T0199271

Geochemical and Hydrological Reactivity of Heavy Metals in Soils

Geochemical and Hydrological Reactivity of Heavy Metals in Soils

Edited by
H. Magdi Selim
William L. Kingery

LEWIS PUBLISHERS

A CRC Press Company
Boca Raton London New York Washington, D.C.

Library of Congress Cataloging-in-Publication Data

Geochemical and hydrological reactivity of heavy metals in soils / edited by H. Magdi Selim, William L. Kingery.
p. cm.
Includes bibliographical references and index.
ISBN 1-56670-623-8 (alk. paper)
1. Soils—Heavy metal content. 2. Havy metals—Environmental aspects. 3. Soil moisture. 4. Soil chemistry. I. Kingery, William L. II. Title.

S592.6.H43S44 2003
628.5′5—dc21 2002043282

Lewis Publishers is an imprint of CRC Press LLC

No claim to original U.S. Government works
International Standard Book Number 1-56670-623-8
Library of Congress Card Number 2002043282
Printed in the United States of America 1 2 3 4 5 6 7 8 9 0
Printed on acid-free paper

Preface

This volume aims at providing a coherent presentation of recent developments and understanding of heavy metal reactivity in soils. Such an understanding is necessary in addressing heavy metals concerns in the environment. The implicit framework of multiple reactivity acknowledges the widely known role played by the various colloidal surface functional groups in concomitant reactions. This overarching frame of reference allows unification between molecular structure-reactivity relationships at one scale and transport processes at the other.

The broader concepts of geochemical and hydrological reactivities are coupled in the first three chapters. Chapter 1 addresses hydrological and geochemical processes controlling the fate and transport of low-level radioactive waste in a subsurface media comprised of fractured saprolite and interbedded fractured limestone and shale bedrock. Use is made of a wealth of observations using multiple tracer techniques in the field to explain the impact of transport processes on the nature and extent of secondary contaminant sources. The effect of colloid mineralogy on their capacity to mediate the transport of heavy metals is presented in Chapter 2 through a series of experimental results employing metal-contaminated suspensions of *ex-situ* soil colloids with montmorillonitic, illitic, and kaolinitic mineralogy applied to undisturbed soil monoliths. Eluent, colloid, and metal recoveries varied with metal, colloid, and soil properties. Mechanisms responsible for colloidal involvement in metal mobility are discussed. Appearing in Chapter 3 is a description of the extension of a multireaction/transport simulation system from one-dimension to three-dimensional applications, using the alternating-direction-implicit numerical algorithm. Also discussed is its implementation in a web-based system with a user-friendly interface providing client-side and server-side computing applications and web-based visualization functionality.

Chapters 4 through 10 collectively treat various aspects of geochemical reactivity on a variety of levels. Fundamental approaches using both experimental observations and theoretical analysis of mineral and organic colloidal surface complexation of contaminants are described in Chapters 4, 5, and 6, while microbial processes are the focus of Chapter 7. Chapter 4 points to mechanisms of metal binding at mineral–water interfaces through an examination of proton adsorption. The authors primarily center on the protonating-deprotonating properties of phyllosilicate clays by stressing the key role played by lattice charges in determining sorption at both basal planes and particle edges. Results are given in Chapter 5 of attempts to combine computational chemistry and molecular modeling to provide a molecular-scale picture of the relation between the complex structural framework of soil organic matter and metal complexation. In the process, the conceptual underpinnings of the application of quantum and classical mechanical simulations to the study of geochemical reactivity are provided. In Chapter 6, the author speaks to the current emphasis on the environmental significance of connections among inorganic species, mineral surfaces, organic molecules, and microbes. Employing the metalloid silicon, the second-most abundant element in the Earth's crust, an examination is given of the possible existence of hypercoordinated silicon-organic ligand-silicon complexes in ambient aqueous solutions using classical electrostatic and solvation theory, crystal chemistry, thermodynamics, and *ab initio* molecular orbital calculations. The kinetics of biological Mn oxidation and the reactivity of biogenic Mn oxides are developed in Chapter 7. Biological Mn oxidation is important because it can exert an important influence on the cycling and bioavailability of trace metals. In addition, the authors describe details and implications of recent studies demonstrating the greater reactivity of biogenic Mn oxides as compared to those formed under abiotic conditions.

Chapters 8 through 10 form a related series addressing geochemical reactivity in soils. In Chapter 8, the authors use a combination of analytical techniques, including x-ray diffraction, electron microscopy, and x-ray absorption, for speciation and quantification of Zn in a soil contaminated as a result of smelting operations. They unite these with chemical extraction and leaching experiments in order to relate metal speciation to bioavailability. Trace metal solubility is explained in Chapter 9 in terms of the release associated with soil Mn-oxides reduction and by an associated displacement of exchangeable metals by high solution concentrations of the dissolved Mn oxide. The authors describe the coupling of these processes using empirical relationships and ion exchange equilibria. The properties and behavior in soils of dissolved organic matter (DOM) and its effect on the environmental behavior of metals is given in Chapter 10. Here the authors discuss the relation of DOM hydrophilic and hydrophobic characteristics and the influence of DOM on trace metal sorption.

The concept of hydrologic reactivity is emphasized in Chapters 11 and 12. In Chapter 11, problems associated with analytical techniques that lead to an inadequate assessment of a complex array of natural surfaces are discussed. The aim is to present some of the key instrumental analysis methods that have wide application to the study of mobile colloids, including light-scattering methods (i.e., photon correlation spectroscopy); acoustic/electroacoustic methods; field flow fractionation; and scanning and transmission electron microscopy. Chapter 12 presents a general-purpose transport model of the multireaction type to describe sulfate transport in forest soils. Discussion is provided on the use of various versions of the multireaction model (equilibrium and kinetic) in describing effluent results from different soil layers. In Chapter 13, a brief treatment is given of fluoride solubility in arid-zone soils contaminated by a nearby phosphate fertilizer production plant.

We wish to sincerely thank the contributors of this book for their diligence and cooperation in achieving our goal and making this volume a reality. We are most grateful for their time and effort in critiquing the various chapters, and in keeping with the focus on the reactivity of heavy metals in soils. Special thanks are due to the reviewers for their help in reviewing individual chapter contributions. Without the support of Louisiana State University and Mississippi State University, this project could not have been realized. We also express our thanks to Randi Cohen, Erika Dery, and the CRC staff for their help and cooperation in the publication of this book.

H. Magdi Selim
Louisiana State University

William L. Kingery
Mississippi State University

Editors

H. Magdi Selim is Professor of Soil Physics at Louisiana State University, Baton Rouge, Louisiana. Dr. Selim received his M.S. and Ph.D. in Soil Physics from Iowa State University, Ames, Iowa, in 1969 and 1971, respectively, and his B.S. in Soil Science from Alexandria University in 1964. Professor Selim has published extensively in scientific journals, bulletins, and monographs, and has also authored and edited several books. His research interests focus on modeling of the mobility of dissolved chemicals and their reactivity in soils and groundwater, and also include saturated and unsaturated water flow in multilayered soils. He is the recipient of several awards including the Phi Kappa Phi Award, the First Mississippi Research Award for Outstanding Research, Gamma Sigma Delta Outstanding Research Award, the Doyle Chambers Achievement Award, and the Sedberry Teaching Award. He served as associate editor of *Water Resources Research* and the *Soil Science Society of America Journal*. Dr. Selim served as chair of the Soil Physics Division of the Soil Science Society of America and as a member of the executive board of the International Society of Trace Element Biogeochemistry (ISTEB). He is a Fellow of the American Society of Agronomy and the Soil Science Society of America.

William L. Kingery is Professor of Agronomy at Mississippi State University. Dr. Kingery received his Ph.D. in Soil Science from Auburn University in 1994, and his M.S. and B.S. in Agronomy from Louisiana State University. Dr. Kingery's research group has published papers in a number of journals including *Naturwissenschaften*. His research interests focus on the application of fundamental chemical techniques to studies in soil science, and to problems in contaminant fate and nutrient management. He and his colleagues have been involved in the application of advanced nuclear magnetic resonance spectroscopy in soils research, animal waste management for land productivity and water quality, and fertility management of soils for rice production. He has co-organized several symposia at the Soil Science Society of America annual meetings and at the International Conference on the Biogeochemistry of Trace Elements. Dr. Kingery served as chair of the Soil Mineralogy Division of the Soil Science Society of America. He is also serving as co-editor of the revision of *Mineralogical Methods* in the book series, *Methods of Soil Analysis*.

Contributors

Vladimir J. Alarcon
Departments of Plant and Soil Sciences and Civil Engineering
Mississippi State University
Mississippi State, Mississippi
valarcon@pss.msstate.edu

Marcelo J. Avena
Departamento de Química
Universidad Nacional del Sur
Bahía Blanca, Argentina
mavena@uns.edu.ar

P.M. Bertsch
Advanced Analytical Center for Environmental Sciences
Savannah River Ecology Laboratory
The University of Georgia
Aiken, South Carolina
bertsch@srel.edu

Grant E. Cardon
Soil and Crop Sciences Department
Colorado State University
Fort Collins, Colorado
gcardon@agsci.colostate.edu

Nicholas Clarke
Norwegian Forest Research Institute
Ås, Norway
nicholas.clarke@nisk.no

Carlos P. De Pauli
Departamento de Fisicoquímica
Universidad Nacional de Córdoba
Córdoba, Argentina
depauli@mail.fcq.unc.edu.ar

George R. Gobran
Department of Ecology and Environmental Research
The Swedish University of Agricultural Sciences
Uppsala, Sweden
george.gobran@eom.slu.se

Colleen H. Green
Soil and Crop Sciences Department
Colorado State University
Fort Collins, Colorado
greenco@lamar.colostate.edu

Ximing Guan
Department of Ecology and Environmental Research
The Swedish University of Agricultural Sciences
Uppsala, Sweden
ximing.guan@eom.slu.se

M. Guerin
Advanced Analytical Center for Environmental Sciences
Savannah River Ecology Laboratory
The University of Georgia
Aiken, South Carolina
guerin@srel.edu

Dean M. Heil
Soil and Crop Sciences Department
Colorado State University
Fort Collins, Colorado
dheil@lamar.colostate.edu

B.P. Jackson
Advanced Analytical Center for Environmental Sciences
Savannah River Ecology Laboratory
The University of Georgia
Aiken, South Carolina
jackson@srel.edu

Richard E. Jackson
Department of Renewable Resources
University of Wyoming
Laramie, Wyoming
oct17@uwyo.edu

Philip M. Jardine
Oak Ridge National Laboratory
Oak Ridge, Tennessee
jardinepm@ornl.gov

A.D. Karathanasis
Department of Agronomy
University of Kentucky
Lexington, Kentucky
akaratha@uky.edu

William L. Kingery
Department of Plant and Soil Sciences
Mississippi State University
Mississippi State, Mississippi
wkingery@pss.msstate.edu

James D. Kubicki
Department of Geosciences
The Pennsylvania State University
University Park, Pennsylvania
kubicki@geosc.psu.edu

Leonard W. Lion
School of Civil and Environmental Engineering
Cornell University
Ithaca, New York
LWL3@cornell.edu

Tonia L. Mehlhorn
Oak Ridge National Laboratory
Oak Ridge, Tennessee
mehlhorntl@ornl.gov

Yarrow M. Nelson
Department of Civil and Environmental Engineering
California Polytechnic State University
San Luis Obispo, California
ynelson@calpoly.edu

Michelle M. Patterson
Department of Renewable Resources
University of Wyoming
Laramie, Wyoming
mickeyp@uwyo.edu

Barry L. Perryman
School of Veterinary Medicine
University of Nevada at Reno
Reno, Nevada
bperryman@cabnr.unr.edu

J.F. Ranville
Chemistry and Geochemistry Department
Colorado School of Mines
Golden, Colorado
jranvill@mines.edu

K.J. Reddy
Department of Renewable Resources
University of Wyoming
Laramie, Wyoming
katta@uwyo.edu

Darryl Roberts
Department of Physics
University of Ottawa
Ottawa, Canada
droberts@uottawa.ca

J. Daniel Rodgers
Department of Renewable Resources
University of Wyoming
Laramie, Wyoming
drodgers@uwyo.edu

Y. Roh
Oak Ridge National Laboratory
Oak Ridge, Tennessee
rohy@ornl.gov

Nita Sahai
Department of Geology and Geophysics
University of Wisconsin
Madison, Wisconsin
sahai@geology.wisc.edu

William E. Sanford
Department of Earth Resources
Colorado State University
310 Natural Resources
Fort Collins, Colorado
William.Sanford@colostate.edu

Andreas C. Scheinost
Department of Environmental Sciences
ETHZ
Schlieren, Switzerland
scheinost@ito.umnw.ethz.ch

John C. Seaman
Advanced Analytical Center for
Environmental Sciences
Savannah River Ecology Laboratory
University of Georgia
Aiken, South Carolina
seaman@srel.edu

H. Magdi Selim
Agronomy Department
Louisiana State University
Baton Rouge, Louisiana
mselim@acgtr.lsu.edu

Donald L. Sparks
Department of Plant and Soil Sciences
University of Delaware
Newark, Delaware
dlsparks@udel.edu

Chad C. Trout
Department of Geosciences
Pennsylvania State University
University Park, Pennsylvania
cct107@psu.edu

J.W.C. Wong
Department of Biology
Hong Kong Baptist University
Hong Kong, China
jwcwong@njau.edu.cn

Honghai Zeng
NSF Engineering Research Center
Mississippi State University
Mississippi State, Mississippi
zeng@erc.msstate.edu

Lixiang Zhou
Department of Environmental Science
and Engineering
Nanjing Agricultural University
Nanjing, China
lxzhou@njau.edu.cn

Jianping Zhu
Department of Theoretical and Applied
Mathematics
The University of Akron
Akron, Ohio
jzhu@math.uakron.edu

Contents

Chapter 1
Hydrological and Geochemical Processes Controlling the Fate and Transport of Contaminants in Fractured Bedrock 1
Philip M. Jardine, Tonia L. Mehlhorn, Y. Roh, and William E. Sanford

Chapter 2
Mineral Controls in Colloid-Mediated Transport of Metals in Soil Environments 25
A.D. Karathanasis

Chapter 3
A Web-Based Three-Dimensional Simulation System for Solute Transport in Soil 51
Honghai Zeng, Vladimir J. Alarcon, William L. Kingery, H. Magdi Selim, and Jianping Zhu

Chapter 4
Effect of Structural Charges on Proton Adsorption at Clay Surfaces 79
Marcelo J. Avena and Carlos P. De Pauli

Chapter 5
Molecular Modeling of Fulvic and Humic Acids: Charging Effects and Interactions with Al^{3+}, Benzene, and Pyridine 113
James D. Kubicki and Chad C. Trout

Chapter 6
Silicon-Organic Interactions in the Environment and in Organisms 145
Nita Sahai

Chapter 7
Formation of Biogenic Manganese Oxides and Their Influence on the Scavenging of Toxic Trace Elements 169
Yarrow M. Nelson and Leonard W. Lion

Chapter 8
Zinc Speciation in Contaminated Soils Combining Direct and Indirect Characterization Methods 187
Darryl Roberts, Andreas Scheinost, and Donald L. Sparks

Chapter 9
Reduction/Cation Exchange Model of the Coincident Release of Manganese and Trace Metals following Soil Reduction 229
Dean M. Heil, Grant E. Cardon, and Colleen H. Green

Chapter 10
Behavior of Heavy Metals in Soil: Effect of Dissolved Organic Matter 245
Lixiang X. Zhou and J.W.C. Wong

Chapter 11
Analytical Techniques for Characterizing Complex Mineral Assemblages: Mobile Soil and Groundwater Colloids 271
John C. Seaman, M. Guerin, B.P. Jackson, P.M. Bertsch, and J.F. Ranville

Chapter 12
Kinetic Modeling of Sulfate Transport in a Forest Soil 311
H. Magdi Selim, George R. Gobran, Ximing Guan, and Nicholas Clarke

Chapter 13
Solubility of Fluoride in Semi-Arid Environments 331
K.J. Reddy, Michelle M. Patterson, J. Daniel Rodgers, Richard E. Jackson, and Barry L. Perryman

Index 351

1 Hydrological and Geochemical Processes Controlling the Fate and Transport of Contaminants in Fractured Bedrock

Philip M. Jardine, Tonia L. Mehlhorn, Y. Roh, and William E. Sanford

CONTENTS

1.1 Introduction 1
1.2 Case Study 2
1.2.1 Multiple Nonreactive Tracer Transport 6
1.2.1.1 Preferential Flow and Matrix Diffusion 6
1.2.1.2 Modeling 10
1.2.2 Chelated Radionuclide Transport 13
1.2.2.1 Chelated Radionuclide Dissociation 13
1.2.2.2 Chelated Radionuclide Oxidation 16
1.2.2.3 Modeling 18
1.3 Summary and a Look Ahead 21
Acknowledgments 22
References 22

1.1 INTRODUCTION

The disposal of low-level radioactive waste generated by the U.S. Department of Energy (DOE) during the Cold War era has historically involved shallow land burial in unconfined pits and trenches. The lack of physical or chemical barriers to impede waste migration has resulted in the formation of secondary contaminant sources where

1-56670-623-8/03/$0.00+$1.50

radionuclides have moved into the surrounding soil and bedrock, as well as groundwater and surface water sources. At certain DOE facilities, such as the Oak Ridge National Laboratory (ORNL) located in eastern Tennessee, the extent of the problem is massive, where thousands of underground disposal trenches have contributed to the spread of radioactive contaminants across tens of kilometers of landscape. The subsurface media at ORNL is comprised of fractured saprolite and interbedded fractured limestone and shale bedrock which are conducive to rapid preferential flow coupled with significant matrix storage (Jardine et al., 2001). Fractures are highly interconnected and surround low permeability, high porosity matrix blocks. Subsurface transport processes are driven by large annual rainfall inputs (~1400 mm/year) where as much as 50% of the infiltrating precipitation results in groundwater and surface water recharge (10 and 40%, respectively). Thus, storm-flow infiltration into the media often results in large physical and geochemical gradients among the various flow regimes, which causes nonequilibrium conditions during solute transport.

At ORNL, waste trenches were often excavated to depths that approached the bedrock-saprolite interface. Seasonally fluctuating groundwater levels coupled with storm-derived, perched water tables allow contaminants to easily access the underlying bedrock. Further, the very nature of the trench design allows for rapid vertical infiltration of storm water and direct connection with the fracture network of the underlying limestone–shale bedrock. In these systems, the bedrock matrix has been exposed to migrating contaminants for many decades, and thus accounts for a significant inventory of the total waste (i.e., secondary sources). A limitation in defining remediation needs of the secondary sources results from an insufficient understanding of the transport processes that control contaminant migration. Without this knowledge base, it is impossible to assess the risk associated with the secondary source contribution to the total off-site migration of contaminants. The objectives of our research were to help resolve this dilemma by providing an improved understanding of contaminant transport processes in highly structured, heterogeneous subsurface environments that are complicated by fracture flow and matrix diffusion. The investigations described here address coupled hydrological and geochemical processes controlling the fate and transport of contaminants in fractured shale bedrock. A heavily instrumented field facility was used in conjunction with multiple tracer techniques to unravel the impact of coupled transport processes on the nature and extent of secondary contaminant sources.

1.2 CASE STUDY

In an effort to understand the importance of secondary source formation and contaminant reactivity in the fractured limestone–shale bedrock that is commonplace at ORNL, a field facility was established within the saturated zone of Waste Area Grouping 5 (WAG 5) on the Oak Ridge Reservation (Figure 1.1(a) and (b); Jardine et al., 1999). The field site resides down gradient to numerous waste disposal trenches that contain ill-defined mixtures of DNAPL, ^{3}H, and ^{90}Sr (Figure 1.1(c) and (d)). A 35-m transect of 33 multilevel groundwater monitoring wells was established along the geologic strike within a fast-flowing fracture regime and a slow-flowing matrix regime (Figure 1.1(d) and Figure 1.2). The slower flowing intervals are referred to

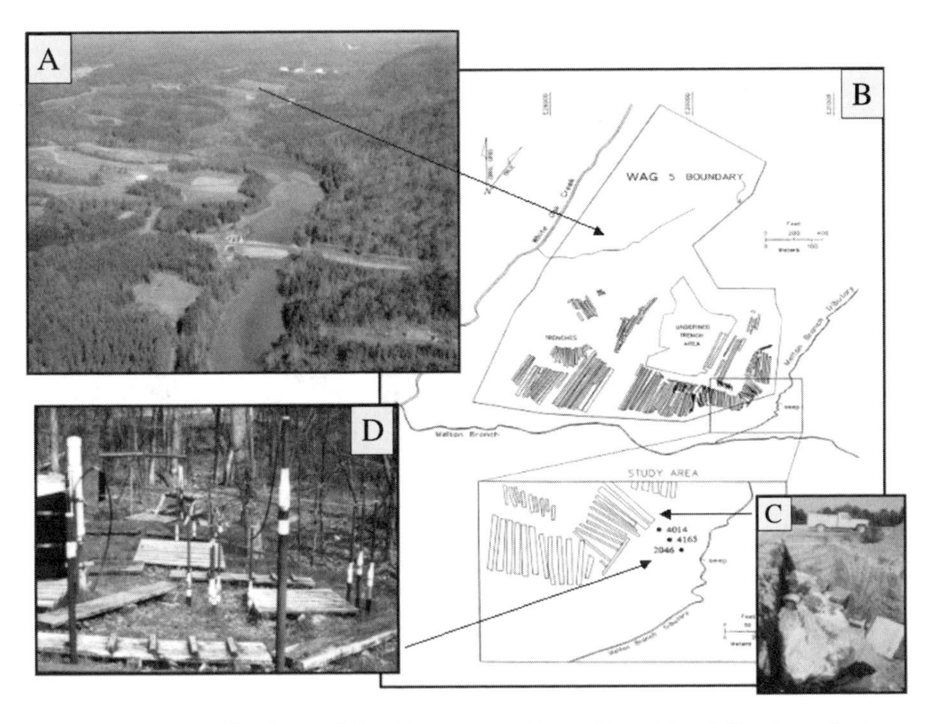

FIGURE 1.1 (See color insert following page 112) (a) Pictorial and (b) schematic views of Waste Area Grouping 5 showing (c) the location of buried waste trenches and the location of the experimental field facility, with a few examples of groundwater monitoring wells (4014, 4165, 2065) that have been strategically placed at strike parallel. The (d) primary well field contains 33 groundwater monitoring positions and a designated tracer injection well.

as "matrix zones" within the bedrock since fracture interconnectedness is sparse. The faster-flowing interval is referred to as the "fractured zone" since fracture orientation and connectivity result in rapid preferential flow. The hydraulic conductivity of the fracture zone is at least an order of magnitude larger than the matrix zone. Each well resides below the water table and within the interbedded fractured shale and limestone bedrock (Figure 1.3). An additional well (4014) was also installed up gradient from the primary well field, and it served as the tracer injection well (Figure 1.2).

Groundwater at the site is generally anaerobic with typical dissolved oxygen (DO) concentrations of 0.016 to 1.0 mg O_2/L. It is highly buffered with HCO_3^- (670–900 mg/L^{-1}) resulting in a consistent groundwater pH of 6.7 to 7.0. The redox environment of the groundwater is Fe-reducing with a few locations that are SO_4-reducing. The media are completely devoid of Mn and Fe bearing mineral oxides, and x-ray diffraction analysis coupled with scanning electron microscopy with energy-dispersive analysis suggested that the solid phase is composed of calcite, quartz, K-feldspar, biotite, and small quantities of kaolinite.

An automated tracer injection system was utilized to dispense tracers into the fracture regime of well 4014 under natural gradient conditions (Figure 1.2(b)). Two long-term, steady-state natural gradient experiments, each with a duration of 1.5 to 2

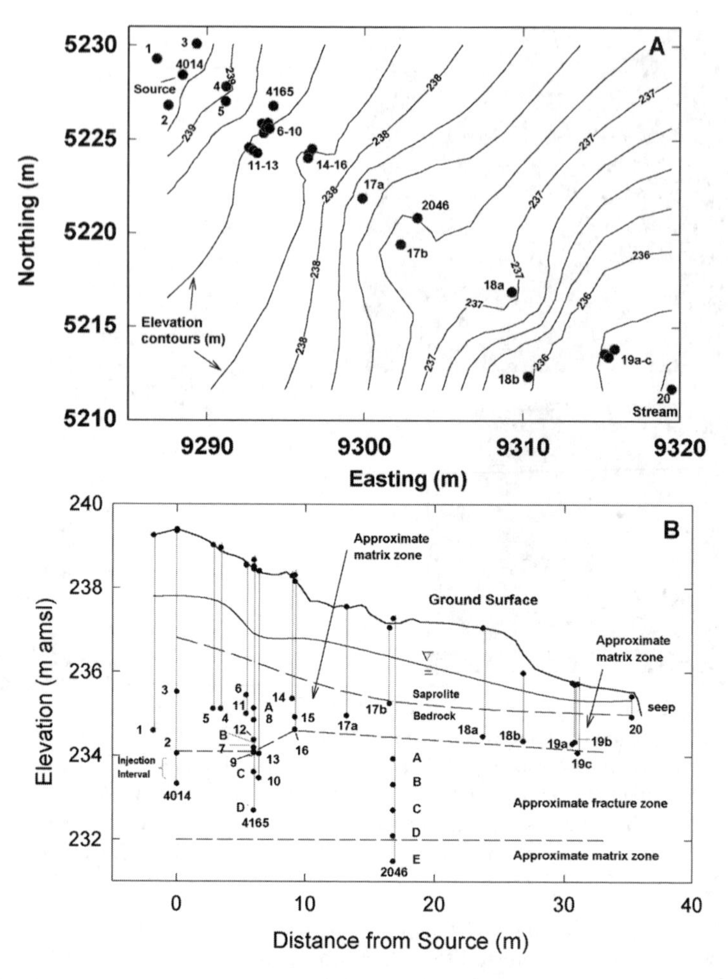

FIGURE 1.2 (a) Map view and (b) cross-section of the experimental field facility showing the location and sampling depth of all groundwater monitoring wells, with well 4014 serving as the tracer injection well. Specific information about each well and the labeling strategy is provided in Jardine, P.M., et al., *Water Resour. Res.,* 35, 2015, 1999. The wells form a strike parallel transect from the waste trenches to a seep that drains into a cross-cutting tributary. The approximate locations of the fracture and matrix regimes are also illustrated. (Modified from Jardine, P.M., et al., *Water Resour. Res.*, 35, 2015, 1999. With permission.)

years, were conducted using multiple nonreactive tracers (He, Ne, Br) and multiple reactive tracers (^{57}Co(II)EDTA and ^{109}CdEDTA) coupled with nonreactive Br^- (Jardine et al., 1999, 2002). Chelated reactive tracers were used to simulate historical waste removal practices that involved co-disposal of radioactive inorganic fission byproducts with various chelating agents such as ethylenediaminetetraacetic acid (EDTA) (Ayres, 1971; Toste and Lechner-Fish, 1989; Riley and Zachara, 1992). Details of the tracer characteristics and injection concentrations can be found in Jardine et al. (1999, 2002). The multiple tracer strategy was designed to take advantage of differences in the

FIGURE 1.3 (See color insert) Pictorial example of bedrock core obtained near the experimental site showing the interbedded fractured shale (black) and limestone (white/gray) that dominates the saturated zone.

molecular diffusion coefficients and geochemical reactivity between the tracers. Thus, the experiments sought to quantify the significance of fracture flow, matrix diffusion, and chemical reactivity on the formation and longevity of secondary contaminant sources in the interbedded shale and limestone bedrock. Each of the two experiments involved the natural gradient injection of tracer for 6 months, followed by a 12-month washout. Spatial and temporal monitoring of the tracers was performed in the matrix and fracture regimes of the bedrock using the multilevel sampling wells instrumented down gradient from the injection source (Figure 1.2).

1.2.1 Multiple Nonreactive Tracer Transport

1.2.1.1 Preferential Flow and Matrix Diffusion

A multiple nonreactive tracer technique involving He, Ne, and Br^- was used to assess the significance of physical nonequilibrium processes on the transport of contaminants at ORNL and comparable geologic formations. The dissolved gas tracers He and Ne, and the dissolved solute tracer Br, were chosen since they differ only in their free-water molecular diffusion coefficients with values of 6.0×10^{-4}, 3.5×10^{-4}, and 2.0×10^{-4} m^2/d for He, Ne, and Br, respectively (Sanford et al., 1996; Sanford and Solomon, 1998). Thus, if physical nonequilibrium processes such as preferential flow and matrix diffusion are significant, differences in tracer breakthrough profiles should be observed. Tracers migrated preferentially along the geologic strike, and their concentrations in the fracture regime quickly reached a near steady-state value close to the source (Figure 1.4(a) and (b)) and a more gradual attainment of near steady-state conditions at greater distances from the source (Figure 1.4(c)). Breakthrough profiles were similar for the three tracers at 6 m and 9 m from the source, due to rapid movement of the tracers along fractures. The earlier washout of gas tracers relative to Br was due to a premature termination of the gas tracer injection prior to the 178-d target date. For conditions farther from the source (e.g., 17 m and Figure 1.4(c)), the breakthrough of dissolved gas tracers in the fracture regime was significantly delayed relative to Br^-, which suggests appreciable up-gradient (toward source) mass loss of the former as the tracer residence times increased. Since the tracers are nonreactive, the observed delay in gas breakthrough within the fracture regime may be the result of greater mass loss to the matrix relative to Br^-. This scenario is consistent with the larger diffusion coefficients of the gas tracers compared to Br^-.

Tracer breakthrough profiles within the matrix regime confirmed that the dissolved gases and Br^- were slowly moving into the bedrock matrix and at different rates. Concentration profiles of the three tracers 6 m from the source and 0.8 m into the matrix relative to the fracture zone are shown in Figure 1.5(a). The movement of He and Ne into and from the matrix was more rapid than Br^-, which is consistent with the larger molecular diffusion coefficients for the dissolved gases relative to Br^-. These results support the notion that matrix diffusion contributed to the overall physical nonequilibrium process that controls solute transport in bedrock at this site (Maloszewski and Zuber, 1990, 1993).

At greater distances from the source, the contribution of matrix interactions is still prominent, and tracer breakthrough profiles remain suggestive of a diffusion mechanism, although at first glance this may not be apparent (Figure 1.5(b), (c)). At 13 m from the source and 0.6 m into the matrix, the three tracers break through nearly simultaneously, with the concentration of the gas tracers eventually surpassing Br^- (Figure 1.5(b)). This is followed by tracer washout after the input pulse was terminated at 178 d. At 23 m from the source and 0.1 m into the matrix, the movement of Br^- into the matrix is actually more rapid than that of the noble gas tracers, which is exactly opposite of what was observed 6 m from the source. This apparent paradox is caused by the preferential loss of gas tracers to the rock matrix closer to the source. Thus, Br^- remains within the advective flow field (fracture regime) for a

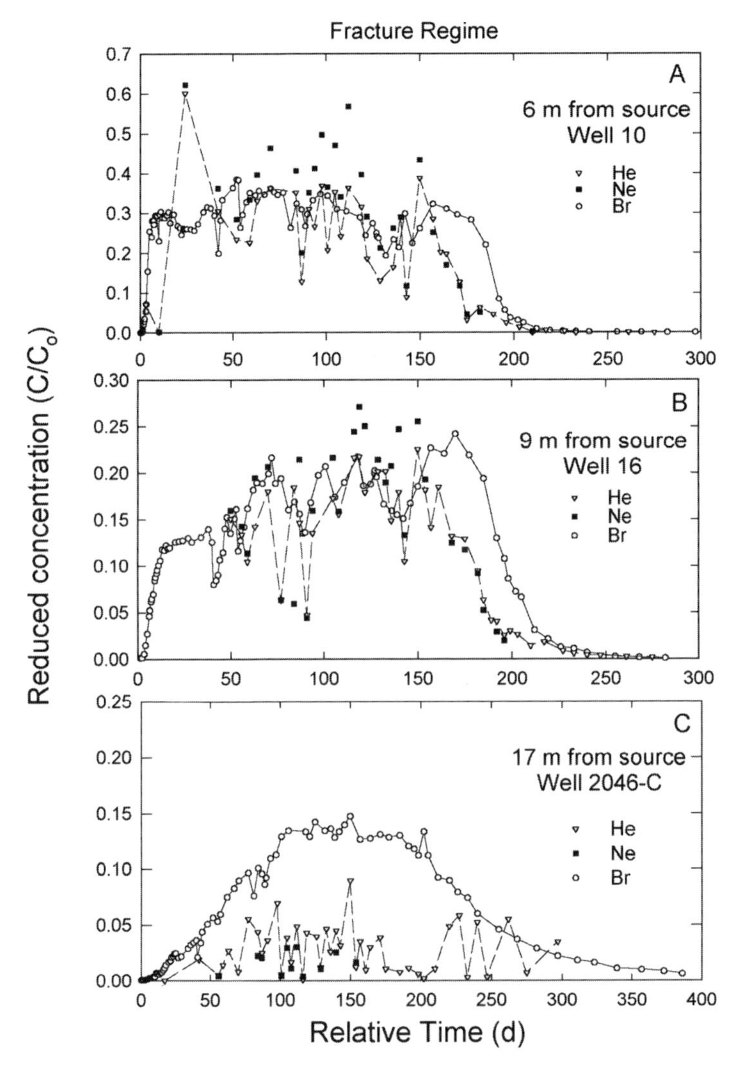

FIGURE 1.4 Observed breakthrough of He, Ne, and Br within the fracture regime, at (a) 6 m (well 10); (b) 9 m (well 16); and (c) 17 m (well 2046-C) from the source.

longer time period, allowing it to be transported greater distances (e.g., Figure 1.4(c)). Having been transported farther down gradient, Br^- begins diffusing into the matrix ahead of He and Ne at greater distances from the source. Eventually, He and Ne arrive at the same locations and also begin to diffuse into the matrix, lagging behind Br^- (Figure 1.5 (b) and (c)). Because of the larger diffusion coefficients of the noble gases, the movement of He and Ne into the matrix is more rapid and, if given time, the He and Ne breakthrough curves will eventually cross over and surpass the Br^- breakthrough curves (see Figure 1.5(b) as an example).

At 31 m and 35 m from the source, separation of the three tracers was even more dramatic (Figure 1.5(d) and (e)) and followed the same trends observed in

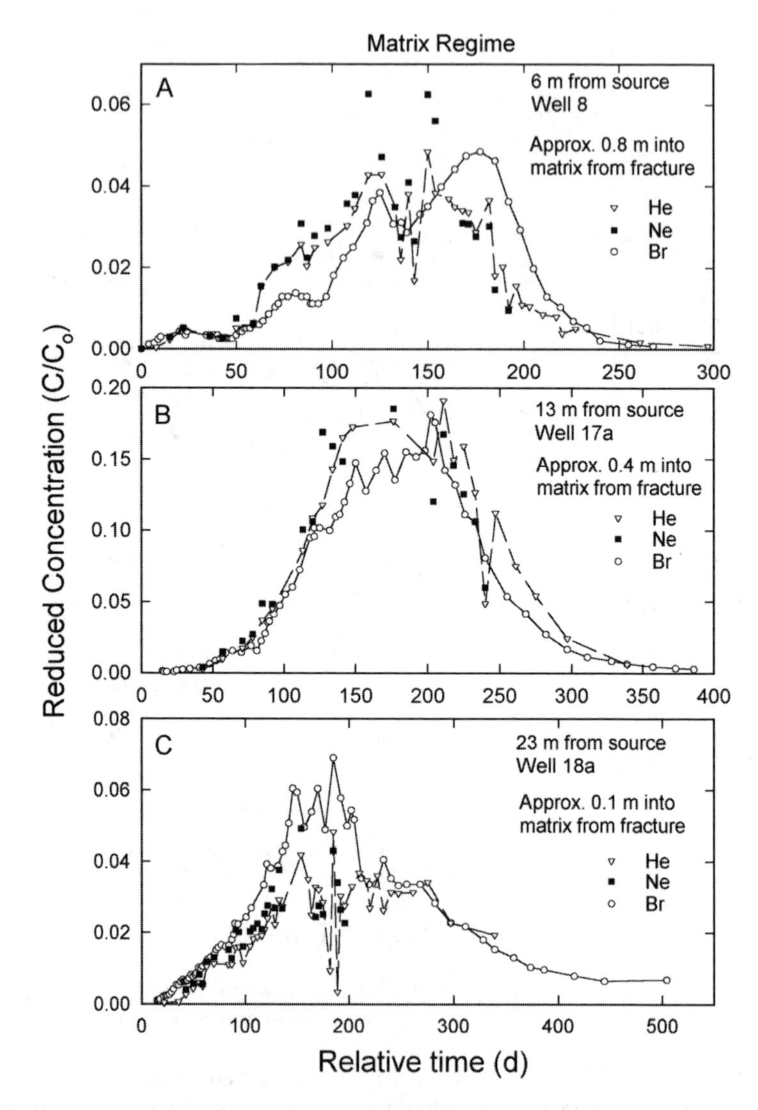

FIGURE 1.5 Observed breakthrough of He, Ne, and Br within the matrix regime, at (a) 6 m (well 8); (b) 13 m (well 17a); and (c) 23 m (well 18a) from the source. (Modified from Jardine, P.M., et al., *Water Resour. Res.*, 35, 2015, 1999. With permission.)

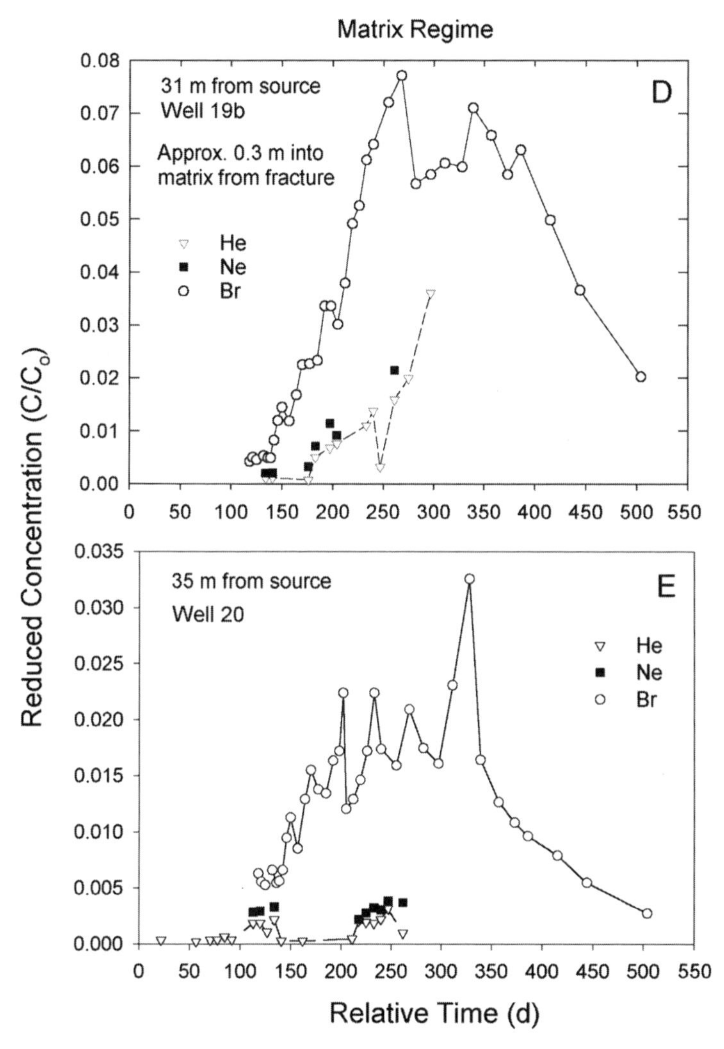

FIGURE 1.5 CONTINUED Observed breakthrough of He, Ne, and Br within the matrix regime, at (d) 31 m (well 19 b); and (e) 35 m (well 20) from the source.

Figure 1.5(c); however, dissolved gas tracer data from 35 m (Figure 1.5(e)) may have been impacted by volatilization since this monitoring well is situated near a groundwater discharge point at the end of the sampling transect (Figure 1.2). Nevertheless, results from 31 m clearly show that solute diffusion into the bedrock matrix is a significant process contributing to contaminant storage in this subsurface media, and that the diffusion process becomes increasingly important as the residence time of the solute increases in the system.

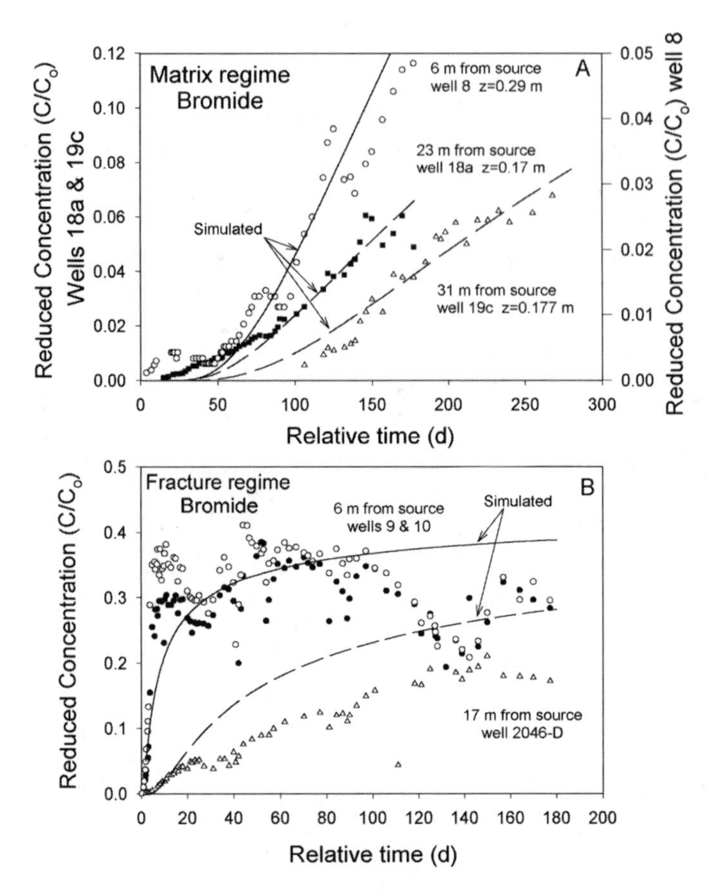

FIGURE 1.6 Observed and model-simulated breakthrough curves for Br within the (a) matrix and (b) fracture regimes at various distances from the source, where $\theta = 0.20$ is the porosity, 2b = 8.4 E-5 m is the fracture aperture, 2B = 2.0 m is the fracture spacing, v = 100 m d^{-1} is the mean pore water velocity within the fracture regime, $\alpha = 0.1$m is the dispersivity, $\tau = 0.6$ is the tortuosity, and z is the distance into the matrix. (From Jardine, P.M., et al., *Water Resour. Res.*, 35, 2015, 1999. With permission.)

1.2.1.2 Modeling

In order to further test the hypothesis that diffusion may be controlling the migration of tracer into the matrix, the one-dimensional, fracture-flow model, CRAFLUSH (Sudicky and Frind, 1982), was used to simulate tracer migration within the fracture and matrix regime along the entire transect of sampling wells. Only the ascending limbs of the tracer breakthrough curves were simulated since the model assumes complete recovery of mass from the matrix upon tracer washout, and this was not observed over the timeframe of our experiment. Full mass recoveries from the matrix would require significantly longer time periods than those studied here.

Model input parameters were based on independent field and laboratory observations that used subsurface media similar to that at the WAG 5 field facility. The mean pore-water velocity (v) of the fracture regime was estimated from measured

TABLE 1.1
Retardation Coefficients for Chelated Metal Migration at Various Distances from Source for Both Fracture and Matrix Regimes

Distance from Source (m)	Fracture Wells	Matrix Wells	R_f for Total EDTA Species	R_f for Co- and Cd-EDTA Species	R_m for Total EDTA Species	R_m for Co- and Cd-EDTA Species
6.0	9, 10	7, 8	260	500	1.1–1.5	1.7
9.2	16	14, 15	330	500	1.8	3.5
13.2	—	17a	360	450	2.1	3.5
16.8	2046	—	360	500	3.3	9.5
23.7	—	18a	260	NA	1.5	NA

NA, not applicable since Co- and Cd-EDTA species never arrived at this location.
R_f and R_m, retardation coefficients in fracture and matrix regimes, respectively.

field-scale advective flow rates in fractured saprolite using bacteriophage (McKay et al., 2000). Porosity (θ) was obtained from direct measurements on similar subsurface material (Wilson et al., 1992; Dorsch et al., 1996), and fracture spacing (2B) and aperture (2b) were estimated from measured values in shale bedrock on an adjacent field site (Dreier et al., 1987). Parameter values used in all simulations are provided in the figure captions and Table 1.1. The only unknown was the solute travel distance from the fracture regime to the various sampling wells within the matrix regime (z). Observed distances based on measured well depths are potentially subject to large errors since it is impossible to know where the fracture regime ends and the matrix regime begins throughout the site. Therefore, simulations of tracer transport throughout the entire site involved holding all model parameters constant except the distance into the matrix z, where all three tracers used the same z at any physical point. Values of z were varied until an adequate simulation of the observed Br^- movement into the matrix was achieved (e.g., Figure 1.6(a)). The magnitude of optimized z values obtained from model simulations was found to be on average 47% of the measured z values across the entire site.

Simulations of Br^- mobility in the matrix regime simultaneously generated simulations of Br^- transport in the fracture regime (z = 0). It was therefore possible to check the adequacy of the model for describing tracer mobility at the field site since we experimentally measure tracer within both the fracture and matrix regimes. Matching the model to observed Br^- mobility within the matrix (Figure 1.6a) produced reasonable simulations of Br^- transport through the fracture regime (Figure 1.6b) without prior knowledge of the observed fracture regime data. This finding suggests that the model, to a certain extent, is accurately describing tracer mobility through the bedrock and that diffusion may be an important mechanism in the storage of solutes in the shale–limestone bedrock.

The migration of the three tracers into the bedrock matrix was reasonably well described using CRAFLUSH (Figure 1.7). All model parameters were held constant as described earlier and z was matched to the observed Br breakthrough as described

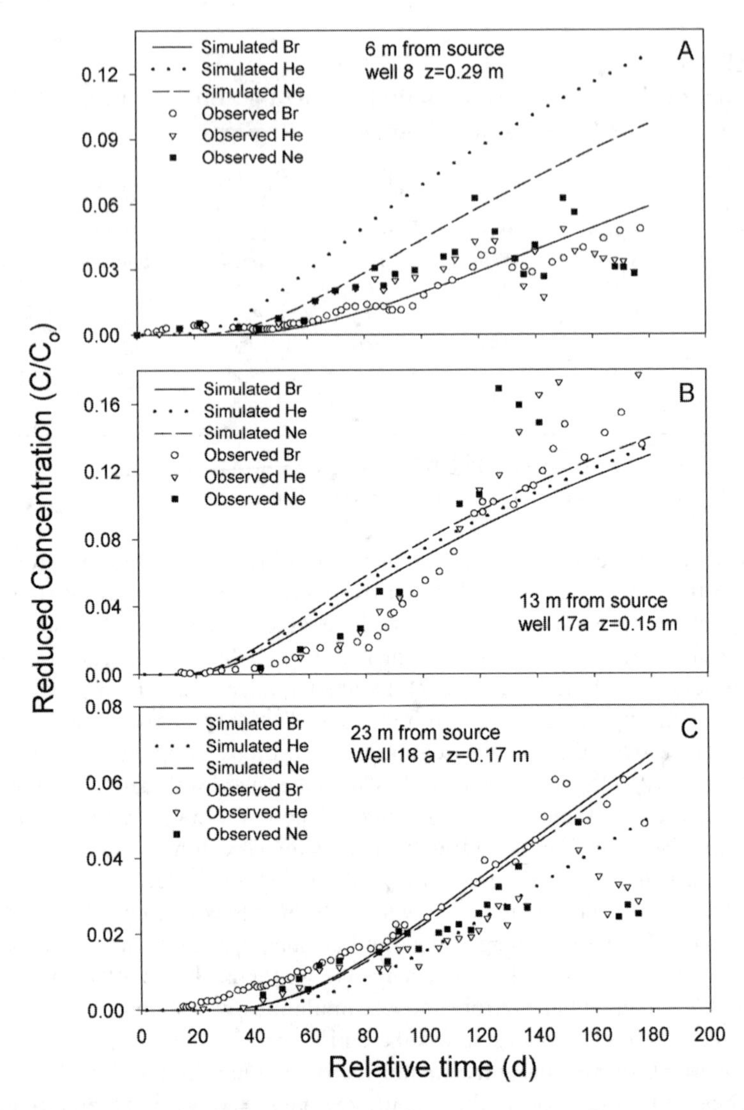

FIGURE 1.7 Observed and model simulated breakthrough curves for He, Ne, and Br within the matrix regime, (a) 6 m, (b) 13 m , and (c) 23 m from the source, where $\theta = 0.20$ is the porosity, $2b = 8.4$ E-5m is the fracture aperture, $2B = 2.0$ m is the fracture spacing, $v = 100$ m d^{-1} is the mean pore water velocity within the fracture regime, $\alpha = 0.1$ m is the dispersivity, $\tau = 0.6$ is the tortuosity, and z is the distance into the matrix. (Modified from Jardine, P.M., et al., *Water Resour. Res.*, 35, 2015, 1999. With permission.)

above. The only model parameter that differed among the three tracers was their free-water molecular diffusion coefficient (D_w). Model results showed that for conditions close to the source, the migration of Br^- into the matrix was slower relative to the gas tracers, and the migration of Ne was slower than that of He (Figure 1.7(a)). These results are partially consistent with the observed data since the tracer diffusion coefficients follow He > Ne > Br. With increasing distance from the source, model simulated curves, on average, more adequately described the observed migration of the three tracers into the matrix (Figure 1.7(b) and (c)). At 13 m from the source, model simulated curves merged in a similar manner as the observed data (Figure 1.7(b)). Further, the simulated curves reproduced the observed early arrival of Br^- into the matrix followed by He and then Ne. For conditions far from the source (i.e., 23 m), model results again matched the observed tracer concentration trends and showed that Br^- migrated into the matrix ahead of the He and Ne, with Ne migrating more rapidly than He (Figure 1.7(c)). Numerical confirmation of the observed multi-tracer breakthrough trends further supports the mechanism of diffusion as an important process controlling the fate and transport of solutes in this shale bedrock. The multiple-tracer technique and ability to monitor both the fracture and matrix regimes provided the necessary experimental constraints for the accurate numerical quantification of the diffusion process.

The results of the multiple nonreactive tracer study showed that secondary contaminant sources may form within the bedrock matrix, and that the importance of this source increases with continued contaminant discharge through the bedrock fracture network. This may be particularly important for reactive contaminants such as radionuclides, where matrix diffusion can enhance solute retardation by many orders of magnitude.

1.2.2 Chelated Radionuclide Transport

1.2.2.1 Chelated Radionuclide Dissociation

Since geochemical processes may also influence the behavior of radioactive contaminants in subsurface environments, field investigations were conducted involving the reactive tracers $^{57}Co(II)EDTA^{2-}$ and $^{109}CdEDTA^{2-}$. The effort was designed to assess the significance of fracture flow, matrix diffusion, and chemical reactivity on the migration of chelated radionuclides from waste trenches into the underlying bedrock. The natural gradient injection of $^{57}Co(II)EDTA^{2-}$, $^{109}CdEDTA^{2-}$, and Br^- was initiated at the WAG 5 field facility for 6 months, followed by a 12-month washout. The breakthrough and washout of the chelated radionuclides were significantly delayed relative to the nonreactive Br^- tracer, suggesting that the former were reactive with the solid phase (Figure 1.8). Down-gradient first-arrival times were two to more than ten times larger for the chelated radionuclides and metals relative to Br^-. Within the fracture regime, the retarded breakthrough of $Co(II)EDTA^{2-}$ and $CdEDTA^{2-}$ was preceded by the breakthrough of $Fe(III)EDTA^-$ (Figure 1.8). These results suggest that the formation of $Fe(III)EDTA^-$ is coupled to the dissociation and subsequent delayed breakthrough of the Co- and Cd- EDTA complexes. This phenomenon has been observed in oxic subsurface environments where Fe-oxyhydroxides are known to effectively dissociate complexes such as Co(II)- and Cd-EDTA with the subsequent formation of $Fe(III)EDTA^-$ (Kent et al., 1991, 1992; Jardine et al., 1993; Zachara et al., 1995; Davis et al., 2000; Mayes et al., 2000). At the current field site, however, these results were somewhat unexpected since the subsurface environment was Fe-reducing and devoid of

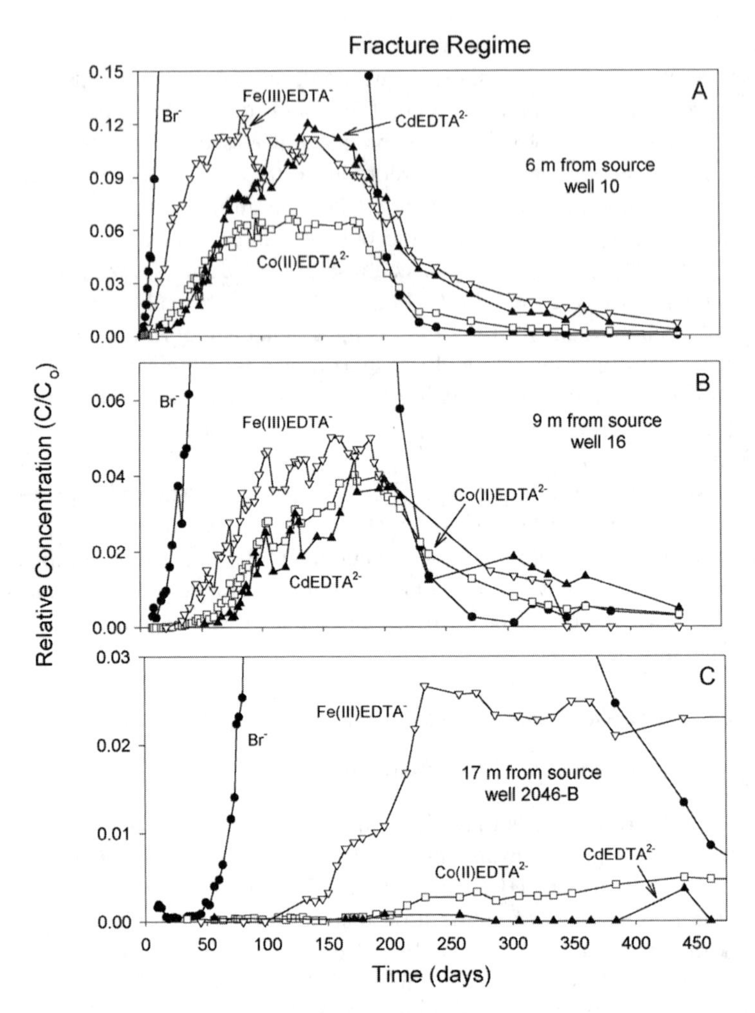

FIGURE 1.8 Breakthrough curves for Br^-, $Co(II)EDTA^{2-}$, $CdEDTA^{2-}$, and $Fe(III)EDTA^-$ in the fracture regime (a) 6 m (well 10), (b) 9 m (well 16), and (c) 17 m (well 2046-B) from the source. For the chelated metals, C_0 is the concentration of all EDTA species in the groundwater injection interval (i.e., Co and Cd). (From Jardine, P.M., et al., *J. Contamin. Hydrol.*, 55, 137, 2002. With permission.)

mineral oxides, such as Fe-oxyhydroxides. Mineralogical analysis of the bedrock, however, revealed small quantities of the primary mineral biotite, $K(Mg,Fe^{2+})_3(Al,Fe^{3+})Si_3O_{10}(OH)_2$, which was the only detectable Fe(III) containing mineral in the subsurface (Figure 1.9). Although Fe(II) dominates the octahedral layers of biotite, sufficient Fe(III) impurity also resides within the mineral lattice. The Fe(III) in biotite is highly unstable and is readily removable from the lattice structure if an appropriate ligand forms a surface complex. Rath and Subramanian (1997) have shown that EDTA is readily adsorbed by biotite. Thus, it is likely that Co- and Cd-EDTA form surface complexes with the biotite. Since the

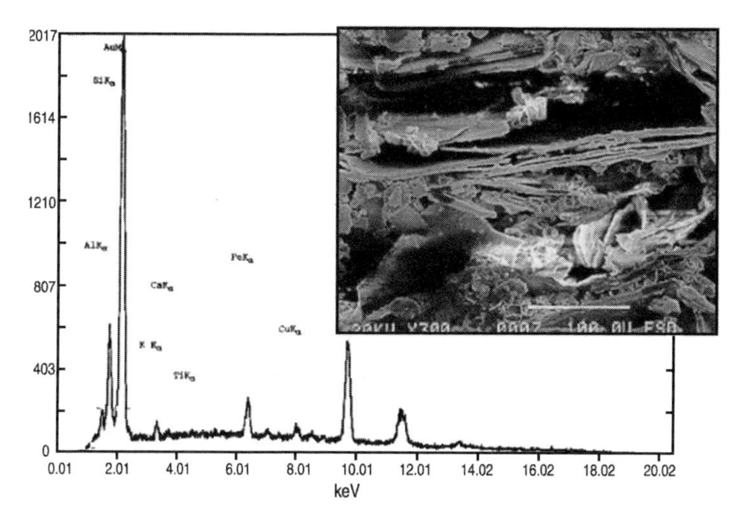

FIGURE 1.9 Mineralogical characterization of bedrock core from the experimental site using scanning electron microscopy with energy-dispersive analysis showing the frayed edge sites of the 2:1 phyllosilicate biotite (i.e., interlayer spaces between sheet silicates in photograph).

stability constant of Fe(III)EDTA$^-$ (log K = 25.1) is significantly greater than Co- and Cd-EDTA (log K = 16.3 and 16.5, respectively), the structural Fe(III) in biotite can successfully dissociate the metal-EDTA complexes, forming aqueous Fe(III)EDTA$^-$.

The dissociation of Co- and Cd-EDTA, with subsequent formation of Fe(III)EDTA$^-$, becomes much more pronounced with increasing distance from the injection source (Figure 1.8(a), (b), and (c)). At 17 m from the source (Figure 1.8(c)), Co(II)EDTA^{2-}, and in particular CdEDTA^{2-}, are nearly absent in the groundwater, with Fe(III)EDTA$^-$ dominating the metal–chelate consortium. Geochemical speciation modeling suggested that the liberated Co and Cd metals are precipitated as $CoCO_3$ and $CdCO_3$, which are not readily mobile in the groundwater. These results are significant since they counter current conceptual models that suggest chelated metals and radionuclides do not interact with the solid phase in Fe-reducing environments (Means and Alexander, 1981; Williams et al., 1991).

Although solid-phase geochemical processes significantly influenced the mobility of Co(II)- and Cd-EDTA in the fracture regime, these processes were not demonstrated to a significant extent within the matrix regime (Figure 1.10). The movement of Co(II)EDTA^{2-} and CdEDTA^{2-} into the matrix did not result in an appreciable increase in Fe(III)EDTA$^-$ that could not be attributed to physical diffusion processes that act to drive mass from the fracture regime into the matrix regime. This is evident when comparing the decreasing mass of Br$^-$ versus metal-EDTA with increasing distance into the matrix from the fracture (e.g., Figures 1.8(a) and 1.10(a) and (b)) and will become more apparent when the modeling results of this data are discussed below. These results suggest that physical retardation mechanisms dominate within the matrix regime, whereas geochemical retardation mechanisms are dominant within the fracture regime.

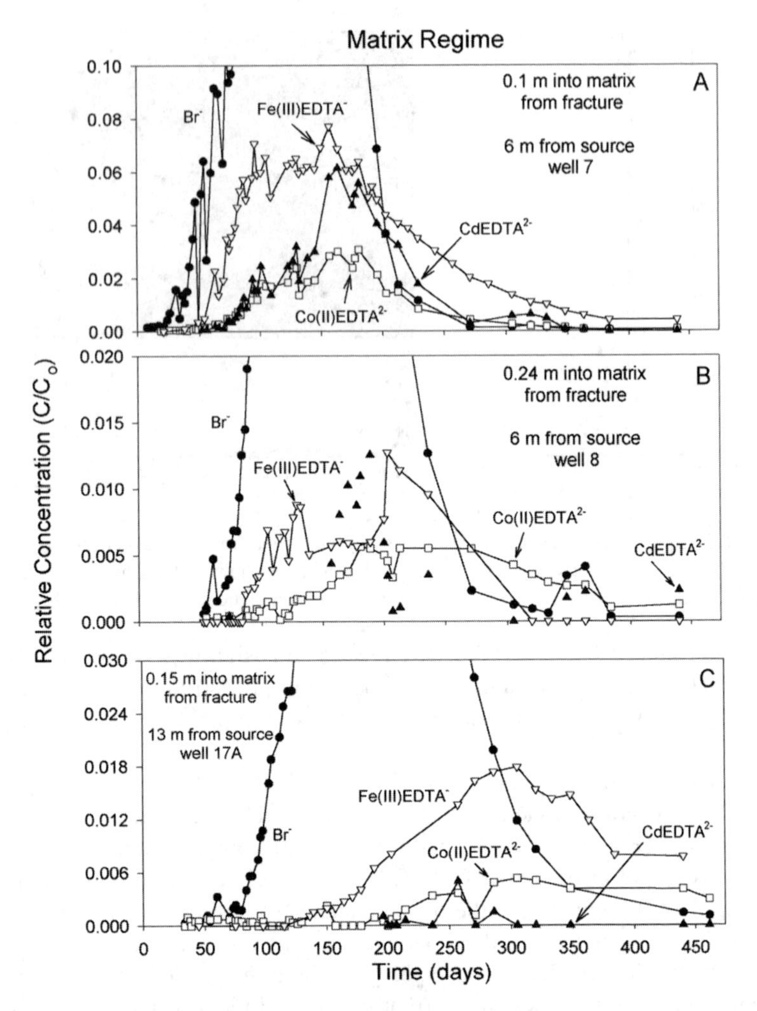

FIGURE 1.10 Breakthrough curves for Br^-, $Co(II)EDTA^{2-}$, $CdEDTA^{2-}$, and $Fe(III)EDTA^-$ in the matrix regime (a and b) 6 m (wells 7 and 8), and (c) 13 m (well 17a) from the source. For the chelated metals, C_0 is the concentration of all EDTA species in the groundwater injection interval (i.e. Co and Cd). (From Jardine, P.M., et al., *J. Contamin. Hydrol.*, 55, 137, 2002. With permission.)

1.2.2.2 Chelated Radionuclide Oxidation

The oxidation of $Co(II)EDTA^{2-}$ to $Co(III)EDTA^-$ was generally not significant under the conditions present at the site since groundwater concentrations of $Co(III)EDTA^-$ were typically 20 to 40 times lower than $Co(II)EDTA^{2-}$. This scenario is in contrast to oxic environments that typically contain appreciable Fe- and Mn- oxides that have been shown to catalyze this redox reaction resulting in appreciable production of $Co(III)EDTA^-$ (Jardine et al., 1993; Jardine and Taylor, 1995a, 1995b; Zachara et al., 1995; Brooks et al., 1996, 1999; Szecsody et al., 1998a, 1998b; Mayes et al., 2000). The adverse environmental implications of this redox reaction are pronounced

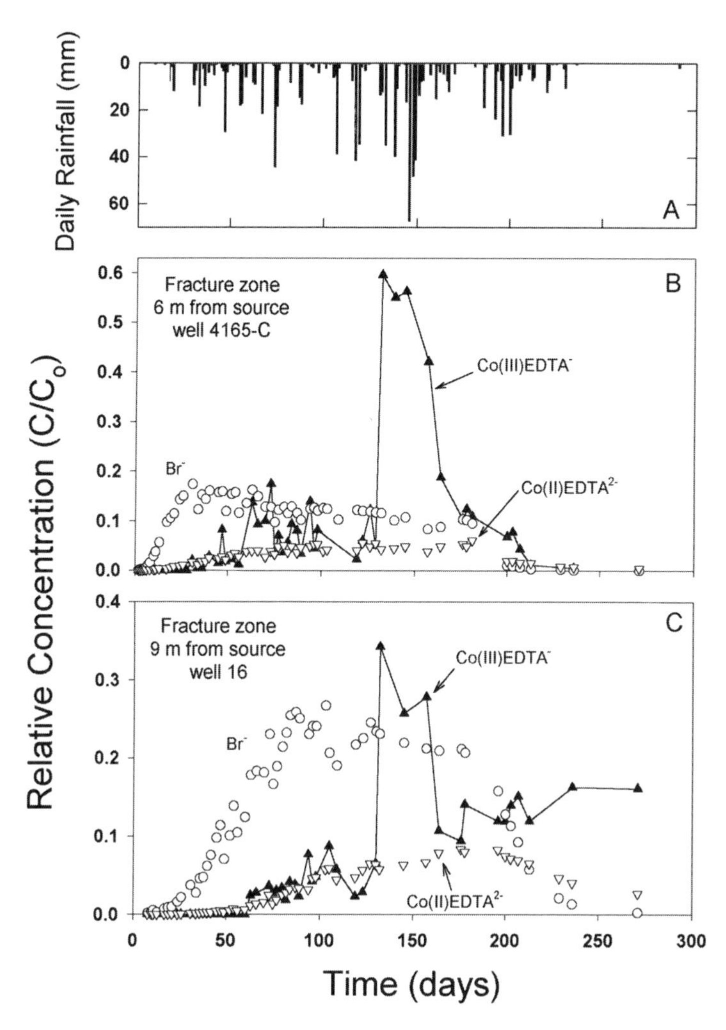

FIGURE 1.11 (a) Daily rainfall breakthrough curves for Br^-, $Co(II)EDTA^{2-}$, and $Co(III)EDTA^-$ in the fracture regime (b) 6 m (well 4165-C), and (c) 9 m (well 16) from the source. The y-axis is the relative concentration of each species with respect to itself.

since $Co(III)EDTA^-$ is extremely stable (log K = 10^{40}), which enhances its persistence and transport potential in subsurface environments. Thus, under the Fe-reducing conditions of this field site, limited $Co(III)EDTA^-$ production can be attributed to the lack of mineral oxides such as Fe- and Mn- oxides. It is noteworthy, however, that small increases in the concentration of $Co(III)EDTA^-$ were detected in the fracture regime during large storm events when groundwater recharge was accelerated (Figure 1.11). Although the relative concentrations of $Co(III)EDTA^-$ are large, the actual molar concentrations are 20 to 40 times less than $Co(II)EDTA^{2-}$. Nevertheless, during large storm events, dissolved oxygen concentrations in the fracture regime were found to increase dramatically from sub-part-per-billion levels to as high as 4 mg O_2/L^{-1} for numerous days. Laboratory studies have shown that

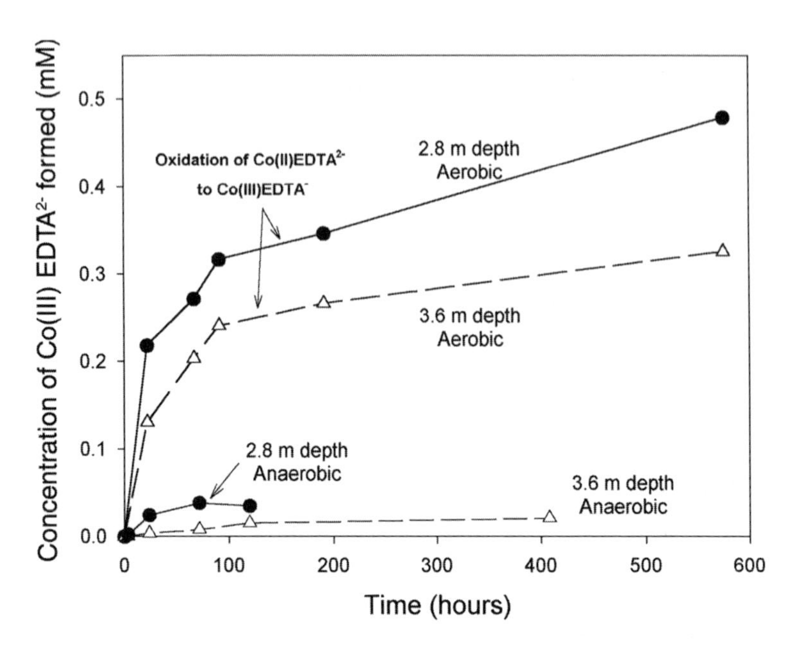

FIGURE 1.12 Batch kinetic studies investigating the oxidation of 1.0 mM Co(II)EDTA^{2-} to Co(III)EDTA^{-} by bedrock from the experimental site in the presence and absence of dissolved oxygen. Experimental conditions were similar to those of the site groundwater and included a constant ionic strength of 13 mM using $NaHCO_3$ and a pH of 6.8.

aqueous Co(II)EDTA^{2-} is stable for months in the presence of dissolved O_2. However, batch kinetic experiments using crushed bedrock from the site have shown that Co(II)EDTA^{2-} is oxidized to Co(III)EDTA^{-} under aerobic conditions where O_2 is present, but not under anaerobic, O_2-free conditions (Figure 1.12). These results suggest that dissolved O_2 interactions with the bedrock serve to catalyze, to a limited extent, the oxidation of Co(II)EDTA^{2-} to Co(III)EDTA^{-} during extreme recharge events. This process may be significant during the fate and transport of dilute concentrations of ^{60}Co-EDTA (μCi levels) that typically seep from shallow, land waste trenches, such as those present at the Oak Ridge National Laboratory (Means et al., 1978; Olsen et al., 1986).

1.2.2.3 Modeling

As in the multiple nonreactive tracer study described above, the one-dimensional fracture flow model CRAFLUSH (Sudicky and Frind, 1982) was used to simulate Br^{-} and chelated-metal migration within the fracture and matrix regime along the entire transect of sampling wells. The model was used to quantify bulk retardation (sorption/dissociation reactions) and diffusion parameters for describing the transport behavior of chelated contaminants in these systems. Only the ascending limbs of the Br^{-} and chelated-metal breakthrough curves were simulated, since the model assumes complete recovery of mass from the matrix upon tracer washout, and this was not observed over the timeframe of our experiment. Full mass recoveries from the matrix would require time periods significantly longer than those studied here.

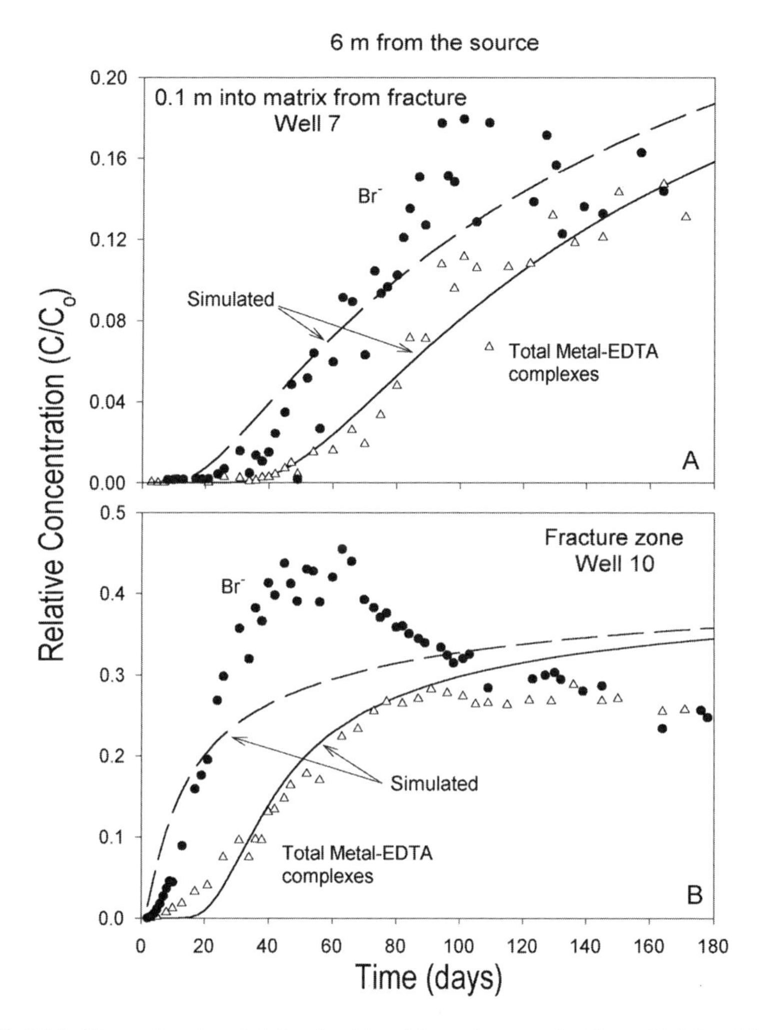

FIGURE 1.13 Observed and model-simulated breakthrough curves for Br^- and total metal-EDTA complexes within the (a) matrix and (b) fracture regimes at 6 m from the source, where $\theta = 0.20$ $cm^3\ cm^{-3}$ is the porosity, $2b = 8.4 \times 10^{-5}$ m is the fracture aperture, $2B = 2.0$ m is the fracture spacing, v=55 m d^{-1} is the mean pore water velocity within the fracture regime, $\alpha = 0.24$ m is the dispersivity, $De = 2.0 \times 10^{-4}$ $m^2\ d^{-1}$ is the molecular diffusion coefficient for Br^- and assume equivalent for the chelated metals, $\tau = 0.45$ is the tortuosity, and $z = 0.1$ m is the distance into the matrix. (From Jardine, P.M., et al., *J. Contamin. Hydrol.*, 55, 137, 2002. With permission.)

Matching simulated to observed Br^- mobility in the matrix (Figures 1.13(a) and 1.14(a)) produced reasonable simulations of Br^- transport through the fracture regime (Figures 1.13(b) and 1.14(b)) with no prior knowledge of the observed fracture regime data. Thus, a single set of model parameters could be used to assess solute diffusion processes from the fracture regime into the matrix regime across the entire field facility. These results are consistent with those obtained for the multiple nonreactive tracer study described above.

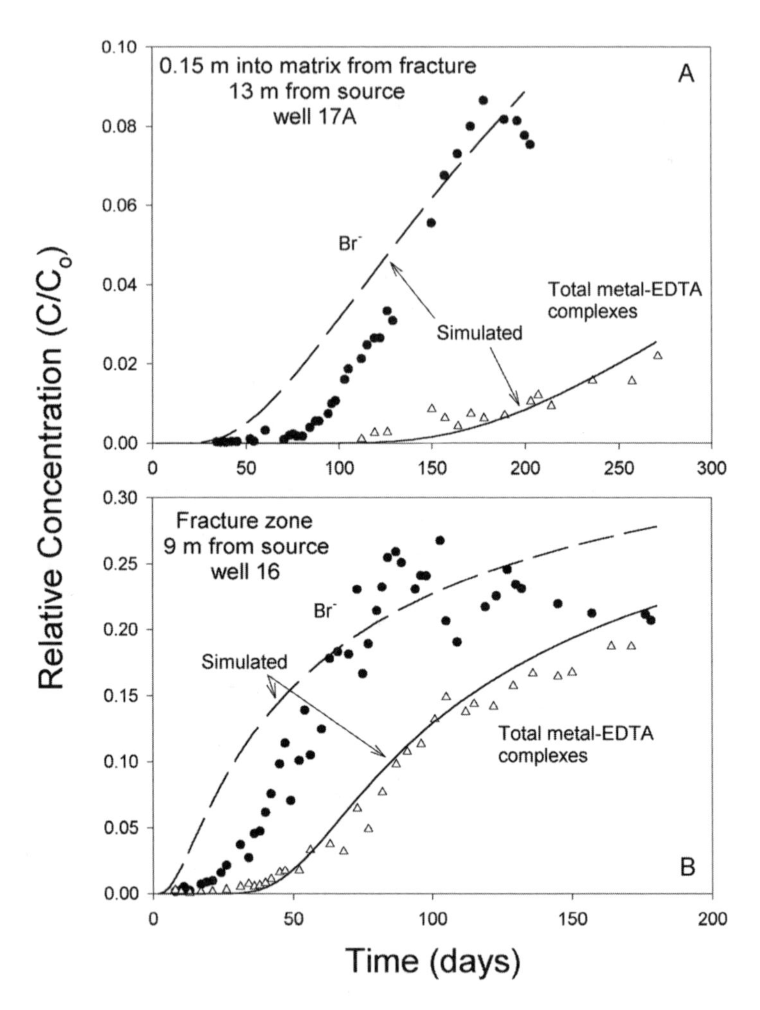

FIGURE 1.14 Observed and model-simulated breakthrough curves for Br^- and total metal-EDTA complexes within the matrix and fracture regimes (a and b, respectively) at either 9 m or 13 m from the source where $\theta = 0.20$ cm^3 cm^{-3} is the porosity, $2b = 8.4 \times 10^{-5}$ m is the fracture aperture, $2B = 2.0$ m is the fracture spacing, $v = 55$ m d^{-1} is the mean pore water velocity within the fracture regime, $\alpha = 0.24$ m is the dispersivity, $De = 2.0 \times 10^{-4}$ m^2 d^{-1} is the molecular diffusion coefficient for Br^- and assumed equivalent for the chelated metals, $\tau = 0.45$ is the tortuosity, and $z = 0.15$ m is the distance into the matrix. (From Jardine, P.M., et al., *J. Contamin. Hydrol.*, 55, 137, 2002. With permission.)

Model simulations involving chelated metals used the same parameters as those for Br^-, and the retardation coefficients in the fracture and matrix regime (R_f and R_m, respectively; see Sudicky and Frind, 1982) were varied to best match chelated metal movement into the matrix (Figures 1.13(a) and 1.14(a) and Table 1.1). Model simulations of total metal-EDTA complexes in the matrix (i.e., Co-, Cd-, and Fe-EDTA) produced good simulations of total chelated-metal migration of these complexes through the fracture regime (Figures 1.13(b) and 1.14(b)). The model also provided a good description of Co- and Cd-EDTA (original contaminants without Fe(III)EDTA byproduct) migration through both the fracture and matrix regimes (not shown). Model-simulated retardation coefficients suggested that metal-EDTA attenuation processes were dominant in the fracture regime, which was consistent with earlier observations of the data (Table 1.1) and suggested that physical retardation mechanisms dominated within the matrix regime. As expected, values of R_f and R_m were larger for Co- and Cd-EDTA species versus Co-, Cd-, and Fe-EDTA together, since $Fe(III)EDTA^-$ appeared to migrate conservatively through this subsurface environment. It should be noted that a single set of values for R_f and R_m could not be used to describe chelated-metal migration across the entire site, thus somewhat limiting the utility of this model for describing larger-scale reactive-contaminant migration in these environments. Variability in R_f and R_m is most likely related to spatially heterogeneous mineralogic properties that control the interfacial geochemical reactions that are responsible for the attenuation of the chelated metals and radionuclides. The model does, however, confirm the importance of chelated-metal attenuation along fractures. Since the fracture regime is the most hydrologically active zone, weathering processes are most likely accelerated and the exposure of primary minerals, such as Fe(III)-enriched biotite, is abundant. Certain metal-EDTA complexes that interact with the biotite are likely to be dissociated due to the strong competition of mineral Fe(III) for the EDTA chelate. These findings serve to enhance current conceptual models that previously assumed that chelated contaminants moved conservatively through these Fe-reducing subsurface environments.

1.3 SUMMARY AND A LOOK AHEAD

The research described above is motivated by the desire to provide an improved conceptual understanding and predictive capability of the geochemical and hydrological processes controlling contaminant migration from secondary sources (storage in soil/rock matrix) that plague waste disposal sites throughout the United States. Such research can unravel complex, time-dependent coupled processes, such as preferential flow, matrix diffusion, sorption, and redox transformations in heterogeneous media. Basic research strategies designed around novel tracer techniques and experimental manipulations not only improve our conceptual understanding of time-dependent contaminant migration in subsurface media, they also provide the necessary experimental constraints needed for the accurate numerical quantification of the coupled nonequilibrium processes. Such information is critical to contaminant fate and transport modeling and risk assessment modeling. Too often risk assessment models treat soil and bedrock as inert media or assume that they are in equilibrium with migrating contaminants. Failure to consider the time-dependent significance of secondary contaminant sources will greatly overpredict the off-site contribution of contaminants from the primary trench sources and thus provide an inaccurate assessment of pending risk. Basic research provides the necessary information needed for more appropriate risk modeling strategies that can potentially translate into multimillion-dollar savings by eliminating the need for certain inappropriate remedial efforts. These research endeavors also provide information that is necessary for improving our decision-making strategies regarding the selection of effective remedial actions and the interpretation of monitoring results after remediation is complete. Simple distinctions of how fracture networks operate to disseminate contaminants versus matrix storage mechanistic are basic research issues that can drastically influence the type and cost of mandated remedial efforts. Such research endeavors provide a more thorough understanding of coupled transport mechanisms so that remediation concepts are not limited to a purely empirical approach. This will allow engineers to develop remediation strategies targeted at specific problems and with a higher probability of success.

ACKNOWLEDGMENTS

This research was supported by the U.S. Department of Energy's Office of Science, and is a contribution from the Natural and Accelerated Bioremediation (NABIR) Program of the Office of Biological and Environmental Research. The authors would like to thank Mr. Paul Bayer and Dr. Anna Palmisano, program managers of DOE, for financially supporting this research. Oak Ridge National Laboratory is managed by the University of Tennessee — Battelle, LLC, under contract DE-AC05-00OR22725 with the U.S. Department of Energy.

REFERENCES

Ayres, J.A., Equipment Decontamination with Special Attention to Solid Waste Treatment, Survey Report BNWL-B-90. Battelle Northwest Laboratories, Richland, WA, 1971.

Brooks, S.C., Taylor, D.L., and Jardine, P.M., Reactive transport of EDTA-complexed cobalt in the presence of ferrihydrite, *Geochim. Cosmochim. Acta*, 60, 1899, 1996.

Brooks, S.C., Carroll, S.L., and Jardine, P.M., Sustained bacterial reduction of Co(III)EDTA - in the presence of competing geochemical oxidation during dynamic flow, *Environ. Sci. Technol.*, 33, 3002, 1999.

Davis, J.A. et al., Multispecies reactive tracer test in an aquifer with spatially variable chemical conditions, *Water Resour. Res.*, 36, 119, 2000.

Dorsch, J. et al., Effective Porosity and Pore-Throat Size of Conasauga Group Mudrock: Application, Test and Evaluation of Petrophysical Techniques, Oak Ridge National Laboratory, Oak Ridge, TN, ORNL/GWPO-021, 1996.

Dreier, R.B., Solomon, D.K., and Beaudoin, C.M., Fracture characterization in the unsaturated zone of a shallow land burial facility, in Flow and Transport Through Unsaturated Rock, Evans, D.D. and Nicholson, T.J., Eds., *Geophys. Monogr.*, 42, 51, 1987.

Jardine, P.M., Jacobs, G.K., and O'Dell, J.D., Unsaturated transport processes in undisturbed heterogeneous porous media II: Co-contaminants, *Soil Sci. Soc. Am. J.*, 57, 954, 1993.

Jardine, P.M. and Taylor, D.L., Kinetics and mechanisms of Co(II)EDTA oxidation by pyrolusite, *Geochim. Cosmochim. Acta*, 59, 4193, 1995a.

Jardine, P.M. and Taylor, D.L., Fate and Transport of Ethylenediaminetetraacetate Chelated Contaminants in Subsurface Environments, in *Soil Environmental Chemistry*, D.L. Sparks, Ed., Elsevier Science Publishers, Amsterdam (*Geoderma*, 67, 125), 1995b.

Jardine, P.M. et al., Quantifying diffusive mass transfer in fractured shale bedrock, *Water Resour. Res.*, 35, 2015, 1999.

Jardine, P.M. et al., Conceptual model of vadose-zone transport in fractured weathered shales. In *Conceptual Models of Flow and Transport in the Fractured Vadose Zone*, U.S. National Committee for Rock Mechanics, National Research Council. National Academy Press, Washington DC, 2001, p. 87.

Jardine, P.M. et al., Influence of hydrological and geochemical processes on the transport of chelated-metals and chromate in fractured shale bedrock, *J. Contamin. Hydrol.*, 55, 137, 2002.

Kent, D.B. et al., Transport of Zinc in the Presence of a Strong Complexing Agent in a Shallow Aquifer on Cape Cod, Massachusetts, U.S. Geological Survey Water-Resources Investigations, Rep. 91-4034, 1991, p. 78.

Kent, D.B. et al., Ligand-Enhanced Transport of Strongly Adsorbing Metal Ions in the Groundwater Environment, in Proceedings of the 7th International Symposium on Water-Rock Interaction, WRI-7, Kharaka, Y.K. and Maest, A.S., Eds., Park City, Utah, 13–18 July, 1992, 805.

Maloszewski, P. and Zuber, A., Mathematical modeling of tracer behavior in short-term experiments in fissured rocks, *Water Resour. Res.*, 26, 1517, 1990.

Maloszewski, P. and Zuber, A., Tracer experiments in fractured rocks: Matrix diffusion and the validity of models, *Water Resour. Res.*, 29, 2723, 1993.

Mayes, M.A. et al., Multispecies transport of metal-EDTA complexes and chromate through undisturbed columns of weathered, fractured saprolite, *J. Contam. Hydrol.*, 45, 265, 2000.

McKay, L.D., Sanford, W.E., and Strong, J.M., Field scale migration of colloidal tracers in a fractured shale saprolite, *Ground Water*, 38, 139, 2000.

Means, J.L. and Alexander, C.A., The environmental biogeochemistry of chelating agents and recommendations for the disposal of chelated radioactive wastes, *Nucl.Chem. Waste Manage.* 2, 196, 1981.

Means, J.L., Crerar, D.A., and Duguid, J.O., Migration of radioactive wastes: Radionuclide mobilization by complexing agents, *Science* (Washington, DC), 200, 1477, 1978.

Olsen, C.R. et al., Geochemical and environmental processes affecting radionuclide migration from a formerly used seepage trench, *Geochim. Cosmochim. Acta*, 50, 607, 1986.

Rath, R.K. and S. Subramanian, Studies on adsorption of guar gum onto biotite mica, *Miner. Eng.*, 10, 1405, 1997.

Riley, R.G. and Zachara, J.M., Chemical Contaminants on DOE Lands and Selection of Contaminant Mixtures for Subsurface Science Research, DOE/ER-0547T, U.S. Government Printing Office, Washington, DC, 1992.

Sanford, W.E., Shropshire, R.G., and Solomon, D.K., Dissolved gas tracers in groundwater: Simplified injection, sampling, and analysis. *Water Resour. Res.*, 32, 1635, 1996.

Sanford, W.E. and Solomon, D.K., Using dissolved gases for site characterization and containment assessment, *J. Environ. Eng.*, 124, 572, 1998.

Sudicky, E.A. and Frind, E.O., Contaminant transport in fractured porous media: Analytical solutions for a system of parallel fractures. *Water Resour. Res.*, 18, 1634, 1982.

Szecsody, J.E. et al., Importance of flow and particle-scale heterogeneity on Co(II/III)EDTA reactive transport, *J. Hydrol.*, 209, 112, 1998a.

Szecsody, J.E. et al., Influence of iron oxide inclusion shape on CoII/IIIEDTA reactive transport through spatially heterogeneous sediment, *Water Resour. Res.*, 34, 2501, 1998b.

Toste, A.P. and Lechner-Fish, T.J., Organic digenesis in commercial, low-level nuclear wastes. *Radioactive Waste Manage. Nucl. Fuel Cycle*, 12, 291, 1989.

Williams, G.M. et al., The influence of organics in field migration experiments. Part 1. In situ tracer tests and preliminary modeling, *Radiochimica Acta*, 52/53, 463, 1991.

Wilson, G.V., Jardine, P.M., and Gwo, J.P., Modeling the hydraulic properties of a multiregion soil, *Soil Sci. Soc. Am. J.*, 56, 1731, 1992.

Zachara, J.M. et al., Oxidation and adsorption of Co(II)EDTA2- complexes in subsurface materials with iron and manganese oxide grain coatings, *Geochim. Cosmochim. Acta*, 59, 4449, 1995.

2 Mineral Controls in Colloid-Mediated Transport of Metals in Soil Environments

A.D. Karathanasis

CONTENTS

2.1 Introduction 26
2.2 Case Study 1 27
  2.2.1 Metal Solutions and Colloid Suspensions 27
  2.2.2 Soil Monoliths 28
  2.2.3 Leaching Experiments 28
  2.2.4 Eluent Characterization 29
  2.2.5 Colloid Elution 29
  2.2.6 Metal Transport 30
2.3 Case Study 2 34
  2.3.1 Metal Solutions and Colloid Suspensions 35
  2.3.2 Soil Monoliths 35
  2.3.3 Leaching Experiments 35
  2.3.4 Colloid Elution 36
  2.3.5 Elution of Desorbed Pb 37
2.4 Case Study 3 39
  2.4.1 Metal Solutions and Biosolid Colloid Fractions 39
  2.4.2 Leaching Experiments 40
  2.4.3 Biosolid Colloid Elution 40
  2.4.4 Metal Elution 42
2.5 Summary 45
2.6 Conclusions 46
References 47

1-56670-623-8/03/$0.00+$1.50

2.1 INTRODUCTION

In recent years, improper disposal of various waste materials has posed serious threats to surface and groundwater supplies and developed into a global scale soil and water pollution problem [1]. Heavy metals account for much of the contamination found at hazardous waste sites in the United States, and have been detected in the soil and groundwater at approximately 65% of the U.S. Environmental Protection Agency Superfund sites [2]. Dramatic increases in land application of agricultural and municipal biosolids have accentuated the problem. In spite of their beneficial contributions as nutrient sources and soil conditioners, these amendments, if not monitored, pose a considerable environmental risk because of their high heavy-metal concentrations [3].

Traditionally, hydrophobic environmental contaminants such as heavy metals were assumed to be relatively immobile in subsurface soil environments because they are strongly sorbed by the soil matrix. However, under certain conditions colloid particles may exceed ordinary transport rates and pose a significant threat to surface and groundwater quality. This threat has been substantiated by recent research evidence showing that water-dispersed colloidal particles migrating through soil macropores and fractures can significantly enhance metal mobility, causing dramatic increases in transported metal load and migration distances [4–8]. Due to a large surface area (100 to 500 m^2g^{-1}) [6] and potentially high surface charge [9], partition coefficients and sorption energies of the colloidal phase may be sufficiently high to exhibit preferential sorption for soluble metals over that of the immobile solid phase [10]. In highly contaminated sites, colloids may even strip metals from the soil matrix to establish a new equilibrium between the two solid phases [4].

Laboratory-scale research experiments with packed or undisturbed soil columns have clearly demonstrated significant colloid-mediated transport of herbicides [11] and heavy metals [12–15] with or without association of organic coatings. Colloid-facilitated transport has been documented as the dominant transport pathway for strongly sorbing metal contaminants, with solute model–predicted amounts being underestimated by several orders of magnitude [16]. Some mineralogical preferences in colloid generation and mobility in reconstructed soil pedons have also been documented, but no association trends with contaminants were established [9]. Colloid-facilitated transport of contaminants has also been reported in several field scale investigations. In groundwater samples of underground nuclear test cavities at the Nevada site, virtually all the activity of Mn, Co, Sb, Cs, Ce, and Cu was associated with colloidal particles [17]. Significant associations of Cr, Ni, Cu, Cd, Pb, and U with groundwater colloids were also found in an acidified sandy aquifer [18]. Organic colloid migration following humus disintegration has been found to be the main transport mechanism for Pb in subsoils of forested ecosystems in Switzerland affected by the nearby aluminum industry [19]. Similarly, the degree of metal-colloid association in pineland streams in New Jersey was controlled by the metal affinity for humic materials [20]. However, other studies have reported metal partitioning and binding potential differences between suspended particulate material and dissolved organic carbon (DOC) carried in two contrasting Wisconsin watersheds due to variability in their composition [21]. Similarly, Fe-and Al-rich colloids were found to play a significant role in transporting Cu, Pb, and Zn in stream discharges affected by AMD in Colorado, depending on pH and colloid concentration [22]. Other studies have suggested that sludge particulates have strong affinity toward metal ions, with the carboxyl moiety being the major surface functional group controlling the association as a function of pH [23].

Although the potential role of colloid particles as carriers or facilitators of contaminants has been well documented, most of the research findings have emphasized the importance of organic constituents or organic coatings on colloid particles as major contributors in the co-transport process, while paying very little attention to contributions of associated mineral colloids with variable composition [24–28]. However, in many cases the generally higher binding energies of trace metals to mineral- rather than organic-colloid surfaces may render high-surface-charge mineral colloids more potent carriers of metal contaminants [29]. Recent studies demonstrated that colloid generation and associated contaminant transport processes in surface and subsurface environments may be significantly affected by complex couplings and reactivity modifications of permanent charge phyllosilicates and variable charge Fe-oxyhydroxide phases [30]. Furthermore, information on contaminant–mineral interactions and colloid-mediated transport derived from model mineral systems cannot be readily extrapolated to complex mineral assemblages of natural systems without adequate experimentation.

The objectives of this study were (1) to assess the effect of colloid mineralogical composition on colloid-mediated transport of metals in subsurface soil environments, and (2) to establish physicochemical gradients and conditions enhancing or inhibiting colloid-mediated transport. The following case studies were used to demonstrate the effects of mineralogy on colloid-mediated transport of metals.

2.2 CASE STUDY 1

In this experiment, *ex situ* soil colloids with diverse mineralogical composition after equilibration with metal solutions of known concentrations were leached through undisturbed soil monoliths exhibiting considerable macroporosity. The colloids (<2 μm) were separated from upper-soil Bt horizons with montmorillonitic, illitic, and kaolinitic mineralogy. The equilibration metal solutions contained Cu, Zn, and Pb. Eluents were monitored over ten pore volumes for colloid and metal concentrations.

2.2.1 Metal Solutions and Colloid Suspensions

Aqueous solutions (10 mg/l^{-1}) of Cu, Zn, and Pb were prepared from $CuCl_2$, $ZnCl_2$, and $PbCl_2$ reagents (>99% purity, Aldrich Chemicals, Milwaukee, WI). These solutions were used as controls and in mixtures with 300 mg/l^{-1} colloid suspensions in the leaching experiments. The same metal chloride reagents were used to prepare the equilibrium solutions in adsorption isotherm experiments for metal affinity determinations.

Water-dispersible colloids were fractionated from upper Bt horizons of three soils representing the series: Beasley (fine, smectitic, mesic Typic Hapludalfs), Shrouts (fine, illitic, mesic Typic Hapludalfs), and Waynesboro (fine, kaolinitic, thermic Typic Paleudults). The extraction of the WDC fractions (<2 μm) was accomplished by mixing ~10 g of soil with 200 ml of deionized H_2O (without addition of dispersing agent) in plastic bottles, shaking overnight, centrifuging at 750 rpm (× 130 *g*) for 3.5 min, and decanting. The concentration of the colloid fraction was determined gravimetrically. Physicochemical and mineralogical properties of the colloid fractions were determined following methods of the U.S. Department of Agriculture-National Soil Survey Center [31] (Table 2.1). Metal-colloid adsorption isotherms were constructed following batch equilibrium experiments to determine Freundlich metal distribution coefficients (K_f) [29].

2.2.2 Soil Monoliths

Upper Bt horizons of a Maury (fine, mixed, semiactive, mesic Typic Paleudalf) and a Loradale (fine, mixed, semi-active, mesic Typic Argiudoll) soil, which in previous studies had exhibited considerable macroporosity and preferential flow, were used for the leaching experiments. Undisturbed soil monoliths of 15-cm diameter and 20 cm length were prepared in the field by carving cylindrically shaped pedestals and encasing them with a PVC pipe of a slightly larger diameter. The annulus was sealed with expansible polyurethane foam to prevent preferential flow along the PVC walls. Physicochemical and mineralogical properties of the soils [29] are shown in Table 2.1. Freundlich metal distribution coefficients (K_f) for the two soils were determined from adsorption isotherms, following the same procedure used for the colloids [29].

2.2.3 Leaching Experiments

Prior to setting up the leaching experiment, four undisturbed soil monoliths from each soil were saturated from the bottom upward with deionized water (D-H_2O) to remove air pockets. Then, about three pore volumes of D-H_2O containing 0.002% NaN_3 were introduced into each monolith (downward vertical gravity flow) using a peristaltic pump at a constant flux (2.2 cm/h^{-1}) to remove loose material from the pores of the soil monoliths. One of the monoliths was used to evaluate the elution of a conservative tracer (1 mM of $CaCl_2$) for comparison with the colloid elution patterns. A metal solution containing 10 mg/l^{-1} of Cu, Zn, and Pb (without colloids) was passed through the second monolith, representing the control treatment. Each one of the other two monoliths received a mixture of 300 mg/l^{-1} colloid suspension and 10 mg/l^{-1} metal solution, following a 24-h equilibration period. All solutions and suspensions were applied to the top of the monoliths with a continuous step input of 2.2 cm/h^{-1}, controlled by the peristaltic pump. Eluents were monitored periodically with respect to volume, Cl^-, colloid, and metal concentration. Breakthrough curves (BTCs) were constructed based on reduced concentrations (ratio of effluent concentration to influent concentration, C/C_o) and pore volumes (flux averaged volume of solution pumped per monolith pore volume).

2.2.4 Eluent Characterization

Colloid concentrations in the eluent were determined with a Bio-Tek multichannel (optical densitometer with fiber-optics technology; Bio-Tek Instruments, Winooski, VT) microplate reader, precalibrated with known concentrations of each colloid at 540 nm. Total metal concentration in the eluents was allocated to solution phase and colloidal phase (colloid-bound contaminant). The eluent samples were centrifuged for 30 min at 3500 rpm (× 2750 *g*) to separate the soluble contaminant fraction from the colloid-bound contaminant fraction. The absence of colloidal material in the supernatant solution was verified by filtration through a 0.2-μm membrane filter. The soluble metal (Cu, Zn, Pb) fractions were analyzed by atomic absorption (concentrations >0.5 mg/l^{-1}) or inductively coupled plasma (ICP) spectrometry (concentrations <0.5 mg/l^{-1}). The colloid fraction was extracted with 1 *M* HNO_3-HCl [32] solution and analyzed with the same methodology used for the soluble fraction. The results for the duplicate soil monoliths and for the two soils were combined for practical purposes, because the reproducibility between soil monoliths was within ±15%.

2.2.5 Colloid Elution

In spite of some tailing in the BTCs of the conservative Cl^- tracer, suggesting some preferential flow, Cl^- elution was generally symmetrical. In contrast, the colloid breakthrough was gradual and somewhat irregular, indicative of the dynamic interactions between matrix, colloids, and solutes occurring during the leaching process (Figure 2.1). Colloid recovery maxima varied by metal saturation and colloid mineralogy, ranging from a high of about 1.00 C/C_0 for the Zn-saturated montmorillonitic colloids to a low of about 0.20 C/C_0 for the Zn-saturated kaolinitic colloids. Generally, colloid breakthrough decreased according to the metal saturation sequence Zn > Cu > Pb, and the mineralogy sequence montmorillonitic > illitic > kaolinitic. The somewhat higher recovery maxima for the Zn colloids are attributed to the lower affinity (K_f) of Zn for the soil matrix (Table 2.1). The greater overall mobility of the montmorillonitic and illitic colloids is consistent with their lower mean size diameter and the more negative electrophoretic mobility, which limited particle filtration by the soil matrix. The elevated pH associated with the colloids (Table 2.1) may have also enhanced their stability and transportability. Settling rate experiments (Figure 2.2) indicated a decline in the concentration of kaolinitic colloids remaining in suspension at pH <5.5 compared to the illitic and montmorillonitic colloids, in spite of high stability at pH levels >6.0. The reduced stability of the kaolinitic colloids is associated with their low pH (5.2), which is closer to their pH_{zpc} range compared to the illitic or montmorillonitic colloids (Table 2.1). Metal saturation is expected to induce easier coagulation and flocculation of the kaolinitic colloids due to a significant reduction in the net surface potential. It is also likely that the stability of the montmorillonitic and illitic colloids was enhanced by their higher OC content (Table 2.1). According to Kretzschmar et al. [26], organic coatings promote colloid stability through steric hindrance effects. In contrast, the mobility of the kaolinitic colloids may have been deterred further by their high Fe and Al hydroxide content; Fe and

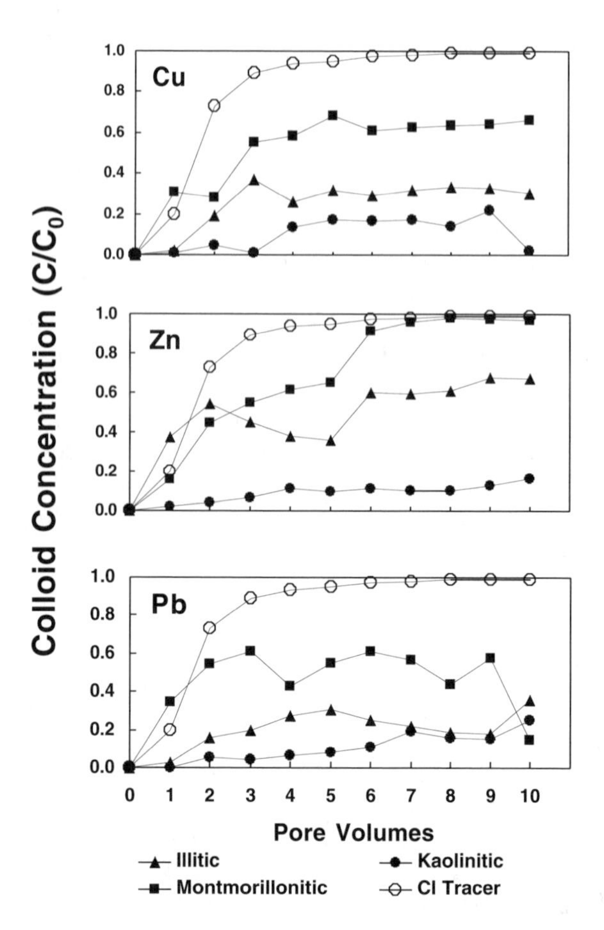

FIGURE 2.1 BTCs for Cu, Zn, and Pb soil colloids with montmorillonitic, illitic, or kaolinitic mineralogy eluted from the soil monoliths.

Al hydroxide are known to act as binding agents and induce flocculation [33]. In all cases, eluent electrical conductivity values (EC), and therefore ionic strength, remained low (50–100 μS cm^{-1}) during the course of the leaching experiment, suggesting that the electrochemical conditions were not conducive for adequate suppression of the thickness of the double layer that would sufficiently reduce the electrostatic repulsive forces between colloid particles and cause flocculation [34].

2.2.6 Metal Transport

Figures 2.3 and 2.4 show breakthrough curves for total and soluble metal fractions, respectively, eluted in the absence and presence of colloids. In the absence of colloids (controls) practically none of the metals exhibited any meaningful breakthrough, suggesting nearly complete sorption by the soil matrix (Figure 2.3). The presence of colloids enhanced considerably total metal elution and in most cases even soluble metal elution, thus providing strong evidence for colloid-mediated metal transport.

TABLE 2.1
Physicochemical and Mineralogical Properties of Soils and Colloids Used in the Case Studies

	Soils		Colloids					
Properties	Loradale	Maury	Montmorillonitic	Mixed	Illitic	Kaolinitic	LSB+$CaCO_3$	LSB–$CaCO_3$
Clay (%)	21	35	—	—	—	—	—	—
Hydraulic conductivity (cm min^{-1})	1.3 ± 0.5	2.6 ± 1.7	—	—	—	—	—	—
Bulk density (g cm^{-3})	1.5	1.6	—	—	—	—	—	—
Mean colloid diameter (nm)[a]	—	—	220	300	270	1050	410	360
Organic C (%)	2.1	0.5	0.8	3.4	0.8	0.4	20	38
pH	6.3	5.8	6.2	6.7	5.8	5.2	~11.0	~7.0
CEC (cmol kg^{-1-})	25.2	21.9	63.4	81.8	46.4	29.0	32.0	60.0
Extractable bases (cmol kg^{-1-})	15.0	10.1	26.5	29.2	17.3	8.1	—	—
Dithionite extractable Fe (mg g^{-1})	6.5	8.3	15.9	15.9	16.4	75.7	—	—
Dithionite extractable Al (mg g^{-1})	4.4	2.8	6.1	5.2	9.2	61.3	—	—
Surface area (m^2g^{-1})	83	65	386	186	123	114	360	400
Electrophoretic mobility (μm cm $v^{-1}s^{-1}$)	—	—	−1.8	−1.9	−1.6	−0.8	—	—
Smectite+vermiculite (%)	—	—	60	—	17	—	—	—
HISM+HIV (%)	10	15	—	44	—	21	—	—
Mica (%)	30	20	20	15	60	11	{5}	{15}
Kaolinite (%)	15	20	16	35	20	56	—	—
Quartz (%)	40	40	4	6	3	12	—	—
$CaCO_3$	—	—	—	—	—	—	55	5
K_f(Cu)	1.99	1.14	2.82	3.93	0.83	0.55	1.29	1.56
K_f(Zn)	1.17	0.78	1.95	3.22	1.19	0.93	7.44	2.90
K_f(Pb)	0.60	1.75	11.43	15.29	4.15	2.69	6.61	5.53

[a] Mean colloid diameter is expressed on a mass basis as measured by a microscan particle-size analyzer.

Note: CEC = cation exchange capacity; HISM = hydroxyinterlayered smectite; HIV = hydroxyinterlayered vermiculite; K_f = Freundlich metal distribution coefficients; LSB = lime stabilized biosolids; { } = total aluminosilicates.

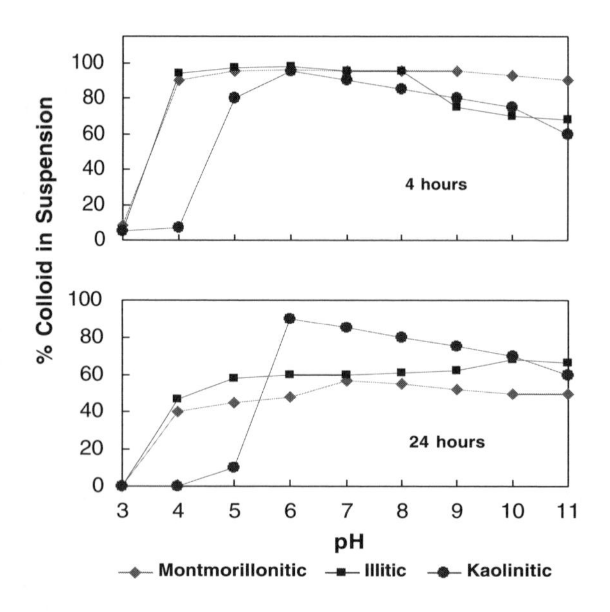

FIGURE 2.2 Settling kinetics curves for soil colloids with montmorillonitic, illitic, or kaolinitic mineralogy.

Most BTCs showed considerable asymmetry, attributed to partial clogging and flushing cycles and/or chemical interactions among solutes, colloids, and soil matrix. These interactions are anticipated considering colloid attachment/detachment phases and the different affinities of metals for colloid and soil surfaces (Table 2.1). Generally, total metal elution was higher than soluble metal elution. Considering that the difference between total metal and soluble metal load represents the colloid-bound fraction and given the strong correlation between total metal and colloid elution, it could be rationalized that the colloids are acting as carriers of the majority of the metal load. As was the case with the colloid elution, the metal load carrying efficiency followed the sequence montmorillonitic > illitic > kaolinitic, indicating a strong relationship with colloid surface charge properties. Therefore, this provides compelling evidence that the primary mechanism for the enhanced metal transport is mainly metal chemisorption to reactive colloid surfaces, especially in cases where metal affinity for colloid sites is greater than that for soil matrix sites. However, competitive metal sorption between colloid and soil matrix may also occur during the leaching cycle, in spite of metal affinities, in order to establish local equilibrium between the two solid phases.

Metal transport increases were also metal specific, following the sequence Zn > Pb > Cu for total metal elution and Zn > Cu > Pb for soluble metal elution. Overall, however, between 30 and 90% of Cu was transported in the soluble fraction, while >60% of Zn and Pb were transported in the colloid-sorbed fraction. This is generally consistent with the metal affinities of the different colloids in conjunction with OC content and colloid size differences. Average increases of total Cu transport in the presence of colloids were three-fold for kaolinitic, five-fold for illitic, and six-fold for montmorillonitic colloids compared to the controls. The respective average

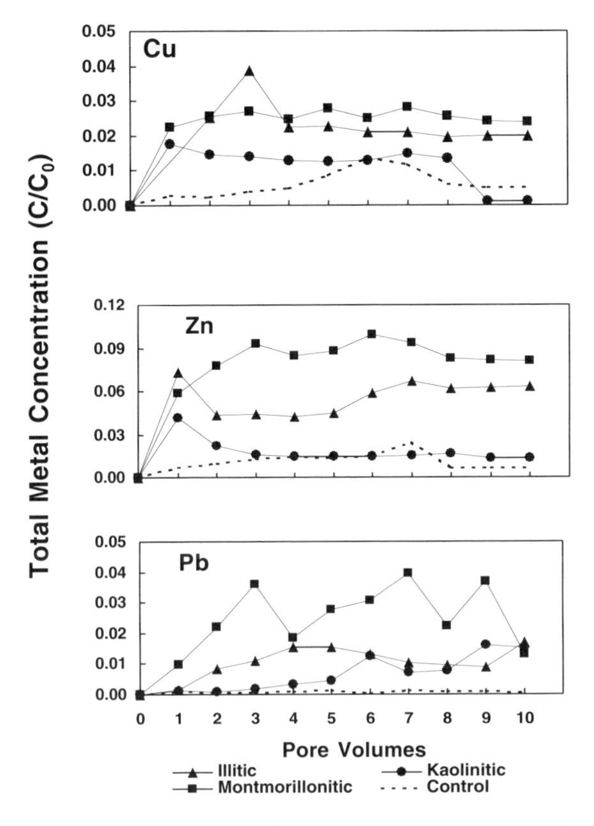

FIGURE 2.3 BTCs for total Cu, Zn, and Pb eluted in the presence or absence (control) of soil colloids with montmorillonitic, illitic, or kaolinitic mineralogy.

increases for Zn transport were 1.5-fold for kaolinitic, six-fold for illitic, and nine-fold for montmorillonitic colloids. Average increases for total Pb were the highest, ranging from seven-fold for kaolinitic up to 30-fold for montmorillonitic colloids. Average soluble metal elution increases were not as dramatic for Cu and Zn (up to three-fold), but more substantial for Pb (up to 11-fold), with the maxima being associated with either montmorillonitic or illitic colloids. The similar soluble Cu load transported by all colloids regardless of mineralogy is attributed to the strong affinity of this metal to form organic complexes. This mechanism may also be partially responsible for the additional soluble metal loads of Zn and Pb recovered in the presence of colloids. Furthermore, exclusion of soluble metal species from soil matrix sites blocked by colloids and elution of metal ions associated with the diffuse layer of colloid particles may have increased the soluble metal load.

These findings clearly demonstrate the role of colloid mineralogical composition on their ability to induce and mediate the transport of heavy metals in subsurface soil environments. In all treatments, the magnitude of colloid-mediated metal transport decreased according to the sequence montmorillonitic > illitic > kaolinitic. In spite of considerable differences between the two soils in terms of physical and

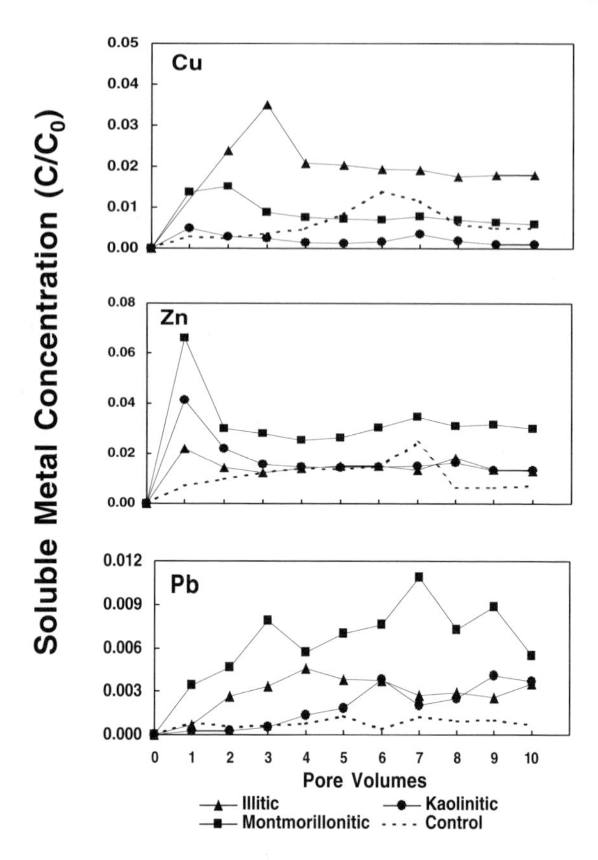

FIGURE 2.4 BTCs for soluble Cu, Zn, and Pb eluted in the presence or absence (control) of soil colloids with montmorillonitic, illitic, or kaolinitic mineralogy.

chemical properties, these trends remained consistent, with <15% variability in metal elution. These relationships appear to be influenced primarily by inherent and/or accessory mineralogical and physicochemical properties of the colloids, such as surface charge, surface area, electrophoretic mobility, and mean colloid diameter, and much less by coincidental factors, such as OC, pH, Fe-Al hydroxides, and ionic strengths, normally encountered in soil environments.

2.3 CASE STUDY 2

This study investigated the potential of *ex situ* water-dispersible colloids with diverse mineralogical composition to desorb Pb from the contaminated soil matrix of undisturbed soil monoliths and co-transport it to groundwater. The study employed intact monoliths contaminated by Pb, which were flushed with colloid suspensions of different mineralogical composition and D-H_2O, used as a control. The soil monoliths represented upper solum horizons of the soils used in Case Study 1 (Maury and Loradale). The soil colloids were fractionated from low ionic strength Bt horizons of Alfisols with montmorillonitic, mixed, and illitic mineralogy.

2.3.1 Metal Solutions and Colloid Suspensions

An aqueous solution of 100 mg/l^{-1} was prepared from a $PbCl_2$ reagent (>99% purity, Aldrich Chemicals, Milwaukee, WI). This solution was used in the contamination phase of the leaching experiments. The same $PbCl_2$ reagent was used to prepare the equilibrium solutions for the adsorption isotherm experiments [29] from which the K_f values were determined (Table 2.1). Water-dispersible colloids (WDCs) were fractionated from upper Bt horizons of three soils representing the series: Beasley (fine, smectitic, mesic Typic Hapludalf), Loradale (fine, mixed, semiactive, mesic Typic Argiudoll), and Shrouts (fine, illitic, mesic Typic Hapludalf), using the procedure described in Case Study 1. Physicochemical and mineralogical properties of the colloid fractions are shown in Table 2.1.

2.3.2 Soil Monoliths

The same soils and the same procedure used in Case Study 1 were used to prepare the undisturbed soil monoliths used in this experiment. Their physicochemical and mineralogical properties are also reported in Table 2.1.

2.3.3 Leaching Experiments

Four soil monoliths from each soil were used in the leaching experiment. Before initiating the contamination phase, the monoliths were saturated from the bottom up with D-H_2O to remove air pockets, and then leached with about three pore volumes of D-H_2O containing 0.002% NaN_3 to remove loose material from the pores of the soil monoliths and suppress biological activity. Subsequently, the monoliths were leached with a 100 mg/l^{-1} Pb flushing solution at a rate of 2.2 cm/h^{-1} for 350 to 400 pore volumes to achieve a certain level of Pb contamination. The target level of contamination was considered reached when the eluted Pb attained a concentration of about 5 mg/l^{-1}, which corresponded to about 40% saturation of the soil matrix as determined at the end of the experiment. At that point, a flushing solution consisting of D-H_2O was applied to a replicate set of monoliths from each soil (controls) at a constant flux of 2.2 cm/h^{-1} for the next 25 to 28 pore volumes. Each of the remaining replicate monolith sets received a flushing suspension consisting of 300 mg/l^{-1} colloid (one for each soil and colloid type) in D-H_2O. Eluents were monitored periodically with respect to volume, colloid, and Pb concentration. Breakthrough curves (BTC) were constructed based on normalized Pb and colloid conccntrations (C/C_o) and pore volumes. Λ value of C_o -·5 mg/l^{-1} was used for Pb, and C_o = 300 mg/l^{-1} was used for colloids.

Colloid concentrations in the eluent were determined by placing 200 ml of the sample into a Bio-Tek multichannel (optical densitometer with fiber-optics technology; Bio-Tek Instruments, Inc., Winooski, VT) microplate reader and scanning at 540 nm. Total Pb concentration in the eluents was allocated to solution phase and colloidal phase. The eluent samples were centrifuged for 30 min at 3500 rpm to separate the soluble contaminant fraction from the colloid-bound contaminant fraction. The colloid-bound Pb was extracted with 1 N HCl-HNO_3 solution, and along with the soluble Pb fraction, was analyzed by ICP spectrometry.

2.3.4 Colloid Elution

Colloid breakthrough in flushing suspensions was irregular but greater than anticipated, considering that the soil monoliths were nearly 40% saturated with Pb (Figure 2.5). Apparently, very little soluble Pb remained in the pore space of the saturated monoliths to cause sufficient colloid flocculation and filtration, while most was tightly held by soil matrix sites. Colloid elution in the presence of all three colloids increased sharply during the first five pore volumes of leaching, thereafter experiencing a more gradual increase before tailing off between 0.55 and 0.80 C/C_o. No colloid elution was observed in D-H_2O (control) flushing solutions. No significant differences in the breakthrough of montmorillonitic and mixed colloids were observed during the first 10 to 15 pore volumes, but the illitic colloid maintained a lower elution throughout the leaching cycle. After the 15th pore volume, the colloids experienced another surge during which they reached maxima of 0.90 C/C_o for montmorillonitic, 0.70 for mixed, and 0.60 for illitic. These differences are associated with the lowest mean colloid diameter of the montmorillonitic colloids, the highest electrophoretic mobility of the illitic colloids, and the higher pH and organic carbon content of the mixed colloids, which probably makes up for their larger overall mean colloid diameter. The irregular colloid breakthrough pattern is indicative of the dynamic nature of the leaching process and the physical and chemical interactions occurring within the soil matrix. The observed colloid breakthrough thresholds are attributed to steady state porosities reached by the monoliths as a function of colloid flux and colloid filtration rates that compromised a portion of the originally available colloid flow paths. The elution resurgence after the 15th pore volume is probably reflecting flow path rearrangements, due to flushing of partially clogged pores, colloid detachment, or some biological activity within the monoliths.

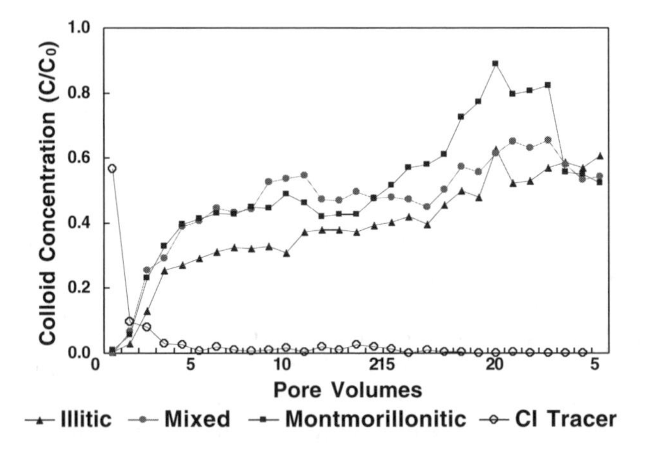

FIGURE 2.5 BTCs for soil colloids with montmorillonitic, mixed, or illitic mineralogy eluted through Pb-contaminated soil monoliths during the colloid-flushing phase.

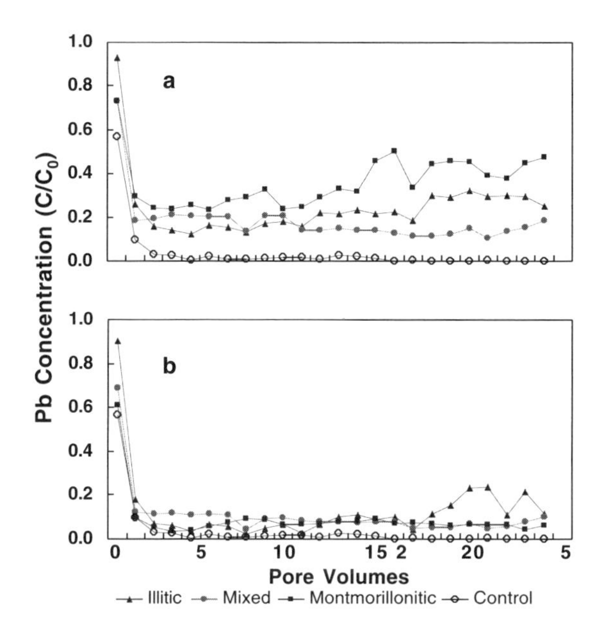

FIGURE 2.6 Desorbed (a) total and (b) soluble Pb elution in D-H2O (control) and colloid suspensions with montmorillonitic, mixed, or illitic mineralogy flushed through the soil monoliths.

2.3.5 Elution of Desorbed Pb

Total Pb elution by D-H_2O flushing solutions (controls) decreased drastically to near 0 after six pore volumes, suggesting absence of soluble Pb in the macropore space and total inability of D-H_2O to desorb Pb, previously attenuated by the soil by matrix (Figure 2.6). In contrast, Pb elution as soluble or total (soluble plus colloid-bound) by colloid-flushing suspensions continued throughout the leaching cycle in all soil monoliths. While the soluble Pb fraction in the eluents was maintained relatively stable, averaging <0.1 C/C_o (Figure 2.6(b)), the total Pb, and therefore, the colloid-bound fraction, varied significantly among colloids averaging between 0.2 and 0.5 C/C_o (Figure 2.6(a)). Most BTCs showed considerable asymmetry, which is attributed to the variable affinity of soil matrices and colloids for Pb (Table 2.1); and the variable filtration rate of the colloids by the soil matrix, which may alter flow paths and soil hydraulic conductivity [35]. Temporary decreases in flow velocity within the matrix may have also enhanced the formation of soluble metal-organic complexes, due to increased interaction time and reduced mass transfer resistance for Pb dissolution (Figure 2.6(b)).

In all cases, the total Pb fraction was considerably higher than the soluble Pb fraction during the colloid application cycles, showing good correlation with colloid breakthrough trends. Since there was no soluble source of Pb in the macropore space, this is the strongest evidence yet that the higher affinity of the colloids for Pb over that of the soil matrix resulted in competitive sorption between the two solid phases, which allowed Pb to be stripped from the soil matrix and adsorbed onto migrating colloids. This mechanism was supported by the identical mineralogical composition of the eluted compared to the input colloids, suggesting that *in situ* colloid generation and detachment within the soil monoliths, and therefore, contribution to the eluted Pb was negligible. Mills et al. [4] suggested that competitive sorption exchanges of a metal between two solid phases might continue until a state of equilibrium is established. Exchange equilibrium rates are relatively fast, and likely within the range of residence times spent by the colloids within the monoliths. Since the number of interactive exchange sites available on the eluting colloids is limited compared to the sites available within the matrix of the entire soil monolith, the extent of Pb desorption is controlled by the concentration of the colloid eluted through the soil monolith, and the accessibility of interactive sites within the soil matrix [36]. The association of the eluted Pb with the eluted colloids was assessed with HCl-HNO_3 extractions, which indicated 35% to 60% colloid saturation with Pb.

Total Pb desorption and remobilization was highest by the montmorillonitic and generally lowest by the mixed colloids (Figures 2.6(a) and 2.7(a)), in spite of the high affinity of the latter for Pb (Table 2.1). Apparently, the high pH of these colloids (6.7) did not induce sufficient Pb solubilization from the soil matrix. Also, the binding energy of Pb to surface sites of mixed colloids may be compromised by the presence of considerable amounts of organic particles, which are known to form weaker outer-sphere metal complexes or soluble organic complexes [37]. Indeed, with the exception of a few elution points, the mixed colloids exhibited the most consistent elution of soluble Pb, with nearly 50% of the total Pb transported being in the soluble fraction (Figure 2.7(b)). The greater potential of the illitic colloids to desorb Pb from the soil

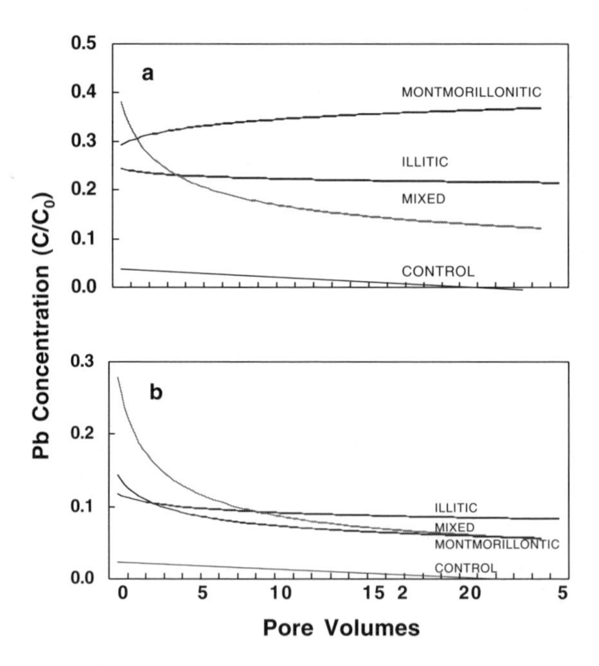

FIGURE 2.7 Power-function–fitted BTCs for desorbed total and soluble Pb eluted through soil monoliths in the presence or absence of soil colloids.

matrix compared to the mixed colloids, in spite of their four-fold lower K_f and smaller surface charge, is surprising, but may be related to their low pH, which is conducive to greater solubilization potential of Pb attached to the soil matrix. Alternatively, other physicochemical factors, including physical exclusion mechanisms, may have exerted considerable influence on the overall colloid behavior.

The presence of colloids in the flushing suspensions enhanced the transport of both soluble and colloid-bound Pb fractions. Since the soluble source of the Pb in the macropore space was negligible, as indicated by the control solutions (D-H_2O), the additional soluble Pb eluted in the presence of colloids must have been caused by colloid-induced desorption from the soil matrix. Weakly held outer-sphere Pb complexes sorbed on the colloids or soil matrix may be easily converted to soluble forms through ionic strength changes or organo-metallic interactions induced by continuous flow rate and flow path changes within the coil matrix [38]. Furthermore, direct ion-exchange reactions between soluble cations present in the colloid suspensions and Pb sorbed in the soil matrix may also contribute a portion to the eluted soluble Pb fraction. Even though the differences in the eluted soluble Pb fraction between colloids were small, elution was highest in the presence of illitic colloids, which had the lowest (K_f) or the mixed colloids, which had the highest pH and OC content (Figure 2.7(b)).

The findings of this experiment clearly demonstrate the potential of *ex situ* colloids with diverse mineralogical composition to desorb and remobilize Pb from contaminated soils. The magnitude of desorption and remobilization appears to be strongly related to the mineralogical composition and inherent or accessory physicochemical characteristics of the colloids. The high sorptive affinity for Pb and the small particle size diameter of the montmorillonitic colloids contributed to significant enhancement in desorption and transport of soluble and colloid-bound Pb compared to that generated by D-H_2O flushing solutions or other colloid types. Surface charge and colloid size similarities between mixed and illitic colloids caused subtle differences in Pb desorption and remobilization potential, which were controlled by pH and OC changes.

2.4 CASE STUDY 3

This study investigated the role of colloid particles with or without carbonates dispersed from lime-stabilized biosolids to mediate the transport of associated metals through intact soil monoliths in laboratory leaching experiments. The biosolid colloids were applied to undisturbed soil monoliths of a Maury soil under steady rate (2.2 cm/h^{-1}) gravity flow conditions. Deionized water spiked with metals at levels similar to the total load (soluble plus sorbed) carried by the colloids was used as a control leaching treatment. The eluents were monitored for colloid, Cu, Zn, and Pb breakthrough concentrations.

2.4.1 Metal Solutions and Biosolid Colloid Fractions

An aqueous solution containing 10 mg/l^{-1} Cu, Zn, and Pb as chloride salts (>99% purity, Aldrich Chemicals, Milwaukee, WI) was used as a control leaching treatment. A composted, municipal lime-stabilized biosolid material was dispersed in D-H_2O (1:20 ratio), shaken for 1 h, and then centrifuged at 750 rpm × 130 *g* for 3.5 min to collect the colloid fraction in a stock suspension. The concentration of the colloid fraction was determined gravimetrically and turbidimetrically. Two biosolid colloid fractions of 100-mg/l^{-1} concentration were prepared from the stock suspension. One of the fractions was diluted with D-H_2O to 100 mg/l^{-1}, while maintaining the original carbonates in its composition. A second biosolid fraction was made up to 100 mg/l^{-1} following three cycles of washing with deionized water to remove most of the carbonates present. The purpose of these biosolid colloid treatments was to assess their behavior under different carbonate content, pH, DOC, and ionic strength conditions. Subsamples of the two biosolid colloid fractions were collected for physicochemical and mineralogical characterization. The results are reported in Table 2.1.

2.4.2 Leaching Experiments

Six undisturbed soil monoliths taken from the upper Bt horizon of the Maury soil (fine, mixed, semiactive, mesic Typic Paleudalf) utilized in the previous case studies were employed in the leaching experiment. The preparation and conditions were similar to those described earlier. A set of replicate soil monoliths was used for each of the following treatments:

1. Control with metal solution containing 10 mg/l^{-1} Cu, Zn, and Pb as chloride salts.
2. Unwashed biosolid colloids (LSB+$CaCO_3$) with concentrations of 100 mg/l^{-1}, pH ~11, EC ~1500 μS cm^{-1}, DOC ~180 mg/l^{-1}, and total metal load of about 10 mg/l^{-1}.
3. D-H_2O washed biosolid colloids (LSB–$CaCO_3$) with concentration of 100 mg/l^{-1}, pH ~8.0, EC ~150 μS cm^{-1}, DOC ~46 mg/l^{-1}, and total metal load of about 10 mg/l^{-1}.

The leaching solutions/suspensions were applied to the top of each monolith through a continuous step input of 2.2 cm/h^{-1} controlled with a peristaltic pump. This rate was tested in earlier experiments and found to provide consistent free-flow conditions without ponding on the top of the monoliths. All input mixtures were allowed to equilibrate for 24 h before application. For ~10 days, eluents were monitored with respect to volume, colloid, and metal concentration. BTCs were constructed based on reduced metal and colloid concentrations (C/C_o) and pore volumes.

2.4.3 Biosolid Colloid Elution

Biosolid colloid breakthrough was highly irregular with several maxima and minima of different intensity throughout the experiment. This pattern is typical of alternating convective cycles, during which colloids are transported by preferential mass flow through soil macropores, and diffusion cycles, during which colloid

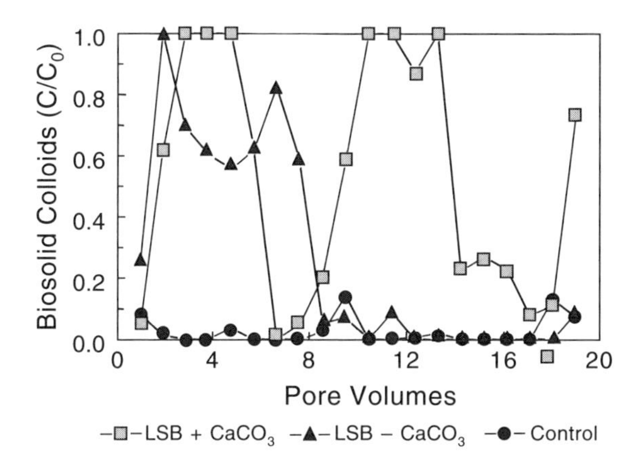

FIGURE 2.8 BTCs for LSB+$CaCO_3$ and LSB–$CaCO_3$ biosolid colloids and a metal solution control eluted through the soil monoliths.

elution is limited or restricted by physical filtration and/or chemical interaction within the soil matrix. Multiple colloid elution maxima were observed, with the highest and longer-duration events being associated with the biosolid colloids containing carbonates (LSB+$CaCO_3$) (Figure 2.8). Elution maxima of C/C_0 exhibited by these colloids between 3 to 6 and again between 9 to 14 pore volumes are probably the result of indigenous soil colloid mobilization or remobilization of already filtered biosolid colloids within the soil matrix as a result of dispersion phenomena caused by the high pH of the biosolid suspensions. This is evident from the pH and EC BTCs (Figure 2.9), showing good correlation with colloid elution maxima. For the carbonate-free biosolid colloids (LSB–$CaCO_3$), convective preferential flow maxima were more prominent during the first eight pore volumes of elution, with only minor resurgences afterward. In contrast, there was essentially no elution of colloids in control treatments in the absence of colloids, involving metal chloride-H_2O solutions.

The pH of the eluted suspensions appeared to be the dominant factor controlling these elution patterns, since it was maintained around 10 to 11 throughout the leaching experiment for the LSB+$CaCO_3$ colloids, while it ranged between 7 and 8.5 for the LSB–$CaCO_3$ (Figure 2.9). These relationships are consistent with the colloid stability patterns shown by the settling kinetics experiments (Figure 2.10). It is interesting that in spite of the high EC and, therefore, increased ionic strength of the LSB+$CaCO_3$ colloids, their high pH was able to maintain a dispersive environment that promoted greater stability and mobility, while the lower buffered pH of the LSB–$CaCO_3$ colloids apparently induced coagulation, and thus easier filtration by the soil matrix.

These findings indicate that even though the lime stabilization process of the biosolid waste may contribute to increased immobilization of soluble metals by sorption or precipitation onto the solid phase, in retrospect, it could create conditions favorable for increased dispersion and mobility of colloid particles and their metal load.

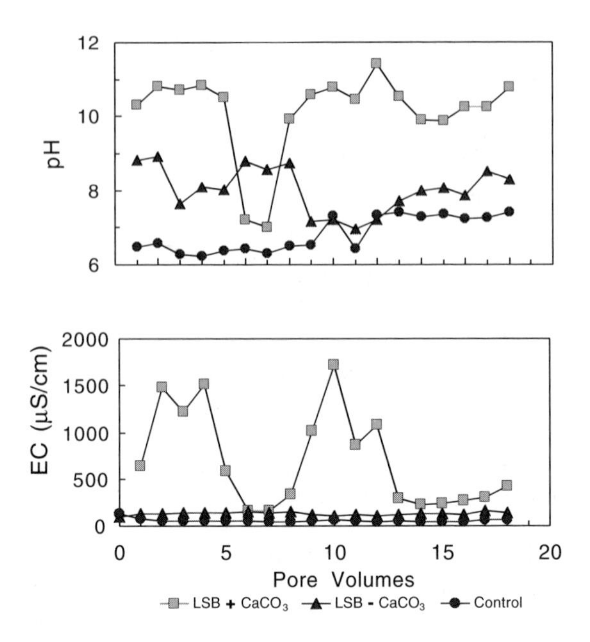

FIGURE 2.9 Breakthrough curves for pH and EC of the eluted control solutions and biosolid colloid suspensions through the soil monoliths.

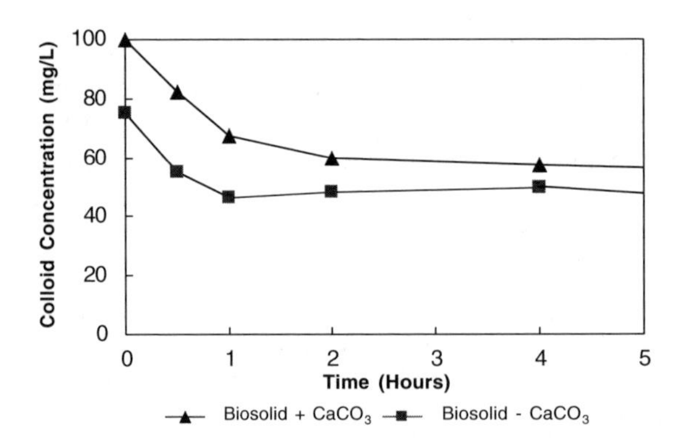

FIGURE 2.10 Settling kinetics curves for LSB+$CaCO_3$ and LSB–$CaCO_3$ biosolid colloids.

2.4.4 Metal Elution

Metal elution was essentially zero for all control treatments, suggesting total attenuation by the soils matrix (Figures 2.11 and 2.12). The presence of colloids enhanced drastically the elution of both soluble and total metal levels, showing an excellent correlation with colloid elution patterns. This confirms the strong association between metals and colloids and their role as carriers or facilitators in the transport process.

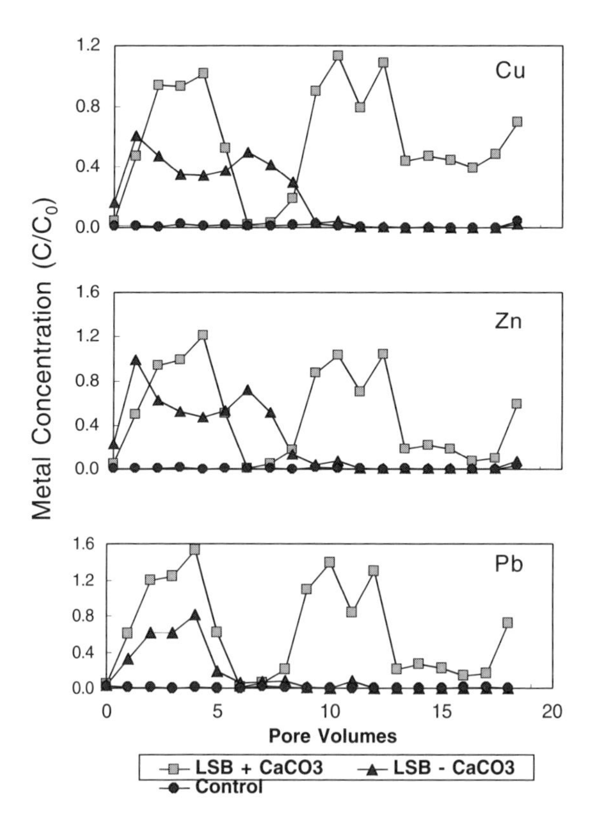

FIGURE 2.11 BTCs for total Cu, Zn, or Pb eluted in the presence or absence (control) of biosolid colloids through the soil monoliths.

Total Cu elution was highest with the LSB+$CaCO_3$ biosolid colloids, exceeding by several orders of magnitude the control, and being more than two-fold greater than the LSB–$CaCO_3$ colloids (Figure 2.11). The high pH of the LSB+$CaCO_3$ colloids apparently was responsible for the enhanced mobility of Cu. The high pH not only increased colloid stability but may have also enhanced the solubility of low molecular weight organic complexes associated with the biosolid colloids [28]. The mobilization of organic complexes and their high affinity for Cu could account for the drastic increases in Cu elution [39]. Nearly 50% of the total eluted Cu was in the soluble form, thus providing strong evidence of the role played by the DOC in the Cu transport process (Figure 2.12). The similar (1:1) ratio of the eluted soluble to colloid-bound Cu for both colloids is consistent with their K_f values, which are not very different (Table 2.1). Due to its high binding potential with DOC complexes and relatively low K_f for the soil matrix, Cu has been documented to be one of the most mobile metals from land-applied biosolids [40]. This mobility has been found to increase by nearly 50% at elevated pH levels, such as those generated during the lime stabilization process [41].

Total Zn elution patterns were very similar to those observed for total Cu (Figure 2.11). The eluted soluble Zn load was similar for the two colloids, but much lower

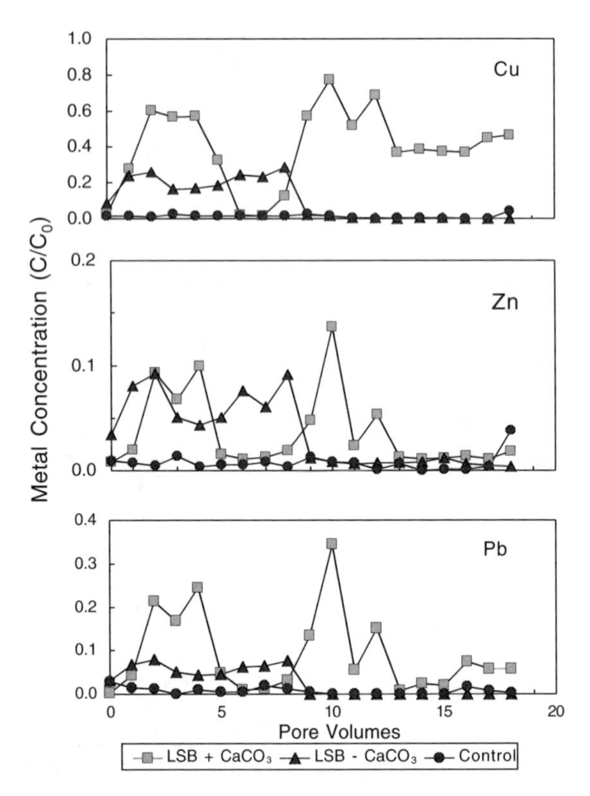

FIGURE 2.12 BTCs for soluble Cu, Zn, or Pb eluted in the presence or absence (control) of biosolid colloids through the soil monoliths.

than that for Cu, amounting to <20% of total Zn. There was an excellent agreement between colloid and Zn elution with >80% of the eluted Zn being in the colloid bound phase. Therefore, chemisorption appears to be the dominant Zn-transport mechanism. This mechanism is corroborated by the strong affinity of Zn for the biosolid colloids over that of the soil matrix, especially for the LSB+$CaCO_3$ colloids, which is attributed to the presence of carbonate precipitates. Carbonates have high sorption capacity for Zn, particularly at elevated pH's, through initial formation of surface-hydrated complexes and eventual co-precipitation and incorporation of the metal into the carbonate structure [42]. Therefore, chemisorption processes, involving ion exchange at organic particle surfaces and co-precipitation on carbonate particle surfaces are responsible for the majority of the total eluted Zn. The small contribution by DOC complexes is consistent with the generally low complexation affinity of Zn compared to that of Cu and other metals [43].

Total Pb elution peaked briefly around 0.8 C/C_o in the presence of LSB–$CaCO_3$ colloids during early stages of leaching and dropped to near 0 afterward (Figure 2.11). In contrast, total Pb breakthrough exceeded 1.0 C/C_o in the presence of LSB+$CaCO_3$ colloids. Total Pb elution patterns were nearly identical to colloid breakthrough patterns, showing the strong sorption affinity of Pb for colloid surfaces. The soluble Pb fraction eluted in the presence of LSB+$CaCO_3$ colloids was five to six times greater than that of the LSB–$CaCO_3$ colloids, but considerably lower than the soluble Cu fraction (Figure 2.12). This suggests that although the transport of small Pb fractions may have been facilitated through DOC-Pb complexes, the largest fraction was transported through chemisorption to colloid particles. The high, but similar K_f values of both colloids for Pb (Table 2.1) compared to the low affinity for Pb of the soil matrix are consistent with these elution patterns. The greater affinity of the LSB+$CaCO_3$ colloids for Pb may be due to carbonate co-precipitation phenomena, and the higher overall Pb sorption energy associated with the carbonate sites rather than the organic sites of the LSB–$CaCO_3$ colloids. The increased soluble Pb fraction eluted in the presence of the LSB+$CaCO_3$ colloids is probably the result of the elevated pH that mobilized additional DOC complexes.

The findings of this study provide strong evidence for the increased migration potential of biosolid colloids and associated metals in subsurface soil environments following land application of lime-stabilized biosolid wastes. The increased mobilization of metals appears to be the result of chemisorption or co-precipitation onto carbonate colloids, which were generated under the highly dispersive alkaline environment of the lime stabilization process, and co-transport through preferential flow processes. A secondary mechanism for additional transport of soluble metal loads involves metal complexation with organic ligands, which become more abundant under the prevailing high pH conditions.

2.5 SUMMARY

This study investigated the effect of colloid mineralogy on its capacity to mediate the transport of heavy metals through undisturbed soil monoliths (15 cm × 20 cm), representing upper Bt horizons of a Typic Paleudalf and/or a Typic Argiudoll. In the first experiment, uncontaminated soil monoliths were leached with metal-contaminated (Cu, Zn, or Pb) suspensions of *ex situ* soil colloids, fractionated from low ionic strength Bt horizons with montmorillonitic, illitic, and kaolinitic mineralogy. In the second experiment, uncontaminated suspensions of the same colloids were leached through undisturbed soil monoliths contaminated with Pb (~40% saturation). In a third experiment, uncontaminated soil monoliths were leached with metal-contaminated biosolid colloids fractionated from a lime-stabilized municipal waste. Metal solutions containing soluble metal concentrations equivalent to the colloid suspensions were used as controls. Eluent colloid and metal recoveries varied with metal, colloid, and soil properties. In the presence of contaminated colloids total metal transport through the undisturbed soil monoliths increased up to 50-fold for Cu and Zn and up to 3000-fold for Pb compared to control treatments, which showed a ~100% attenuation by the soil matrix. The presence of colloids increased the transport of both the soluble and colloid-sorbed metal fractions. Colloid-mediated transport increased with colloid surface charge and electrophoretic mobility, and was inhibited by increasing colloid size, and Fe/Al-sesquioxide content. Metal mobility and load-carrying capacity followed the mineralogical sequence montmorillonitic>illitic>kaolinitic, although increased soil organic carbon content appeared to partially compensate for mineral charge deficiencies. Montmorillonitic and to a lesser extent mixed or illitic colloids leached through Pb-contaminated monoliths were able to desorb and remobilize previously retained Pb of levels up to 50 times greater than control soluble metal solutions. Organically enriched colloids were less effective metal desorbers than mineral colloids due to their lower binding energy, but caused significant soluble Pb mobilization through formation of organo-metallic complexes. Lime-stabilized biosolid colloids also enhanced significantly greater metal transport than carbonate-free biosolid colloids or soluble metal controls in spite of the higher ionic strength, following the sequence Pb>Zn>Cu. The alkaline conditions generated during the lime stabilization process apparently created a dispersive environment conducive to formation of carbonate colloids, which mobilized greater metal loads through chemisorption and co-precipitation processes.

A stronger specific metal sorption affinity and/or energy for the mineral colloid surface than the soil matrix appeared to be the dominant mechanism facilitating metal transport. However, physical exclusion, organic complexation, competitive adsorption, co-precipitation, and metal solubility enhancement in the presence of colloids were also important. The findings of this study document the important role of mineral colloid particles as metal carriers and facilitators in subsurface soil environments. Depending on colloid mineralogical composition, metal loads transported in the presence of colloids may be several orders of magnitude greater than those transported by the soluble phase alone. These results have important ramifications on modeling and prediction of contaminant transport, and the application of suitable remediation technologies.

2.6 CONCLUSIONS

In all experiments, the presence of mineral colloids of diverse mineralogical composition in leaching solutions enhanced the elution of colloid-bound and soluble metal load up to several orders of magnitude compared to that of soluble metal controls. Generally, minerals of higher surface charge and electrophoretic mobility or smaller colloid size diameter induced greater metal transportability. The magnitude of colloid-mediated metal transport was drastically reduced in the presence of larger size and low surface charge mineral colloids with Fe-Al-oxyhydroxide coatings. A stronger specific metal sorption affinity for the mineral colloid surface than the soil matrix appears to be the dominant mechanism facilitating metal transport, but physical exclusion, competitive sorption, and metal solubility enhancement in the presence of colloids are also important processes. The chemisorption affinity of the metals for some mineral colloids appears to be high enough to enable desorption of already retained metals by the soil matrix and remobilization into subsurface environments. Colloid-mediated metal desorption and remobilization potential is not only a function of total surface charge, or metal distribution coefficients, but also a function of the bonding energy between metal-colloid surfaces, which appears to be higher in mineral than organic colloids.

Colloid-induced metal transport is also possible in soil environments receiving biosolid applications. Although a moderate metal mobilization is expected in these cases through colloid transport or soluble metal complexation with organic ligands, an even higher metal load could be mobilized with lime-stabilized biosolids. In spite of the supposed beneficial reduction of the soluble metal load through the lime stabilization process, the resulting alkaline dispersive environment may mobilize additional metal pools associated with organic and/or carbonate colloid particles through complexation or co-precipitation mechanisms. These findings strongly suggest that the role of mineral colloids as potential carriers and facilitators of metals in leaching solutions should not be underestimated, even in cases where the influence of the organic phase appears to be quite dominant.

REFERENCES

1. Alloway, B.J., Ed., *Heavy Metals in Soils*, Blackie Academic & Professional, London, 1995.
2. U.S. Environmental Protection Agency, Recent Developments for In situ Treatment of Metal Contaminated Soils, EPA-542-R-97–004, USEPA, Washington, D.C., 1997.
3. Forstner, U., Land contamination by metals: Global scope and magnitude of problem, in *Metal Speciation and Contamination of Soil*, Allen, H.E. et al., Eds., Lewis Publishers, Boca Raton, FL, 1995, p. 1.
4. Mills, W.B., Liu, S., and Fong, F.K., Literature review and model (COMET) for colloid/metals transport in porous media, *Ground Water*, 29, 199, 1991.
5. Puls, R.W. and Powell, R.M., Transport of inorganic colloids through natural aquifer material: Implications for contaminant transport, *Environ. Sci. Technol.*, 26, 614, 1992.
6. Liang.L. and McCarthy, J.F., Colloidal transport of metal contaminants in groundwater, *Metal Speciation and Contamination of Soil*, Allen, H.E. et al., Eds., Lewis Publishers, Boca Raton, FL, 1995, p. 87.
7. Ouyang, Y. et al., Colloid-enhanced transport of chemicals in subsurface environments: A review, *Crit. Rev. Environ. Sci. Technol.*, 26, 189, 1996.
8. Ryan, J.N. and Elimelech, M., Colloid mobilization and transport in groundwater, *Colloids Surfaces A Physicochem. Eng. Aspects*, 107, 1, 1996.
9. Kaplan, D.I., Bertsch, P.M., and Adriano, D.C., Mineralogical and physicochemical differences between mobile and nonmobile colloidal phases in reconstructed pedons, *Soil Sci. Soc. Am. J.*, 61, 641, 1997.
10. Tessier, A., Carnigan, R., and Belzile, N., Reactions of trace metals near the sediment-water interface in lakes, in *Transport and Transformations of Contaminants Near the Sediment Water Interface*, DePinto, J.V., Lick, V., and Paul, J.F., Eds., Lewis, Chelsea, MI, 1994, p. 129.
11. Seta, A.K. and Karathanasis, A.D., Atrazine adsorption by soil colloids and co-transport through subsurface environments, *Soil Sci. Soc. Am.J.*, 61, 612, 1997.
12. Roy, S.C. and Dzombak, D.A., Chemical factors influencing colloid-facilitated transport of contaminants in porous media, *Environ. Sci. Technol.*, 37, 656, 1997.
13. Karathanasis, A.D., Subsurface migration of Cu and Zn mediated by soil colloids, *Soil Sci. Soc. Am. J.*, 63, 830, 1999.
14. Karathanasis, A.D., Colloid-mediated transport of Pb through soil porous media, *Int. J. Environ. Stud.*, 57, 579, 2000.
15. Kretzschmar, R. and Sticher, H., Transport of humic-coated iron oxide colloids in a sandy soil: Influence of Ca^{2+} and trace metals, *Environ. Sci. Technol.*, 31, 3497, 1997.
16. Grolimund, D. et al., Colloid-facilitated transport of strongly sorbing contaminants in natural porous media: A laboratory column study, *Environ. Sci. Technol.*, 30, 3118, 1996.

17. Buddemeier, R.W. and Hunt, J.R., Transport of colloidal contaminants in groundwater: Radionuclide migration at the Nevada Test Site, *Appl. Geochem.*, 3, 535, 1988.
18. Kaplan, D.I., Bertsch, P.M., and Adriano, D.A., Facilitated transport of contaminant metals through an acidified aquifer, *Ground Water*, 33, 708, 1995.
19. Egli, M., Fitze, P., and Oswald, M., Changes in heavy metal contents in an acidic forest soil affected by depletion of soil organic matter within the time span 1969–93, *Environ. Pollut.*, 105, 367, 1999.
20. Ross, J.M. and Sherrell, R.M., The rate of colloids in trace metal transport and adsorption behavior in New Jersey pinelands streams, *Limnol. Oceanogr.*, 44, 1019, 1999.
21. Shafer, M.M. et al., The influence of dissolved organic carbon, suspended particulates, and hydrology on the concentration, partitioning and variability of trace metals in two contrasting Wisconsin watersheds, *Chem. Geol.*, 136, 71, 1997.
22. Schemel, L.E., Kimball, B.A., and Bencala, K.E., Colloid formation and metal transport through two mixing zones affected by acid mine drainage near Silverton, Colorado, *Appl. Geochem.*, 15, 1003, 2000.
23. Wang, J.M., Huang, C.P., and Allen, H.E., Surface physical-chemical characteristics of sludge particulates, *Water Environ. Res.*, 72, 545, 2000.
24. Johnson, W.P. and Avery, G.L., Facilitated transport and enhanced desorption of polycyclic aromatic hydrocarbons by natural organic matter in aquifer sediments, *Environ. Sci. Technol.*, 29, 807, 1995.
25. McCarthy, J.F. et al., Mobility of natural organic matter in a sandy aquifer, *Environ. Sci. Technol.*, 27, 667, 1993.
26. Kretzschmar, R., Robarge, W.P., and Amoozegar, A., Influence of natural organic matter on colloid transport through saprolite, *Water Resour. Res.*, 31, 435, 1995.
27. Kaplan, D.J. et al., Soil-borne mobile colloids as influenced by water flow and organic carbon, *Environ. Sci. Technol.*, 27, 1193, 1993.
28. Han, N. and Thompson, M.L., Copper-binding ability of dissolved organic matter derived from anaerobically digested biosolids, *J. Environ. Qual.*, 28, 939, 1999.
29. Sparks, D.L., Kinetics of metal sorption reactions, in *Metal Speciation and Contamination of Soil*, Allen, H.E. et al., Eds., Lewis Publishers, Boca Raton, FL, 1995, p. 35.
30. Bertsch, P.M. and Seaman, J.C., Characterization of complex mineral assemblages: Implications for contaminant transport and environmental remediation, *Proc. Natl. Acad. Sci. USA*, 96, 3350, 1999.
31. U.S. Department of Agriculture-National Soil Survey Center, Soil Survey Laboratory Methods Manual, Soil Survey Investigations Report No. 42, Version 3.0, National Soil Survey Center, Lincoln, NE, 1996.
32. U.S. Environmental Protection Agency, Methods for the Determination of Metals in Environmental Samples, Method 200.2, EPA/600/R-94/111, USEPA, Washington, D.C., 1994.
33. Goldberg, S., Kapoor, B.D., and Rhoades, J.D., Effect of aluminum and iron oxides and organic matter on flocculation and dispersion of arid zone soils, *Soil Sci.*, 150, 588, 1990.
34. Jekel, M.R., The stabilization of dispersed mineral particles by adsorption of humic substances, *Water Resour. Res.*, 20, 1543, 1986.
35. Davis, A. and Singh, I., Washing of zinc (II) from contaminated soil column, *J. Environ. Eng.*, 121, 174, 1995.
36. Elliott, H.A., Liberati, M.R., and Huang, C.P., Competitive adsorption of heavy metals by soils, *J. Environ. Qual.*, 15, 214, 1986.

37. Verloo, M. and Cottenie, A., Stability and behavior of complexes of Cu, Zn, Fe, Mn, and Pb with humic substances of soils, *Pedobiology*, 22, 174, 1972.
38. Ji, G.L. and Li, H.Y., Electrostatic adsorption of cations, in *Chemistry of Variable Charged Soils*, Yu, T.T., Ed., Oxford University Press, Oxford, 1997, p. 64.
39. Temminghoff, E.J.M., van Der Zee, S.E.A.T.M., and DeHaan, F.A.M., Copper mobility in a Cu-contaminated sandy soil as affected by pH and solid and dissolved organic matter, *Environ. Sci. Technol.*, 31, 1109, 1997.
40. McBride, M.B. et al., Long-term leaching of trace elements in a heavily sludge-amended silty clay loam soil, *Soil Sci.*, 164, 613, 1999.
41. Richards, B.K. et al., Effect of processing mode on trace elements in dewatered sludge products, *J. Environ. Qual.*, 26, 782, 1997.
42. Zachara, J.M., Cowan, C.E., and Resch, C.T., Sorption of divalent metals on calcite, *Geochim. Cosmochim. Acta*, 55, 1549, 1991.
43. Harter, R.D., Effect of soil pH on adsorption of Pb, Cu, Zn, and Ni, *Soil Sci. Soc. Am. J.*, 47, 47, 1983.

3 A Web-Based Three-Dimensional Simulation System for Solute Transport in Soil

Honghai Zeng, Vladimir J. Alarcon, William L. Kingery, H. Magdi Selim, and Jianping Zhu

CONTENTS

3.1 Introduction 52
3.2 Three-Dimensional MRTM Model and ADI Method 55
  3.2.1 Three-Dimensional MRTM Model 55
  3.2.2 Boundary Conditions 56
  3.2.3 Three-Step ADI Method 57
3.3 System Implementation 61
  3.3.1 Introduction 61
  3.3.2 Three-Dimensional Simulation System and Related Technologies 63
    3.3.2.1 Traditional Computing Model 63
    3.3.2.2 Web-Based High-Performance Computing Model 64
  3.3.3 J2EE Technologies 66
    3.3.3.1 Client Tier 66
    3.3.3.2 Web Tier 67
    3.3.3.3 Business Tier 68
    3.3.3.4 Resource Tier 68
3.4 Results 69
3.5 Conclusions 76
Acknowledgment 77
References 77

1-56670-623/03/$0.00+$1.50

3.1 INTRODUCTION

Concerns regarding heavy metal contamination in the environment affecting all ecosystem components including aquatic and terrestrial systems have been identified, with increasing attention to limiting their bioavailability in the soil and groundwater. Metal mining, smelting, and processing throughout the world have contaminated soils with heavy metals in excess of natural soil background concentrations. Such activities introduce metal contaminants into the environment through gaseous and particulate emissions, waste liquids, and solid wastes. In addition to the soil contamination, many mining and smelting sites have considerable surface water and groundwater contamination from heavy metals released and transported from contaminated soils. This contamination endangers water supply resources as well as the soil quality.

Different processes, including advection, dispersion, diffusion, and adsorption, work together or separately to determine the transport of contaminants in the groundwater. The movement of contaminants through the soil matrix to groundwater is primarily a liquid phase process (Figure 3.1), but the partitioning of the chemicals between sorbed and dissolved phases is a critical factor in determining how rapidly the contaminants leach (Selim, 1992). The adsorption of contaminants by soil constituents is an important mechanism of retention. The ability for contaminants to attach to the soil is determined by the properties of both the soil and the contaminants (Figure 3.2). The term "adsorption" is often used to include all retention and release reactions in soils, including precipitation, dissolution, ion exchange, and adsorption–desorption mechanisms (Selim and Amacher, 1997). The usual conceptualization of retention mechanisms in soil often includes (a) equilibrium models, in which it is assumed that the reaction of an individual solute species is sufficiently fast or instantaneous; and (b) kinetic models, in which the amount of solute retained or released from the soil solution is time dependent. Multireaction models are those that deal with multiple interactions of one species in the soil environment (Selim

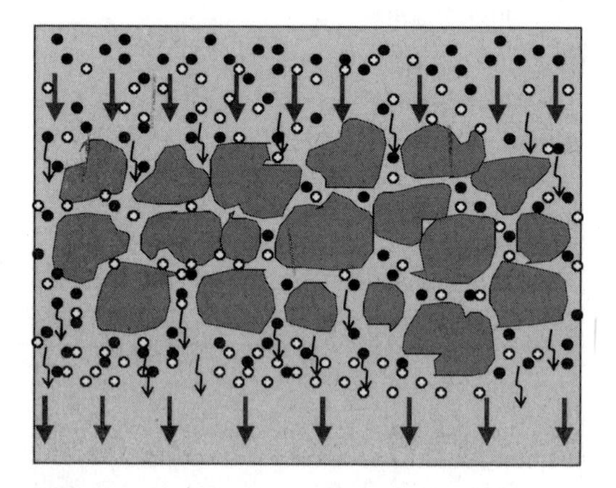

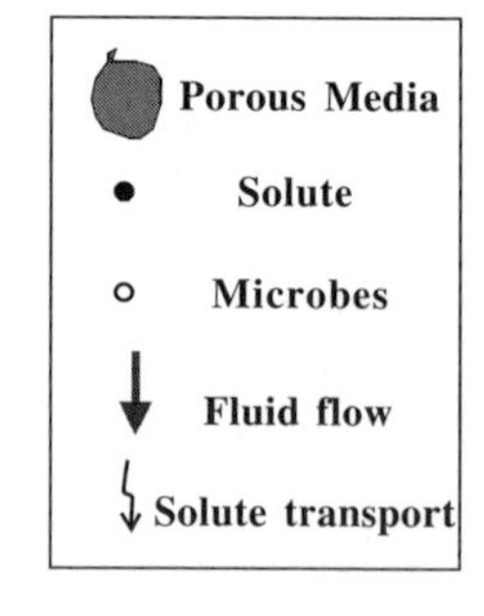

FIGURE 3.1 Migration of contaminants through the soil. The soil is conceptualized as a porous medium.

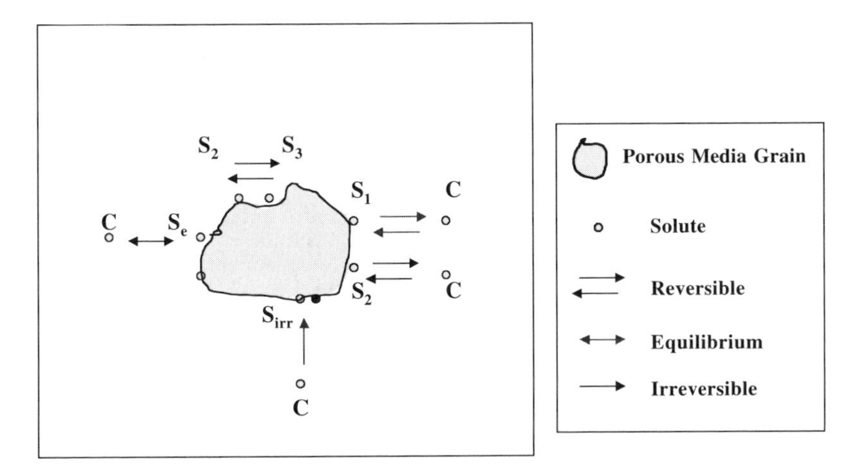

FIGURE 3.2 Solute phases and reactions in the soil postulated by the model. The solute can be in solution (C); sorbed reversibly in the soil (S_e) in equilibrium with C; sorbed reversibly and reacting kinetically (S_1); sorbed reversibly and slowly reacting kinetically (S_2); strongly and reversibly sorbed slowly reacting kinetically (S_3); or sorbed irreversibly in the soil (S_{irr}).

and Amacher, 1997), accounting for reversible as well as irreversible processes of solutes in the soil environment (Selim et al., 1990).

To analyze the transport and retention of chemical contaminants in groundwater flowing through soils, experimental and theoretical studies generated several reliable models. Diverse numerical methods have been applied to solve the governing equations efficiently. Some computer models include the simulation of physical and chemical processes.

Among several analytical methods for the prediction of movement of dissolved substances in soils, one model (Leij et al., 1993) was developed for three-dimensional nonequilibrium transport with one-dimensional steady flow in a semi-infinite soil system. In this model, the solute movement was treated as one-dimensional downward flow with three-dimensional dispersion to simplify the analytical solution. Another model (Rudakov and Rudakov, 1999) analyzed the risk of groundwater pollution caused by leaks from surface depositories containing water-soluble toxic substances. In this analytical model, the pollutant migration was also simplified into two stages: predominantly vertical (one-dimensional) advection and three-dimensional dispersion of the pollutants in the groundwater. Typically, analytical methods have many restrictions when dealing with three-dimensional models and do not include complicated boundary conditions.

Due to the difficulties of getting analytical solutions, many numerical methods were developed to simulate the solute transport and retention processes in the soil. Deane et al. (1999) analyzed the transport and fate of hydrophobic organic chemicals (HOCs) in consolidated sediments and saturated soils. Walter et al. (1994) developed a model for simulating transport of multiple thermodynamically reacting chemical substances in groundwater systems. Islam et al. (1999) presented a modeling approach to simulate the complex biogeochemical interactions in landfill leachate-contaminated soils. However, none of these models deal with three-dimensional domains. In numerical models, higher dimensionality increases greatly the computational complexity. Therefore, previous research usually uses simplified models to decrease the level of computational difficulty.

Manguerra and Garcia (1997) introduced a modified linked approach to solve the governing partial differential equations (PDEs) for subsurface flow and solute transport using finite difference methods. A variant of the strongly implicit procedure (SIP), one of the most popular methods for solving matrix equations by iteration, was adopted. However, when the desired application required a full three-dimensional implementation of the model, innovative modifications were applied to reduce the computations involved. In order to circumvent the difficulties in fully three-dimensional, saturated-unsaturated flow and transport models, Yakirevich et al. (1998) reduced the governing equations to quasi-three-dimensional formulations. This model coupled the one-dimensional Richards equation for vertical flow in the unsaturated zone and the two-dimensional equation for horizontal flow in the saturated zone. Simulations for the quasi-two-dimensional case, using finite difference numerical schemes, proved to be computationally efficient. However, the method was unstable for large time steps, and could not be used for the cases in which horizontal water fluxes were significant. Some other numerical schemes, such as the Hopscotch algorithm (Segol, 1992) and the Galerkin finite element technique (Srivastava and Yeh, 1992), were applied to the three-dimensional transport and retention problem. They all suffered the efficiency problem resulting from the computational complexity in three-dimensional applications.

The three-dimensional method of characteristics (MOC3D) (Konikow et al., 1996; Kipp et al., 1998; Heberton et al., 2000) is a transport model that calculates transient changes in the concentration of a single solute in a three-dimensional groundwater flow field. The groundwater flow equation describes the head distribution in the aquifer. The solute transport equation describes the solute concentration within the flow system. The MOC3D coupled the flow equation with the solute-transport equation, so that this model can be applied to both steady-state and transient groundwater-flow problems. Instead of solving the advection-dispersion governing PDEs directly, the MOC3D solves an equivalent system of ordinary differential equations to increase computational efficiency. This approximation inevitably decreases the accuracy and precision of the numerical results. MOC3D programs, developed in FORTRAN 77, are restricted to mathematically simple retention reactions, such as first-order adsorption.

In order to avoid the restrictions to complicated adsorptive reactions in the MOC3D, Selim et al. (1990) developed a simulation system based on the multireaction model (MRM) and multireaction transport model (MRTM). The MRM model includes concurrent and concurrent-consecutive retention processes of the nonlinear kinetic type. It accounts for equilibrium (Freundlich) sorption and irreversible reactions. The processes considered are based on linear (first order) and nonlinear kinetic reactions. The MRM model assumes that the solute in the soil environment is present in the soil solution and in several phases representing retention by various soil constituents and mechanisms. This model is capable of describing chemicals under batch (kinetic) conditions without considering the water flow. The MRTM model represents an extension of the MRM model because it includes transport processes in addition to the adsorption behavior of chemicals in the soil environment (Selim et al., 1990). The kinetic retention–reaction equations and advection–dispersion equations are solved in explicit-implicit, finite-difference methods in both the MRM and MRTM models. However, only one-dimensional problems were considered due to the complexity of higher dimensional problems.

This review shows that due to the constraints of the analytical three-dimensional models in dealing with realistic boundary, flow, and transport conditions, and the computational complexity involved in numerical schemes, previous multiple-reaction models simplified or avoided three-dimensional models. Additionally, they do not fully utilize the current power of remote computing services provided through the Internet. Due to the lack of web-based remote computing capability, users usually need to install the computational software on their local computers before running them.

In this chapter, we extend the one-dimensional MRTM transport and adsorption model (Zeng et al., 2002; Selim et al., 1990) to three-dimensional applications, using the alternating-direction-implicit (ADI) numerical algorithm to reduce computational complexity. We also discuss the development of a web-based system with a user-friendly interface providing client-side and server-side computing applications and web-based visualization functionality. The accuracy and efficiency issues of solution algorithms will be addressed in future publications.

3.2 THREE-DIMENSIONAL MRTM MODEL AND ADI METHOD

3.2.1 THREE-DIMENSIONAL MRTM MODEL

If all advection, dispersion, diffusion, and adsorption processes in three dimensions are considered, the three-dimensional MRTM governing equation can be expressed in the following form:

$$\theta\frac{\partial C}{\partial t}+\rho\frac{\partial S}{\partial t}=\theta D_x\frac{\partial^2 C}{\partial x^2}-q_x\frac{\partial C}{\partial x}+\theta D_y\frac{\partial^2 C}{\partial y^2}-q_y\frac{\partial C}{\partial y}+\theta D_z\frac{\partial^2 C}{\partial z^2}-q_z\frac{\partial C}{\partial z}-Q \quad (3.1)$$

where S is the solute concentration associated with the solid phase in the soil; ρ is the soil bulk density; D_x, D_y, and D_z are the hydrodynamic dispersion coefficients in different directions; q_x, q_y, and q_z are the Darcy's water flux densities in different directions; Q is a sink/source term; x, y, and z are the soil domain length, width, depth, respectively; and t is time. R is a retardation term that accounts for equilibrium-reversible and/or Freundlich solute retention in the soil. It is explicitly introduced as

$$. R=1+\frac{b\rho K_d}{\theta}C^{b-1} \quad (3.2)$$

The MRM model is connected to the transport model through S, the same as in the one-dimensional MRTM model (Selim et al., 1990):

$$S = S_e + S_1 + S_2 + S_3 \rightarrow \frac{\partial S}{\partial t} = \frac{\partial S_e}{\partial t} + \frac{\partial S_1}{\partial t} + \frac{\partial S_2}{\partial t} + \frac{\partial S_3}{\partial t} \tag{3.3}$$

and

$$Q = \rho \frac{\partial S_{irr}}{\partial t} = \theta K_s C \tag{3.4}$$

where S_e, S_1, S_2, S_3, and S_{irr} are the different phases of solute interacting with the soil (Selim et al., 1990).

3.2.2 Boundary Conditions

In order to simplify discussions, Equation (3.1) is re-written as:

$$\frac{\partial C}{\partial t} + d\frac{\partial C}{\partial x} + e\frac{\partial C}{\partial y} + f\frac{\partial C}{\partial z} = a\frac{\partial^2 C}{\partial x^2} + b\frac{\partial^2 C}{\partial y^2} + c\frac{\partial^2 C}{\partial z^2} + F(x,y,z,t) \tag{3.5}$$

or

$$C_t + dC_x + eC_y + fC_z = aC_{xx} + bC_{yy} + cC_{zz} + F(x,y,z,t) \tag{3.6}$$

where $d = \frac{q_x}{\theta}$, $e = \frac{q_y}{\theta}$, $f = \frac{q_z}{\theta}$, $a = D_x$, $b = D_y$, $c = D_z$, and $F = -\frac{Q}{\theta} - \frac{\rho}{\theta}\frac{\partial S}{\partial t}$

The computational domain is normalized into a cube with $(x, y, z) \in [0,1]\times[0,1]\times[0,1]$. Initial conditions are imposed by the equations below. It is assumed that the soil contains a uniform initial concentration C_i, in the solution and the soil matrix is devoid of sorbed phases at time zero:

$$C(x,y,z,0) = C_i \tag{3.7}$$

and

$$S_e(x,y,z,0) = S_1(x,y,z,0) = S_2(x,y,z,0) = S_3(x,y,z,0) = 0 \tag{3.8}$$

where

$$(x,y,z) \in [0,1]\times[0,1]\times[0,1].$$

The Dirichlet-type boundary conditions assume that a solution of known concentration (C_0) is applied at the soil top surface for a given duration t_p. This solute pulse-type input is assumed to be followed by a solute-free solution application at the soil top surface:

$$q_z C_0 = -\theta D_z \frac{\partial C}{\partial z} + q_z C, \ z = 0, \ t < t_p \tag{3.9}$$

$$0 = -\theta D_z \frac{\partial C}{\partial z} + q_z C, \ z = 0, \ t > t_p \tag{3.10}$$

At the bottom of the soil profile, a Neumann-type no-flow boundary condition is specified as

$$\frac{\partial C}{\partial z} = 0, \ z = 1, \ t > 0 \tag{3.11}$$

On the other sides of the soil profile, Neumann-type no-flow boundary conditions are also specified as

$$\frac{\partial C}{\partial x} = 0, \ x = 0, \ t > 0 \tag{3.12}$$

$$\frac{\partial C}{\partial x} = 0, \ x = 1, \ t > 0 \tag{3.13}$$

$$\frac{\partial C}{\partial y} = 0, \ y = 1, \ t > 0 \tag{3.14}$$

and

$$\frac{\partial C}{\partial y} = 0, \ y = 1, \ t > 0 \tag{3.15}$$

3.2.3 Three-Step ADI Method

The numerical solution to the advection-dispersion equation and associated adsorption equations can be performed using finite difference schemes, either in their implicit and/or explicit form. In the one-dimensional MRTM model (Selim et al., 1990), the Crank–Nicholson algorithm was applied to solve the governing equations of the chemical transport and retention in soils. The web-based simulation system for the one-dimensional MRTM model is detailed in Zeng et al. (2002). The alternating direction-implicit (ADI) method is used here to solve the three-dimensional models.

In the three-dimensional advection-dispersion-adsorption equation, the adsorptive reaction part $\frac{\partial S}{\partial t}$ is treated explicitly, so that it can be merged into $F(x, y, z, t)$ together with the sink/source term Q. The finite-difference form for the adsorptive reaction part is the same as in the one-dimensional numerical method. The numerical method used in the one-dimensional case makes the computational complexity unacceptable in the three-dimensional case, because the solution to the multidiagonal matrix–vector equation systems is too expensive. In order to reduce computational complexity, we use the alternating direction-implicit (ADI) algorithm on a rectangular grid (x_i, y_j, z_k), where $i = 1,2,3,...,NI$, $j = 1,2,3,...,NJ$, and $k = 1,2,3,...NK$. The solute concentration is expressed as

$$C(x,y,z,t) = C(i\Delta x, j\Delta y, k\Delta z, n\Delta t) = C^n_{i,j,k}$$

$$i = 1,2,3,...,NI,\ \ j = 1,2,3,...NJ,\ \ k = 1,2,3,...,NK,\ \ t = 0,1,2,... \tag{3.16}$$

where $\Delta x = \frac{1}{NI}$, $\Delta y = \frac{1}{NJ}$, $\Delta z = \frac{1}{NK}$.

The ADI method, first used by Peaceman and Rachford (1955) for solving parabolic PDEs, can also be derived from the Crank-Nicholson algorithm. In the three-dimensional MRTM model, the governing equation can be discretized by the Crank-Nicholson algorithm as

$$\frac{C^{n+1}_{i,j,k} - C^n_{i,j,k}}{\Delta t} + \frac{d}{2}((C_x)^{n+1}_{i,j,k} + (C_x)^n_{i,j,k}) + \frac{e}{2}((C_y)^{n+1}_{i,j,k} + (C_y)^n_{i,j,k}) + \frac{f}{2}((C_z)^{n+1}_{i,j,k} + (C_z)^n_{i,j,k})$$

$$= \frac{a}{2}((C_{xx})^{n+1}_{i,j,k} + (C_{sx})^n_{i,j,k}) + \frac{b}{2}((C_{yy})^{n+1}_{i,j,k} + (C_{yy})^n_{i,j,k}) + \frac{c}{2}((C_{zz})^{n+1}_{i,j,k} + (C_{zz})^n_{i,j,k})$$

$$+ \frac{F^{n+1}_{i,j,k} + F^n_{i,j,k}}{2} \tag{3.17}$$

or

$$\frac{C^{n+1}_{i,j,k} - C^n_{i,j,k}}{\Delta t} + d\delta_x(\frac{C^{n+1}_{i,j,k} + C^n_{i,j,k}}{2}) + e\delta_y(\frac{C^{n+1}_{i,j,k} + C^n_{i,j,k}}{2}) + f\delta_z(\frac{C^{n+1}_{i,j,k} + C^n_{i,j,k}}{2})$$

$$= a\delta_{xx}(\frac{C^{n+1}_{i,j,k} + C^n_{i,j,k}}{2}) + b\delta_{yy}(\frac{C^{n+1}_{i,j,k} + C^n_{i,j,k}}{2}) + c\delta_{zz}(\frac{C^{n+1}_{i,j,k} + C^n_{i,j,k}}{2}) + \frac{F^{n+1}_{i,j,k} + F^n_{i,j,k}}{2} \tag{3.18}$$

where

$$\delta_x C^n_{i,j,k} = \frac{C^n_{i+1,j,k} - C^n_{i-1,j,k}}{2\Delta x}$$

$$\delta_y C^n_{i,j,k} = \frac{C^n_{i,j+1,k} - C^n_{i,j-1,k}}{2\Delta y}$$

$$\delta_z C^n_{i,j,k} = \frac{C^n_{i,j,k+1} - C^n_{i,j,k-1}}{2\Delta z}$$

$$\delta^2_x C^n_{i,j,k} = \frac{C^n_{i+1,j,k} - 2C^n_{i,j,k} + C^n_{i-1,j,k}}{\Delta x^2}$$

$$\delta^2_y C^n_{i,j,k} = \frac{C^n_{i,j+1,k} - 2C^n_{i,j,k} + C^n_{i,j-1,k}}{\Delta y^2}$$

$$\delta^2_z C^n_{i,j,k} = \frac{C^n_{i,j,k+1} - 2C^n_{i,j,k} + C^n_{i,j,k-1}}{\Delta z^2}$$

This discretization is known to be second order accurate in both time and space. Grouping the unknown terms (at time step n+1) to the left and the known terms (at time step n) to the right, we have:

$$(I + \frac{\Delta td}{2}\delta_x - \frac{\Delta ta}{2}\delta^2_x + \frac{\Delta te}{2}\delta_y - \frac{\Delta tb}{2}\delta^2_y + \frac{\Delta tf}{2}\delta_z - \frac{\Delta tc}{2}\delta^2_z)C^{n+1}_{i,j,k} =$$

$$(I - \frac{\Delta td}{2}\delta_x + \frac{\Delta ta}{2}\delta^2_x - \frac{\Delta te}{2}\delta_y + \frac{\Delta tb}{2}\delta^2_y - \frac{\Delta tf}{2}\delta_z + \frac{\Delta tc}{2}\delta^2_z)C^n_{i,j,k} + \frac{F^{n+1}_{i,j,k} + F^n_{i,j,k}}{2}\Delta t \quad (3.19)$$

The above expression can be approximately factorized as:

$$(I + \frac{\Delta td}{2}\delta_x - \frac{\Delta ta}{2}\delta^2_x)(I + \frac{\Delta te}{2}\delta_y - \frac{\Delta tb}{2}\delta^2_y)(I + \frac{\Delta tf}{2}\delta_z - \frac{\Delta tc}{2}\delta^2_z)C^{n+1}_{i,j,k} =$$

$$(I - \frac{\Delta td}{2}\delta_x + \frac{\Delta ta}{2}\delta^2_x)(I - \frac{\Delta te}{2}\delta_y + \frac{\Delta tb}{2}\delta^2_y)(I - \frac{\Delta tf}{2}\delta_z + \frac{\Delta tc}{2}\delta^2_z)C^n_{i,j,k}$$

$$+ \frac{F^{n+1}_{i,j,k} + F^n_{i,j,k}}{2}\Delta t \quad (3.20)$$

The difference between the last two equations, or the additional error of the factorization, is $O(\Delta t^3)$. Therefore, the following three-step ADI method still has second-order accuracy in both time and space.

$$(I+\frac{\Delta td}{2}\delta_x-\frac{\Delta ta}{2}\delta_x^2)C^*_{i,j,k}=$$

$$(I-\frac{\Delta td}{2}\delta_x+\frac{\Delta ta}{2}\delta_x^2)(I-\frac{\Delta te}{2}\delta_y+\frac{\Delta tb}{2}\delta_y^2)(I-\frac{\Delta tf}{2}\delta_z+\frac{\Delta tc}{2}\delta_z^2)C^n_{i,j,k}$$

$$+\frac{F^{n+1}_{i,j,k}+F^n_{i,j,k}}{2}\Delta t \tag{3.21}$$

$$(I+\frac{\Delta te}{2}\delta_y-\frac{\Delta tb}{2}\delta_y^2)C^{**}_{i,j,k}=C^*_{i,j,k} \tag{3.22}$$

$$(I+\frac{\Delta tf}{2}\delta_z-\frac{\Delta tc}{2}\delta_z^2)C^{n+1}_{i,j,k}=C^{**}_{i,j,k} \tag{3.23}$$

The solutions to the three separate equations 3.2, 3.3, and 3.4 can be computed by solving only tridiagonal equations since the left sides of the equations involve only three-point central difference operators $\delta_x, \delta_x^2, \delta_y, \delta_y^2, \delta_z, \delta_z^2$ and δ. Although the right side of equation 3.3 involves the product of these operators, it does not complicate the solution process since they are applied to the known solution values from the previous time step.

While solving the second-step equation 3.3, boundary conditions for $C^{**}_{i,0,k}$ and $C^{**}_{i,NJ,k}$ (i = *1,2,...NI* and i = *1,2,...NK*) need to be calculated reversely from the third equation by setting $j = 0$ and $j = NJ$, respectively:

$$C^{**}_{i,0,k}=(I+\frac{\Delta tf}{2}\delta_z-\frac{\Delta tc}{2}\delta_z^2)C^{n+1}_{i,0,k} \tag{3.24}$$

$$C^{**}_{i,,NJ,k}=(I+\frac{\Delta tf}{2}\delta_z-\frac{\Delta tc}{2}\delta_z^2)C^{n+1}_{i,NJ,k} \tag{3.25}$$

Similar reverse calculations of boundary conditions for $C^*_{0,j,k}$ and $C^*_{NI,j,k}$ (j = *1,2,...NJ* and i = *1,2,...NK*) are needed while solving the first-step equation 3.2:

$$C^*_{0,j,k}=(I+\frac{\Delta te}{2}\delta_y-\frac{\Delta tb}{2}\delta_y^2)(I+\frac{\Delta tf}{2}\delta_z-\frac{\Delta tc}{2}\delta_z^2)C^{n+1}_{0,j,k} \tag{3.26}$$

$$C^*_{NI,j,k} = (I + \frac{\Delta te}{2}\delta_y - \frac{\Delta tb}{2}\delta_y^2)(I + \frac{\Delta tf}{2}\delta_z - \frac{\Delta tc}{2}\delta_z^2)C^{n+1}_{NI,j,k} \tag{3.27}$$

The three-step ADI method greatly reduces the computational complexity by solving independent equation systems with tridiagonal coefficient matrices. These independent computations also provide enough parallelism for parallel computing (Zhu, 1994).

3.3 SYSTEM IMPLEMENTATION

3.3.1 INTRODUCTION

With the development of the Internet, online access to scientific computing resources became possible. Diverse web technologies are used in this research to separately implement the one- and three-dimensional MRTM simulation systems. The web technologies used in the one-dimensional simulation system are discussed in Zeng et al. (2002). This section covers details of web technologies, especially Java technologies, that are used in the three-dimensional simulation system. The major goals and system requirements of the two simulation systems are also compared.

The one-dimensional MRTM simulation system makes use of client-side computing resources. Since the one-dimensional computational loads are not heavy, they can be handled by the client-side computers. The major goal is to keep the legacy code written in C or FORTRAN untouched as much as possible. The computations wrapped within the Java code are executed in the web browser using the Java Native Interface (JNI) technology. The JNI allows the Java code that runs inside a Java virtual machine (JVM) of the web browser to interoperate with applications and libraries written in other programming languages, such as C, C++, and FORTRAN. This makes it possible to reuse legacy code (i.e., existing code) written in other programming languages and enable researchers from any other discipline, in collaboration with computer scientists and mathematicians, to develop large code blocks and add new functions. The code writing can be done independently according to the researcher's preferred programming language. Those programs can be linked afterward and become, as a whole, the desired application.

The legacy code is packaged into dynamic libraries, which are inevitably platform dependent. In other words, the one-dimensional simulation library running on the Windows platform is different from that on the Unix platform. The client–server relationship is relatively simple in the one-dimensional simulation system (Figure 3.3). The simulation programs are packaged into libraries and uploaded (through FTP) to a web server supporting basic HTTP protocols, so that users can access the web server (or website) with their favorite web browsers. The libraries are downloaded (through HTTP) to the web browser on the client (or user) side and run inside the browser's JVM. The computations in the one-dimensional simulation system execute on the client side, and the incurred Java client-side security problem has been considered and solved (Zeng et al., 2002).

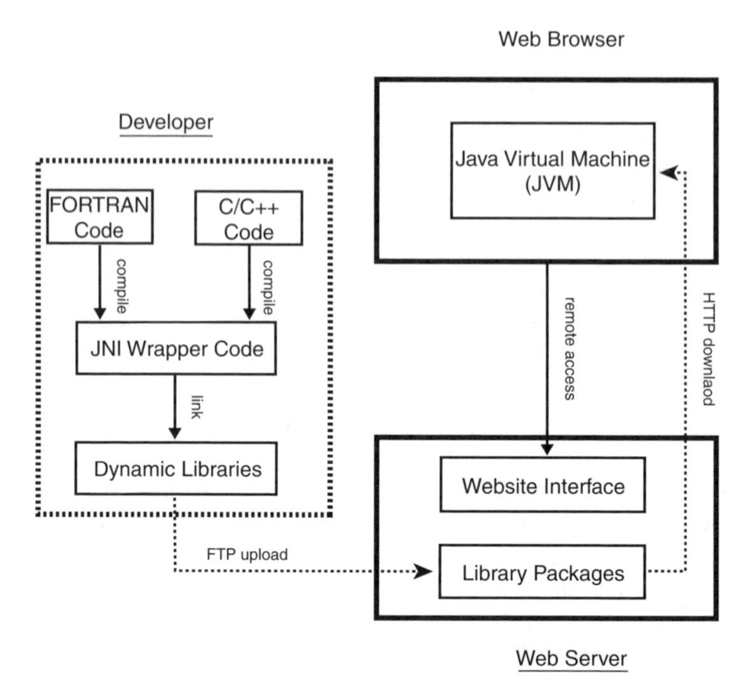

FIGURE 3.3 One-dimensional MRTM simulation architecture. Simulation programs packaged into libraries are uploaded to a web server. Users access the web server with their favorite web browser.

The computations in the three-dimensional MRTM simulation system are much more time consuming than those in the one-dimensional MRTM simulation system. This kind of workload is usually too heavy for client-side computing resources. Therefore, the client-side computing model used in the one-dimensional MRTM simulation system does not fit the three-dimensional case. High-performance (parallel) computing technologies should be used to achieve better performance. This research aims to transfer the traditional high-performance computing model to a web-based one, so that users are relieved of the tedious text-based terminal operations and manual maintenance of result data files. Java 2 Enterprise Edition (J2EE) technologies enable the server-side computing features for the three-dimensional MRTM simulation system. In the multitiered computing architecture (Figure 3.4), the middleware program runs inside a J2EE server all the time, and provides web services for the requests to execute the three-dimensional simulation computations. The J2EE server connects the client and server sides smoothly. Computing requests are submitted by the client-side web browser, arranged by the J2EE middleware, and finally sent to the server-side computing resources. The high-performance computing is executed on the server-side parallel computers. These parallel computers may be the same machines where the J2EE server runs, but can also be different machines. The computing job information and result data are maintained inside a database that resides on the additional database server. Although the system architecture becomes much more complicated than the simple client–server model in the

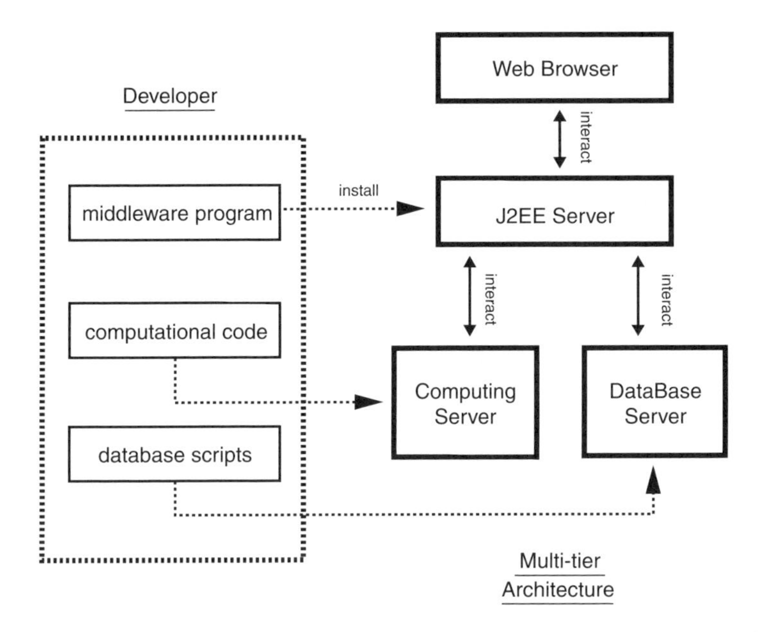

FIGURE 3.4 Three-dimensional MRTM simulation architecture. The multitiered architecture allows middleware programs to run inside a J2EE server and also provides web services to the users of the system.

one-dimensional system, the three-dimensional MRTM simulation system is more flexible, portable, and secure. Besides the usual workload of the server that might delay processing, there are no other factors that would affect server performance.

3.3.2 Three-Dimensional Simulation System and Related Technologies

As mentioned previously, the three-dimensional MRTM model is computation intensive. The JNI technology, which wraps Java code with the legacy code and executes the computations on the client side usually cannot meet the requirement of such computing workloads. High-performance (parallel) computing is also difficult to implement inside the client-side web browsers. Furthermore, the web-based programs with JNI blending technology are platform dependent and require users either to know how to configure security on their local computers, or trust Java applet providers completely. All these restrictions lead developers to the server-side computing model to fulfill the requirements of high-performance computations.

3.3.2.1 Traditional Computing Model

Traditional programs in high-performance computational field simulations are often platform dependent, stand-alone, and lack embedded visualization tools. Platform dependence requires users to compile programs again before running them on a new platform. The source code requires some changes if the software packages have platform-dependent features. Stand-alone applications require users to keep a copy of compiled programs and install all necessary supporting software on their local computers. The traditional solution of using parallel resources often entails allocating accounts to various users, and permitting them to log on to the parallel system and then submit computational jobs. Such solutions add more administrative loads to the parallel system, and apparently do work efficiently for large numbers of users. Furthermore, visualization of the computational results requires separate operations without the help of embedded visualization tools. The computational results are traditionally saved as data files after programs terminate. Users can view the results with their favorite visualization software, but this means that they must install the visualization software locally and maintain their data files manually.

Figure 3.5 shows the traditional procedure of executing high-performance computational programs (e.g., three-dimensional MRTM parallel programs) and viewing the results afterward. Authorized users log in to the parallel computing system remotely, and submit their computational jobs. The results are saved on the remote system disks after the jobs are done. Then, users may download the data files to their local computers (manually through FTP), and separately launch visualization software to view the results.

3.3.2.2 Web-Based High-Performance Computing Model

With the development of the Internet, online access to computer programs seems to be a natural extension of the previous traditional computational model. This research implements a web-based, three-dimensional MRTM simulation system, which takes full advantage of the most popular J2EE technologies, and links all previous standalone procedures smoothly as shown in Figure 3.6. The J2EE server transmits the computing requests from the client-side web browser to the server-side, high-performance computing facilities, and feeds back the responses. Web-based graphics packages relieve users from the manual visualization operations in the traditional computing model. The web-based system also makes use of a database to store computing job information and link data files automatically. Since Java technologies have the "compile once, run everywhere" feature (Sun Microsystems, 1997), systems built on J2EE are also platform independent. In general, this new web-based system makes users feel as though they are shopping online when they submit computing jobs and view the graphical result data. The new computing model also manages and uses expensive computing resources (that a regular user cannot afford) more efficiently and economically.

The Reference Implementation of J2EE (downloadable free from http://java.sun.com/j2ee) was chosen for the J2EE server of the three-dimensional MRTM simulation system. The three-dimensional simulation programs, developed as a multitiered J2EE application or product, is assembled and deployed inside the J2EE server. The J2EE server, which contains the running environment for the J2EE application, connects the client and server sides smoothly. The Enterprise JavaBeans (EJB) container and web container, which reside in the J2EE server, provide the low-level platform-specific functionality that supports the J2EE components. The EJB container manages the execution of EJB components for the J2EE application. The web container manages the execution of the Java Server Page (JSP) and Java

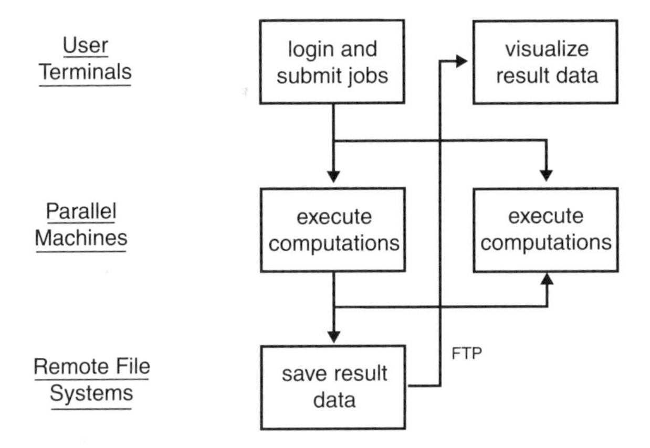

FIGURE 3.5 Traditional high-performance computing model. Authorized users log in and submit jobs to the parallel computing system remotely. The remote system executes computations and saves the results. Users may download the results to their local machines.

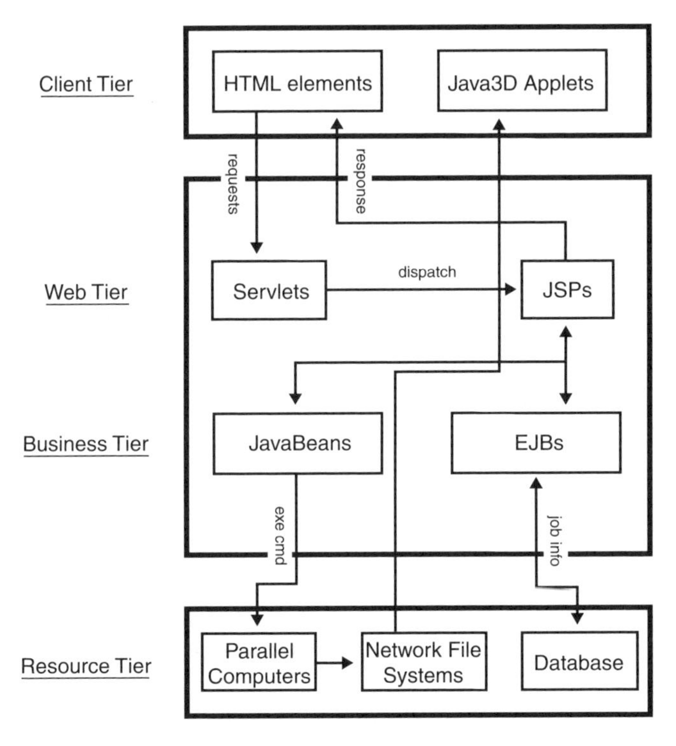

FIGURE 3.6 J2EE multitiered high-performance computing model. The J2EE server transmits the computing requests from the client-side web browser to the server-side, high-performance computing facilities, and feeds back the responses. Web-based graphics packages relieve users from the manual visualization operations in the traditional computing model.

servlet components for the J2EE application. The J2EE components are discussed in detail below. The existence of containers relieves component developers from low-level system implementations, such as system security and database transaction management. In conclusion, the multitiered computing architecture makes the three-dimensional MRTM simulation system more flexible, portable, and secure.

3.3.3 J2EE Technologies

The J2EE technologies, used in the three-dimensional MRTM simulation, provide a component-based approach to the design, development, assembly, and deployment of enterprise applications. The simulation system adopts a multitiered distributed-application model. The system logic is divided into various components in different tiers that run on different machines. Components are used in different tiers of the three-dimensional MRTM system, including an applet in the client tier, JSP in the web tier, and EJB in the business tier. The component paradigm, as the foundation of J2EE technologies, greatly simplifies enterprise application development (Perrone and Chaganti, 2000). Thus, people with different skills working on different components or tiers can collaborate during the whole development process. Therefore, the simulation system based on J2EE technologies, often called the enterprise application, is featured by scalability, portability, security, and flexibility.

3.3.3.1 Client Tier

In multitiered architecture, the client tier is the top layer of the three-dimensional MRTM simulation system. The components inside the client tier run inside the web browsers on the client machines, or users' local computers. In the client tier, users via web browsers access the enterprise application on the MRTM simulation-system website, input parameters for the simulation (including the thread number to use for parallelization), and submit computing jobs to parallel computers on the server side. Other simulation system information, retrieved from the database through the underlying tiers, can also dynamically display on the web pages. Thus, users may check their job information (view the computing results), and the simulation administrator may change the user information (add or delete users). All these presentations are implemented by components called web clients, which are dynamic web pages written in various types of markup language (HTML, XML, and so on). The web browser transforms requests in the client tier into HTTP requests, sends them to the next tier (web tier), and later receives and transforms HTTP responses from the web tier into graphic-user-interface contents to render in the web clients.

The web clients are often called "thin clients" in the enterprise application, because they do not directly query databases, execute complex business logic, or connect to legacy applications. Such heavyweight operations are off-loaded to the components in the web tier or business tier underneath in order to utilize the security, speed, service, and reliability provided by J2EE server-side technologies. The thin-client design makes the enterprise application scalable, and the development process simple and flexible.

Java applets, another kind of J2EE components, are also embedded in the client tier. These applets are the containers holding visualization programs written in Java3D graphical packages. The Java Swing technology combined with Java3D supports the visualization interface for the MRTM computational result data. As discussed in the one-dimensional MRTM simulation system, Java applets with security constraints execute in the JVM installed in the web browser. Therefore, the client tier needs the Java Plug-in (Sun Microsystems, 2001a) and possibly a security policy file (Sun Microsystems, 2001b) for Java applets to safely execute in the client-side web browser. The security restrictions for the one-dimensional simulation system are solved in this enterprise application, and discussed below in the resource tier section.

3.3.3.2 Web Tier

As the second tier in the multitiered enterprise simulation system, the web tier communicates with both the upper client tier and lower business tier. Inside this tier, the information from the client-tier requests are processed and forwarded to the business tier. The dynamically generated responses by the web tier, together with the feedback from the business tier, are transmitted back to the client tier for display in the web pages. Web components separate the web logic control from the web page design, enabling concise and modular development.

The logic of the web tier is implemented by the web-tier components running on the J2EE server. Java servlets and Java Server Pages (JSPs) are the major components to generate dynamical responses for the client-tier requests. Servlets written as Java classes dynamically process HTTP requests and construct HTTP responses for web applications. Due to the object-oriented features of Java language, servlets have the advantages of maintainability and reusability, and hence are used as the central controlling units in the web tier of the three-dimensional, MRTM enterprise-simulation system. According to the client-tier requests, the controlling servlets forward the logic flow to different JSPs, or the processing units.

JSPs are the processing units in the web tier of the three-dimensional, MRTM enterprise-simulation system. Compared to the tedious pure-Java programming for servlet development, JSP technology allows a natural approach to creating web contents because of its simplicity, flexibility, and compatibility. JSPs are text-based documents with two types of contents: the static scripts expressed in HTML or XML format, and the JSP elements, which construct dynamic responses. With the help of HTML or XML scripts, generating dynamical web responses according to client-side requests becomes very straightforward. The JSP elements contain the Java code that retrieves, processes, and forwards information between the client and business tiers. Custom tags, for example, realize loop operations in dynamical web-content generation in the MRTM simulation system.

JavaBeans, another type of web-tier component, retrieve the input information from client-side HTTP requests, such as user account and job management information. Since JavaBean components run on the J2EE server, they can access the runtime Java Virtual Machine (JVM) on the server side. Compiled binary code written in native languages such as C/C++ and FORTRAN can be called directly by the server-side, runtime JVM through JavaBeans. Therefore, the client-side JNI technology used in the one-dimensional simulation system is replaced by server-side JavaBean component technology. The binary code directly called by JavaBeans in the three-dimensional MRTM enterprise system is more efficient than the JNI wrapper code in the one-dimensional system.

3.3.3.3 Business Tier

In order to improve performance, business logic, such as processing user and job information, is removed from the web tier and handled by the business tier. In the business tier, EJB components receive information from the web-tier components, process the format, and send the information to the database in the resource tier for storage. EJBs also retrieve data from the database in the resource tier, process them if necessary, and send them back to the web tier for display.

In the MRTM enterprise system, two kinds of EJBs are used: session beans (session EJB) and entity beans (entity EJB). A session bean represents a transient conversation with the client side. When the client finishes executing (e.g., the user logs off of the simulation system), the session bean and the transient data are gone. In contrast, an entity bean represents persistent data stored in one row of a database table. If the client terminates or if the server shuts down, the underlying J2EE services ensure that the entity bean data are saved in the database. The session EJBs retrieve information directly from the web tier, process the business logic, and then call the entity EJBs. The entity EJBs directly connect the database in the resource tier to load or store the information processed by the session EJBs. Therefore, the system logic flows through servlets, JSPs, session EJBs, and entity EJBs, and finally arrives at the database in the resource tier.

The enterprise system can improve performance by dividing the business tier to session EJBs and entity EJBs. Since the J2EE server can only accommodate a limited number of concurrent database connections, excessive interactions between JSPs and entity EJBs (e.g., when many users access the MRTM system simultaneously) can overburden the server system resources. The session EJBs, acting as a buffer, process the business logic from the web tiers first, and call the entity EJBs to connect to the database only if necessary. Thus, the number of concurrent database connections is greatly reduced, and system performance improves. This further multitiered design also makes the J2EE application more scalable. The source code of servlets, JSPs, session EJBs, and entity EJBs in the enterprise system is clearly modulated and greatly simplified. Future enhancement in different tiers only causes minimum changes to the entire enterprise system.

3.3.3.4 Resource Tier

The resource tier in the J2EE-based MRTM simulation system contains the database server, parallel computing resources, and network file systems. The database server—for example, an Oracle server—can run on a separate machine. This enterprise system using the Cloudscape database server embedded in the J2EE server is used to manage the database. Therefore, both the database server and J2EE server run on the same machine. The parallel computing resources, or the parallel computers, are eight Sun SuperMSPARC processors with the shared-memory environment. The network file systems store the computing results from the MRTM simulations.

Maintaining enterprise data, such as the user information and the computing results, is one of the most important steps involved in building the three-dimensional, MRTM enterprise simulation system. Data collection, often referred to as a database, can be managed by a database management system (DBMS). The DBMS is the information technology used by enterprise system to efficiently retrieve, update, and manage enterprise data. Java Database Connectivity architecture (JDBC) is the standard means for connecting to a database from Java applications. All three-dimensional MRTM simulation information is stored and managed in a database through JDBC connections.

In this enterprise system, the entity EJBs in the business tier access the database through JDBC connections. JavaBeans in the web tier submit parallel computing jobs directly to the resource tier. The parallel-computing job queues are managed by the Portable Batch System (PBS), which provides much flexibility and functionality. The PBS operates on multiple Unix network environments, including heterogeneous clusters of workstations, supercomputers, and other parallel systems. Users submit jobs to the PBS, specifying the number and type of CPUs. The manual operations are automated in the web-based, enterprise MRTM simulation system. With a specified email address in the PBS job submission, the job execution acknowledgment can be sent to users synchronously.

Although the parallel computers and the J2EE server run in different places, they access the same network file systems in the resource tier. The result data files of the MRTM simulation system are saved by the parallel computers to the *public_html* directory of the J2EE server, and are directly read and rendered by the Java3D applets in the client tier. URL connections help Java3D applets read data from the resource tier. Policy files for security configuration used in the one-dimensional simulation system are not needed in the three-dimensional enterprise simulation. The J2EE administrator, instead of the system users, configures the access authority of the *public_html* directory, so as to clear the security restrictions for the applets in the client tier. Since the system logic is performed transparently to users in the J2EE server, this server-side security configuration makes the enterprise system more transparent, simple, and secure.

The one-dimensional MRTM system uses Java2D technology, whereas the three-dimensional simulation system uses more powerful Java3D technology. Java2D and Java3D are parts of the graphics and multimedia components for J2EE user interfacing. Java2D enables sophisticated two-dimensional rendering of objects and images. Java3D provides interfaces for creating and manipulating three-dimensional shapes, as well as adding animation, texture, lighting, and shading to these objects. These graphics technologies help the simulation system visualize the result data online interactively.

3.4 RESULTS

The three-dimensional model was compared to the one-dimensional MRTM model (Selim et al., 1990). To create comparable results, a nominal case was analyzed in which all spatial and temporal variables and parameters were set to a range of 0 to 1. This normalization step provided the means to compare results while avoiding

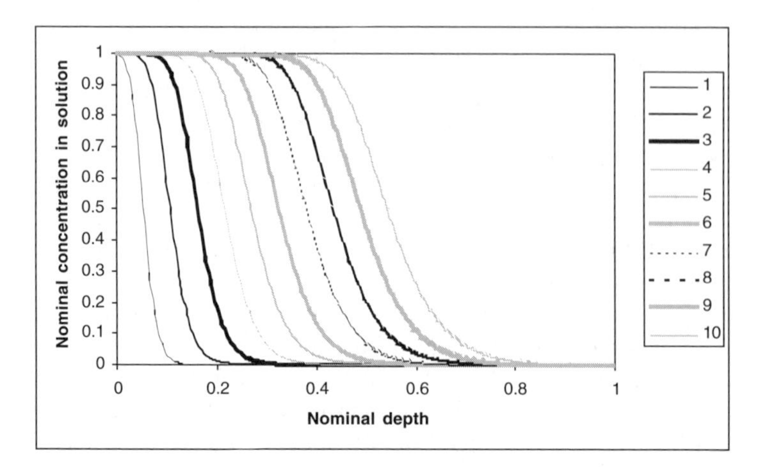

FIGURE 3.7 Concentration-depth curves generated by the one-dimensional MRTM model. Ten consecutive time steps are shown. Note how the contaminant front travels towards the bottom of the soil column, generating sigmoidal-shape curves with a strong slope.

the dimensional analysis that is usually involved. To perform the comparison, contaminant concentrations are plotted in the form of concentration-depth curves.

Figure 3.7 shows results generated by the one-dimensional MRTM model for ten consecutive time steps. The one-dimensional model predicts that the contaminant concentration in the soil solution is almost null at a nominal depth of 0.1 after one time step. After ten time steps, the contaminant concentration almost disappears at 0.8 of nominal depth. The concentration-depth values through the soil profile, for all time-steps, decrease smoothly from 1.0 (at the top soil layer) to zero. The curves are of sigmoidal shape, meaning that the fastest concentration depletion occurs in the middle soil layers. The model predicts that the topsoil layers will be saturated with the contaminant for time steps four to ten from 0.2 to 0.5 of nominal depth. Even the first three time steps show saturation at smaller portions of topsoil layers.

Figure 3.8 shows results from the three-dimensional simulation. In this simulation, the contaminant concentration depletion does not occur up to 0.2 of nominal depth after one time step. After ten time steps, the contaminant concentration only disappears at around 0.9 of nominal depth. For most of the time steps, the concentration-depth curve is of a sigmoidal shape but the slope of the linear portion of the curve (corresponding to the middle soil layers) is less strong than the slope of the one-dimensional MRTM simulation. This means that the three-dimensional MRTM estimates that the middle soil layers are less efficient in retaining the contaminant; hence, the soil solution carries the contaminant farther down the soil depth. The first three curves are of an exponential shape, meaning that the middle soil layers, initially devoid of any solute presence, are very efficient in retaining the contaminant. This efficiency decreases with time as mentioned above. The top layers show that, in almost all cases, maximum concentration in the soil solution does not reach total saturation. The model predicts that those soil top layers will only be partially saturated with the contaminant after the third time step (>90% saturation) from 0.1 to 0.3 of nominal depth.

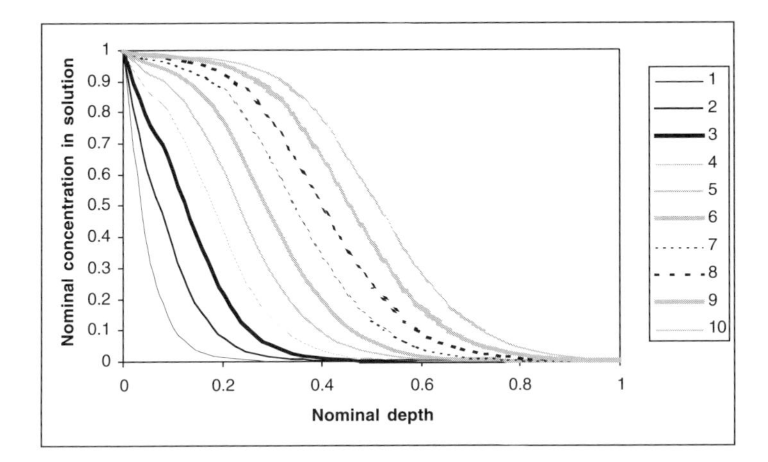

FIGURE 3.8 Concentration-depth curves generated by the three-dimensional MRTM model. Ten consecutive time steps are shown. Overall, the concentration-depth curves generated by the three-dimensional model are consistent with the estimations of the one-dimensional MRTM. Note, however, the lack of total saturation (in terms of concentration) for the top soil layers at any time step. The curves are also smoother than those generated by the one-dimensional model.

Despite some different results between the one- and three-dimensional models that might be due to the fact that simple models (such as the one-dimensional MRTM) provide averaged values, results from both models seem to be consistent. The concentrations estimated by the three-dimensional MRTM seem to better reflect the actual phenomena since this MRTM shows that immediate saturation of the top soil layers does not occur (this would occur only if flow and dispersion in the top horizontal planes would have unrealistic high values). Furthermore, even after ten time steps the topsoil layers are only partially saturated.

One of the immediate applications of this model is to simulate advection and dispersion of contaminants in soil columns with low permeability and strong retention mechanisms. In these cases, the contaminant would not be expected to travel to the lowest layers of the soil profile very soon. In this chapter, results from two common contaminant-input conceptualizations were simulated. The first scenario consisted of a localized contaminant load (point source) located at the center of the topsoil column layer (point-source contaminant inputs may simulate localized leakages of chemicals on the surface of the earth). The second simulation consists of a linearly distributed contaminant load (line source) over the top surface of a soil column. Line sources are commonly used in groundwater contamination problems (e.g., see Mulligan and Ahlfeld, 2001). Additionally, a two-point source simulation is included for purposes of comparison.

The contaminant chosen for this application is a trace compound, such as mercury (Hg). The U.S. Environmental Protection Agency (1999) sets the Criteria Maximum Concentration (CMC) for mercury in fresh water at 1.4 μg/l and the Criterion Continuous Concentration (CCC) at 0.77 μg/l. Given the three types of contaminant sources available in three-dimensional MRTM, this application seeks to know how different the three-dimensional distribution of the contaminant (in the soil solution) would be for each of those input sources.

In all simulations, the column consists of a contaminant-free soil of 1-cm depth, having a volumetric moisture content of 0.4 cm^3/cm^3 and a bulk density of 1.25 g/cm^3. It receives a load of 1 mg/l (i.e., 1000 μg/l) of Hg in solution during 1 h. The dispersion coefficient of the aqueous solution in the soil is isotropic and estimated to be 0.01 cm^2/h in all directions. The Darcy flux is anisotropic with $q_x = 0.05\ cm/h$, $q_y = 0.05\ cm/h$, and $q_z = 0.1\ cm/h$.

The compound–soil interaction is reflected in the following parameters: $k_d = 1.0\ cm^3/g$ (distribution coefficient); NEQ = 1.1 (Freundlich parameter); $k_1 = 0.01\ h^{-1}$ (forward kinetic reaction rate); $k_2 = 0.02\ h^{-1}$ (backward kinetic reaction rate); U = 1.2 (nonlinear kinetic parameter); $k_3 = 0.01\ h^{-1}$ (forward kinetic reaction rate); $k_4 = 0.02\ h^{-1}$ (backward kinetic reaction rate); W = 1.3 (non-linear kinetic parameter); $k_5 = 0.01\ h$ (forward kinetic reaction rate); $k_6 = 0.02\ h^{-1}$ (backward kinetic reaction rate), k_s = 0.005 (irreversible reaction rate).

The spatial size of the grid used for all simulations was 20×20×20. To avoid confusion, a percentage format for the dimension is used in the following comments (e.g., 50% depth corresponds to horizontal plane number 10, and so on for other directions). As shown in Figures 3.9, 3.10, and 3.11, three sliding planes are used by the model to show the spatial distribution of the concentration in shades of gray (the maximum concentration corresponds to black). Additionally, the model provides the maximum concentration values and the position of the plane in X (transverse plane), Y (longitudinal plane), or Z (horizontal plane) directions. X-direction is the direction of the arrow that points to the right. Y-direction is the direction of the arrow that points to the horizontal plane. Z-direction is the direction of the arrow that points downwards.

The three-dimensional distribution of the contaminant concentration (after 1 h of simulation) for the point source is shown in Figure 3.9. At the top, the maximum concentration is clearly equal to the input source concentration (1 μg/l Hg). The anisotropic advection and isotropic dispersion distributes the load asymmetrically around the point of application because the contaminant plume travels following a line that dissects the XY plane at 45 degrees. This can be seen in the dim plume shadow shown at the top horizontal plane, but it is more evident at the horizontal plane that corresponds to about 55% depth. Also at this plane the solute concentration is being reduced below to the CMC and CCC criteria for Hg, 0.0005 mg/l (0.5 μg/l). The model also shows that in a transverse plane (perpendicular to the X axis) at 30% of the distance from the leftmost plane, the concentration (0.0003 mg/l) also meets both criteria. Sliding the plane farther to the right, the plume only meets the criteria again at the rightmost transverse plane.

The transport associated with a line source of contaminant was also studied for the same data set (anisotropic media, isotropic dispersion). The model, besides providing numerical details of the simulation (maximum concentration value) also shows (Figure 3.10) the three-dimensional distribution of the contaminant for two cutting planes. The migration of the contaminant concentration for the line source has a wider distribution in the horizontal and transversal planes as shown in the figure. The contaminant concentration meets the CMC and CCC criteria just at 50%

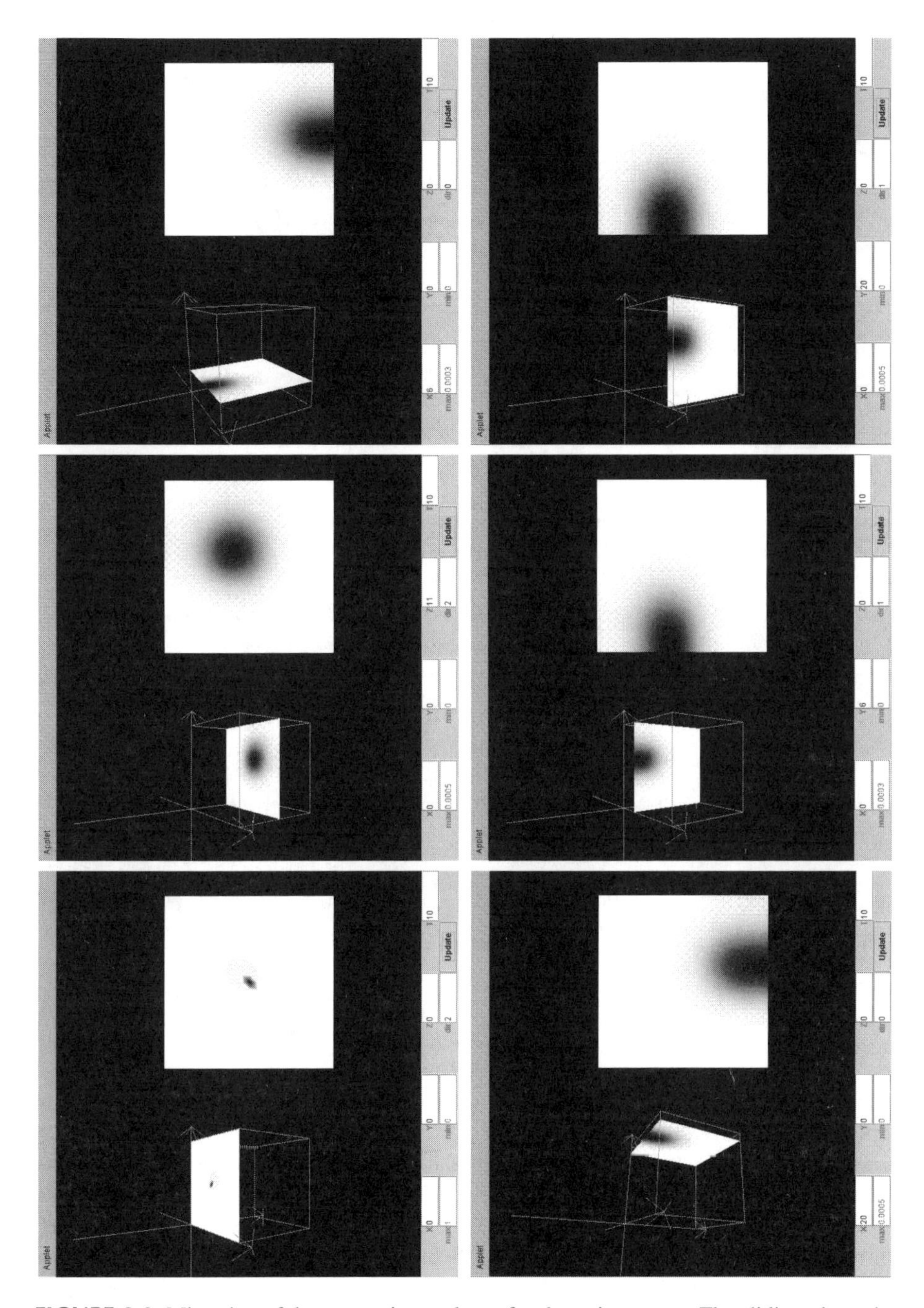

FIGURE 3.9 Migration of the contaminant plume for the point source. The sliding planes in the X, Y, and Z directions show the position of the soil column regions in which the solute concentration complies with EPA criteria for Hg in fresh water. The numerical value of the maximum concentration in those planes is also shown.

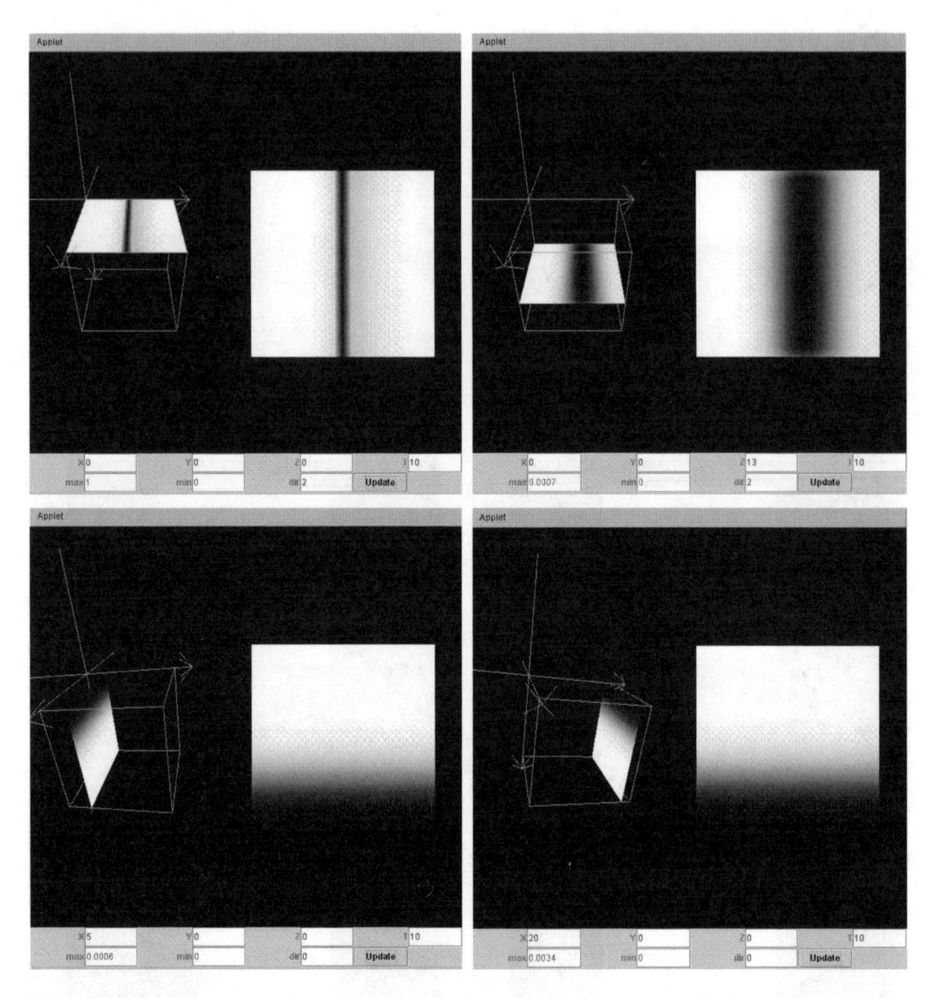

FIGURE 3.10 Migration of the contaminant plume for the line source. Only two visualization planes are shown (horizontal and transverse) because of the almost symmetric contaminant distribution in the longitudinal plane.

of the depth, meaning that the solute has traveled deeper into the soil. The same is true for the transverse plane because the EPA criteria are met at 25% distance from the leftmost transverse plane (against 30% as estimated in the previous simulation). Since there is no advection in the negative X-direction (to the left), the plume has traveled up to this plane only due to dispersion, that is, the line source spreads wider in the X-direction. The migration in the positive X-direction (to the right) is also stronger. The plume is still slightly above the CMC criteria and doubles the CCC criterion at the rightmost transverse plane.

The simulation results for the two-point source (Figure 3.11) are similar to the one-point source case in the transverse and horizontal planes. The solute concentration in soil meets the EPA criteria at 55% depth and at 30% distance from the leftmost transverse plane. The concentration values are also below the criteria at the

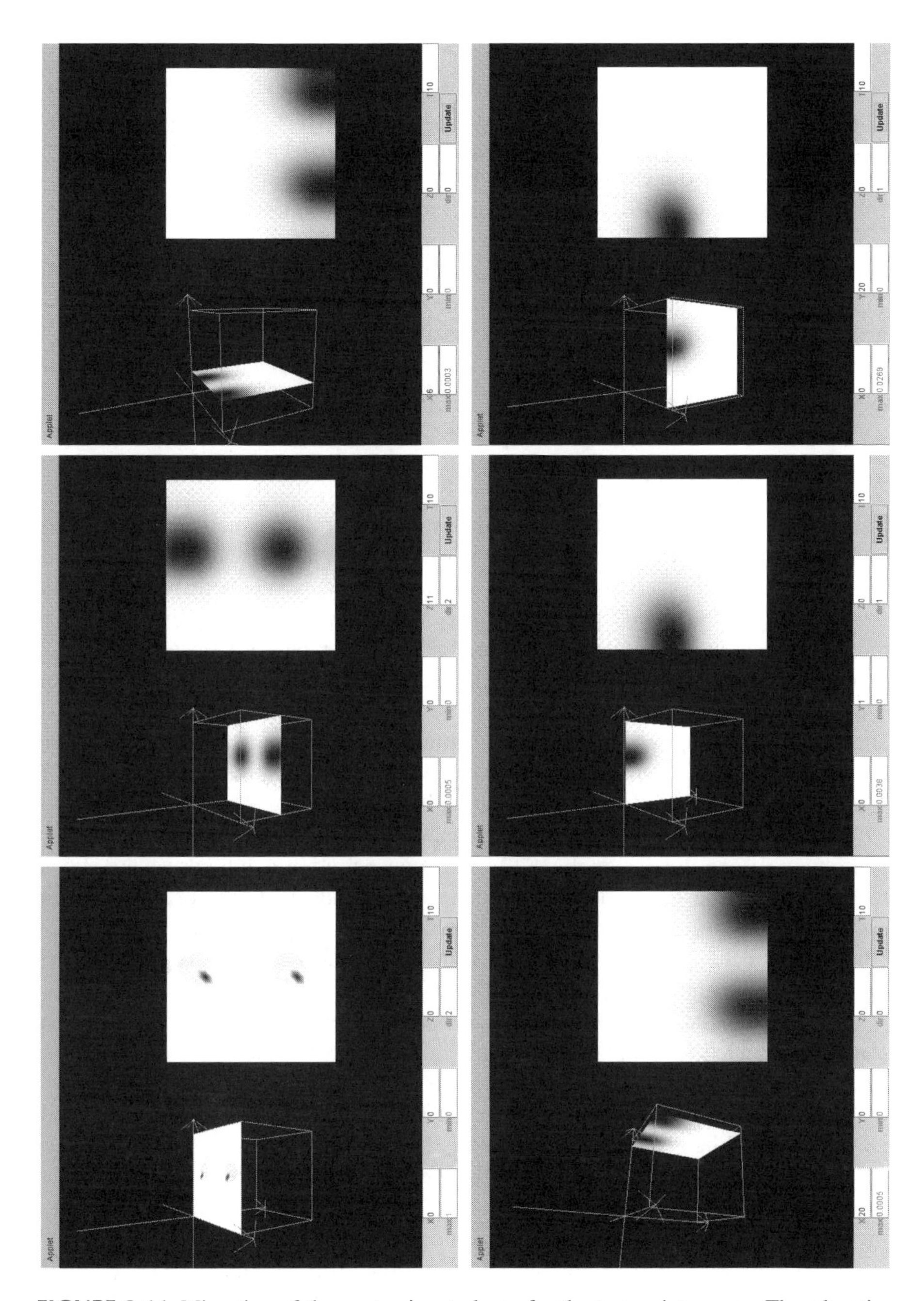

FIGURE 3.11 Migration of the contaminant plume for the two-point source. The advective and dispersive contaminant front is visualized at planes in which the solute concentration complies with the EPA criteria. Concentrations at the frontmost and backmost planes, however, are shown to be over the CMC and CCC limits.

rightmost transverse plane. This was expected due to the relatively wide separation between both source points, that is, combined dispersion and advection effects are absent. But, the longitudinal plane shows that the solute concentrations in this direction are not met even at the border frontmost and backmost planes. Solute concentrations are shown to be higher than 3.8 μg/l (0.0038 mg/l).

It is interesting to see how the visualization part of three-dimensional MRTM scales the concentration values between 0.0 to the maximum concentration value at the plane in which the results are visualized. This could be very useful when the solute is a trace compound. When the trace compound is hazardous (e.g., a heavy metal such as mercury), it is also necessary to monitor the spatial distribution, particularly of very low concentrations. The current three-dimensional MRTM visualization provides a means to track these types of compounds.

3.5 CONCLUSIONS

This three-dimensional MRTM simulation system sets up a generic framework for high-performance computing applications using J2EE enterprise technologies. Platform independence, web accessibility, and multitiered architecture provide a thin front-end to the back-end high-performance computing resources. Given access to the Internet, users can create and execute their own computing jobs using adequate computational resources anywhere anytime, even from a laptop personal computer. The web and business tiers, often called the middle tier, have the responsibility of allocating computing resources and managing result data. As the system grows, the multitier design also has the advantage of allocating programming tasks more efficiently to different people with different skills.

The results provided by three-dimensional MRTM are consistent with the numerical output of one-dimensional MRTM. The concentration-depth curves are shown to be similar for a nominal test case that is independent of temporal and spatial scales. Besides the numerical output that the model generates, the visualization component of the model gives an almost instantaneous look into the spatial distribution of the contaminant. This visualization is made by sliding three planes (horizontal, longitudinal, and transversal) across the entire simulation domain. Concentrations are scaled from 0.0 to the maximum values so that the trace concentrations can be easily visualized. The numerical value of the maximum concentration is also output in the visualization window, together with the current position of the visualization plane. When the trace compound is hazardous (e.g., a heavy metal such as mercury), it is also necessary to monitor the spatial distribution of very low concentrations. The current three-dimensional, MRTM visualization method provides the means to track these types of trace concentrations.

The model was used to study the advection and dispersion of a trace contaminant load into a low-permeability soil column. An anisotropic case was simulated. The model estimated and showed the spatial distribution of the contaminant plume and visually depicted the concentration values in grayscale. The three-dimensional visualization provided by the model was shown to be very useful for identifying the extent and severity of the soil contamination due to the trace compound load under three different types of input load distribution (point source, line source, and two-point sources). The quantification and visualization shown in the example could be used for remediation purposes. Models of lower dimensions might not provide comparable information, especially on the spatial distribution of the contaminant plume.

The efficiency problem caused by the three-dimensional intensive computations can be further improved by applying parallel computing technologies, since the ADI method provides enough parallelism. Details of parallelization will be reported in a future paper. The current three-dimensional visualization demonstrates the distribution of solution concentrations. With further extensions of the web-based simulation system, other solute distributions in different phases will be included.

ACKNOWLEDGMENT

This research was partially supported by U.S. National Science Foundation grants 0075009 and 0082979.

REFERENCES

Deane, G., Chroneer, Z., and Lick., W., Diffusion and sorption of hexachlorobenzene in sediments and saturated soil, *J. Environ. Eng.*, 125, 689, 1999.

Heberton, C.I., Russel, T.F., Konikow, L.F., and Hornberger, G.Z., A Three-Dimensional Finite-Volume Eulerian-Lagrangian Localized Adjoint Method (ELLAM) for Solute-Transport Modeling, Water Resources Investigations Report 00-4087, U.S. Geological Survey, Reston, VA, 2000.

Islam, J., Singhal, N., and Jaffe, P., Modeling biogeochemical interactions in soils under leaking landfills, Proceedings of the 1999 Contaminated Site Remediation Conference, Fremantle, Western Australia, 1999.

Kipp, K.L. Jr., Konikow, L.F., and Hornberger, G.Z., An Implicit Dispersive Transport Algorithm for the U.S. Geological Survey MOC3D Solute-Transport Model, Water Resources Investigations Report 98-4234, U.S. Geological Survey, Reston, VA, 1998.

Konikow, L.F., Goode, D.J., and Hornberger, G.Z., A Three-Dimensional Method-of-Characteristics Solute-Transport Model (MOC3D), Water Resources Investigations Report 96-4267, U.S. Geological Survey, Reston, VA, 1996.

Ladd, S.R., *Java Algorithms*, McGraw-Hill, New York, 1998.

Leij, F.J., Toride, N., and van Genuchten, M.T., Analytical solutions for non-equilibrium solute transport in three-dimensional porous media, *J. Hydrol.*, 151, 193, 1993.

Manguerra, H.B. and Garcia, L.A., Modeling flow and transport in drainage areas with shallow ground water, *J.Irrigation Drainage Eng.*, 123, 185, 1997.

Mulligan, A.E. and Ahlfeld, D.P., Optimal plume capture in unconfined aquifers, in *Physicochemical Groundwater Remediation*, Smith, J.A. and Burns, S.E., Eds., Kluwer Academics, New York, 2001, p. 23.

Peaceman, D.W. and Rachford, H.H., The numerical solution of parabolic and elliptic differential equations, *SIAM J.*, 3, 28, 1955.

Perrone, P.J. and Chaganti, V.S.R.K.R., Building Java Enterprise Systems with J2EE, SAMS, Indianapolis, IN, 2000.

Rudakov, D. and Rudakov, V.C., Analytical modeling of aquifer pollution caused by solid waste depositories, *Ground Water*, 37, 352, 1999.

Segol, G., Hopscotch algorithm for three-dimensional simulation, *J. Hydraulic Eng.*, 118, 385, 1992.

Selim, H.M., Modeling the transport and retention of inorganics in soils, *Adv. Agron.*, 47, 331, 1992.

Selim, H.M. and Amacher, M.C., *Reactivity and Transport of Heavy Metals in Soils*, CRC/Lewis Publishers, Boca Raton, FL, 1997.

Selim, H.M., Amacher, M.C., and Iskandar, I.K., Modeling the Transport of Heavy Metals in Soil, Monograph 90-2, U.S. Army Cold Regions Research and Engineering Laboratory, Hanover, NH, 1990.

Srivastava, R. and Jim Yeh, T.C., A three-dimensional numerical model for water flow and transport of chemically reactive solute through porous media under variably saturated conditions, *Adv. Water Resour.*, 15, 275, 1992.

Sun Microsystems, Java Native Interface Specification, available at http://java.sun.com/j2se/1.3/docs/guide/jni/spec/jniTOC.doc.html, 1997.

Sun Microsystems, Java Plug-in Product, available at http://java.sun.com/products/plugin, 2001a.

Sun Microsystems, Security in Java 2 SDK 1.2, available at http://java.sun.com/docs/books/tutorial/security1.2/index.html, 2001b.

U.S. Environmental Protection Agency, National Recommended Water Quality Criteria—Correction, EPA 822-Z-99-001, USEPA, Washington, D.C., 1999.

Walter A.L. et al., Modeling of multicomponent reactive transport in groundwater: 1. Model development and evaluation, *Water Resour. Res.*, 30, 3137, 1994.

Yakirevich, A., Borisov, V., and Sorek, S., A quasi three-dimensional model for flow and transport in unsaturated and saturated zones: 1. Implementation of the quasi two-dimensional case, *Adv. Water Resour.*, 21, 679, 1998.

Zeng, H. et al., A web-based simulation system for transport and retention of dissolved contaminants in soil, *Comput. Electron. Agric.*, 33, 105, 2002.

Zhu, J., *Solving Partial Differential Equations on Parallel Computers*, World Scientific Publishing, Singapore, 1994.

4 Effect of Structural Charges on Proton Adsorption at Clay Surfaces

Marcelo J. Avena and Carlos P. De Pauli

CONTENTS

4.1 Introduction 79
4.2 Structure of Clays 80
4.2.1 The Surface of a Clay Layer 82
4.2.2 Proton Adsorption 83
4.2.3 The Intrinsic Component of K_H^{eff} : The Bond-Valence Principle 86
4.2.4 The Electrostatic Component of K_H^{eff} 90
4.3 Case Study 93
4.3.1 Modeling Proton Adsorption 99
4.3.2 Choosing the Model 99
4.3.3 Application to Montmorillonite 103
4.3.4 Application to Illite 104
4.3.5 Application to a Kaolinitic Soil 105
4.3.6 Differences in Behavior of Clays and Metal Oxides 106
4.4 Summary and Concluding Remarks 107
4.5 Acknowledgments 107
4.6 Appendix: Isolated Layer Model 107
References 109

4.1 INTRODUCTION

Ion adsorption and desorption at the mineral–water interface are important processes in soils, sediments, surface waters, and groundwater. By capturing or releasing ions, mineral surfaces play key roles in soil fertility, soil aggregation, chemical speciation, weathering, and the transport and fate of nutrients and pollutants in the environment. Proton adsorption is a very specific form of ion adsorption. This area is so important

1-56670-623-8/03/$0.00+$1.50

that it is usually treated separately from other forms. Most minerals have reactive surface groups that are capable of binding or releasing protons. This leads to the development of electrical charges at the surface and the ability to control the attachment of metal complexes, ions different from protons, organic molecules, polymers, microorganisms, and particles. According to Brady et al.,[1] understanding proton adsorption is a necessary first step to unraveling the affinity of mineral surfaces for both inorganic and organic species.

The main effects of proton adsorption-desorption on the adsorption of metals in general and heavy metals in particular have been recognized for many decades. Protons can be exchanged by metal ions at exchange sites on the mineral surface; desorbing protons can leave negatively charged groups at the surface, which act as Lewis bases that coordinate metal ions; adsorbed protons can form proton bonds between surface groups and metal complexes; and adsorbed protons can also generate positive charges at the surface repelling or attracting respectively positively or negatively charged metal complexes. A good understanding of proton adsorption is essential to learn more about metal adsorption at the mineral–water interface.

Most scientific articles on proton adsorption focus discussion on the oxide–water interface. Although it has been intensively studied, proton adsorption at the clay–water interface has been addressed with less detail or was taken as a particular case of adsorption on oxides. This chapter deals mainly with the protonating-deprotonating properties of phyllosilicate clays. It stresses the key role that the presence of structural charges (one of the most important differences between clays and normal metal oxides) plays in determining the adsorption, not only at the basal planes but also at the edges of the particles. A brief description of both the bulk and surface structure of phyllosilicate clays is given, and conventional models for proton adsorption and the electric double layer especially developed for clays are used. The chapter is based on a recent review,[2] which in turn is based on older articles by numerous authors who performed a great deal of work since Pauling[3] introduced the basis for explaining mineral structure and reactivity. Our aims are to provide insight into the main processes that control the proton adsorption at a phyllosilicate surface and to highlight the differences between the behavior of clays and oxides.

4.2 STRUCTURE OF CLAYS

Books describing clay structure are numerous.[4–7] Thus, only a brief description is given here. The basic building bricks of phyllosilicate clays are tetrahedrons with Si^{4+} in the center and four O^{2-} in the corners, and octahedrons with a metal cation Me^{m+} (usually Al^{3+} or Mg^{2+}) in the center and six O^{2-} and/or OH^- in the corners. The tetrahedrons share oxygens to form hexagonal rings, and the combination of rings lead to the formation of a flat tetrahedral sheet. Similarly, the octahedrons share oxygens to form a flat octahedral sheet. If only two-thirds of the octahedral sites are occupied by cations the sheet is termed octahedral. If all possible sites are occupied the sheet is termed trioctahedral.

The tetrahedral and octahedral sheets can be stacked on top of each other to form a phyllosilicate layer. Indeed, the first classification of phyllosilicate clay minerals is based on the type and number of sheets that form the layer. The superposition of one tetrahedral and one octahedral sheet results in a 1:1 layer. This layer type is represented in soils by the kaolin group, kaolinite being the most common mineral of the group. On the other hand, the superposition of two tetrahedral sheets with one octahedral sheet between them results in a 2:1 layer. There are three clay groups with the 2:1 structure: illitic (mica), vermiculite, and smectite (montmorillonite). Schematic representations of sheets, layers, and stacks of layers are given in Figure 4.1.

In many phyllosilicate layers there are isomorphic substitutions. These substitutions occur when a Si^{4+} or Me^{m+} ion in the ideal phyllosilicate structure is substituted by another cation. Since the valence of the new cation is usually lower than that of Si^{4+} or Me^{m+}, the layer structure has a shortage of positive charges, which is interpreted as a net negative charge. The negative charges originated by isomorphic substitution within the structure of a clay layer are usually called structural charges or permanent charges. Structural charges can be tetrahedral or octahedral, depending on the layer where isomorphic substitution took place.

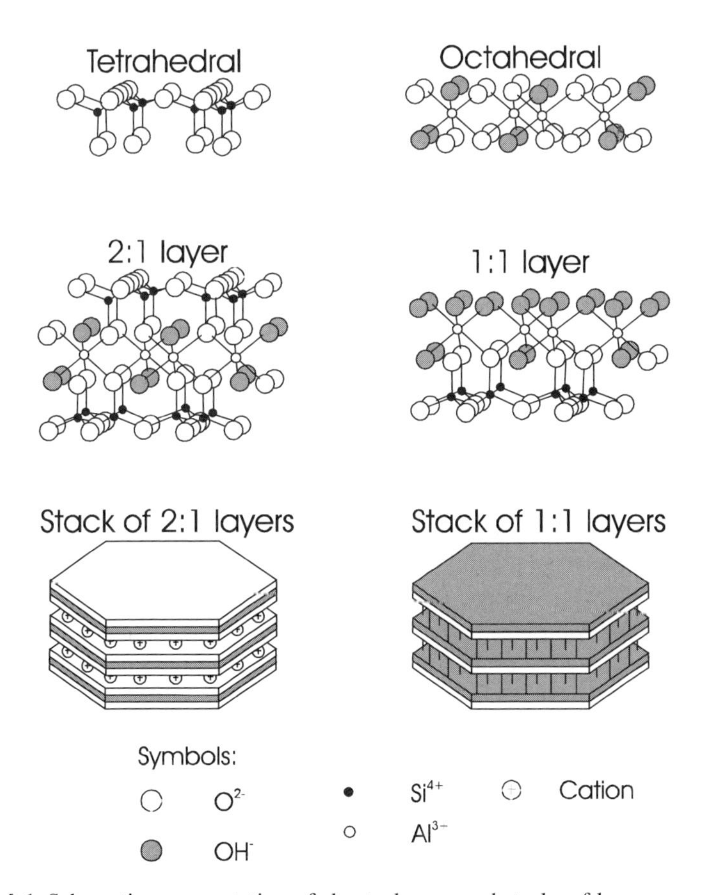

FIGURE 4.1 Schematic representation of sheets, layers, and stacks of layers.

The tetrahedral and octahedral sheets are held together in a 1:1 or 2:1 layer through the sharing of oxygens belonging to the joined faces; very strong bonds keep the sheets together. In addition, layers can associate face-to-face among them to form stacks or platelets (Figure 4.1). Layers having negative structural charges are held together in a platelet by cations intercalated in the interlayer spacing. These cations neutralize the structural charge and serve as an electrostatic binder. In 2:1 layers, when structural charges are located mainly in the tetrahedral sheets, the short distance between interlayer cation and the structural charge sites results in a relatively strong interaction, which impedes complete delamination. This is the case for micas and vermiculites. On the contrary, when structural charges reside mainly in the octahedral sheet the electrostatic interaction is weaker. This is the case for montmorillonite, where the attractive forces are weak enough to allow water to enter the interlayer spacing, produce swelling, and lead to a completely delaminated system under certain conditions. In 1:1 phyllosilicate layers, besides electrostatic interactions there are hydrogen bonds between the octahedral sheet of a layer and the tetrahedral sheet of another layer. Hydrogen bonds are strong enough to keep the layers firmly together and to produce non-delaminating systems, even in the absence of electrostatic attraction between structural charges and interlayer cations.

Layers or platelets can also be associated into aggregates or flocs, and in the extreme case of very concentrated solutions, a gel can also be formed.[8–11] This is especially the case of montmorillonite, where edge-to-face interactions can lead to the formation of a gel at concentrations higher than about 4%.

From the discussion above, it can be seen how the atomic structure of phyllosilicate clays plays a key role in determining the final state of clay particles in aqueous media. The presence of structural charges, neutralizing cations, and the capacity of forming hydrogen bonds between different layers produces a system that can be completely delaminated, completely flocculated, or in an intermediate state having flocs mixed with isolated layers. Whether the more stable situation corresponds to isolated layers, flocs, or a mixture depends on the type of clay, its concentration, pH, concentration and type of supporting electrolyte, and so on.

4.2.1 The Surface of a Clay Layer

Figure 4.2 shows a drawing of a 2:1 layer with the two tetrahedral sheets sandwiching the octahedral sheet. The layers are very thin and flat — the thickness is only 9.6Å whereas the length and width can be several micrometers. The layers are so thin that a large fraction of atoms is located at the surface. The oxygens at the top of the upper tetrahedral sheet and those at the bottom of the lower tetrahedral sheet are at the surface planes (basal planes) of the layer. The rest of the atoms can be considered in the layer bulk. A simple counting reveals that surface oxygens represent 35 to 40% of the atoms forming part of the solid. This extremely high fraction of surface atoms, as compared to that of bulky oxides explains why reactions at a clay surface may become important. Figure 4.2 is accurate for a complete delaminated system. However, if the solid is formed mainly by thick platelets, such as in micas, vermiculites, or illites, the fraction of atoms exposed to the solution is much lower.

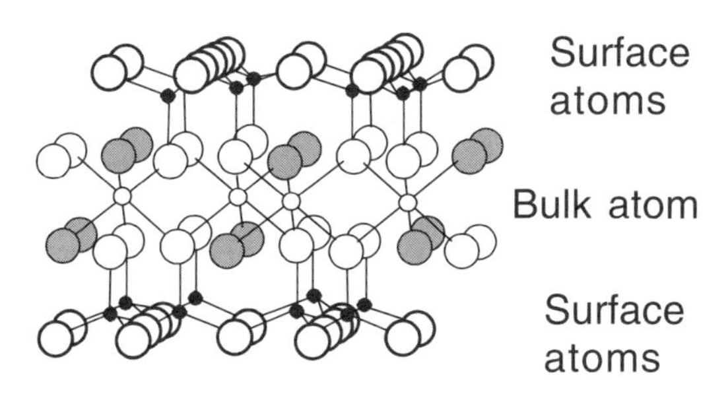

FIGURE 4.2 2:1 Layer showing the oxygen atoms at basal surfaces. Symbols used are the same as in Figure 4.1.

When a clay is dispersed in an aqueous solution, surface oxygens or surface hydroxyls become potentially reactive. Their reactivity depends on the type and spatial distribution of the atoms surrounding them. Oxygens at the siloxane surface, for example, are bonded to two Si^{4+}, and it is customary to define the species Si_2-O as a surface group called the siloxane group.[12] Hydroxyls belonging to the basal surface of 1:1 layers are bonded to two Al^{3+}, and the formed Al_2-OH group is sometimes called a gibbsite surface group,[2] because it is the same as the group located at the basal surface of gibbsite. Besides the Si_2-O and Al_2-OH groups, the broken edges of the layers contain other surface groups. According to White and Zelazni,[13] there are three main groups at the edges of a phyllosilicate surface: tetrahedral ^{IV}Si-OH, octahedral ^{IV}Al-OH, and transitional ^{IV}Si^{VI}Al-OH groups. The presence of irregularities in the structure, such as steps and ledges, may add other surface groups at the edges. Drawings representing typical basal and edge groups of 2:1 and 1:1 layers are schematized in Figure 4.3.

4.2.2 Proton Adsorption

Proton adsorption on a clay particle is the process of transferring the proton from the solution bulk to the surface of the particle. The term "particle" is used here generically to represent a phyllosilicate layer, platelet, or floc. There are several possible states for adsorbed protons, depending on the aggregation state of the clays (Figure 4.4). In an isolated layer, protons can adsorb either at the basal surface or edge surface. Besides these adsorption modes, protons can be adsorbed in a stack of layers, either between two basal surfaces or at the edge of a layer, but existing in very close contact with a nearby basal surface belonging to another layer. When protons are "absorbed" into a floc, they can be attached to edges, at basal surfaces, to both edges and basal surfaces, or can even be dissolved in the aqueous solution trapped within the floc. An exact mathematical treatment of the proton adsorption process, taking into account all these possible adsorption states, seems to be quite difficult. Thus, a simplified treatment is usually applied. The mathematics of proton adsorption considered here follows that presented by Avena,[2] Avena et al.,[14] and Borkovec et al.[15] The treatment is not new, and is written in a general way so that it can be applied to the adsorption of protons to the surface of metal oxides, clays, and other minerals, such as carbonates and sulfurs. Moreover, it can also be used

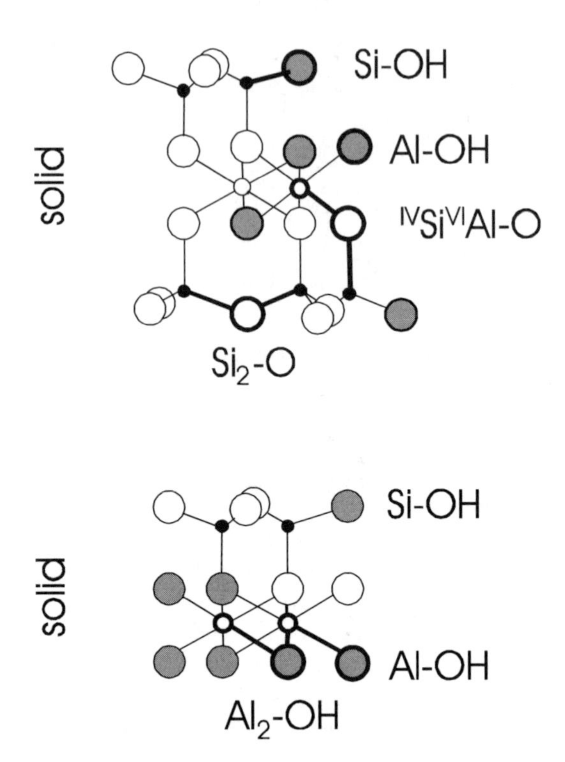

FIGURE 4.3 Phyllosilicate layers showing different surface groups. Symbols used are the same as in Figure 4.1.

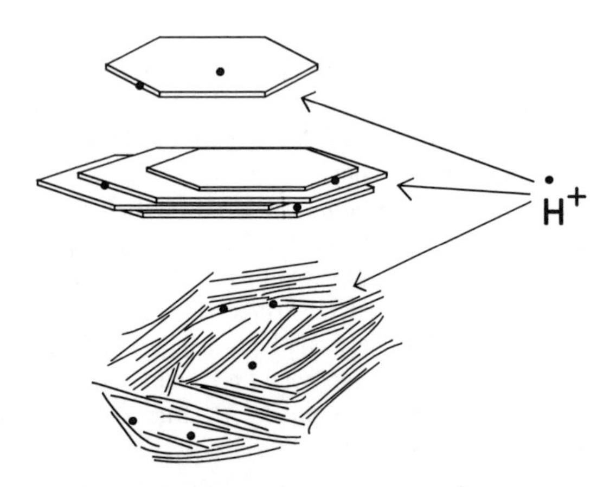

FIGURE 4.4 Schematic representation of the proton adsorption process on layers, platelets, and flocs. (Reprinted from Ref. 2, p. 109 by courtesy of Marcel Dekker, Inc.)

to describe the adsorption of any other ion at the solid liquid interface if the proper charge of the ion is used in the equations.

The binding of a proton ion to a surface group is represented by

$$A^x + H^+ = AH^{x+1} \tag{4.1}$$

where A^x denotes a functional surface group carrying a charge x and AH^{x+1} is the protonated group. The mass action law of this reaction is

$$K_H^{\text{int}} = \frac{\Gamma_{AH}}{\Gamma_A a_{H,0}} \tag{4.2}$$

where Γ_{AH} and Γ_A are the surface densities of the protonated group (AH^{x+1} in this case) and of the unprotonated group (A^x), respectively; K_H^{int} is the protonation constant; and $a_{H,0}$ is an expression for the proton activity at the location of the adsorption site. $a_{H,0}$ is defined as

$$a_{H,0} = a_H e^{-\frac{F\psi_0}{RT}} \tag{4.3}$$

where Ψ_0 is the smeared-out surface potential and represents the difference in the electrical potential between the surface and the bulk solution, and a_H represents the activity of protons in the bulk.

The combination of equations 4.2 and 4.3 gives

$$K_H^{\text{int}} e^{-F\psi_0/RT} = \frac{\Gamma_{AH}}{\Gamma_A a_H} \tag{4.4}$$

In dilute solutions K_H^{int} is independent of the electric potential, and is called the intrinsic protonation constant. The left side of equation 4.4 depends on the magnitude of the surface potential, and is called the effective or apparent constant, K_H^{eff},

$$K_H^{eff} = K_H^{\text{int}} e^{-F\psi_0/RT} \tag{4.5}$$

In logarithmic form, the above equation is

$$Log K_H^{eff} = Log K_H^{\text{int}} - \frac{F\psi_0}{2.303RT} \tag{4.6}$$

According to equations 4.5 and 4.6, the effective affinity of a group for protons results from two different contributions: a chemical or intrinsic contribution, given by $Log K_H^{\text{int}}$, and an electrostatic contribution, given by the term containing the surface potential. Any factor or process affecting either $Log K_H^{\text{int}}$ or Ψ_0 affects the effective affinity of the reactive group for protons. These factors are discussed in the next sections.

4.2.3 The Intrinsic Component of K_H^{eff}: The Bond-Valence Principle

Atoms and ions located at the surface of a solid are characterized by an imbalance of chemical forces because they usually have a lower coordination number than equivalent atoms in the bulk. The undercoordinated cations are Lewis bases and the undercoordinated anions are Lewis acids, and both are unstable in the presence of water. The tendency to restore the balance of chemical forces drives the reactivity of surface groups.

The charge of ions in ionic crystals is neutralized by the surrounding ions of opposite charge. According to Pauling's principle of electroneutrality,[3] the charge of a cation is compensated by the charge of the surrounding anions and vice versa. Thus, the charge of an anion, for example, is only partially compensated by one surrounding cation, and the magnitude of this partial compensation is given by the bond valence, v, defined as the charge z of a cation divided by its coordination number in the solid, CN:

$$v = \frac{z}{CN}$$

In the case of $Al(OH)_3$, for example, where $z = 3$ and $CN = 6$, the value $v = 0.5$ implies that each aluminum atom neutralizes on average half the unit charge of OH^- per Al-OH bond. Then, two aluminum atoms need to be coordinated per OH^- in order to compensate for the hydroxyl charge and to achieve electroneutrality in the structure.

Hiemstra et al.[16,17] applied this concept of local neutralization of charge and geometrical considerations to develop the MUSIC model, which permits estimation of the intrinsic protonation constant of various surface groups. The model was proposed to explain the reactivity of surface groups in metal oxides, but it can also be applied to evaluate the reactivity of groups belonging to clay surfaces. Indeed, Pauling's concepts, on which the MUSIC model is based, were developed for minerals and clays.

The MUSIC model relates the intrinsic affinity for protons of any surface oxygen to the degree of charge neutralization that the surrounding cations achieve. Strictly speaking, the model is only applicable to ionic solids. Bleam[18] re-analyzed the MUSIC model and formulated a similar one: the crystallochemical model, which is based on the bond-valence principle used by crystallographers to predict structure and properties of solids. The crystallochemical model relates the affinity for protons of surface groups to the Lewis basicity of surface oxygens, and is more general than the MUSIC model because it can be also applied to nonionic materials. More recently, Hiemstra et al.[19] combined the MUSIC model and the crystallochemical model, proposing a modified version of MUSIC, including the main concepts of the bond-valence principle.

The modified MUSIC model formulates the protonation reactions as

$$Me_n - O^{nv-2} + H^+ = Me_n - OH^{nv-1} \tag{4.7}$$

$$Me_n - OH^{nv-1} + H^+ = Me_n - OH_2^{nv} \tag{4.8}$$

where $Me_n\text{–}O^{nv-2}$, $Me_n\text{–}OH^{nv-1}$ and $Me_n\text{–}OH_2^{ny}$ are oxo- and hydroxo-surface groups; n is an integer that represents the number of metal ions (Me) bonded to the protonating oxygen; and v is the bond valence. Equations 4.7 and 4.8 show that the protonating entity of the group is an oxygen. The model states that the valence V of an oxygen atom in the bulk of a solid is neutralized by j bonds with the surrounding atoms, yielding $V = -\sum s_j$, where s is the actual bond valence.[19] At the surface, part of the bonds are missing but new bonds, either covalent bonds with protons or hydrogen bonds with water molecules, are formed. At the surface, this usually means that the valence of the surface oxygen is either undersaturated or oversaturated. Therefore, the oxygen will react, tending to restore the equality between V and $-\sum s_j$, which gives a more stable bond arrangement.

Mathematically, the model is formulated as

$$LogK_H^{\text{int}} = -19.8\left(\sum s_j + V\right) \tag{4.9}$$

where $LogK_H^{\text{int}}$ is the logarithm of the protonating constant of equations 4.7 and 4.8, and the value 19.8 results from a model calibration using protonation constants of oxo- and hydroxo-solution complexes.[19] To solve equation 4.9, it is necessary to know V (–2) and $\sum s_j$. This last formula can be expressed as

$$\sum s_j = ns_{Me} + ms_H + i(1 - s_H) \tag{4.10}$$

where n, m, and i are integers; s_{Me} is the actual bond valence of the Me-O bonds (there are different ways of calculating this bond valence,[19] although as a first approximation the value of v will be used here); s_H is the bond valence of the H donating bond and $(1-s_H)$ the bond valence of H accepting bond. The coordination with H donating or accepting bonds is taken into account with appropriate values of s_H and $(1-s_H)$. Bleam[18] and Hiemstra, Venema, and Van Riemsdijk et al.[19] used values of s_H corresponding to hydrogen bonds between water molecules in pure water. In this case, about 0.2 valence units are transferred per bond. Thus, s_H is about 0.8 and $(1-s_H)$ is about 0.2.

As a calculation example, Figure 4.5 shows a siloxane group (Si_2-O) and a protonated siloxane group (Si_2-OH^{+1}) at the basal surface of a layer. In the oxo group, there are two Si-O bonds, each with $s_{Me} \approx v$; and two H accepting bonds, each with $s_H = 0.8$. The resulting value for $\sum s_j$ is 2.4 and thus $LogK_H^{\text{int}} = -7.9$. In the case of the hydroxo group, there are two Si-O bonds, and one H donating and one H accepting bond. Therefore, $\sum s_j = 1$ and $LogK_H^{\text{int}} = -19.8$. The same procedure is applied to calculate $LogK_H^{\text{int}}$ of other groups.

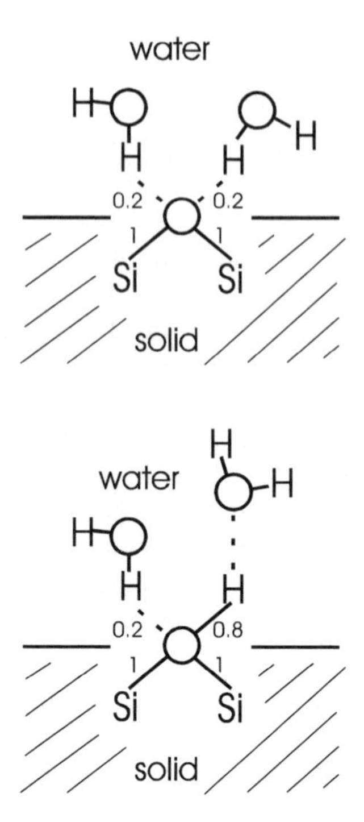

FIGURE 4.5 Clay surface with accepting and donating H bonds.

TABLE 4.1
$LogK_H^{int}$ of Different Groups at Clay Surfaces According to Modified MUSIC Model

Group	$S_{Me} \approx v$	n	m	i	$V + \sum s_j$	$LogK_H^{int}$
Si_2-O	1	2	0	2	0.4	–7.9
Si_2-OH	1	2	1	1	1	–19.8
Si-O	1	1	0	3	–0.4	7.92
Si-OH	1	1	1	2	0.2	–4.0
Al-O	0.5	1	0	2	–1.1	21.8
Al-OH	0.5	1	1	1	–0.5	9.9
Al_2-O	0.5	2	0	1 (2)	–0.8 (–0.6)	15.8 (11.9)
Al_2-OH	0.5	2	1	0 (1)	–0.2 (0)	4.0 (0.0)
$^{IV}Si^{VI}Al$-O	$s_{Al} = 0.5$, $s_{Si} = 1$	$n_{Al} = 1$, $n_{Si} = 1$	0	1 (2)	–0.3 (–0.1)	5.9 (2.0)
$^{IV}Si^{VI}Al$-OH	$s_{Al} = 0.5$, $s_{Si} = 1$	$n_{Al} = 1$, $n_{Si} = 1$	1	0 (1)	0.3 (0.5)	–5.9 (–9.9)

Table 4.1 shows $LogK_H^{int}$ values estimated for different groups at clay surfaces. The model predicts a very low affinity of Si_2-O groups for protons, which will never be protonated in a normal pH range in an aqueous solution. Even less likely is the probability of protonating a hydroxo Si_2-OH^{+1} group, to give Si_2-OH_2^{+2}. Conversely, $LogK_H^{int} = 7.92$ indicates that the Si-O^{-1} group undergoes protonation at an intermediate pH. A second protonation of the group is impossible in aqueous media ($LogK_H^{int} = -4$ for Si-OH). For surface groups containing aluminum atoms, $LogK_H^{int} = 9.9$ for Al-$OH^{-1/2}$ indicates that this group can become protonated to give Al-$OH_2^{+1/2}$, but that Al-$O^{-3/2}$ will not be present in water because its affinity for protons is so high ($LogK_H^{int} = 21.8$) that it is always protonated. Some doubts remain about Al_2-O^{-1} and $^{IV}Si^{VI}Al$-$O^{-1/2}$ groups because the total number of H bonds ($m+i$) that they form in contact with water is not well established. The actual number $m+i$ depends on the surface structure and steric factors.[19] If $m+i = 1$ for Al_2-O^{-1}, $LogK_H^{int} =$ 4, indicating that the group could become significantly protonated in relatively acidic media (pH < 4). If $m+i = 2$, $LogK_H^{int} = 0$ and the group does not become protonated in a normal pH range. As for the group $^{IV}Si^{VI}Al$-$O^{-1/2}$, $LogK_H^{int} = 5.9$ or $LogK_H^{int} =$ 2 for $m+i = 1$ or $m+i = 2$, respectively. In the first case, significant protonation can take place in acidic media.

The simple estimation of the affinity constant of different surface groups is a significant advantage of the model. The advantage becomes very important when the distribution of groups at the surfaces is also known because the protonation-deprotonation behavior of each surface can be predicted. In the case of 2:1 clays, for example, the only groups located at the basal surfaces are the siloxane groups. Since these groups have a very low affinity for protons, it can be predicted that the whole surface does not protonate in a normal pH range. Only proton bonds with surrounding water molecules will be established. Similarly, the only groups located at the gibbsite surface of 1:1 layers are Al_2-OH groups, which are quite unreactive. Thus, the surface seldom acquires extra protons. It only can become protonated under relatively acidic conditions if $LogK_H^{int} = 4.0$.

It is more difficult to understand the group distribution at edges because this depends on the type of crystallographic planes exposed to the aqueous solution and on the presence of irregularities in the structure. In spite of this, an estimation of the behavior can be made because $^{IV}Si^{VI}Al$-$O^{-1/2}$, Al-$OH^{-1/2}$, Si-O^-, and/or their protonated species undoubtedly are present at edges. Since all these groups undergo protonation-deprotonation in a normal pH range, the edge surface will have acid-base properties.

Although the inertness of basal surfaces and the reactivity of edge surfaces have been recognized for a long time, the modified MUSIC model provides a clear explanation of this behavior. Another important prediction of the model is that the difference between the affinity of an oxo group and the conjugated hydroxo group is very large, about 12 LogK units. This fact is also predicted by the MUSIC model[16] and the crystallochemical model.[18] The difference between the two consecutive protonation constants is so large that only one reaction (either protonation of the oxo group or protonation of the hydroxo group) operates in a normal pH range.[15] This situation appears to be the case in aqueous media for any oxygen-containing group located at the surface of a solid, forming part of an organic molecule[14] or belonging to a water molecule or to an aqueous species.[16] Finally, another advantage of the model with respect to the classical MUSIC model is that it can predict the protonation constant of transitional sites such as $^{IV}Si^{VI}Al$-O. This could not be estimated before.

4.2.4 The Electrostatic Component of K_H^{eff}

Besides $LogK_H^{int}$, the potential at the adsorption site needs to be evaluated in order to know the effective protonation constant of a reactive group (equations 4.5 and 4.6). An accurate evaluation is very difficult because of the complexity of the charge distribution in clay layers and clay aggregates. Consider proton ions represented in Figure 4.4. Under most conditions edge surfaces have a different surface potential than basal surfaces because of their different charging behavior. Therefore, a proton attaching an edge "feels" a different electric field than a proton attaching a basal surface. In addition, since isolated layers, platelets, and flocs may coexist in clay–water dispersions, layer–layer interactions may significantly affect the electrostatic environment of an adsorbing ion. Each of the protons represented in Figure 4.4 is in a particular electrostatic environment and the electric potential is different at each position. A very good knowledge of the size and shape of the layers and of the structure of platelets and flocs is required in order to get a good mapping of the electric potential along the surface or within a floc. The mapping of ideal isolated layers was performed by Bleam[20] and Bleam, et al.,[21] but the mapping of a mix of flocs and isolated particles is impossible thus far, and approximations are usually made. We shall treat these electrostatic approximations in sections 4.3.2 and 4.6. Only qualitative analysis is investigated here.

The electric potential at a clay particle can be evaluated from the charge density if a charge-potential relationship is known. The charge density, in turn, depends on the distribution of surface sites and structural charges. Si_2-O and Al_2-OH groups carry no net charge and a surface populated with these groups only is also electrically neutral. This is the case of siloxane and gibbsite surfaces in 2:1 and 1:1 phyllosilicates; therefore they are uncharged (a gibbsite surface could acquire a positive charge in very acidic media). The electroneutrality of basal surfaces is seldom addressed in the literature, and in most articles it is assumed that basal planes carry a net negative charge because of the presence of structural charges. A better interpretation of the charge distribution in clays is that basal surfaces are neutral and that structural charges are within the clay structure. Charges belonging to the octahedral sheet are at about 5Å from both basal surfaces. Charges belonging to the tetrahedral sheet are located very close to the surface in a plane beneath the surface plane, but still within the layer. On the other hand, edges are populated with groups that are able to become protonated or deprotonated. In summary, in a phyllosilicate layer basal surfaces are normally neutral, structural charges reside within the layer, and edge surfaces are positive, neutral, or negative, depending on the degree of protonation.

Although structural charges do not reside at the surface, they produce an electric field that emanates in all the directions affecting the surface potential.[20] This potential, which is negative with respect to that of the solution bulk, drives the particular cation adsorption properties of basal surfaces. As mentioned previously,[2] Figure 4.6

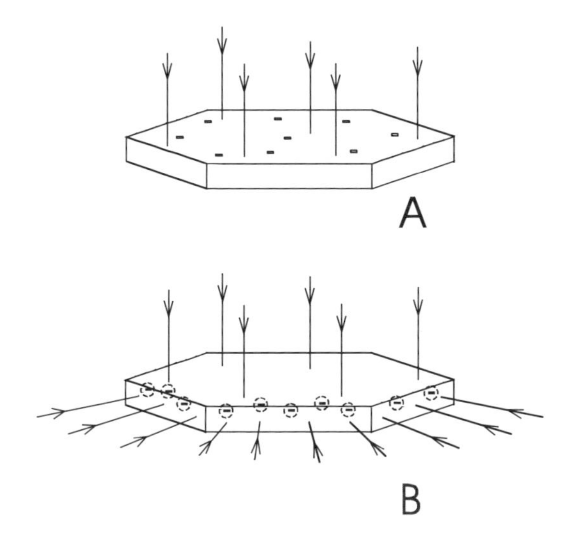

FIGURE 4.6 Phyllosilicate layers having structural charges at (A) the basal planes and (B) within the layer. The layers are assumed to be at a pH where edges carry no net charge. Arrows schematize the electric field generated by the structural charges.

shows two different drawings of a 2:1 phyllosilicate layer, which is assumed to be at a pH where the net charge at the edge surface is zero. Arrows represent the electric field emanating from the structural charges. In drawing A, the charges are located at the basal plane, which is the common assumption. This drawing can lead to the wrong conclusion that structural charges affect only the electric potential at the basal surface without affecting the electric potential at edges. In drawing B, structural charges are represented within the clay layer, which is a better interpretation. Since the electric field generated by the charges emanates in all directions, structural charges affect not only the potential at the basal surface but also at the edge surfaces.

The magnitude of the electric potential at the edges depends on the edge charge and the magnitude and location of structural charges. Güven[22,23] calculated the separation distance between structural charge sites in a hypothetical dioctahedral smectite where the sites are assumed to be regularly separated in a plane parallel to the basal plane. In such smectite, each substitution site is between 7 and 9 Å (average 8Å) apart from the next one, and at an average distance of about 4 Å from the edge surface. The same type of calculation applied to a hypothetical 2:1 dioctahedral layer with tetrahedral substitutions, indicates that each substitution site is between 10Å and 9 Å (average 9.5 Å) apart[23] and at an average distance of about 4.7 Å from the edges.

The situation for both hypothetical layers is schematized in Figure 4.7. In the layer with octahedral substitutions, the distance between structural charges and edges (4Å) is similar to the distance between the structural charge and the basal plane (4.8 Å). This means that the electric field emanating from the charges may similarly affect the electric potential at the basal surface and the edge surface. Thus, under pH conditions where no net charge resides at edges, both basal and edge surfaces are uncharged, and the electric potential is negative and of similar magnitude at both

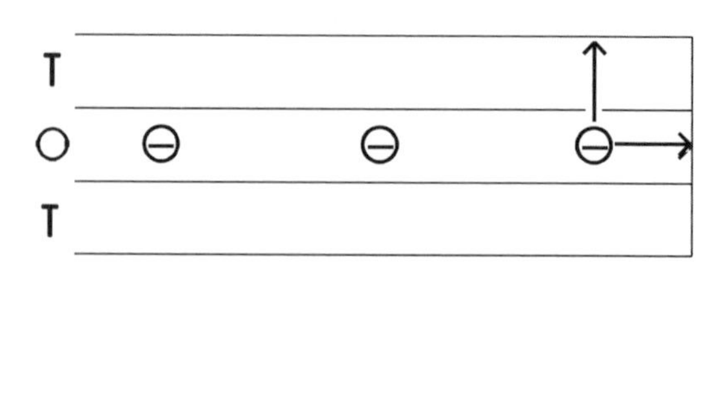

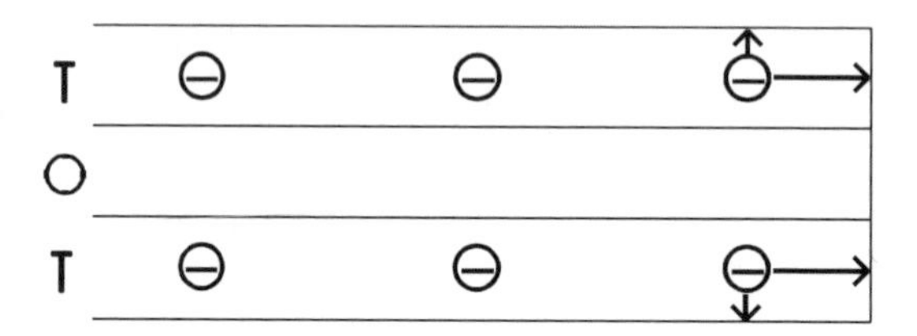

FIGURE 4.7 2:1 Layers showing structural charges in the tetrahedral layers (T) and the octahedral layers (O). Arrows represent the distances between the charges and the basal surface or the edge surface.

surfaces. Even for a low positive-edge charge, the edge potential will be negative if enough structural charges are present.

The negative electric potential generated at the basal surface by structural charges is high enough to induce cation adsorption in positions normal to the basal plane, and to control the distribution of interlayer species. Monte Carlo (MC) and molecular dynamic (MD) simulations show that in Li^+-hectorite (substitution sites in the octahedral sheet), interlayer Li^+ is coordinated to water molecules and its equilibrium position oscillates around a location that is directly on top of the structural charge.[24,25] This localized arrangement of cations in the interlayer demonstrates that the electric field emanating from the charges is readily felt by cations approaching the basal surface. Unfortunately, there is no MC or MD information about the effects of structural charges on the behavior of ions adsorbing at edges. However, due to the vicinity of structural charges, it is expected that an important attractive field is felt by a proton ion or any other cation approaching a surface group at edges.

In the layer with tetrahedral substitution, the average distance between structural charges and edges is again 4Å, but is larger than the distance between the structural charge and the basal planes. The electric field emanating from the structural charges is stronger at basal surfaces than at the edges. The field is so strong that MC and MD simulations predict that in hydrated Li^+-beidellite (substitution in tetrahedral sheets), the interlayer Li^+ also resides in a position normal to the structural charge, but at a smaller distance from the basal surface, directly coordinated to surface oxygens of Si_2-O groups.[24] At the edges, any approaching proton or other cation must also experience an electrostatic attraction. The strength of this attraction should be smaller than that received by a cation approaching the basal surface.

Besides the electric field developed by structural charges, charged groups generated by protonation or deprotonation reactions also contribute to the edge surface potential. When a surface group is protonated at an edge surface, the charge changes and the edge potential is modified. This change in potential changes the effective constant for protonation of the next groups. This effect is seen as an electrostatic interaction between protonating sites, because the proton affinity of a given group changes when another group is being protonated in its neighborhood.[15]

4.3 CASE STUDY

H^+ adsorption at the mineral–water interface modifies the surface charge, the surface potential, and the distribution of ions in the solution surrounding the solid. Experiments aiming to quantify H^+ adsorption and its effects usually combine techniques that are indicators of charge development (H^+ and ion adsorption, ion exchange)[26–29] and techniques that are indicators of electric potential (electrophoresis, streaming potential).[26,30,31] The first group of techniques is by far the most employed in clay and soil systems. A brief description of the techniques is given below.

Proton adsorption. Either potentiometric titrations or batch methods are used for proton adsorption measurements. In the first case, the solid sample is dispersed in an aqueous electrolyte and the dispersion is titrated with a strong acid or a strong base and the H^+ addition is recorded as a function of pH (an OH^- addition is recorded as a negative H^+ addition). A blank, prepared with the supporting electrolyte, is also titrated. The difference in the amount of H^+ necessary to reach a desired pH between the dispersion and the blank provides the H^+ consumption by the solid, and a curve as a function of pH can be constructed. This consumption is not an absolute value. It is relative to the initial state of the solid. The H^+ consumption versus pH curve reflects the absolute proton adsorption only if the sample studied is initially free of acid or base impurities. Otherwise, correction is required. Potentiometric titration is straightforward and can be performed in a relatively short time. Unfortunately, lateral reactions such as solid dissolution can complicate data analysis, because they modify the overall proton consumption and cannot be corrected by performing the blank titration. In the second case (batch methods), H^+ consumption at each pH is evaluated separately. The dispersion is equilibrated in a flask at the desired pH, and the electrolyte concentration and amount of H^+ or OH^- needed to reach that pH is quantified. The same procedure is followed with a blank solution. The difference in H^+ consumption by the dispersion and the blank represents the H^+ consumed by the solid. This technique is more time and sample consuming than potentiometric titration because it is necessary to prepare as many flasks with dispersion and blanks as the number of data points desired. It has the advantage that the supernatant of every sample can be analyzed, and thus dissolution processes or other H^+-consuming processes can be detected and accounted for. Another important advantage is that the technique can be used to perform ion adsorption measurements simultaneously.[29] Like potentiometric titrations, the batch technique also measures relative H^+ consumption.[29]

Ion adsorption. Ions of the supporting electrolyte adsorb to neutralize the surface charge and the structural charge of the solid. Batch methods are usually employed to measure this adsorption. Once the solid and the electrolyte solution at the desired pH are at equilibrium, the dispersion is centrifuged or filtrated. The supernatant is separated from the solid, and the latter is treated with a concentrated electrolyte (different from the supporting electrolyte) to displace adsorbed ions. These ions are then quantified in the extracts and they reflect the net amount of adsorbed ions. If ions are only electrostatically adsorbed, ion adsorption data provide the net charge of the studied material. Cation exchange capacity[32] and inner sphere Cs^+ adsorption[29] are specific cases of ion adsorption.

Zeta potential. Zeta potential measurements are used to investigate the electric potential of the interface. It is usually measured with electrophoresis or streaming potential.[30] These techniques are less commonly applied than the two described above. It is difficult to use electrophoresis with heterogeneous samples such as soil. This technique has to be used with more homogeneous samples such as purified clay.

Chorover and Sposito[29] and Sposito et al.[33] classified the charge of solid particles in three operational categories or components: (1) the structural charge density, σ_{str}, originating from isomorphous substitutions; (2) the H^+ adsorption charge density; σ_H, originating from proton adsorption-desorption processes; and (3) the ion adsorption charge density, which is the difference between the adsorption of cations and anions. Δq includes all the adsorption modes (electrostatic, inner sphere, outer sphere) but excludes H^+ adsorption that is already accounted for in σ_H. The three components of charge can be measured by Cs^+ adsorption for σ_{str}, proton adsorption for σ_H, and ion adsorption for $\Delta q = q_+ - q_-$. Although cation exchange capacity (CEC) measurements are usually used to estimate σ_{str}, it depends on both the structural charge and the proton charge. CEC is only a good estimate of the structural charge when the sample contains negligible surface groups that can become protonated or deprotonated, or when measurements are performed at the pH where $\sigma_H = 0$.

Any clay or soil dispersion in water is electrically neutral. The electroneutrality condition is

$$\sigma_{str} + \sigma_H + \Delta q = 0$$

The equation can be rewritten in the form

$$\Delta q = -(\sigma_{str} + \sigma_H),$$

indicating that a plot Δq versus σ_H should be a straight line with x- and y-intercepts both equal to σ_{str} and slope equal to -1. Since the electroneutrality condition is completely general, the mentioned plot should be independent of ionic strength.

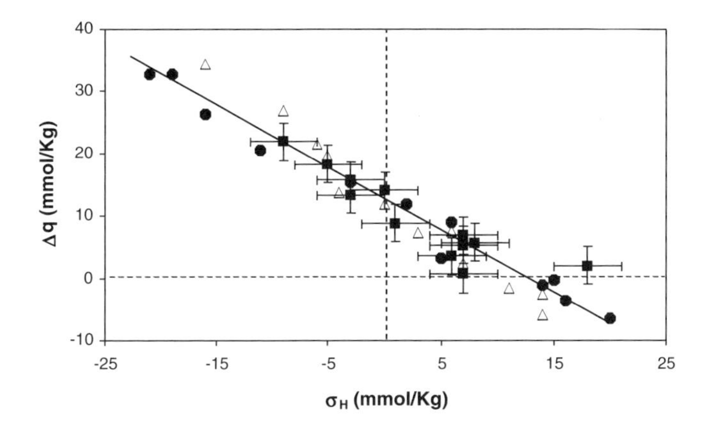

FIGURE 4.8 Charge balance test for a kaolinitic soil in LiCl solutions. Electrolyte concentration: circles, 0.001 M; triangles, 0.005 M; and squares, 0.001 M. Error bars are only given for 0.001-M electrolyte. Bars are similar in the other two cases. Data provided by J. Chorover.

Examples of this behavior were given by Chorover and Sposito[29] for kaolinitic soils from Brazil and by Schroth and Sposito[34] for two Georgia kaolinites, shown in Figure 4.8. Equation 4.10 is a test for charge balance and consistency. Deviations from the behavior described by the equation reveal data inconsistency and indicate inaccuracy or inappropriateness of any of the methods used to measure the charge components. The equation is also very useful to correct relative σ_H versus pH curves.[29] If σ_{str} and Δq are known, at least one pH value, the absolute σ_H , can be calculated at that pH, and the whole relative σ_H versus pH curve can be corrected.

Very few articles have appeared in the literature in which the three components of charge were measured.[29,34] Most other articles report only on σ_H or Δq and give little or no information about the means of establishing the absolute σ_H data. It appears that in these cases only relative σ_H versus pH curves are presented. The authors stated that relative data were analyzed in only a few cases.[35,36]

Several sets of experimental data providing proton adsorption and ion adsorption on clays and clay-containing soils are described below and modeled later. The description starts with two monmorillonite samples, continues with illite, and finishes with a kaolinitic soil.

Baeyens and Bradbury[28] presented a complete and reliable set of proton adsorption data corresponding to the <0.5 μm fraction of a SWy-1 Na-montmorillonite (Crook County, Wyoming). Proton adsorption at edges, proton adsorption at exchanging sites, and total proton adsorption were given, and they are shown in Figure 4.9. Proton adsorption at edges increases by decreasing pH at a given ionic strength. The curves at different electrolyte concentration do not intersect each other, and at constant pH the adsorption decreases by increasing the electrolyte concentration. It is evident that the pH at zero adsorption decreases as the electrolyte concentration increases.

Proton adsorption by Na^+-H^+ exchange is also pH- and ionic-strength dependent. There is a negligible adsorption at relatively high pH, but the adsorption suddenly increases at a given pH. This pH value decreases by increasing the electrolyte concentration.

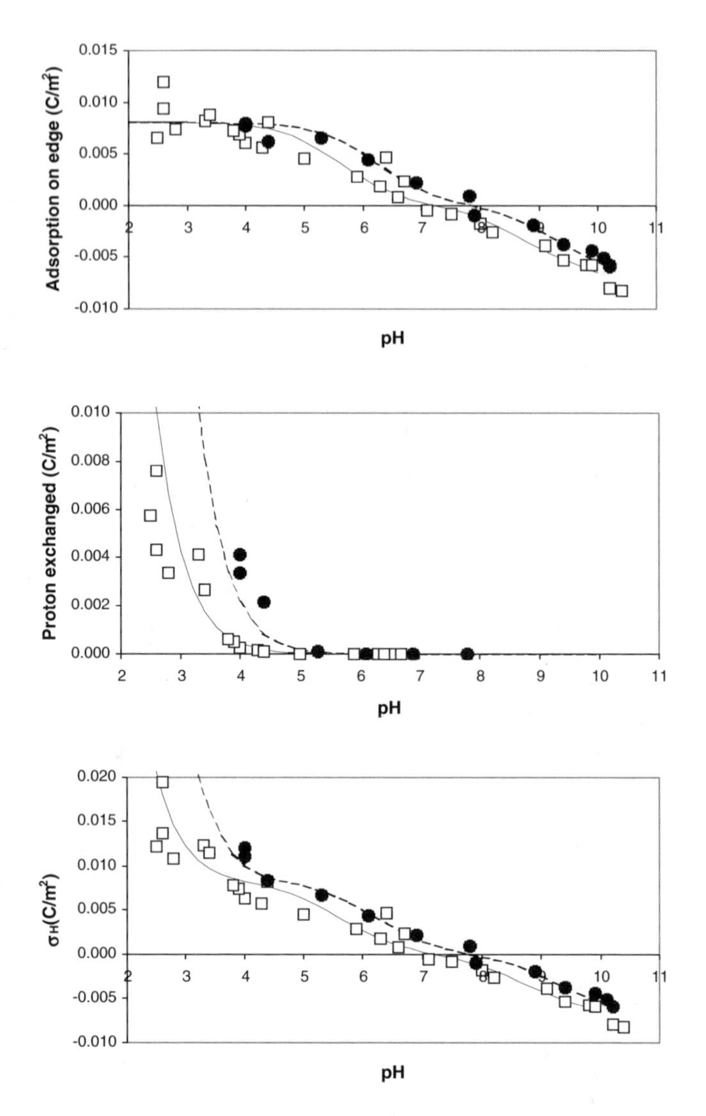

FIGURE 4.9 Proton adsorption on SWy-1 Na-montmorillonite in NaCl solutions. Symbols correspond to experimental data. Lines correspond to model predictions. Electrolyte concentration: squares and solid line, 0.1 M; circles and dashed line, 0.05 M.

Total proton adsorption, which can be identified with σ_H, is the sum of adsorption at edges and at exchange sites. The curves run almost parallel to each other, and there is an increase in proton adsorption on decreasing the pH and the electrolyte concentration. The pH where $\sigma_H = 0$ (pH_0) decreases when the electrolyte concentration increases. This behavior is markedly different from the behavior exhibited by metal oxides having no structural charge. In these oxides, pH_0 does not depend on the electrolyte concentration and appears as a common intersection point of the curves resulting from experiments performed at different electrolyte concentrations.

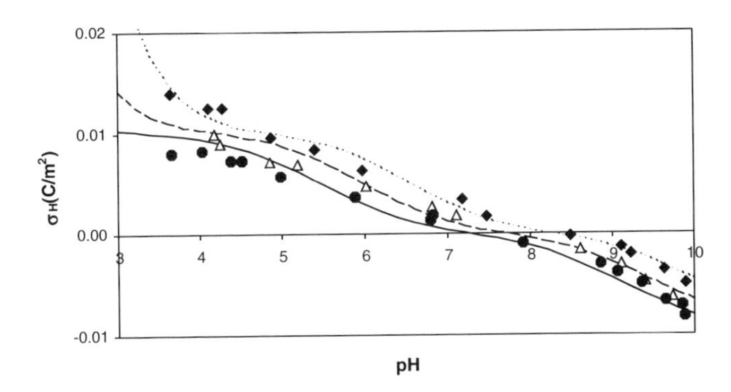

FIGURE 4.10 Proton adsorption on an Argentinean montmorillonite in NaCl solutions. Symbols correspond to experimental data. Lines correspond to model predictions. Electrolyte concentration: circles and solid line, 0.12 M; triangles and dashed line, 0.01 M; diamonds and dotted line, 0.002 M.

This common intersection point in oxides corresponds to the point of zero charge, PZC, which is the pH where the net surface charge is zero.[37,38]

The absence of intersection points in σ_H versus pH curves is a typical characteristic of clays with a high content of structural charges. Avena and De Pauli[26] presented titration curves for three NaCl concentrations using an Argentinean Na-montmorillonite (data shown in Figure 4.10); Turner at al.[39]presented titration curves for two electrolyte concentrations of a smectitic clay; and Madrid and Diaz-Barrientos[40] presented titration curves for an Arizona montmorillonite (Apache County, sample SAZ-1, Clays Mineral Society Source Clays Repository). In all these cases σ_H versus pH curves showed the same features as those in Figure 4.9.

Illitic soils also exhibit a behavior different from oxides and similar to montmorillonites. The experimental data shown in Figure 4.11 corresponds to those published by Hendershot and Lavkulich.[32] Other illites also behave in the same way.[41]

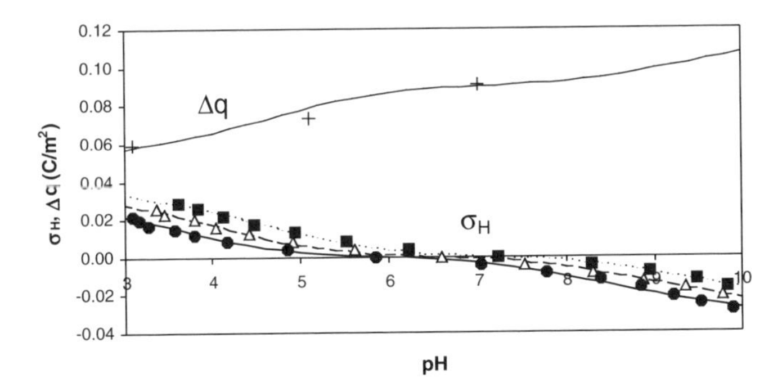

FIGURE 4.11 Proton and ion adsorption on a Na illite in NaCl solutions. Symbols correspond to experimental data. Lines correspond to model predictions. Electrolyte concentration: circles and solid line, 0.1 M; triangles and dashed line, 0.01 M; squares and dotted line, 0.001 M.

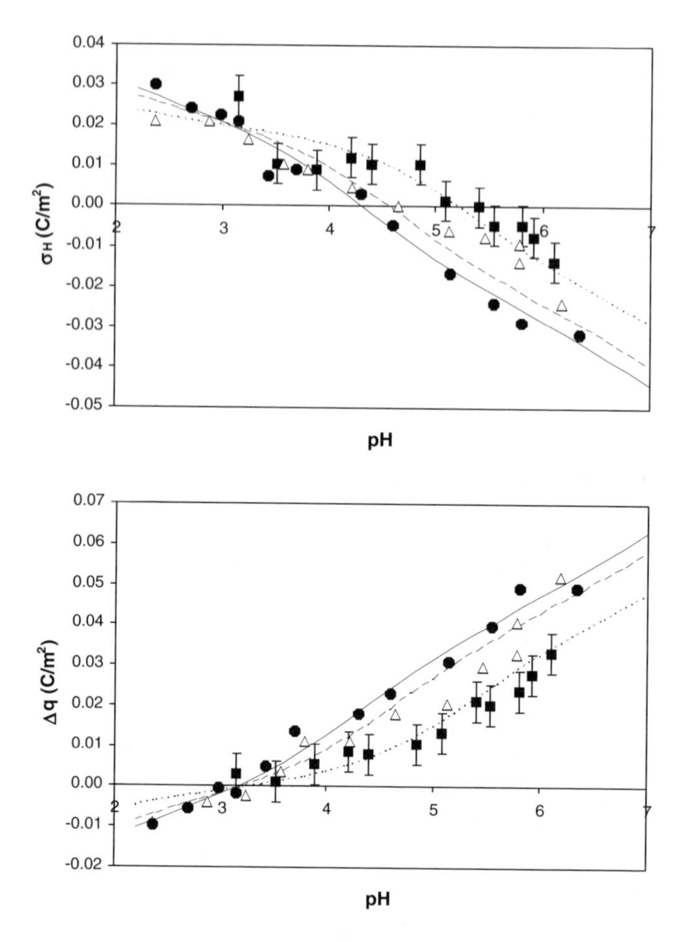

FIGURE 4.12 Proton and ion adsorption on a kaolinitic soil in LiCl solutions. Symbols correspond to experimental data. Lines correspond to model predictions. Electrolyte concentration: circles and solid line, 0.01 M; triangles and dashed lines, 0.005 M; squares and dotted line, 0.001 M. Error bars are only given for 0.001-M electrolyte. The bars are similar in the other two cases. (Reprinted from Ref. 2, p. 109 by courtesy of Marcel Dekker, Inc.)

σ_H versus pH and Δq versus pH curves for a kaolinitic soil[29] are presented in Figure 4.12. The charge balance test of this sample was given in Figure 4.8. This set of data appears to be one of the most complete and reliable in the literature because of the success in passing the test. The clay fraction (75%) contains mainly kaolinite, and also a small amount of vermiculite (about 1%), which is the source of structural charge.[29] At a given ionic strength, the proton-adsorption charge density decreases with increasing pH. The slope of the curves increases by increasing the ionic strength and there is an apparent data convergence at a pH of approximately 3 to 3.5. Similar to that shown with montmorillonite, pH_0 decreases by increasing the electrolyte concentration.

σ_H versus pH curves are also dependent on pH and ionic strength. Since the structural charge is constant, the variations in Δq should reflect the variations in σ_H if the electroneutrality condition is met. Δq versus pH curves tend to intersect at pH 3 to 3.5 at $\Delta q = 0$.

4.3.1 Modeling Proton Adsorption

Modeling of the clay–water interface began several decades ago. The first models were adapted from those used for metal–water and oxide–water interfaces. Van Raij and Peech[27] carried out pioneering works in applying Gouy-Chapmann and Stern models to kaolinitic soils. Since then, many articles in which modeling is performed have appeared in the literature.[26,40–44] More recently, and due to the development of computers, *ab initio* methods, and MD and MC simulations have appeared.[24,25,45,46] They provide invaluable information at the molecular level but, thus far they can only be applied to extremely small systems such as clusters containing a small number of atoms. They cannot be applied to a complex soil system as yet.

The interest in modeling is driven by several factors. Modeling facilitates better understanding of the mineral–water interface making it possible to recognize the main processes at work. Modeling also allows us to quantitatively account for the chemical processes at the interface and to predict the behavior of a system under different conditions. This section deals with conceptual models of the clay–water interface based on conventional acid–base and electrostatic theories. The acid–base theories applied here rely on equilibrium protonation–deprotonation reactions such as equations 4.1, 4.7, or 4.8, whereas electrostatic theories are based on charge distribution and potential decay in the electrical double layer (EDL). In these models, surfaces are usually considered to have ideal planar or spherical geometries with uniform charge density distributions, and with a mean and smeared-out electric potential (mean field approximation). Although there are rather crude approximations in the models, sometimes they capture the essence of adsorption processes very accurately.

4.3.2 Choosing the Model

Any conventional model of the clay–water interface should take into account certain important characteristics of clays, especially those that make the clay–water interface different from the oxide–water interface. The main characteristics to be considered follow:

1. Presence of structural charges and their effects on the electric potential at both the basal and edge surfaces. This is the main difference between a clay–water system and a metal oxide–water system, which in principle contains no structural charges.
2. Monoprotic protonating-deprotonating surface groups reacting according to equations 4.1, 4.7, or 4.8. Such reactions lead to the development of positive or negative pH-dependent charges.

3. If the model distinguishes between basal surface and edge surface, direct binding of protons to groups belonging to basal surfaces should be avoided. Only protonation-deprotonation of edge groups should be taken into account. In the case of using the modified MUSIC model,[19] these considerations are not necessary because the model by itself predicts reactivity of edges and unreactivity of basal surfaces.
4. Exchange of cations at localized sites driven by the presence of structural charges. MD and MC simulations give clear evidence that cations in the interlayer spacing are localized within a particular energy well caused by structural charges.[24,25]

The equation governing the affinity of protons for surface groups is either equation 4.7 or 4.8. If the type and amount of the group present at the surface is known, values from Table 4.1 can be used as a first approximation. If the population of groups is unknown or some doubts remain in this respect, a surface containing one or two different groups can be considered as adjustable parameters. The following modification of equation 4.1 can be used:

$$SO^{-1/2} + H^{+} = SOH^{+1/2} \tag{4.11}$$

and

$$K_{H}^{\text{int}} e^{-F\psi_0/RT} = \frac{\Gamma_{SOH}}{\Gamma_{SO} a_H} \tag{4.12}$$

where $SO^{-1/2}$ and $SOH^{+1/2}$ are protonated and deprotonated groups, respectively. Equation 4.11 is analogous to equation 4.1 for $x = -1/2$. There is no way of calculating the protonation constant of the reaction in equation 4.11. Its value needs to be evaluated by fitting experimental data.

Even in the case of an isolated layer, the treatment of the electrostatics is complicated and one must select among different possibilities. Consider Figure 4.13. Drawing A depicts a layer containing an uncharged basal surface, an edge surface with a pH-dependent charge, and structural charges within the clay layer affecting

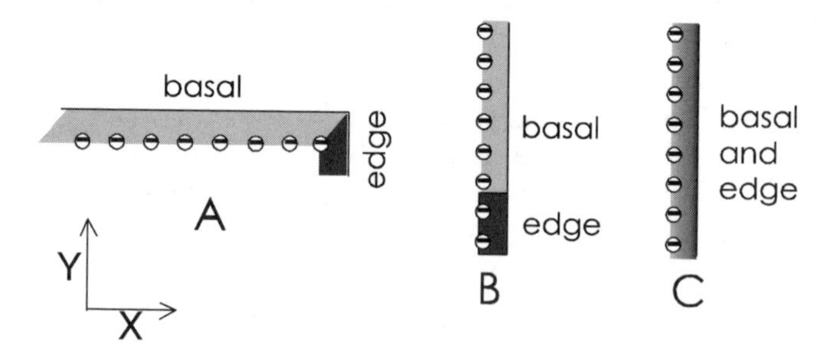

FIGURE 4.13 Three different representations of clay–water interface.

the potential of both surfaces. This arrangement leads to a good description of the charge and potential distribution but is quite difficult to treat mathematically, because two dimensions must be considered (x and y).[47,48] Drawing B is a simplification of A. The edge was located in the direction of the basal plane and thus a patchy surface has been created. Note that structural charges are still considered to be affecting the potential at both surfaces. The last level of approximation is given by drawing C. Here the patchy surface has been "homogenized" in order to have a uniform surface containing basal surface and edge groups. The structural charges consistently affect the whole surface potential. This last drawing is by far the easier to solve and has demonstrated good fitting abilities.[2]

In a previous article, a model similar to the one represented by drawing C in Figure 4.13 was used to model the acid–base properties of clays.[2] The treatment of electrostatics in the model is based on previous work by Avena and De Pauli,[26] which in turn is based on older studies by Kleijn and Oster[43] and Madrid and Diaz-Barrientos.[40] The structure of the clay–water interface according to that model is shown in Figure 4.14. The structural charges are represented as a plane of negative charges inside the solid. The charge density in this plane is the structural charge density. At the surface these charges are considered to express themselves as discrete sites X that can bind cations (either protons or other cations — denoted Cat^+ — from the solution). The edge sites with acid–base properties are located in the same plane and are assumed to "feel" the same potential as X sites. According to Figure

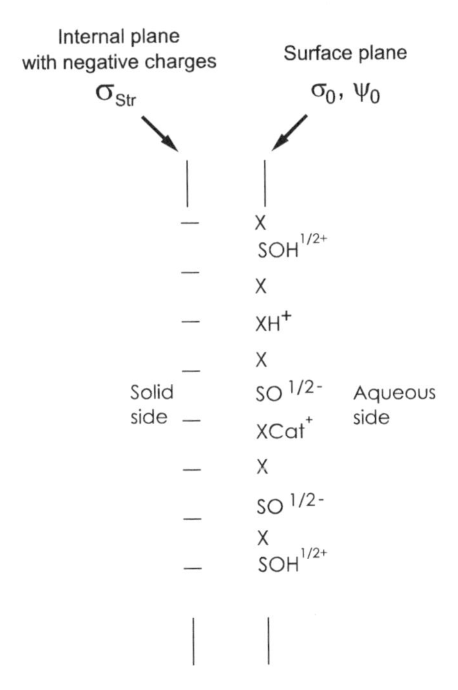

FIGURE 4.14 Representation of clay–water interface according to isolated layer model. (Reprinted from Ref. 2, p. 109 by courtesy of Marcel Dekker, Inc.)

4.7, this is a rather good approximation for 2:1 layers with octahedral substitutions, but it is believed that it underestimates the potential "felt" by X sites in a layer with tetrahedral substitutions. The charge of the hypothetical surface plane can be calculated from the site density of positively charged ($SOH^{1/2+}$, XH^+ and $XCat^+$) and negatively charged species ($SO^{1/2-}$). The proton charge or proton adsorption density can be calculated from the density of species that have gained protons ($SOH^{1/2+}$, XH^+) and those that have lost them ($SO^{1/2-}$). The complete set of equations forming the model is given and analyzed in the appendix, section 4.6. An important prediction of the model is that because of the presence of structural charges in the inner layer, the surface potential is negative even though the net charge at the surface plane is zero, as was concluded in previous sections.

If, instead of isolated layers, there are mainly flocs dispersed in water, a simple approach for modeling is that given by Kraepiel et al.[44] The schematic representation of the clay water interface given by these authors is shown in Figure 4.15. A clay particle (platelet or floc) is imagined as a semi-infinite homogeneous porous solid immersed in an aqueous solution. The solid represents both the crystalline layers of the clay and the aqueous interlayer. It is also assumed that the solid bears a permanent negative charge originating from isomorphic substitutions. This representation is similar to that of a Donnan gel and the clay particle can be seen as a gel particle containing structural charges homogeneously distributed and capable of absorbing electrolyte ions, metal ions, and water within the gel. The interface between the solid and the solution is assumed to be an infinite plane that contains reactive sites that undergo acid–base and surface complexation reactions. The net absorption of ions can be calculated with this model from the difference in ion concentration between the particle bulk and the solution bulk. In the case of protons it is also necessary to consider H^+ attached to the specific surface sites. The charge of the particle is calculated from the density of structural charges and the amount of ab(ad)sorbed ions, and the charge at the particle–water interface is calculated from the densities of protonated and unprotonated sites. A disadvantage of the model as

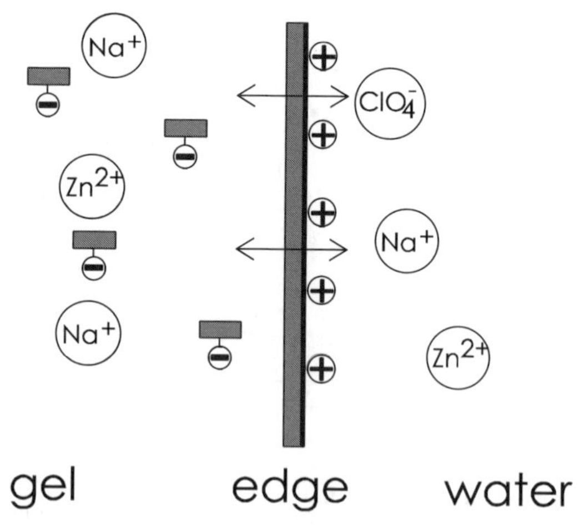

FIGURE 4.15 Representation of clay–water interface according to the Donnan-like model.

presented by Kraepiel et al.[44] is that diprotic groups are considered to be the responsible for the proton adsorption–desorption properties of the interface. As was emphasized in previous sections, this is not the case for groups where the protonating entity is oxygen. However, the consideration of diprotic groups does not complicate the interpretation since in most cases, models with diprotic groups work just as well as models with monoprotic groups.[15] A very important property of the model is that because of the presence of structural charges within the gel, the surface potential is also predicted to be negative even when the net charge at the surface plane is zero, as concluded in previous sections.

Whether the isolated layer model or the Donnan-like model is more appropriate depends on whether the clay particles are considered to be mainly isolated layers or mainly flocs. A montmorillonite dispersion, for example, tends to be formed by isolated layers at high pH and low electrolyte concentrations, especially when the electrolyte cation is Li^+ or Na^+, but it is usually aggregated at low pH and high electrolyte concentrations. Therefore, in a simple titration experiment between pH 3 and 10 at low electrolyte concentration, the aggregation state can change from floc to isolated layer, complicating the choice of model. However, although they are conceptually different, both models should perform similarly with appropriate parameters, and the final model choice is mainly a matter of taste or mathematical convenience.

4.3.3 Application to Montmorillonite

Montmorillonite is a 2:1 clay with isomorphic substitutions mainly in the octahedral sheet and some substitutions in the tetrahedral sheets. When the clay is exchanged with monovalent ions, water and electrolyte ions can enter the interlayer spacing and delaminate the system. With Li^+ or Na^+ as the exchanging cations the delamination is almost complete, whereas with K^+ or Cs^+ the delamination is less effective.[48,49] At low pH, edge-to-face interactions can lead to the formation of aggregates.

The model selected for fitting the experimental data is represented by Figure 4.14 and equations in the appendix (section 4.6). Clearly, montmorillonite does not match exactly with the model: The structural charges are not homogeneously distributed and edge groups are not uniformly mixed with cation exchange sites. However, the essential features of the clay, presence of structural charges, and presence of proton adsorption–desorption groups, are taken into account by the model. In addition, the crucial role of structural charges affecting the reactivity of protonating groups electrostatically (by affecting the electric potential at the location of the groups) is considered by the model.

The fitted data are those of Figure 4.9. The sample is a Na^+-montmorillonite with a CEC = 0.870 meq/g,[50] which is assumed to represent the structural charge. This value, combined with the surface area of 800 m^2/g, leads to $N_{str} = 1.09 \times 10^{-6}$ mol/m^2 and $\sigma_{str} = -FN_{str} = -0.105$ C/m^2. Two different kinds of protonating-deprotonating groups reacting according to equation 4.11 were assumed, with $FN_{edge} = 0.008$C/m^2 (less than 10% of FN_{str}) for each site. The model parameters are listed in Table 4.2 and the fit is shown in Figure 4.9. The data are reasonably well described by the model, especially the absence of an intersection point between the curves,

TABLE 4.2
Parameters used in calculations.

Sample	$LogK^{int}_{H,1}$	$LogK^{int}_{H,2}$	$LogK^{int}_{XH}$	$LogK^{int}_{XCat}$	C_1 (F/m²)	FN_{str} (C/m²)	FN_{edge} (C/m²)
SWy-1 montmorillonite	2.61	5.32	–0.34	–0.73	0.03	0.105 (0.87)	0.008 (0.067)
Argentinean montmorillonite	4.46	6.80	0.72	0.76	0.07	0.096 (0.80)	0.008 (0.067)
Illite	3.98	7.75	1.63	2.47	0.65	0.090 (0.21)	0.029 (0.067)
Kaolinitic soil	3.38	—	1.00	1.15	0.42	0.019 (0.0126)	0.8 (0.53)

Note: Numbers in parentheses are N_{str} or N_{edge} values in meq/g units. *LogK* and C_1 values were optimized in order to obtain the best possible curves fit. FN_{str} and FN_{edge} were kept constant during the optimization procedure.

the decrease in pH_0 by increasing the ionic strength, and the negligible proton adsorption at exchange sites at high pH. The model also correctly predicts that proton adsorption on edges control the acid-base behavior at intermediate and high pH and that adsorption on both edges and cation exchange are important at low pH.

The performance of the model with the Argentinean montmorillonite has been presented previously by Avena.[2] The CEC of the sample is 0.8 meq/g and the surface area 800 m²/g; thus the structural charge is –0.096 C/m². Two different types of protonating groups were also considered. The data fit is satisfactory as shown by Figure 4.10. The model predicts a decrease in proton adsorption and a decrease in pH_0 by increasing the electrolyte concentration. Although it is not shown here, the model could also correctly predict the near independence of the zeta potential with the pH[2].

4.3.4 Application to Illite

Illites belong to the mica group and contain considerable substitution that generates structural negative charges in the tetrahedral layer. Illite platelets are formed by the superposition of many layers kept together by cations, mainly K^+. It is a nonexpandable clay, and cations in the interlayer spacing are not able to leave their position readily and be exchanged by other cations from the aqueous solution. A schematic representation of an illite platelet is given in Figure 4.16. Most of the structural charge residing in the solid is neutralized by the cations trapped in the interlayer spacing. The remaining structural charges, those that are not neutralized by trapped cations, are located in the outermost layers of the platelets or in the vicinities of the edges. Only these non-neutralized charges act as structural charges with cation exchange properties. These charges represent the net structural charge in the model. The drawing also shows protons adsorbing at the edges, which are surely affected by the structural charges located nearby. Neither the isolated layer

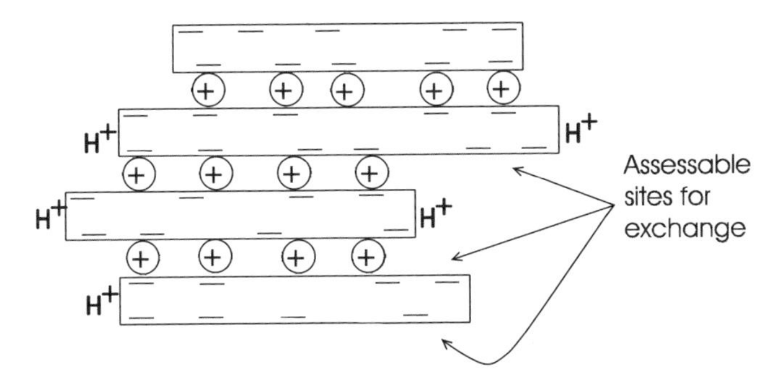

FIGURE 4.16 Scheme of an illite platelet.

model nor the Donnan-like model match this structure. However, both perform well because the effect of structural charges is considered. The data from Hendershot and Lavkulich[32] are compared in Figure 4.11. The CEC of this Na-illite is 0.21 meq/g and the surface area is 224 m^2/g ($\sigma_{str} = -FN_{str} = -0.09$ Cm^{-2}). Two different kinds of protonating groups were also used in this case. The amount of these groups was considered to be equal to that of montmorillonite on a per gram basis. The parameters used in the calculations are also listed in Table 4.2. There is very good agreement between experimental and theoretical data. The model predicts correctly not only the proton adsorption curves at three different ionic strengths, but also the Δq versus pH curve, which can be directly obtained from CEC data because anion adsorption is negligible.[32]

4.3.5 Application to a Kaolinitic Soil

Hydrogen bonds between tetrahedral and octahedral sheets keep the layers strongly attached, impeding an easy delamination of the platelets. The distribution of protonating sites depends on the fraction of basal surfaces and edge surfaces exposed to the solution. Eventually, these fractions depend on the aspect ratio (average diameter to thickness ratio) of the platelet. Large aspect ratios indicate that the system is quite delaminated and then a large fraction of the surface corresponds to basal surfaces. Small aspect ratios correspond to thick platelets with a large fraction of the surface occupied by edges. In addition, the origin and distribution of structural charges can be different in different kaolinite samples. Three different cases can be considered: (1) isomorphic substitutions are real substitutions in the 1:1 layers; (2) structural charges belong to 2:1 layers interstratified with the 1:1 layers; and (3) structural charges belong to 2:1 clays that are mixed with but separated from kaolinite platelets. In the first two cases, the structural charges are included in the platelets and thus it can be assumed that the charges are rather homogeneously distributed within the platelets. In the last case, kaolinite platelets without structural charges and 2:1 clays with structural charges may be considered as independent systems, the adsorptive properties of the whole system being the sum of the individual contributions. Cases 1 or 2 are assumed to occur in this sample.

Only one kind of protonating site is assumed to be present at the surface, together with cation exchange sites. The surface area of the solid was considered to be 64

m^2/g and a site density N_{edge} = 5 sites/nm^2. This last value is lower than the actual site density at edges (about 8 sites/nm^2), and reflects the fact that only a fraction of the surface is reactive (edge surface).[1,2] The structural charge of the sample is $\sigma_{str} = -FN_{str} = -0.019$ C/m^2, as measured by Cs^+ adsorption.[29]

Theoretical data are compared to experimental data in Figure 4.12. The fitting of Δq and σ_H versus pH and electrolyte concentration is rather good.

4.3.6 Differences in Behavior of Clays and Metal Oxides

The main difference between the proton adsorption behavior of clays carrying structural charge and metal oxides with no structural charge is the effect of the electrolyte concentration on the proton adsorption curves. In clays, the pH where proton adsorption is zero decreases as the electrolyte concentration increases, whereas in metal oxides this pH value does not change by changing the electrolyte concentration and a crossing point is observed (PZC).

The model presented here succeeds in predicting the behavior of clay samples because it allows the structural charge to affect the potential at the location of protonating sites. Consider equation 4.12. Using the logarithm and rearranging it results in

$$pH = LogK_H^{\text{int}} - Log\frac{\Gamma_{SOH}}{\Gamma_{SO}} - \frac{F\psi_0}{2.303RT} \tag{4.13}$$

When $\Gamma_{SOH} = \Gamma_{SO}$ (either zero edge charge in clays or zero surface charge in metal oxides), this equation becomes

$$pH_0 = LogK_H^{\text{int}} - \frac{F\psi_0}{2.303RT} \tag{4.14}$$

because pH = pH_0 under this condition.

Oxide surfaces have zero net charge and zero net surface potential at pH_0 when immersed in an indifferent electrolyte solution. Therefore, the second term of the right side of equation 4.14 vanishes. Any change in the concentration of the supporting electrolyte does not change the surface potential and thus pH_0 is independent of the electrolyte concentration. This independence is evidenced by a crossing point of the proton adsorption curves and pH_0 coincides with PZC.

Phyllosilicate layers having structural charges behave in a different way. Although the net charge at the edge surface is zero at pH_0, the net potential is negative because of the presence of structural charges. This potential becomes less negative by increasing the electrolyte concentration, which leads to a decrease in pH_0 (Eq. 4.14). This is evidenced by a shift in the titration curves towards lower pH, such as in Figures 4.9, 4.10, 4.11, and 4.12.

4.4 SUMMARY AND CONCLUDING REMARKS

The proton adsorption properties of clay minerals depend on both, surface groups and structural charges. Surface groups are oxygen-containing groups that can become protonated or deprotonated. The intrinsic affinity of each group for protons is the result of bonds that the oxygen atom establishes with atoms belonging to the solid and water molecules belonging to the solution surrounding it. The unsaturated valence of surface oxygen atoms drives the reactions with protons, and the bond-valence principle seems to predict correctly the acid–base properties of the groups. The principle predicts the unreactivity of basal surfaces populated with Si-O_2 and Al_2-OH groups and the reactivity of basal surfaces populated with Al-$OH^{-1/2}$, Si-O^{-1}, SiAl-$O^{-1/2}$ and other groups. Thus, the unreactivity of basal surfaces does not need to be assumed *a priori* when modeling. It emerges directly by applying the modified MUSIC model.

Structural charges affect proton adsorption by exchanging protons and cations from the supporting electrolyte and by affecting the electric potential at edges. The electric field emanating from structural charges modifies the edge potential and changes the effective affinity of edge groups for protons. This modification in the potential is responsible for the particular protonating behavior of clay surfaces. Simple models considering the mentioned roles of structural charges capture the essence of the proton adsorption process and give a first-order general description of it.

4.5 ACKNOWLEDGMENTS

Financial support for this research was provided by the Fundación Antorchas, CONICET, SECyT-UNC, and Region Rhône-Alpes (France). We are grateful to J. Chorover for providing the kaolinitic soil experimental data.

4.6 APPENDIX: ISOLATED LAYER MODEL

The clay–water interface is represented in Figure 4.14. The structural charge density is given by

$$\sigma_{str} = -FN_{str} \tag{4.A1}$$

where N_{str} represent the moles of substitution sites divided by the total surface area of the solid, and F is the Faraday constant. Only two different cations are considered to be present in the solution: H^+ and the cations of the supporting electrolyte, Cat^+. The adsorption of cations on X sites is then represented by

$$X + Cat^+ = XCat^+ \;\; ; \;\; K_{XCat}^{int}$$

$$X + Cat^+ = XCat^+ \;\; ; \;\; K_{XCat}^{int}$$

Thus,

$$K_{XCat}^{int} e^{-F\psi_0/RT} = \frac{\Gamma_{XCat}}{\Gamma_X a_{Cat}} \tag{4.A2}$$

$$K_{XH}^{int} e^{-F\psi_0/RT} = \frac{\Gamma_{XH}}{\Gamma_X a_H} \tag{4.A3}$$

The mass balance for X sites is expressed by

$$N_{str} = \Gamma_X + \Gamma_{XCat} + \Gamma_{XH} \tag{4.A4}$$

Monoprotic groups are considered to be present at the surface. Equations 4.12 and 4.13 are repeated here:

$$SO^{-1/2} + H^+ = SOH^{+1/2} \tag{4.A5}$$

$$K_H^{int} e^{-F\psi_0/RT} = \frac{\Gamma_{SOH}}{\Gamma_{SO} a_H} \tag{4.A6}$$

The mass balance for these groups is given by

$$N_{edge} = \Gamma_{SO} + \Gamma_{SOH} \tag{4.A7}$$

where N_{edge} represents the moles of groups able to react according to equation 4.A5 divided by the total surface area of the solid. The charge at the hypothetical surface plane results from the contribution of the charged surface species that are located in this plane:

$$\sigma_0 = F\left(\frac{1}{2}\Gamma_{SOH} - \frac{1}{2}\Gamma_{SO} + \Gamma_{XCat} + \Gamma_{XH}\right) \tag{4.A8}$$

The charge potential relationship is given by the Stern-Gouy-Chapmann model. In this model it is assumed that the Stern layer is a region of constant capacitance, C_1, separating the surface plane from the plane where the diffuse layer starts. The charge-potential relationship in the Stern layer is

$$\psi_0 - \psi_d = \frac{\sigma_{str} + \sigma_0}{C_1} \tag{4.A9}$$

where ψ_d is the potential at the beginning of the diffuse layer. The Gouy-Chapmann equation, on the other hand, gives the charge potential relationship in the diffuse layer:

$$\psi_d = \frac{2RT}{zF} \operatorname{arcsin\,h}\left[\frac{-\sigma_d}{\left(8RT\varepsilon_0 \varepsilon c\right)^{1/2}}\right] \qquad (4.A10)$$

where σ_d is the charge of the diffuse layer, ε_0 is the dielectric permitivity of vacuum, ε is the dielectric constant of the liquid medium and c is the concentration of the supporting electrolyte. The model is completed with the electroneutrality condition

$$\sigma_{str} + \sigma_0 + \sigma_d = 0 \qquad (4.A11)$$

Equations 4.A1, 4.A2, 4.A3, 4.A4, and 4.A6 to 4.A11 form the complete set of equations needed for calculations. The set can be solved for any pH and electrolyte concentration if the values for N_{edge}, N_{str}, C_1, K_{XCat}^{int}, K_{XH}^{int}, and K_H^{int} are known. The solution of the equations allows calculation of the surface concentration of the species considered, the potentials ψ_0 and ψ_d and the charges σ_0 and σ_d. The surface concentration of the species allows also the calculation of σ_H, Δq, the proton adsorption at exchange sites and the proton adsorption at the edges, which are the experimentally assessable magnitudes.

The proton adsorption charge is given by

$$\sigma_H = F\left(\frac{1}{2}\Gamma_{SOH} - \frac{1}{2}\Gamma_{SO} + \Gamma_{XH}\right)$$

where $F(1/2\Gamma_{SOH} - 1/2\Gamma_{SO})$ is the proton adsorption on edge groups and $F\Gamma_{XH}$ is the proton adsorption on exchange sites.

Δq is given by the difference between cations and anions adsorbed in the diffuse layer plus the cations adsorbed at the surface:

$$\Delta q = \sigma_d + F\Gamma_{XCat}$$

REFERENCES

1. Brady, P.V., Cygan, R.T., and Nagy, K.L., Molecular controls on kaolinite surface charge, *J. Colloid Interface Sci.*, 183, 356, 1996.
2. Avena, M.J., Acid-base behavior of clay surfaces in aqueous media, in *Encyclopedia of Surface and Colloid Science*, Hubbard, A., Ed., Marcel Dekker, New York, 2002, p. 37.
3. Pauling, L., *The Chemical Bond*, 3rd ed., Cornell University Press, Ithaca, NY, 1960.

4. Bailey, S.W., Ed., *Hydrous Phyllosilicates*, vol. 19 of *Reviews in Mineralogy*, Mineralogical Society of America, Washington, D.C., 1988.
5. Sposito, G., *The Surface Chemistry of Soils*, Oxford University Press, New York, 1984.
6. Greenland, D.J. and Hayes, M.H.B., Eds., *The Chemistry of Soil Constituents*, Wiley, Chichester, England, and New York, 1978.
7. Dixon, J.B. and Weed, S.B., Eds., *Minerals in Soil Environments*, Soil Science Society of America, Madison, WI, 1992.
8. Low, P.F., Interparticle forces in clay suspensions: Flocculation, viscous flow and swelling, in *CMC Workshop Lectures, Clay-Water Interface and Its Reological Implications*, vol. 4, Güven, N. and Polastro, R.M., Eds., Clay Mineral Society, Boulder, CO, 1992, p. 157.
9. Faisandier, K. et al., Structural organization of Na- and K-montmorillonite suspensions in response to osmotic and thermal stresses, *Clays Clay Miner.*, 46, 636, 1998.
10. Derrindinger, L. and Sposito, G., Flocculation kinetics and cluster morphology in illite/NaCl suspensions, *J. Colloid Interface Sci.*, 222, 1, 2000.
11. Lagaly, G., From clay mineral crystals to colloidal clay mineral dispersions, in *Coagulation and Flocculation, Theory and Applications*, Surfactant Science Series, 47, Dobiás, B., Ed., Marcel Dekker, New York, 1993, p. 427.
12. Sposito, G., *The Reactive Solid Surfaces in Soils, The Surface Chemistry of Soils*, Oxford University Press, New York, 1984.
13. White, G.N. and Zelazni, L., Analysis and implications of the edge structures of dioctahedral phyllosilicates, *Clays Clay Miner.*, 36, 141, 1988.
14. Avena, M.J., Mariscal, M., and De Pauli, C.P., Proton binding at clay surfaces in aqueous media, in Proceedings of 12th International Clay Conference, Elsevier, in press.
15. Borkovec, M., Jönsson, B., and Koper, G.J.M., Ionization processes and proton binding in polyprotic systems: Small molecules, proteins, interfaces, and polyelectrolytes, *Surface Colloid Sci.*, 16, 99, 2001.
16. Hiemstra, T., De Wit, J.C.M., and Van Riemsdijk, W.H., Multisite proton adsorption modeling at the solid/solution interface of (hydr)oxides: A new approach. I. Model description and evaluation of intrinsic reaction constants, *J. Colloid Interface Sci.*, 133, 91, 1989.
17. Hiemstra, T., De Wit, J.C.M., and Van Riemsdijk, W.H., Multisite proton adsorption modeling at the solid/solution interface of (hydr)oxides: A new approach. II. Application to various important (hydr)oxides, *J. Colloid Interface Sci.*, 133, 105, 1989.
18. Bleam, W.F., On modeling proton affinity at the oxyde/water interface, *J. Colloid Interface Sci.*, 159, 312, 1993.
19. Hiemstra, T., Venema, P., and Van Riemsdijk, W.H., Intrinsic proton affinity of reactive surface groups of metal (hydroxides: The bond valence principle, *J. Colloid Interface Sci.*, 184, 680, 1996.
20. Bleam, W.F., The nature of cation substitution sites in phyllosilicates, *Clays Clay Miner.*, 38, 527, 1990.
21. Bleam, W.F., Welhouse, G.J., and Janowiak, M.A., The surface coulomb energy and proton coulomb potentials of pyrophillite {010},{110},{100} and {130} edges, *Clays Clay Miner.*, 41, 305, 1993.
22. Güven, N., Hydrous phyllosilicates, in *Reviews in Mineralogy*, vol. 19, Bailey, S.W., Ed., Mineralogical Society of America, Washington, D.C., 1988, p. 497.
23. Güven, N., Molecular aspects of clay/water interactions, in *CMC Workshop Lectures, Clay–Water Interface and Its Reological Implications*, vol. 4, Güven, N. and Polastro, R.M., Eds., Clay Mineral Society, Boulder, CO, 1992, p. 1.

24. Sposito, G. et al., Surface geochemistry of the clay minerals, *Proc. Natl. Acad. Sci. USA*, 96, 3358, 1999.
25. Greathouse, J. and Sposito, G., Monte Carlo and molecular dynamics studies of interlayer structure in Li(H2O)3-smectites, *J. Phys. Chem. B*, 102, 2406, 1998.
26. Avena, M.J. and De Pauli, C.P., Proton adsorption and electrokinetics of an Argentinean montmorillonite, *J. Colloid Interface Sci.*, 202, 195, 1998.
27. Van Raij, B. and Peech, M., Electrochemical properties of some oxisols and alfisols of the tropics, *Soil Sci. Soc. Am. Proc.*, 36, 587, 1972.
28. Baeyens, B. and Bradbury, M., A mechanistic description of Ni and Zn sorption on Na-montmorillonite. Part I: Titration and sorption measurements, *J. Contamin. Hydrol.*, 27, 199, 1997.
29. Chorover, J. and Sposito, G., Surface charge characteristics of kaolinitic tropical soils, *Geochim. Cosmochim. Acta*, 59, 875, 1995.
30. Low, P.F., The clay-water interface, in *Proceedings of the International Clay Conference, Denver*, Schultz, L.G., van Olphen, H., and Mumpton, F.A., Eds., Clay Mineral Society, Bloomington, IN, 1987, p. 247.
31. Sondi, I., Biscan, J., and Pravdic, V., Electrokinetics of pure clay minerals revisited, *J. Colloid Interface Sci.*, 178, 514, 1996.
32. Hendershot, W.H. and Lavkulich, L.M., Effect of sesquioxide coatings on surface charge of standard mineral and soil samples, *Soil. Sci. Soc. Am. J.*, 47, 1252, 1983.
33. Sposito, G., On points of zero charge, *Environ. Sci. Technol.*, 32, 2815, 1998.
34. Schroth, B.K. and Sposito, G., Surface charge properties of kaolinite, *Clays Clay Miner.*, 45, 85, 1997.
35. Thorn, L.H. et al., Polymer adsorption on a patchwise heterogeneous surface, *Progr. Colloid Polym. Sci.*, 109, 153, 1998.
36. Heil, D. and Sposito, G., Organic matter role of illitic soil colloids flocculation: II. Surface charge, *Soil. Sci. Soc. Am. J.*, 57, 1246, 1993.
37. Davis, J.A. and Kent, D.B., Surface complexation models in aqueous chemistry, in *Mineral Water Interface Geochemistry*, vol. 23 of *Reviews in Mineralogy*, Hochella Jr., M.F. and White, A.F., Eds., Mineralogical Society of America, Washington, D.C., 1990, p. 176.
38. Westall, J.C., Adsorption mechanisms in aquatic surface chemistry, in *Aquatic Surface Chemistry*, Stumm, W., Ed., Wiley, New York, 1987, p. 3.
39. Turner, G.D. et al., Surface-charge properties and UO22+ adsorption of a subsurface smectite, *Geochim. Cosmochim. Acta*, 60, 3399, 1996.
40. Madrid, L. and Diaz-Barrientos, E., Description of Titration curves of mixed materials with variable and permanent surface charge by a mathematical model. 1. Theory. 2. Application to mixtures of lepidocrocite and montmorillonite, *J. Soil Sci.*, 39, 215, 1988.
41. Liu, W. et al., A comparative study of surface acid-base characteristics of natural illites from different origins, *J. Colloid Interface Sci.*, 219, 48, 1999.
42. Stadler, M. and Schindler, P.W., *Clays Clay Miner.*, 41, 288, 1993.
43. Kleijn, W.B. and Oster, J.D., Effects of permanent charge on the electrical double-layer properties of clays and oxides, *Soil Sci. Soc. Am. J.*, 47, 821, 1983.
44. Kraepiel, A.M.L., Keller, K., and Morel, F.M.M., On the acid-base chemistry of permanently charged minerals., *Environ. Sci. Technol.*, 32, 2829, 1998.
45. Chang, R.-F.C., Skipper, N.T., and Sposito, G., Monte Carlo and molecular dynamics simulations of interfacial structure in lithium-montmorillonite hydrates., *Langmuir*, 13, 2074, 1997.

46. Chang, R.-F.C., Skipper, N.T., and Sposito, G., Monte Carlo and molecular dynamics simulations of electrical double-layer structure in potassium-montmorillonite hydrates, *Langmuir*, 14, 1201, 1998.
47. Secor, R.B. and Radke, C.J., Spillover of the diffuse double layer on montmorillonite particles, *J. Colloid Interface Sci.*, 103, 237, 1985.
48. Chang, F.-R.C, and Sposito, G., The electrical double layer of a disk-shaped clay mineral particle: Effect of particle size, *J. Colloid Interface Sci.*, 163, 19, 1994.
49. Sposito, G., Surface chemical aspects of soil colloidal stability, in *The Surface Chemistry of Soils*, Oxford University Press, New York, 1984, 189.
50. Baeyens, B. and Bradbury, M.H., A Quantitative Mechanistic Description of Ni, Zn, and Ca Sorption on Na-Montmorillonite, Report number 95-10, Paul Scherrer Institut, Wurenlingen, 1995.

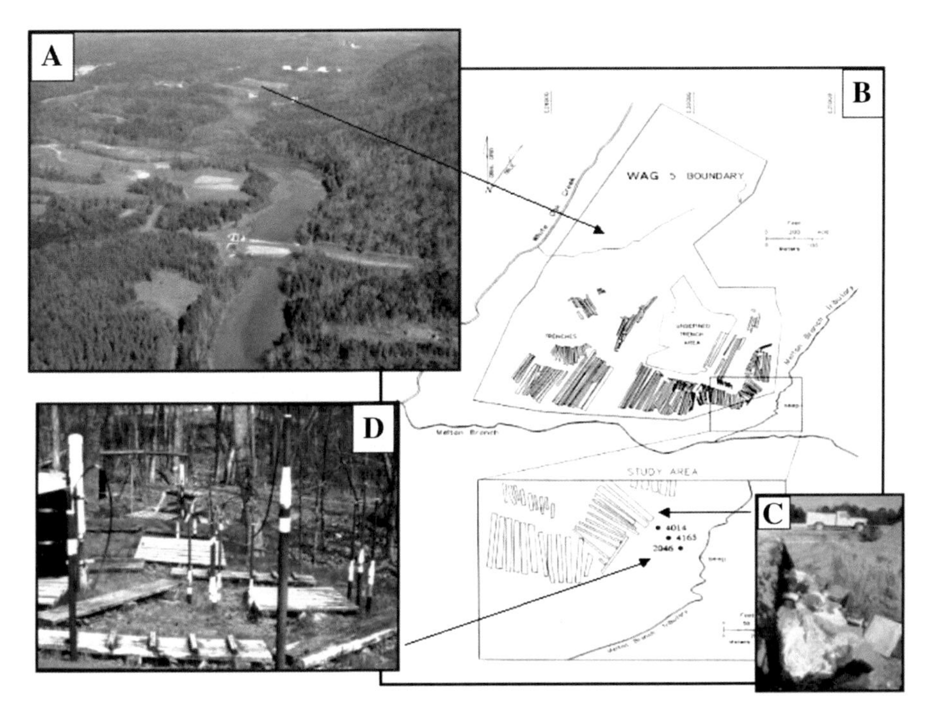

COLOR FIGURE 1.1 (**a**) Pictorial and (**b**) schematic views of Waste Area Grouping 5 showing (**c**) the location of buried waste trenches and the location of the experimental field facility, with a few examples of groundwater monitoring wells (4014, 4165, 2065) that have been strategically placed at strike parallel. The (**d**) primary well field contains 33 groundwater monitoring positions and a designated tracer injection well.

COLOR FIGURE 1.3 Pictorial example of bedrock core obtained near the experimental site showing the interbedded fractured shale (black) and limestone (white/gray) that dominates the saturated zone.

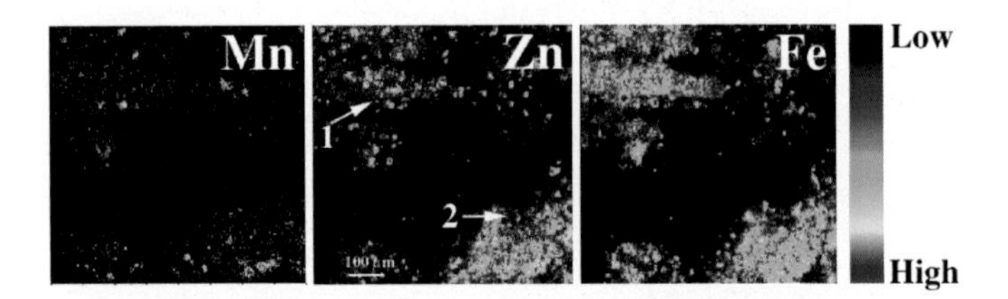

COLOR FIGURE 8.9 Micro-SXRF elemental maps for the surface soil sample. The numbered arrows indicate the micro-EXAFS spectra in Figure 8.11. The color bar gives relative amounts of each element in arbitrary units. (Reprinted with permission from Roberts, D.R. et al., *Environ. Sci. Technol.*, 36, 1742-1750. Copyright 2002. American Chemical Society.)

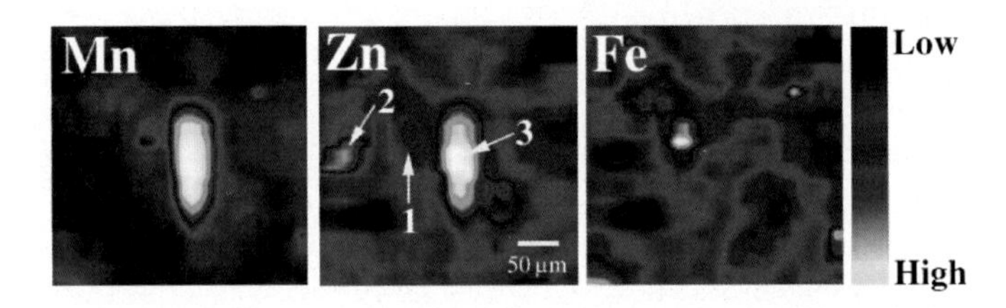

COLOR FIGURE 8.10 Micro-SXRF elemental maps for the subsurface soil sample. The numbered arrows indicate the micro-EXAFS spectra in Figure 8.11. The color bar gives relative amounts of each element in arbitrary units. (Reprinted with permission from Roberts, D.R. et al., *Environ. Sci. Technol.*, 36, 1742-1750. Copyright 2002. American Chemical Society.)

5 Molecular Modeling of Fulvic and Humic Acids: Charging Effects and Interactions with Al^{3+}, Benzene, and Pyridine

James D. Kubicki and Chad C. Trout

CONTENTS

5.1 Introduction 113
5.2 Methods 116
5.2.1 Quantum Calculations 117
5.2.2 Classical Simulations 119
5.3 Results 121
5.3.1 Deprotonation and Complexation of Simple Organic Acids 121
5.3.2 Fulvic Acid: Charging and Solvation Effects on Structure 127
5.3.3 Comparison of Benzene and Pyridine Interactions with Aqueous FA 129
5.3.3.1 Al^{3+} — Complexed Humic Acid 134
5.3.4 Pyridine Interaction with Al-Complexed HA 136
5.4 Conclusions and Future Work 137
5.5 Acknowledgments 138
References 139

5.1 INTRODUCTION

Soils are excellent examples of complex systems. The multitude of feedbacks occurring among the physical, chemical, and biological processes in soils creates an immense challenge for anyone attempting to understand soil formation and behavior. For example, organisms mine soils for essential nutrients, accelerating and modifying

1-56670-623-8/03/$0.00+$1.50

the rate of mineral weathering. In turn, death and decay of organisms leads to development of soil organic matter (SOM). The presence of soil organic matter then affects the soil quality (e.g., water and retention) and the types of weathering products that form because SOM influences dissolution and aqueous speciation.[1] Different mineral or amorphous solid types can then affect the turnover rates of SOM.[2] Such a bewildering interplay of soil components makes understanding the overall behavior of the system extremely difficult; however, important insights can be gleaned from isolating one or two components of the system and determining the key factors that control a given process.

Each component of a soil may have significant complexity. This is especially true for SOM.[3] The result of partial decay of biomolecules, SOM contains numerous types of functional groups that range from hydrophobic to hydrophilic and that can form complexes with various metals.[4–7] Hence, sequestration of organic and metal contaminants is significantly affected by SOM chemistry.[8–11] Although this complexity leads to a wide variation in SOM between soils and even within a single soil, certain important components are common to most SOM.[5] We will never be able to model all this variation, but we can hope to focus on the most important components of the SOM and determine their roles in soil chemistry. Figure 5.1 schematically illustrates the role that molecular modeling can play in soil and environmental science.

Thanks to the efforts of previous researchers, we have begun to see details of SOM molecular structure.[4,12–14] Determining the individual functional groups present with SOM is not a trivial task, but piecing together the larger-scale structure from

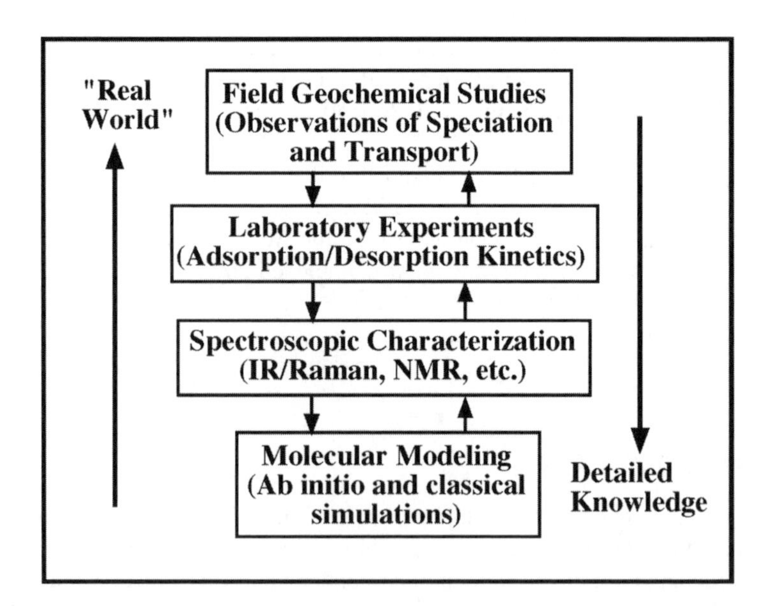

FIGURE 5.1 Schematic representation of the role of molecular modeling in geochemistry shown above. Observations and constraints from field and laboratory studies are key in designing realistic molecular simulations. The feedback among the various approaches adds value to each component of the study.

these puzzle pieces is even more challenging.[12,15,16] Current structural models may not be perfect, and they may not reflect the diversity of SOM, but they are useful starting points for testing hypotheses with regard to SOM chemistry. Use and testing of this first generation of SOM models will lead to new insights and refinements of SOM structure. As molecular modeling techniques become more common among soil scientists, a larger array of model types can be studied and subtle chemical effects investigated.[17] We hope that this chapter will serve as a guidepost to important problems in modeling SOM chemistry and as a roadmap to useful modeling methods.

Important papers have already been published in this area, but this area of research is relatively new and ripe for exploration. Schulten and Schnitzer[13] published some of the first papers on this topic. Early work focused on simple molecular mechanics calculations of neutral, isolated humic acid models. Although this approach neglects important factors such as charging and solvation, simplified models can be used initially as a point of reference for more complex and realistic systems. Recently, inclusion of some of these complicating factors has led to more accurate descriptions of SOM models.[18]

Diallo et al.[15,16] have taken a molecular modeling approach in their attempts to build SOM structural models. The use of new Fourier-transform-ion-cyclotron resonance-mass spectrometry data and NMR spectroscopy has allowed these researchers to piece together a more reliable picture of the large-scale humic acid structure.[19–22] The two most important factors in producing worthwhile molecular simulations are an accurate theoretical model of bonding in the system (discussed in the Methods section) and a realistic description of the system to be modeled. The latter factor should encourage the modeler to use as much experimental data on the structure and chemistry of the system as he or she can. Too often, highly demanding and theoretically accurate computations may be carried out on a model system that does not reflect the true system of interest. Assumptions regarding important structures can lead to useless model predictions. For instance, the catalysis field has long assumed that metal catalysts are controlled by the surface structure of the metal catalyst. Recent research has shown that in many instances, however, oxidation of the metal at the surface occurs before catalytic properties are present.[23] Thus, it is the metal oxide rather than the metal that is the catalyst. Molecular modeling studies that do not include all the important components of a reaction would never be able to predict the behavior of the true system.

Other studies have focused on an important aspect of SOM chemistry: adsorption to mineral surfaces.[13,24,25] Adsorbed SOM is critical to understanding sequestration of contaminants in soils because adsorption can stabilize SOM and affect its sorptive properties.[26–30] Such simulations require knowledge of the SOM structure, the relevant mineral surface structure, and the nature of interaction between the two. Some recent experimental studies have addressed the nature of this interaction, but much more research needs to be performed on this topic.[31–33] Practically speaking, running simulations of a system, which includes a large organic molecule, mineral surface, and water molecules becomes computationally demanding because the number of atoms required to simulate the system will be large (>10,000).

The work discussed in this chapter illustrates one approach to this large, complex problem. First, quantum mechanical calculations are used on small, simplified systems to establish a link between models and experimental spectra (e.g., IR, Raman, NMR, etc.). Although oversimplifying the problem of SOM, this same approach is often used in experiments to gain a handle on the important functional groups involved in a given chemical reaction.[34,35] This step is key because force fields used in classical simulations are not always reliable. Moreover, it can be difficult to know when they are accurate and when they fail. Quantum mechanical results can be carried out with various levels of approximation but are generally more reliable than force fields, especially for unusual chemical bonding situations. When tested against experimental data, a reasonable degree of certainty can be associated with the molecular models used. Once these benchmarks are established, their results can be used to constrain structures of larger-scale classical simulations. In many of the questions regarding metal complexation by SOM, some aspects of the chemistry are more difficult to model than others. For example, if a model fulvic acid is complexed to Al^{3+}, descriptions of the C-C and C-H bonds may be relatively easy to reproduce with classical force fields.[36] This is due to the fact that these bonds have been well studied and accurate parameters describing their interaction are built into the force field. Other interactions, such as Al-O bonding or H-bonding, are not accurately modeled by current force fields because parameterization of these species has not been as well tested. Thus, a combination of quantum mechanical and classical simulations can provide a maximum of information on these complex systems.

5.2 METHODS

Two fundamental types of molecular modeling are discussed in this chapter: quantum mechanical calculations and classical mechanical simulations. The difference between the two is that quantum mechanical (or *ab initio*) calculations describe the electron densities of atoms whereas classical mechanical simulations model atoms as particles connected to others via springs. Description of electron densities is computationally demanding, especially for heavier atoms, so quantum calculations are generally limited to fewer particles than classical simulations (Figure 5.2). The advantage quantum calculations enjoy is flexibility to model systems that are not well understood (i.e., bond lengths, energies, etc., are unknown). The difference between the two is so large that many workers use two different terms to describe these techniques: "computational chemistry" for quantum mechanics and "molecular modeling" for classical simulation. The intent is to associate the former with a more rigorous stature and the latter with more approximate results. In general, this simplified perception is fairly accurate, but quantum mechanical results can be useless and classical simulations can be accurate.

The divide between these two end-members can be fuzzy in practice (Figure 5.2). Development of hybrid codes that employ each method on different components of a model has been a great advance in modeling larger-scale systems.[37] Termed "QM/MM" for quantum mechanics/molecular mechanics, this approach will likely enjoy widespread utilization and success in fields such as soil science, environmental chemistry, and geochemistry due to the nature and complexity of reactions in these fields. Furthermore, as computers become more powerful and software becomes more advanced, it becomes feasible to perform molecular simulations using quantum

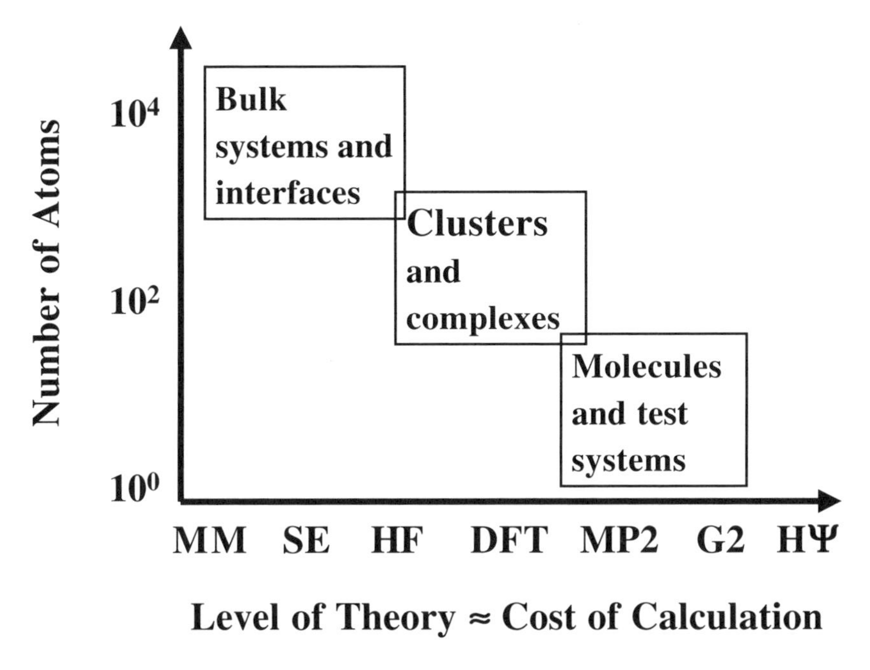

FIGURE 5.2 Matrix representation of a fundamental problem in molecular modeling of geochemical systems. More accurate calculations are computationally more demanding, but larger model systems are needed to account for all the components in a geochemical system. Judicious use of each method can generate accurate and realistic molecular simulations.

mechanics to describe atomic interactions rather than the force field approximation. A few researchers, notably Lubin et al.,[38] Weare et al.,[39] and Iarlori et al.[40] have published excellent studies employing these techniques to systems of geochemical interest. Use of these codes is not yet commonplace, however, because they require a high level of computing power. For example, Hass et al.[41] used 32 nodes of an IBM SP2 for a period of 6 months to perform a simulation. Fortunately, the new generation of PC-based Linux clusters will make this type of simulation affordable for most researchers in the next decade.

5.2.1 Quantum Calculations

All the *ab initio* quantum calculations presented in this chapter were performed with the program Gaussian 98.[42] The Gaussian series of programs has been developed over many years by a large number of researchers adding to and refining the original code. Gaussian was the brainchild of the Nobel laureate John Pople and is a standard program in the field of computational chemistry because of its reliability and flexibility. Other programs available to interested researchers include GAMESS (see Quantum Chemistry Program Exchange, Indiana University), Spartan (Wavefunction, Inc.), Jaguar (Schrodinger, Inc.), Q-CHEM (Q-Chem, Inc.), Parallel Quantum Solutions (PQS, Inc.), HyperChem (Hypercube, Inc.), and DMol^3 (Accelrys, Inc.). Platforms for these calculations can range from a desktop PC to highly parallel supercomputers, but commonly Unix-based workstations are used by molecular modelers because they are fast and affordable. This type of machine is rapidly being replaced by less expensive PC-based Linux (or "Beowulf") clusters.

As mentioned above, the two keys to useful quantum mechanical modeling are constructing an accurate initial model and choosing the appropriate level of theory. The first step will not be discussed here because this procedure varies from problem to problem. In some cases, a number of models may be constructed and tested to determine which one best fits available experimental data. The second step has been addressed with a simple scheme in the research presented in this chapter. One starts with the lowest level of theory possible and tests the results against experimental data and selected higher-level calculations. If the model results are satisfactory for the problem at hand, then the low level of theory is fine. If errors and inconsistencies are found, then higher level calculations must be performed for the suite of models under study.

Two main considerations determine the vague "level of theory" mentioned above. First, the basis set used to describe the electron density must be adequate. Quantum calculations are approximations to the Schrodinger equation, $H\Psi = E\Psi$, where H is the Hamiltonian operator describing the kinetic and potential energy of electrons and nuclei, E is the energy of the system, and Ψ is the electronic wavefunction. Unfortunately, Ψ is not known, so we use various functions to approximate Ψ. Commonly, Gaussian functions, such as $\phi_{1s}(r,\alpha) = (2\,\alpha/\pi)^{3/4}\exp[-\alpha\, r^2]$ where r is the electron-nucleus distance and α is the orbital exponent, are used for computational reasons (which is the origin of the name for the Gaussian program).[43] Basis set notation is obtuse, but a general principle is that the larger number of Gaussian functions used, the more accurate the basis set. In addition, the Gaussian basis set can be split into different sets to describe core and valence electrons. This is helpful because of the different behaviors of electrons near and far from the nucleus. Typically, basis sets are split into sets of two or three, which gives rise to the terminology doubly- and triply-split basis sets. Triply-split basis sets are usually more accurate. For example, one could go from an STO-3G basis set with 3 Gaussians approximating each atomic orbital and no splitting between core and valence electrons to a 6–311G basis set with 6 Gaussians approximating each atomic orbital and a triply-split basis set.

To confuse the issue even more, workers have found that addition of functions to describe formally unfilled atomic orbitals (e.g., *d*-orbitals on Al^{3+}) improves results considerably.[44] Seemingly extraneous orbitals provide for a more accurate description of bonding because they help to account for polarization that occurs between two bonded atoms. A single set of d-orbitals on atoms heavier than H is designated with an asterisk (*); adding p-orbitals to H is designated with two asterisks (**). A more straightforward notation uses the number and type of orbitals included, which leads to a designation such as 6–311G(d,p).

The last point regarding basis sets that will be important for the discussion here is the inclusion of diffuse functions. As the name implies, diffuse functions are used to describe electron density far from a nucleus. The role of electrons far from molecular nuclei is especially important in two cases. Anionic models require diffuse functions because the electron density is spread over a greater volume compared to cations and neutral molecules. Models examining the interaction of two molecules via van der Waal's forces or H-bonding also benefit greatly from the use of diffuse functions. Addition of diffuse functions is designated by a plus sign (+) for heavy atoms only and by two plus signs (++) for heavy atoms and H. For a more complete description of basis sets and their relationship to atomic orbitals, see McQuarrie and Simon.[43]

The second consideration in choosing a method is the level of electron correlation. A range of methods from no electron correlation (Hartree-Fock methods) to full configuration interaction is available; however, the more extensive the electron correlation, the more computationally demanding the calculations become. Some electron correlation methods, such as the Møller-Plesset method, can scale as N^5 where N is the number of electrons.[45] One can imagine that such methods become impractical for larger model systems.

A useful development has been the hybridization of molecular orbital theory and density functional theory.[46] The latter uses a relatively simple equation to estimate the electron correlation as a function of the electronic density. With the electronic density described by the basis sets discussed above, a quicker approximation for electron correlation can be attained. There are numerous exchange and correlation functional pairs, but a commonly used set is the Becke 3-parameter exchange functional and the Lee-Yang-Parr correlation functional.[47,48] This approximation for electron exchange and correlation is simply designated B3LYP in Gaussian 98.[46]

5.2.2 Classical Simulations

Classical mechanical molecular simulations avoid calculation of electron densities altogether. Each atom is given a set of parameters that fit into analytical equations used to describe atomic interactions. For instance, ions affect one another through long-range Coulombic forces described by the equation

$$\phi_{ionic} = Z_i Z_j / \varepsilon r_{ij} \tag{5.1}$$

where ϕ_{ionic} is the ionic potential energy, ε is the dielectric constant of the medium, Z_i is the charge on ion i, and r_{ij} is the distance between ions i and j. Many early simulations were performed with this type of interatomic potential alone (plus repulsion terms and perhaps van der Waal's attraction terms).[49] Today, simulations generally reserve the ionic interaction terms for long-range, nonbonded forces, and any atoms directly bonded to one another interact through covalent terms. Choosing the atomic charges remains an important step in developing an interatomic potential, however. Charges are either determined empirically by adjusting charges within a model to fit experimental data, or they can be determined theoretically by adjusting atomic charges to fit electrostatic potentials around molecules in quantum mechanical calculations.[50]

Other important nonbonded terms are van der Waal's forces and hydrogen bonding. The latter is particularly important in determining the positions of H atoms as solvation energies. A typical description of the van der Waal's forces is the Lennard-Jones or 6–12 potential, so called because of its functional form

$$\phi_{vdW} = A_{ij}/r_{ij}^{12} - B_{ij}/r_{ij}^{6}. \quad (5.2)$$

A_{ij}/r_{ij}^{12} is a repulsive force (i.e., a positive contribution to the potential energy) and B_{ij}/r_{ij}^{6} is an attractive force between nonbonded atoms i and j. Other exponential forms, such as the Buckingham potential, can also be used to describe atomic repulsion.[51] A similar equation can be used to describe H-bonds using different constants:

$$\phi_{H\text{-}bond} = C_{ij}/r_{ij}^{12} - D_{ij}/r_{ij}^{10}. \quad (5.3)$$

Often, H-bonds are treated implicitly by electrostatic interactions; however, for simulations of solutions, clay minerals, and mineral-solution interfaces, explicit consideration of H-bonding should improve results.

As stated above, ionic contributions to the energy are often reserved for nonbonded interactions. Bonded interactions are treated by harmonic approximations. Higher terms can be included and are necessary for configurations deviating from minimum energy structures. For the purpose of this introduction, however, simple harmonic equations will be used to illustrate the concepts behind this type of force field. Bond stretching and bond angle bending can be handled with equations of the form

$$\phi_{bond} = \phi_{ij} = k_{ij}(r_{ij} - r_0)^2 \quad (5.4)$$

$$\phi_{angle} = \phi_{ijk} = k_{ijk}(\theta_{ijk} - \theta_0)^2 \quad (5.5)$$

where the k's are force constants defined by the atom types i, j, and k; r_{ij} and θ_{ijk} are the bond distance and angle in the present configuration; and r_0 and θ_0 are the minimum energy bond length and angle, respectively. Other forms, such as a Morse potential, have also been used successfully.[52]

Often, the potential energy surface is followed until the most energetically stable configuration can be found. These "energy minimizations" occur at 0K and are useful for predicting structures and spectroscopic properties.[53] Energy minimizations are heavily influenced by the starting configuration of the model, however, and can end in local rather than global minima. Molecular dynamics simulations use the interatomic force field to predict positions as a function of time at a finite temperature. Time is explicitly included in the calculation and all the atoms can move in concert according to classical mechanics and their kinetic energies at a given temperature. A Boltzmann distribution of velocities is attained after atomic motions are scaled to a given temperature, which allows for some atoms to be moving with kinetic energies higher than the average value.[54] Molecular dynamics is the method of choice for studying dynamical properties of systems, such as diffusion or other time-dependent reactions.

The COMPASS force field used in the simulations reported here was developed explicitly for condensed-phase systems, which makes it unusual among classical force fields.[55] Generally, gas-phase data are used to constrain force constants and so on, which is a simpler approach, but this method leads to uncertainties regarding application in liquids and solids. The algorithm used to produce COMPASS (Condensed-phase Optimized Molecular Potentials for Atomistic Simulation Studies) begins with gas-phase, *ab initio* calculations to estimate force field parameters, such as atomic charges and force constants. This is a reasonable approach because condensation will act as a perturbation to the atomic parameters in a given molecule.[56] Once the parameters are fit to reproduce the *ab initio* results on gas-phase molecules, the parameters are then refined to fit both gas-phase and liquid-phase properties of the compounds. The process is iterated until a converged set of force field parameters is achieved that fits all the available results (both experimental and *ab initio*).

Another important component of the COMPASS force-field development is the systematic fit to various types of compounds. First, alkane parameters are fit, and then held fixed while alkene parameters are derived. Each functional group is added by fitting to larger arrays of compounds, but the parameters derived in the previous level are not changed, so a self-consistent force field is created that describes a wide variety of functional groups.

COMPASS has been used to model a number of condensed-phase organic systems.[57–59] Hence, we chose COMPASS as a likely candidate for modeling fulvic and humic acids; however, we caution the reader that classical mechanical force fields may be accurate for one system and not for another. The force field must be tested for each new application before the results can be considered reliable.

5.3 RESULTS

5.3.1 DEPROTONATION AND COMPLEXATION OF SIMPLE ORGANIC ACIDS

To begin our investigation of fulvic and humic acid behavior, we first test our methods on simple, well-understood systems. Charging and metal complexation in naturally occurring organic acids is heavily influenced by benzoic acid-type functional groups within the larger acid.[60,61] Thus, we tested our methods on benzoic, salicylic, and phthalic acids, which are commonly used as simple analogs of important functional groups within fulvic and humic acids in experimental studies.[34,35,62] We are interested in modeling the charging behavior of fulvic and humic acids, so we must be able to model the various charged states of the above simple organic acids. To do this, the neutral and deprotonated species of each acid is modeled to predict its energy, structure and vibrational spectrum. (Note: The doubly deprotonated species of salicylic acid, $C_6H_4OCOO^{2-}$, was not modeled because the pK_a for the phenol group is >13, so it should not deprotonate in most natural waters.) To account for solvation, we add H_2O molecules around the organic acids such that the most hydrophilic functional groups are H-bonded. This approach has proven satisfactory for predicting vibrational frequencies of organic acids in aqueous solutions.[63,64] Figure 5.3 illustrates the minimum energy structures

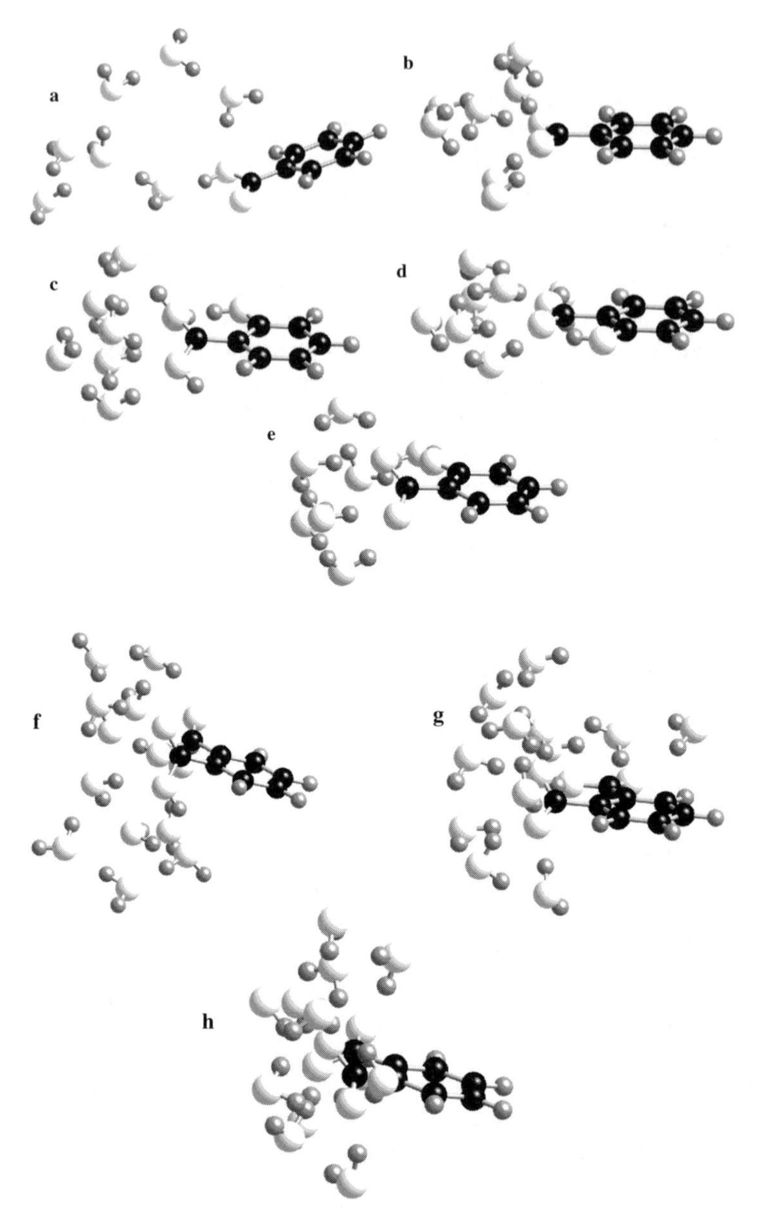

FIGURE 5.3 Model configurations of aqueous organic molecules: (a) benzoic, (b) benzoate, (c) salicylic, (d) salicylate, (e) doubly de-protonated salicylate, (f) phthalic, (g) phthalate, and (h) doubly de-protonated phthalate. A key for complexation to metals is the orientation of the carboxylate groups with respect to the aromatic rings. Ligands with the carboxylate groups oriented in the plane of the aromatic ring (e.g., benzoate, salicylate) tend to be strong ligands. When O-O repulsion exists, then carboxylate groups tend to rotate out of the plane of the aromatic ring (e.g., doubly de-protonatated salicylate, phthalate, and doubly de-protonated phthalate), which limits the ability of the ligand to bind with a metal.

derived using the B3LYP/6–31G* (hybrid DFT/MO) method. An important point to note from these figures is the orientations of the COO^- group relative to the aromatic ring. This will play a role Al^{3+} complexation discussed below. The rotation of carboxylate groups is especially pronounced for the doubly deprotonated phthalic acid. Out-of-plane rotations are explained by the fact that the negatively charged O atoms near each other are attempting to minimize the electrostatic repulsion among them. The out-of-plane orientation causes distortions in the Al-organic complex that reduce the energy of complexation.[65]

A test of the accuracy of these models is to compare the calculated vibrational frequencies against those measured in aqueous solutions. Using literature values for infrared (IR) and Raman spectra as well as our own UV-resonance Raman spectra (UVRR), the model and observed values are compiled in Table 5.1 and an example correlation is plotted in Figure 5.4. Excellent agreement between theory and experiment for these aqueous-phase species suggests that we are adequately representing these organic acids in solution.

The next level of complexity is to add Al^{3+} to the organic acids and model these complexes in aqueous solution. Two problems are presented with the addition of Al^{3+}. First, we must be assured that the bonding mechanism between the organic acid and the Al^{3+} is correct. For example, a common assumption is that

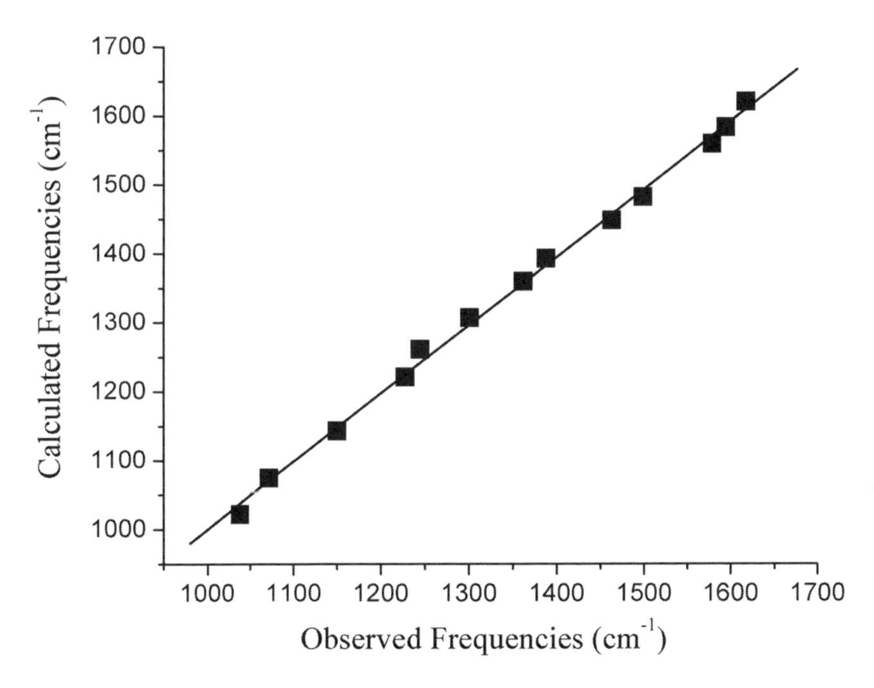

FIGURE 5.4 A strong linear correlation between the calculated frequencies of salicylate (Figure 5.3(d)) and the observed UV resonance Raman frequencies suggests that the molecular modeling is accurately representing this aqueous species. The correlation has a slope of 0.98 and a standard deviation of ±11 cm^{-1}.

TABLE 5.1
Comparison of Observed and Calculated Vibrational Frequencies (cm^{-1}) of Aqueous Organic Acids

Benzoic[a]		Benzoate[a]		Salicylic[b]		Salicylate[b]		Phthalic[b]		Phthalate[b]		Phthalate[b]	
Observed	Calculated	Observed	Calculated	Observed	Calculated	Observed	Calculated	Observed	Calculated	Observed	Calculated	Observed	Calculated
1692	1598	1603	1596	1667	1648	1618	1620	1712	1655	1688	1592	1604	1553
1605	1576	1595	1566	1620	1611	1595	1583	1690	1634	1647	1571	1586	1462
1585	1484	1555	1479	1596	1549	1579	1559	1606	1564	1605	1553	1488	1422
1495	1457	1495	1433	1489	1474	1499	1482	1584	1487	1581	1466	1446	1381
1451	1439	1417	1353	1362	1344	1463	1448	1487	1449	1540	1448	1400	1347
1412	1313	1400	1310	1325	1325	1388	1393	1475	1426	1493	1425	1363	1291
1319	1314	1310	1292	1248	1244	1362	1360	1298	1281	1475	1360	1170	1126
1284	1251	1275	1161	1220	1207	1301	1307	1291	1261	1375	1293	1160	1108
1212	1164	1175	1116	1167	1160	1245	1261	1274	1187	1362	1280	1041	1029
1178	1114	1160	1114	1148	1148	1228	1221	1160	1136	1293	1245	1023	943
1069	1014	1104	1066	1067	1114	1150	1123	1145	1086	1274	1153		
1027	991	1065	1014	1039	1030	1072	1075	1075	1039	1158	1116		
1002	974	1020	981			1038	1022	1045	1023	1050	1033		
1003	966					1035	997						

[a] From Varsányi, G.D., *Assignments for Vibrational Spectra of Seven Hundred Benzene Derivatives*, vol. 1, Wiley, New York, 1974.

[b] From Trout, C.C. and Kubicki, J.D., UV Raman spectroscopy and ab initio calculations of carboxylic acids-Al solutions, *Abstr. Pap. Am. Chem.*, 224:012-Geoc, 2002.

salicylic acid forms a bidentate complex with Al^{3+} as Al bonds to one O of the carboxylate group and to the O of the phenol group.[62] At low pH, however, the structure of the complex may actually be monodentate.[63,66] Consequently, adding these two components together is not a trivial matter because a variety of bonding options are possible in some cases. Second, we must verify that the modeling method is adequate. Reproduction of experimental properties of the organic acids is *not* sufficient to ensure that the Al-organic complexes will be modeled accurately with the B3LYP/6–31G* method. Both of these problems are addressed by testing a variety of possible complex configurations and comparing the results to experimental spectral properties (i.e., vibrational frequencies and NMR $\delta^{27}Al$ values).

Table 5.2 compares the observed and calculated vibrational frequencies for each Al-organic acid complex wherever the experimental data are available. An example correlation is plotted in Figure 5.5. The excellent correlation between theory and experiment substantiates the accuracy of our methodology. When ^{27}Al NMR spectra are available, the same complex that fits the vibrational frequencies also fits the observed ^{27}Al chemical shift.[67] The fact that the same complexes that reproduce vibrational frequencies also reproduce the $\delta^{27}Al$ values is a strong indicator that the complexes are realistically modeled. Consequently, we can use these *ab initio* results

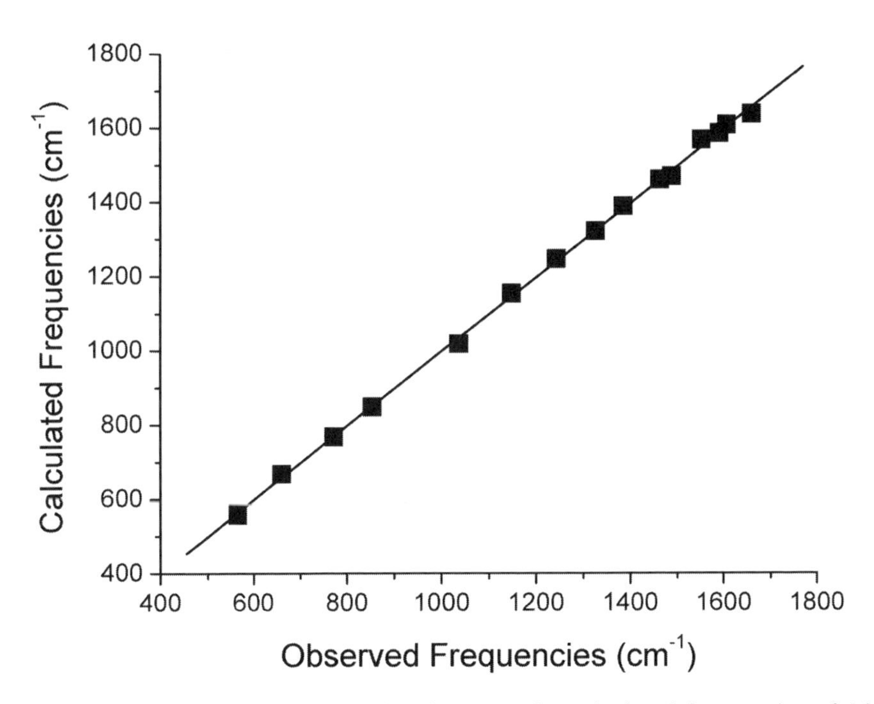

FIGURE 5.5 A strong linear correlation between the calculated frequencies of Al-salicylate complex (Figure 5.6(a)) and the observed UV-resonance Raman frequencies suggests that the molecular modeling is accurately representing this aqueous species. The correlation has a slope of 1.00 and a standard deviation of ±12 cm^{-1}.

TABLE 5.2
Comparison of UV-Resonance Raman and Calculated Vibrational Frequencies (cm^{-1}) of Aqueous Al–Organic Complexes

Phthalic-Al pH 2.5		Phthalate-Al pH 4		Salicylic-Al pH 2.5		Salicylate-Al pH 3.8	
Observed	Monodentate	Observed	Bridging Bidentate	Observed	Monodentate	Observed	Bridging Bidentate
1695	1715	1669	1660	1661	1656	1668	1636
1634	1600	1600	1592	1608	1609	1607	1607
1564	1594	1496	1531	1592	1588	1587	1584
1487	1457	1457	1463	1553	1563	1548	1567
1449	1438	1383	1404	1489	1466	1464	1460
1426	1380	1297	1301	1464	1454	1391	1388
1281	1305	1267	1269	1386	1370	1327	1321
1261	1251	1159	1151	1327	1301	1249	1246
1187	1157	1045	1026	1244	1215	1149	1154
1136	1142	875	870	1149	1151	1086	1088
1039	1061	653	654	1038	1022	1035	1018
781	782			854	862	853	848
661	654			772	769	659	653
				661	663	592	589

Source: Trout, C.C. and Kubicki, J.D., UV Raman spectroscopy and ab initio calculations of carboxylic acids-Al solutions, *Abstr. Pap. Am. Chem.*, 224:012-Geoc, 2002.

on simple systems to test force field results on the same compounds and to constrain the behavior of similar functional groups in fulvic and humic acids during larger-scale classical simulations.

Figure 5.6 compares the energy minimized structures of a bidentate bridging Al-salicylate complex ($Al_2(OH)_4(H_2O)_4C_6H_4OHCOO^+$) using the *ab initio* (B3LYP/6–31G*) and molecular mechanics (COMPASS) methods. Assuming that the *ab initio* structure is close to the actual structure because the experimental frequencies and $\delta^{27}Al$ are reproduced in this model, it is obvious that the molecular mechanics approach does not result in a realistic structure. The mismatch is not surprising in this instance because the COMPASS force field has not been parameterized to account for Al-O bonds in this type of compound. Consequently, in the large-scale molecular mechanics simulations that follow, we will constrain Al-O bonding to values obtained from the *ab initio* calculations. The COMPASS force field should provide an adequate representation of the organic component of the system.[55]

5.3.2 Fulvic Acid: Charging and Solvation Effects on Structure

Some previous models of dissolved natural organic matter (NOM) treated the fulvic or humic acids as charge neutral species.[68] Under most natural conditions, however,

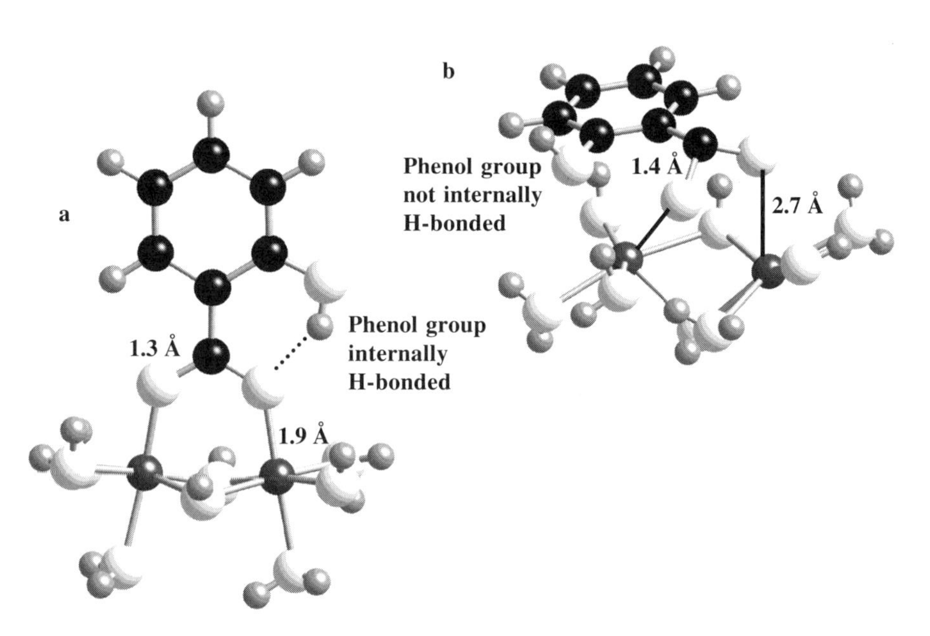

FIGURE 5.6 Comparison of the structures calculated for the salicylate bidentate bridging complex with $[Al_2(OH)_4(H_2O)_4]^{2+}$ calculated with (a) B3LYP/6–31G* in Gaussian 98 and (b) COMPASS in Cerius.[2] The fundamentally different nature of the predicted structures for this complex suggests that the COMPASS force-field parameterization is not capable of modeling Al-organic bonding at this time.

dissolved NOM will be deprotonated because there are abundant functional groups (e.g., carboxylic acids) with pK_a's around 4. Molecular modeling studies conducted on neutral model NOM molecules are of limited value; the charge neutral models were only used because the force fields available did not have atom types defined for the anionic species. In addition, the effect of solvation on molecular conformations has not been extensively studied. Charging and solvation effects are complementary because the presence of water allows the charging to take place and the solvent in turn screens the atomic charges that develop.

In a previous study, the effect of deprotonating carboxylic acid groups on a model of Suwannee fulvic acid was investigated with three modeling techniques: molecular mechanics (MM+); semi-empirical (PM3), and *ab initio* (Hartree-Fock).[65,69–72] One advantage of modeling FA versus HA is that the molecular weight of FA is much lower. Hence, we can treat the average-size FA molecule without truncating the structure as is necessary for HA.[36] Each method resulted in somewhat different conformations of the SFA with the *ab initio* method presumably providing the most reliable result. Based on these energy minimizations, it was concluded that charging and intramolecular H-bonding would have significant effects on the conformation of SFA and by inference other fulvic acids as well.

Solvation effects were neglected in the Kubicki and Apitz study, in part, because of limited computer power.[65] As a matter of general practice, however, one would probably run these types of gas-phase calculations prior to running simulations within a solvent even with unlimited computer resources. This is a common strategy to evaluate the effects of solvation on structure and one that provides an initial guess for the solvation calculations. One advantage of the molecular modeling approach is the ability to add and subtract components at will in order to assess the effects they have on the behavior of the system.

In this chapter, we present new results based on semi-empirical quantum calculations (PM3) that include solvation and charging effects simultaneously on the same model SFA.[71] These calculations were carried out in HyperChem 5.0 (Hypercube, Inc.). Solvation was carried out with two approaches. In the first approach, the neutral, gas-phase SFA model was simulated, then this molecule was deprotonated at each of four carboxylic acid sites. Finally, a solvation sphere of H_2O molecules was used to surround the anionic SFA and the structure obtained via molecular dynamics simulations and energy minimizations as an isolated "nanodroplet." This approach has the advantage of allowing maximum flexibility of the model SFA. Larger model systems may require long simulation runs to sample all available conformations, but isolation of the SFA and water allows each component to move more freely.

The second approach was to employ periodic boundary conditions and molecular mechanics (COMPASS) to model the solvated SFA.[55,73] These simulations were performed with Cerius^2 4.2 (Accelrys, Inc.). Periodic boundary conditions create a bulk system with no surface effects; and hence, this situation is more realistic compared to the experimental system of SFA dissolved in water. H_2O molecules, however, must diffuse to allow motion of the SFA model, so that the SFA model conformations may be restricted due to this limited motion of the surrounding H_2O molecules. Note also that periodic simulations must be charge neutral within the central cell; hence, Na^+ atoms were included to charge balance the negatively charged SFA. Using both approaches in turn allows us to investigate the maximum number of configurations (without periodic boundary conditions) and to approximate a true bulk system (with periodic boundary conditions). With this strategy, we can account for the potential problems inherent in each type of simulation.

Figure 5.7 depicts the changes that occurred to the model SFA as it underwent deprotonation and solvation in the nonperiodic conditions. Structures were generated with 50-ps molecular dynamics simulations at 300K with a time step of 1 fs followed by energy minimization. The neutral, gas-phase model resulted in an open configuration (Figure 5.7(a)). Deprotonation of the carboxylic acid groups causes the structure to coil significantly (Figure 5.7(b)). Intramolecular H-bonding is responsible for this coiling phenomenon because the O atoms in the COO^- groups have excess charge to share with any available H atoms throughout the molecule. The model SFA is long enough that the energy gained forming relatively strong H-bonds (e.g., 1.39 and 1.44 Å) offsets the energy needed to bend the backbone torsion angles. The anion in the gas-phase neglects possible H-bonds to H_2O molecules in the solvent, however. Once these are included (Figure 5.7(c)), the model SFA opens back up to a conformation similar to that observed for the neutral, gas-phase molecule. Thus, the SFA-H_2O H-bonds are able to overcome the intramolecular H-bonding and allow the SFA torsions to relax back toward their preferred values.

An example of a molecular dynamics simulation under periodic boundary conditions is shown in Figure 5.8. The model SFA was solvated with HyperChem 5.0 to generate an input structure for an MD simulation using the COMPASS force field within Cerius^2 4.2. The cell dimensions are ~33 × 34 × 36 Å in this case with 1493 H_2O molecules included. The cell represented in Figure 5.8 is repeated in three dimensions such that there are no surface effects within the simulation. A simulation of 100 ps (100,000 time steps) at 300K was performed on this model system. Although the short duration is not long enough to probe all configuration space for this model, significant conformational changes were observed during the simulation. The overall resulting structure is also similar to that determined with the semi-empirical (PM3) energy minimization on the nonperiodic system (i.e., only one intramolecular H-bond exists as the system uncoils in response to solvation). Longer run times should be performed to better probe configuration space for this system.

5.3.3 Comparison of Benzene and Pyridine Interactions with Aqueous FA

Interest in the interactions of hydrophobic compounds with dissolved NOM is significant because co-solubilization can increase the apparent solubility of hydrophobic contaminants in natural waters dramatically.[74–76] We have chosen two model organic contaminants to represent hydrophobic and hydrophilic organic contaminants interacting with a model SFA. An earlier study[65] examined the interaction of these two compounds with SFA using PM3 energy minimizations, but solvation effects were not included at that time. Here, we present simulation results including H_2O molecules in both periodic and nonperiodic simulations.

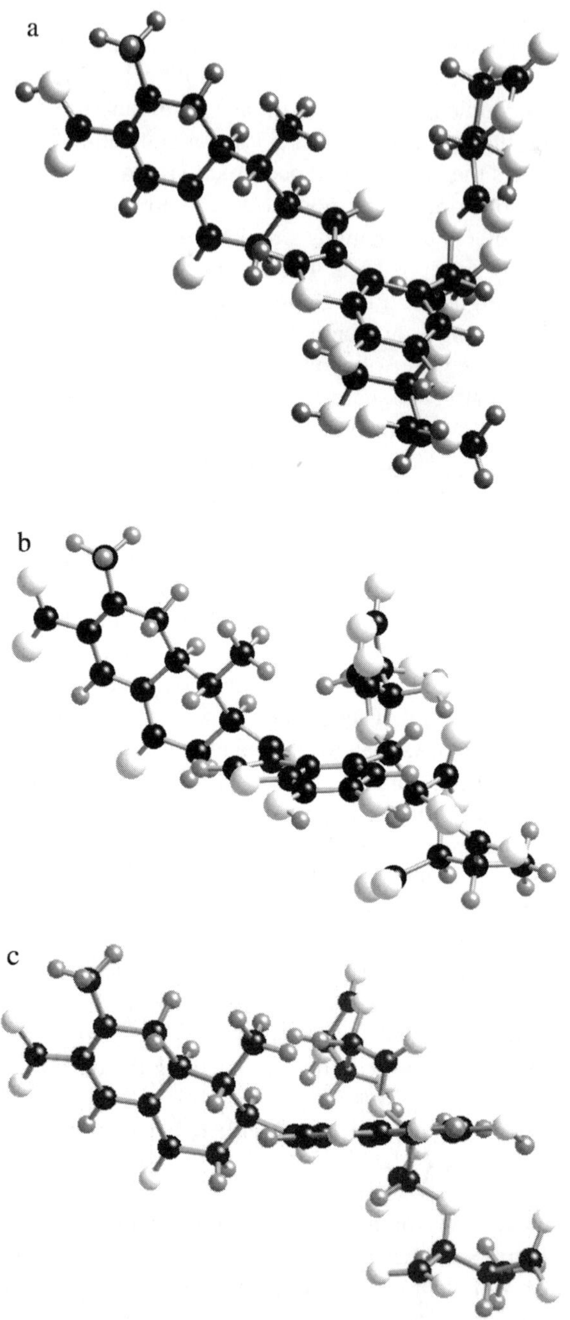

FIGURE 5.7 Semi-empirical PM3 energy-minimized structures of Suwannee fulvic acid.[69] Structure 1 as (a) gas-phase, neutral, (b) gas-phase anion, and (c) aqueous-phase anion. (Note that the H_2O molecules of solvation in the model have been removed from this figure for clarity.) H-bonding affects the conformation of the molecule, and both deprotonation and solvation strongly affect the H-bonding that forms.

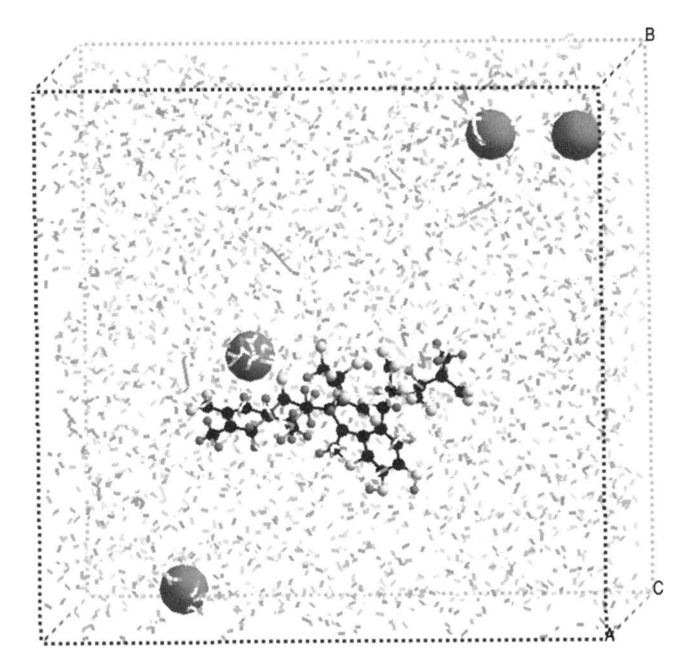

FIGURE 5.8 Model of SFA with a -4 charge in a periodic box of 1493 H_2O molecules and charge balanced with four Na^+ ions after 50 ps of MD simulation with the COMPASS force field in Cerius[2] (Accelrys, Inc.). Results of deprotonation and solvation are similar in these three-dimensional periodic simulations to the nonperiodic energy minimization in Figure 5.7(c).

The nonperiodic model systems (PM3 energy minimizations) presented in Figure 5.9 show that the previous model calculations overpredicted the amount of interaction between H atoms attached to the benzene ring (both in benzene and pyridine) and the O atoms in carboxylate groups (Figure 5.9a).[65] The H-bonding donor capability of H_2O molecules is much greater than that of aromatic H atoms, so the presence of the solvent overwhelms any potential H-bonding between the aromatic H and the carboxylate O atoms. On the other hand, the acid H^+ of the pyridine molecule does remain H-bonded to two O atoms in a carboxylate group (Figure 5.9(b)). Such an interaction is consistent with experimental results of NMR spectroscopy that have been interpreted as association via H-bonding.[76]

A periodic simulation using the same methods outlined above was also performed on the aqueous fulvic acid–benzene system. Figure 5.9(c) illustrates the configuration generated in this model system after 50 ps of simulation. The benzene migrates in this case to associate with the lipid residue of the fulvic acid. Figure 5.9(d) is a snapshot of the simulation at 100 ps. At this point, the benzene molecule is not associated with the SFA at all and is completely solvated by water.

Three points are illustrated by these new configurations generated from the MD simulation. First, in agreement with the PM3 energy minimizations discussed in the paragraph above, the relative strength of the H-bonding between benzene and carboxylate groups is weaker than that between the carboxylate groups and water.

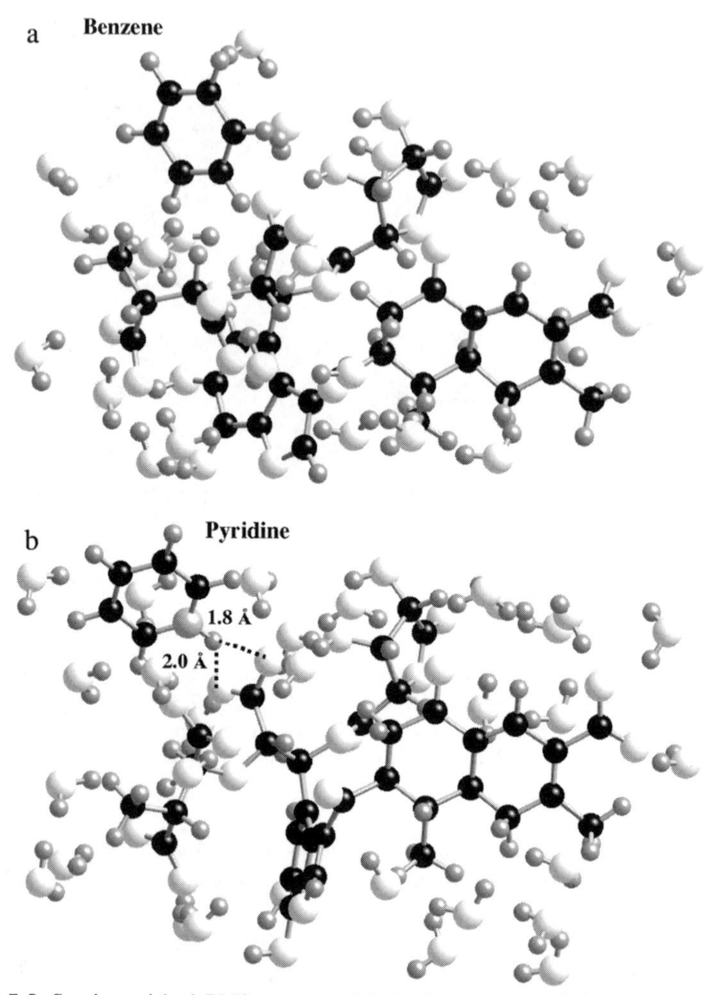

FIGURE 5.9 Semi-empirical PM3 energy minimized structures of Suwannee fulvic acid[69] interacting with (a) benzene and (b) pyridine with explicit solvation by H_2O molecules. The association of these compounds with SFA is severely weakened by the presence of water compared to the gas-phase calculations presented in Kubicki and Apitz.[65] (c) Configuration of benzene and SFA after 50 ps of three-dimensional periodic MD simulation with the COMPASS force field in Cerius^2 (Accelrys, Inc.). The benzene molecule has migrated toward the lipid section of the SFA as defined by Leenheer.[69] (d) Configuration of benzene and SFA after 100 ps of simulation. The benzene molecule has become completely dissociated from the SFA. (Note that the H_2O molecules of solvation in the model have been removed from this figure for clarity.) *(continued)*

Hence, H_2O molecules replace the benzene that had been in proximity to the carboxylate group. One must always remember that actual associations between molecules will be a function of the *relative* energies of all the possible conformations. One cannot predict that a certain association will exist solely on the basis that a negative energy is found for two molecules in proximity. This may seem like an obvious point, but the modeling literature is full of examples where alternative

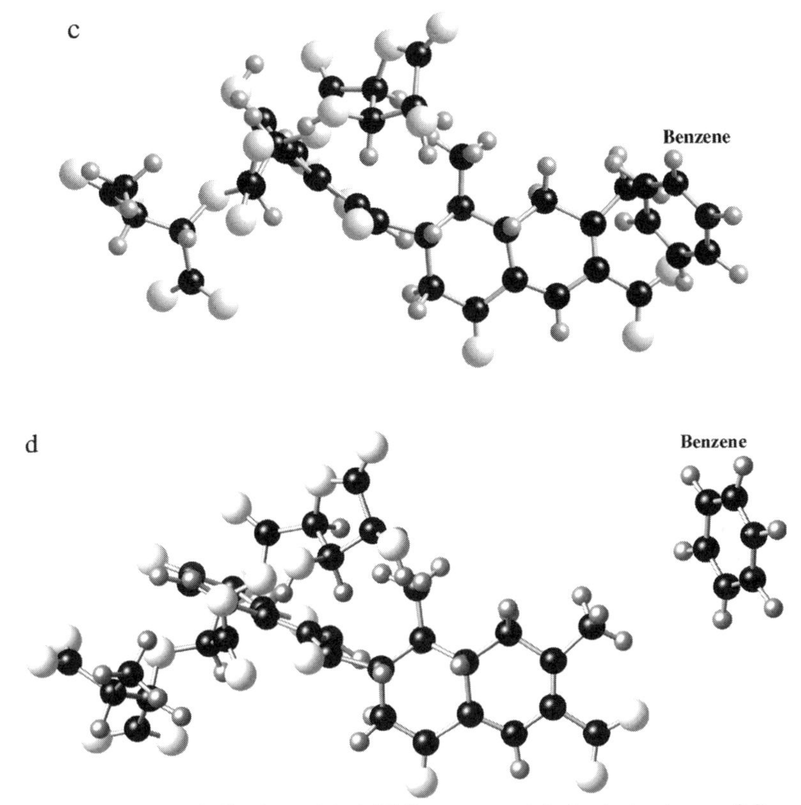

FIGURE 5.9 (*continued*) Semi-empirical PM3 energy minimized structures of Suwannee fulvic acid[69] interacting with (a) benzene and (b) pyridine with explicit solvation by H_2O molecules. The association of these compounds with SFA is severely weakened by the presence of water compared to the gas-phase calculations presented in Kubicki and Apitz.[65,69] (c) Configuration of benzene and SFA after 50 ps of three-dimensional periodic MD simulation with the COMPASS force field in Cerius^2 (Accelrys, Inc.). The benzene molecule has migrated toward the lipid section of the SFA as defined by Leenheer.[69] (d) Configuration of benzene and SFA after 100 ps of simulation. The benzene molecule has become completely dissociated from the SFA. (Note that the H_2O molecules of solvation in the model have been removed from this figure for clarity.)

associations and/or solvation have been neglected in the interest of simplifying the model system. Consequently, we conclude that a particular association is *the* stable one, forgetting that other possibilities have been excluded from the study.

The second point to be reiterated is that energy minimizations often find local minima and not the most stable energy configuration. In this case, the PM3 calculation may be providing a relatively accurate depiction of the benzene–fulvic acid interaction energy, but the energy minimization is not sampling configuration space adequately. Using MD simulations to generate a range of configurations followed by energy minimizations of various time steps throughout the simulation is a good strategy in this case. The MD simulations can be used to overcome the local minima problem, and the energy minimizations can be performed at a higher level of theory than the MD simulations to provide for more accurate energetics.

The third point is that a number of configurations may be energetically possible (see Figures 5.9(c) and 5.9(d)). If the energy difference between configurations is small relative to kT (kT ≈ 2.5 kJ/mol at 298K), then each configuration will be sampled and the occurrence of each configuration should be related to its potential energy. In this manner, a long simulation with enough molecules can be used to predict such properties as partitioning coefficients.[77,78]

5.3.3.1 Al^{3+}-Complexed Humic Acid

The interaction of many metals with dissolved NOM can be complex and irreversible. (Note: The term "irreversible" is used here in the sense that dissociation of the complex does not occur as the reverse process of association.[79] This does *not* mean that the metal cannot be dissociated once it is complexed; it means that thermodynamic equilibrium does not hold due to kinetic barriers preventing dissociation from occurring.) Consequently, the transport, fate, and bioavailability of metals can be strongly influenced by the presence of NOM. In this chapter, we use Al^{3+} as an example of a metal that forms strong complexes with NOM to investigate the nature of this complexation and its effect on the conformation of the NOM itself.[74] This complex interaction has been suggested as a possible source of irreversibility in metal adsorption, but observing the molecular changes that occur in natural systems has been difficult.

In this case, we wanted to model possible long-range effects of Al^{3+} complexation to dissolved NOM, so we chose to use a model humic acid (HA) instead of a fulvic acid.[60,80] The initial bonding of the Al^{3+} to a carboxylate group of the HA was determined from the model Al^{3+}-benzoic acid calculations presented above. The HA-H_2O system was then run through 50 ps of molecular dynamics simulations using PM3 with and without the Al^{3+} bonded to a carboxylic acid group of the HA. Figure 5.10 compares the resulting structures of the model HA. The overall –4 charge on the model HA (five COO^- groups and one NH_3^+ group) causes the structure to uncoil as was observed in the SFA simulations (Figure 5.7(c)). Electron repulsion between the carboxylate groups placed throughout the model HA causes the C backbone to remain relatively open. In addition, the open structure also maximizes the possibilities for H-bonding to the surrounding H_2O molecules. When complexed with Al^{3+}, however, the overall charge on the model HA is almost neutralized and the molecule can begin to coil. Furthermore, an additional carboxylate group is attracted to the Al^{3+} cation and begins to form a bond to the metal ion (Figure 5.10(b)).

The model prediction mentioned above is consistent with the observation of the "salting-out effect" as dissolved NOM enters saline waters.[81–83] As terrestrially derived, dissolved organic matter (DOM) reaches the ocean, much of the organic matter in the river precipitates in the near coastal environment due to the increased ionic strength of seawater. If complexes between humic acids and cations, such as Ca^{2+}, can form similar structures to the conformation in Figure 5.10(b), then the DOM will have fewer hydrophilic groups to interact with the surrounding water, the hydrophobic portions of the DOM will be more exposed at the surface, and the DOM will tend to coil and form colloids. The overall structural rearrangement will

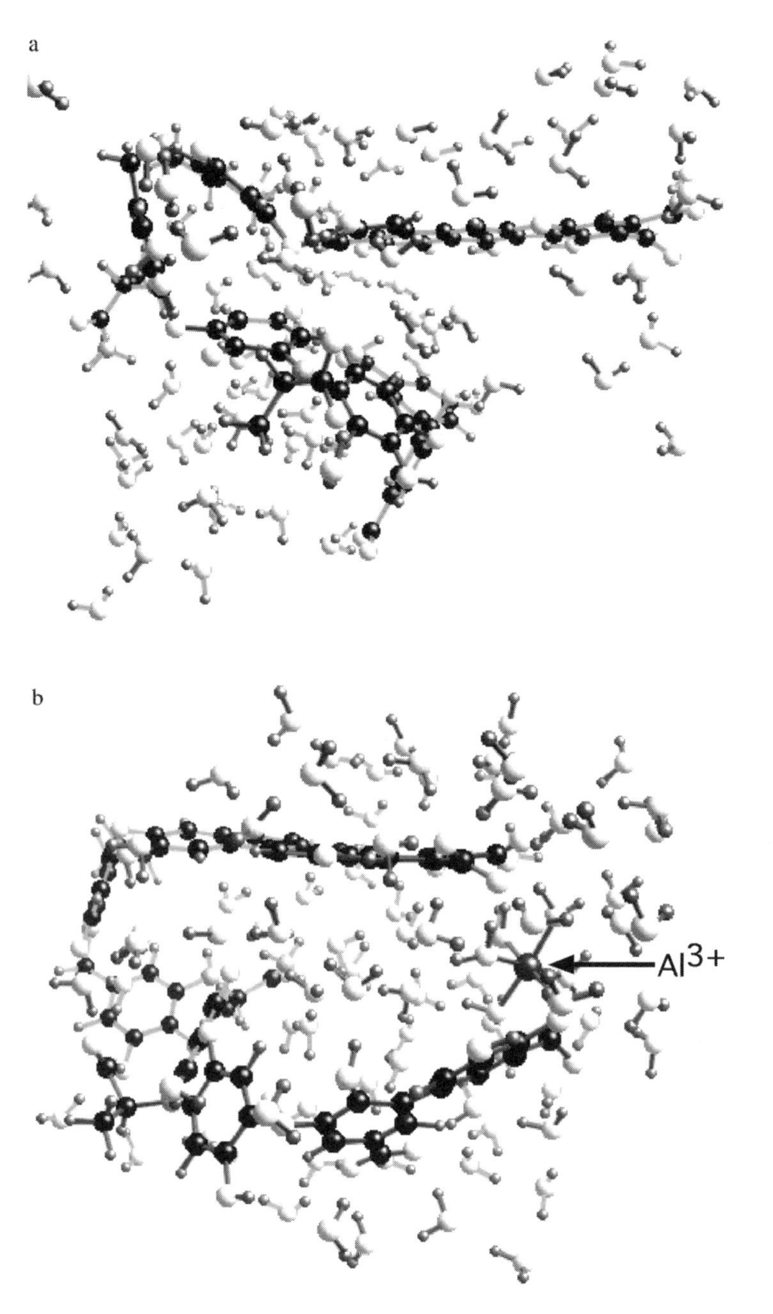

FIGURE 5.10 Models of a simplified humic acid molecule based on the Stevenson structure as pictured in Paul and Clark[80] in (a) a deprotonated and solvated state and (b) deprotonated, solvated, and complexed to Al^{3+}.[60] The force field does not allow for making and breaking covalent bonds within the simulation, but the strong electrostatic attraction between a free COO^- group and the Al^{3+} suggests that the Al^{3+} may ultimately form multiple covalent bonds to the humic acid and thereby affect the conformation of the DOM.

also have a tendency to more strongly sequester any other species (e.g., PAHs, Cu^{2+}) that may be associated with the DOM.

The formation of multiple bonds between carboxylate groups of the model HA and the Al^{3+} can also explain the irreversible adsorption of metals into dissolved NOM. Although the COO-Al bonds may not be extremely strong on an individual basis (as evidenced by the Al-benzoic acid association constants), the formation of multiple bonds creates a significant kinetic barrier to dissociation. Similar to a Fe-siderophore complex, removing a complexed metal from dissolved NOM can be hindered by the fact that numerous bonds need to be broken to release the metal. For instance, if one COO-Al bond should break, the remaining bonds may keep the Al^{3+} associated with the HA long enough that the broken COO-Al bond re-forms before another COO-Al bond is broken. Coiling of the HA upon complexation can increase the effectiveness of this kinetic barrier. Hydrophobic domains may develop within the HA that prevent the free exchange of H_2O molecules.[14,79] Because H_2O exchange is a likely mechanism for dissociating the Al^{3+}-HA complex, diminishing the presence of water locally around the region of the HA bonded to Al^{3+} could severely inhibit the tendency of the complex to dissociate.[84]

5.3.4 Pyridine Interaction with Al-Complexed HA

Lastly, we attempt to put together a model system consisting of each of the components discussed above: a model of HA complexed with Al^{3+} interacting with benzene and pyridine in water. Charging and conformational changes in dissolved NOM discussed above may have a significant effect on the interaction of NOM with hydrophobic organic contaminants. The development of more hydrophobic regions and coiling of the C backbone could increase the sorption of hydrophobic contaminants and increase the sequestration capacity of the NOM. All these components may be available in soils and sediments, so representing this complexity is a significant challenge in molecular modeling studies of NOM.

Figure 5.11 is an example of a complex model system. An energy minimization was performed with the semi-empirical PM3 method without periodic boundary conditions. As in Figure 5.10 above, the Al^{3+} forms a covalent bond to the carboxylate group on the humic acid. A second carboxylate group at the opposite end of the model humic acid is electrostatically attracted to the highly charged Al^{3+} cation that makes the model humic acid loop back on itself. The pyridine molecule is located inside the humic acid loop. Pyridine appears to be interacting with the humic acid via π–π attractions because the aromatic rings are aligned and separated by a distance of approximately 3.5Å. A similar distance is found between carbon rings in graphite. Hence, we suggest that van der Waal's forces between the aromatic rings are favored. In addition, pyridine is able to form a H-bond to an O atom in an H_2O molecule. Although this humic acid model is relatively small compared to natural humic acids, one can envision how organic contaminants could be entrapped within humic acids via the mechanism shown in Figure 5.11. Metals could play an important role in this process because humic-metal complexation could drive conformational changes that develop more hydrophobic regions within the humic acid.

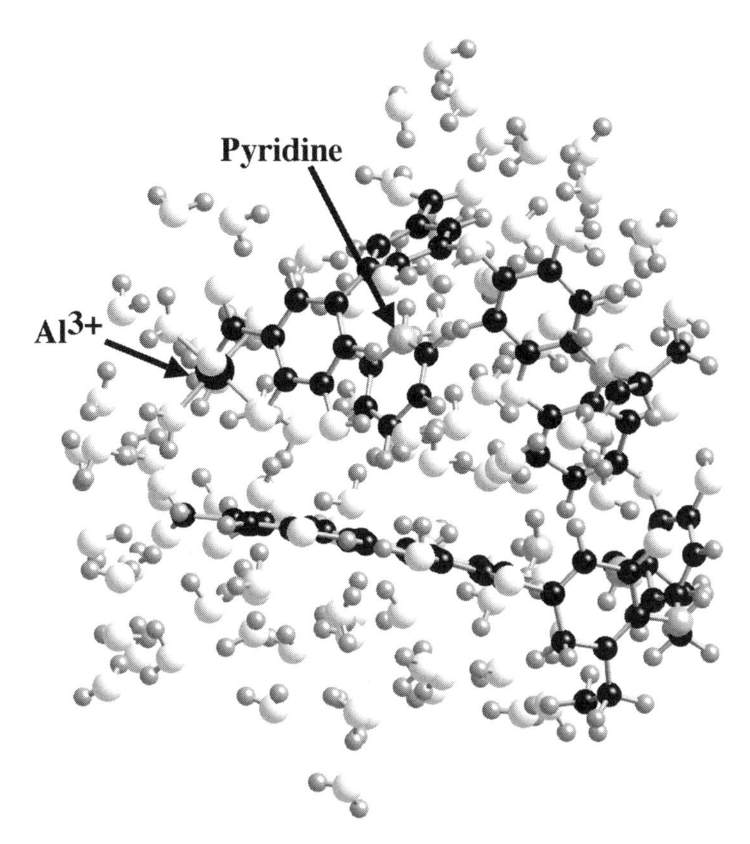

FIGURE 5.11 Model of a simplified humic acid molecule based on the Stevenson structure as pictured in Paul and Clark[80] in deprotonated, solvated and complexed to Al^{3+} demonstrating a possible interaction mechanism with pyridine.[60] Van der Waal's attraction between the π-electrons of the two aromatic rings is a likely driving force for this association.

5.4 CONCLUSIONS AND FUTURE WORK

Due to the complexities of the substances involved and the preliminary nature of the calculations discussed here, we do not make any definitive claims regarding fulvic or humic acid chemistry. The complexity and range of compositions found in fulvic and humic acids makes definitive conclusions based on a few simulations impossible. The value of this chapter hopefully lies in the hypotheses that we have suggested above and in the methodology developed for further simulations. The key elements of our methodology follow:

1. Comparing *ab initio* calculations to experiment to determine proper structural models
2. Testing classical simulations against *ab initio* calculations to verify the accuracy of the force field

3. Performing simulations under a variety of conditions to ensure that results are not model dependent
4. Comparing simulated results and predictions to experimental observations wherever possible

The results of this first step have been successful, so the behavior of simple organic acids can be understood via a combination of spectroscopy and computational chemistry. Steps 2 through 4 are in the initial stages and will require extensive effort before reaching the same level of confidence as the *ab initio* results.

The value of our hypotheses suggested here is that they can be used as motivation for further experimentation and simulation. We believe that most of these hypotheses are consistent with the recent research we are aware of; however, we acknowledge that controversy generally surrounds the interpretation of experimental results involving fulvic and humic acids. The interaction among experimentalists and modelers is of greater value than either activity in isolation because these two independent methods are capable of testing the predictions of the other approach.

Once a methodology is developed for molecular modeling the chemistry of metals and NOM, a vast array of studies become possible. As mentioned above, the complexity of NOM is almost endless. Combined with the large number of metals of geochemical and environmental interest that may interact with NOM, the matrix of possible models is enormous. A primary assumption in tackling this daunting problem is that a number of controlling parameters and guiding principles may determine metal–NOM chemistry. If so, it may be possible to shed some light on how these operate via computational chemistry and thereby eliminate a large number of the possibilities.

In this chapter, we concentrated on aluminum because it is easier to handle theoretically than transition metals and because we are interested in bonding NOM to soil minerals. Al^{3+} and Fe^{3+} are thought to control this strong adsorption interaction, and bonding should be similar between NOM and these two metals. We are currently beginning to study more specific interactions, such as aqueous and surface complexation of Fe^{3+} with siderophores and the biochemistry of silica complexation.[85] Another area of great interest is the complexation of amino acids with metals such as Cu^{2+} because these complexes may play a role in the bioavailability of copper in the environment.[86–88] The main thrust of our research will be trying to incorporate mineral surfaces into the aqueous-phase models we have presented here. Generating accurate models of systems containing water, NOM, metals, and mineral surfaces would begin to give us a molecular-level picture of soil chemistry.

5.5 ACKNOWLEDGMENTS

JDK acknowledges the financial support of the Office of Naval Research and the National Science Foundation through a CRAEMS research grant (CHE-0089156). Computational resources were also provided by the Department of Defense High Performance Computing Initiative through the Aeronautical Systems Center (Dayton, Ohio) and the Space and Naval Warfare Systems Center (San Diego, California).

REFERENCES

1. Drever, J.I. and Stillings, L.L., The role of organic acids in mineral weathering, *Colloid. Surface*, 120, 167, 1997.
2. Torn, M.S. et al., Mineral control of soil organic carbon storage and turnover, *Nature*, 389, 170, 1997.
3. MacCarthy, P., The principles of humic substances, *Soil Sci.*, 166, 738, 2001.
4. Chefetz, B. et al., Structural characterization of soil organic matter and humic acids in particle-size fractions of an agricultural soil, *Soil Sci. Soc. Am. J.*, 66, 129, 2002.
5. Hayes, M.H.B. et al., *Humic Substances II: In Search of Structure*, Wiley & Sons, New York, 1989.
6. Leenheer, J.A. et al., Models of metal-binding structures in fulvic acid from the Suwannee River, Georgia, *Environ. Sci. Technol.*, 32, 2410, 1998.
7. Benedetti, M.F. et al., Metal ion binding to humic substances: Application of the non-ideal competitive adsorption model, *Environ. Sci. Technol.*, 29, 446, 1995.
8. Alexander, M., How toxic are toxic chemicals in soil?, *Environ. Sci. Technol.*, 29, 2713, 1995.
9. Weber, W.J., McGinley, P.M., and Katz, L.E., A distributed reactivity model for sorption by solid and sediments, 1. Conceptual basis and equilibrium assessments, *Environ. Sci. Technol.*, 26, 1955, 1992.
10. Xia, G. and Pignatello, J.J., Detailed sorption isotherms of polar and apolar compounds in a high-organic soil, *Environ. Sci. Technol.*, 35, 84, 2001.
11. Gustafsson, O. et al., Quantification of the dilute sedimentary soot phase: Implications for PAH speciation and bioavailability, *Environ. Sci. Technol.*, 31, 203, 1997.
12. Aiken, G.R., Isolation and concentration techniques for aquatic humic substances, in *Humic Substances in Soil, Sediment, and Water: Geochemistry, Isolation and Characterization*, Aiken, G.R., McKnight, D.M., Wershow, R.L., MacCarthy, P., Eds., John Wiley & Sons, New York, 1988, p. 363.
13. Schulten, H.R. and Schnitzer, M., Chemical model structures for soil organic matter and soils, *Soil Sci.*, 162, 115, 1997.
14. Chen, Y.Y. and Bleam, W.F., Two-dimensional NOESY nuclear magnetic resonance study of pH-dependent changes in humic acid conformation in aqueous solution, *Environ. Sci. Technol.*, 32, 3653, 1998.
15. Diallo M.S. et al., Molecular modeling of Chelsea humic acid: A hierarchical approach based on experimental characterization, computer-assisted structure elucidation, and atomistic simulations, *Abstr. Pap. Am. Chem. Soc.*, 219, 115-ENVR., 2000.
16. Diallo, M. et al., Molecular modeling of humic substances: Binding of hydrophobic organic compounds to Chelsea humic acid, *Abstr. Pap. Am. Chem. Soc.*, 221, 189-ENVR., 2001.
17. Kubicki, J.D. and Bleam, W.F., An introduction to molecular modeling, in *Molecular Modeling of Clays and Mineral Surfaces*, Kubicki, J.D. and Bleam, W.F., Eds., Clay Minerals Society, in press.
18. Schulten H.R., Interactions of dissolved organic matter with xenobiotic compounds: Molecular modeling in water, *Environ. Toxicol. Chem.*, 18, 1643, 1999.
19. Kujawinski, E.B., Hatcher, P.G., and Freitas, M.A., High-resolution Fourier transform ion cyclotron resonance mass spectrometry of humic and fulvic acids: improvements and comparisons, *Anal. Chem.*, 74, 413, 2002.
20. Kujawinski, E.B. et al., The application of electrospray ionization mass spectrometry (ESI MS) to the structural characterization of natural organic matter, *Org. Geochem.*, 33, 171, 2002.

21. Simpson, A.J. et al., Improvements in the two-dimensional nuclear magnetic resonance spectroscopy of humic substances, *J. Environ. Qual.*, 31, 388, 2002.
22. Dria, K.J., Sachleben, J.R., and Hatcher, P.G., Solid-state carbon-13 nuclear magnetic resonance of humic acids at high magnetic field strengths, *J. Environ. Qual.*, 31, 393, 2002.
23. Reuter, K. et al., Atomistic description of oxide formation on metal surfaces: The example of ruthenium, *Chem. Phys. Lett.*, 352, 311, 2002.
24. Shevchenko, S.M. and Bailey, G.W., Non-bonded organo-mineral interactions and sorption of organic compounds on soil surfaces: A model approach, *Theochem. J. Mol. Struc.*, 422, 259, 1998.
25. van Duin, A.C.T. et al., Simulated partitioning of polyaromatic hydrocarbons between mineral surfaces and humic layers in the water phase, *Abstr. Pap. Am. Chem. Soc.*, 219, 71-Geoc, 2000.
26. Mayer, L.M. and Lawrence, M., *Organic Geochemistry*, Plenum Press, New York, 1993, 171.
27. Skjemstad, J.O. et al., High energy ultraviolet photo-oxidation: A novel technique for studying physically protected organic matter in clay and silt-sized particles, *J. Soil Sci.*, 44, 485, 1993.
28. Kohl, S.D. and Rice, J.A., The binding of contaminants to humin: A mass balance, *Chemosphere*, 36, 251, 1998.
29. Murphy, E.M., Zachara, J.M., and Smith, S.C., Influence of mineral-bound humic substances on the sorption of hydrophobic organic compounds, *Environ. Sci. Technol.*, 24, 1507, 1990.
30. Murphy, E.M. et al., The sorption of humic acids to mineral surfaces and their role in contaminant binding, *Sci. Total Environ.*, 118, 413, 1992.
31. Myneni, S.C.B. et al., Imaging of humic substance macromolecular structures in water and soils, *Science*, 286, 1335, 1999.
32. Kubicki, J.D. et al., Attenuated total reflectance Fourier-transform infrared spectroscopy of carboxylic acids adsorbed onto mineral surfaces, *Geochim. Cosmochim. Acta*, 63, 2709, 1999.
33. Namjesnik-Dejanovic, K. and Maurice, P.A., Conformations and aggregate structures of sorbed natural organic matter on muscovite and hematite, *Geochim. Cosmochim Acta*, 65, 1047, 2001.
34. Yost, E.C., Tejedor-Tejedor, M.I., and Anderson, M.A., *In situ* CIR-FTIR characterization of salicylate complexes at the goethite/aqueous solution interface, *Environ. Sci. Technol.*, 24, 822, 1990.
35. Biber, M.V. and Stumm, W., An *in situ* ATR-FTIR study: The surface coordination of salicylic acid on aluminum and iron(III) oxides, *Environ. Sci. Technol.*, 28, 763, 1994.
36. Leenheer, J.A., Wershaw, R.L., and Reddy, M.M., Strong-acid, carboxyl group structures in fulvic acid from the Suwannee River, Georgia, 2. Major structures, *Environ. Sci. Technol.*, 29, 393, 1995.
37. Dapprich, S. et al., A new ONIOM implementation in Gaussian 98, Part I. The calculation of energies, gradients, vibrational frequencies and electric field derivatives, *J. Mol. Struc.-Theochem.*, 462, 1, 1999.
38. Lubin, M.I., Bylaska, E.J., and Weare, J.H., Ab initio molecular dynamics simulations of aluminum ion solvation in water clusters, *Chem. Phys. Lett.*, 322, 447, 2000.
39. Weare, J.H. et al., Thermodynamic models of natural fluids: Theory and practice, *Abstr. Pap. Am. Chem. Soc.*, 221,123-Geoc, 2001.

40. Iarlori, S. et al., Dehydroxylation and silanization of the surfaces of beta-cristobalite silica: An ab initio simulation, *J. Phys. Chem. B*, 105, 8007, 2001.
41. Hass, K.C. et al., The chemistry of water on alumina surfaces: Reaction dynamics from first principles, *Science*, 282, 265, 1998.
42. Frisch, M.J. et al., *Gaussian 98, Revision A.6*, Gaussian, Pittsburgh, 1998.
43. McQuarrie, D.A. and Simon, J.D., *Physical Chemistry: A Molecular Approach*, University Science Books, Sausalito, CA, 1997.
44. Newton, M.D. and Gibbs, G.V., Ab initio calculated geometries and charge distributions for H_4SiO_4 and $H_6Si_2O_7$ compared with experimental values for silicates and siloxanes, *Phys. Chem. Miner.*, 6, 221, 1980.
45. Møller, C. and Plesset, M.S., Note on an approximation treatment for many-electron systems, *Phys. Rev.*, 46, 618, 1934.
46. Foresman, J.B. and Frisch, A.E., *Exploring Chemistry with Electronic Structure Methods*, 2nd ed., Gaussian, Pittsburgh, 1997.
47. Becke, A.D., Density-functional exchange-energy approximation with correct asymptotic-behavior, *Phys. Rev. A*, 38, 3098, 1988.
48. Lee, C.T., Yang, W.T., and Parr, R.G., Development of the Colle-Salvetti correlation energy formula into a functional of the electron density, *Phys. Rev. B*, 37, 785, 1988.
49. Woodcock, L.V., Angell, C.A., and Cheeseman, P., Molecular dynamics studies of the vitreous state: Ionic systems and silica, *J. Chem. Phys.*, 65, 1565, 1976.
50. Breneman, C.M. and Wiberg, K., Determining atom-centered monopoles from molecular electrostatic potentials: The need for high sampling density in formamide conformational analysis, *J. Comput. Chem.*, 11, 361, 1990.
51. Leinenweber, K. and Navrotsky, A., A transferable interatomic potential for crystalline phases in the system MgO-SiO_2, *Phys. Chem. Min.*, 15, 588, 1988.
52. Lasaga, A.C., and Gibbs, G.V., Quantum mechanical potential surfaces and calculations on minerals and molecular clusters, I. STO-3G and 6–31G* results, *Phys. Chem. Miner.*, 16, 29, 1988.
53. Kubicki, J.D., Interpretation of vibrational spectra using molecular orbital theory calculations, in *Molecular Modeling Theory: Applications in the Geosciences*, Cygan, R.T. and Kubicki, J.D., Eds., Geochemical Society of America, Washington, D.C., 2001, p. 459.
54. Kubicki, J.D. and Lasaga, A.C., Molecular dynamics and diffusion in silicate melts, in *Diffusion, Atomic Ordering, and Mass Transport: Selected Problems in Geochemistry*, Ganguly, J., Ed., Springer-Verlag, New York, 1991, p. 1.
55. Sun, H., COMPASS: An ab initio force-field optimized for condensed-phase applications—Overview with details on alkane and benzene compounds, *J. Phys. Chem.*, 102, 7338, 1998.
56. Gibbs, G.V., Molecules as models for bonding in silicates, *Am. Mineral.*, 67, 421, 1982.
57. Fried, J.R., Sadat-Akhavi, M., and Mark, J.E., Molecular simulation of gas permeability: Poly(2,6-dimethyl-1,4-phenylene oxide), *J. Membrane Sci.*, 149, 115, 1998.
58. Kubicki, J.D., Molecular mechanics and quantum mechanical modeling of hexane soot structure and interactions with pyrene, *Geochem. Trans.*, 1, 41, 2000.
59. Launne, T., Neelov, I., and Sundholm, F., Molecular dynamics simulations of polymers of unsubstituted and substituted poly(p-phenylene terephthalate)s in the bulk state, *Macromol. Theor. Simul.*, 10, 137, 2001.
60. Stevenson, F.J., *Humus Chemistry: Genesis, Composition, Reactions*, 2nd ed., John Wiley & Sons, New York, 1994.

61. da Silva, J.C.G.E. and Machado, A.A.S.C., Characterization of the binding sites for Al(III) and Be(II) in a sample of marine fulvic acids, *Mar. Chem.*, 54, 293, 1996.
62. Thomas, F. et al., Aluminum(III) speciation with acetate and oxalate, a potentiometric and ^{27}Al NMR Study, *Environ. Sci. Technol.*, 25, 1553, 1991.
63. Kubicki, J.D. et al., The bonding mechanisms of salicylic acid adsorbed onto illite clay: An ATR-FTIR and MO study, *Environ. Sci. Technol.*, 31, 1151, 1997.
64. Clausén M. et al., Characterisation of gallium(III)-acetate complexes in aqueous solution, *J. Chem. Soc. Dalton*, 12:2559–2564, 2002.
65. Kubicki, J.D. and Apitz, S.E., Models of natural organic matter interactions with organic contaminants, *Org. Geochem.*, 30, 911, 1999.
66. Kummert, R. and Stumm, W., The surface complexation of organic acids on hydrous γ-Al_2O_3, *J. Colloid Interface Sci.*, 75, 373, 1980.
67. Kubicki, J.D., Sykes, D., and Apitz, S.E., Ab initio calculation of aqueous aluminum and aluminum-carboxylate NMR chemical shifts, *J. Phys. Chem. A*, 103, 903, 1999.
68. Schulten, H.R. and Schnitzer, M., Three-dimensional models for humic acids and soil organic matter, *Naturwissenschaften*, 80, 29, 1995.
69. Leenheer, J.A., Chemistry of dissolved organic matter in rivers, lakes, and reservoirs, in *Environmental Chemistry of Lakes and Reservoirs*, Baked, L.A., Ed., American Chemical Society, Washington, D.C., 1994, p. 195.
70. Allinger, N.L., Conformational analysis, 130. MM2, a hydrocarbon force field utilizing V1 and V2 torsional terms, *J. Am. Chem. Soc.*, 99, 8127, 1977.
71. Stewart, J.J.P., Optimization of parameters for semi-empirical methods, 1. Methods, *J. Comput. Chem.*, 10, 209, 1989.
72. Binkley, J.S., Pople, J.A., and Hehre, W.J., Self-consistent molecular orbital methods, 21. Small split-valence basis sets for first row elements, *J. Am. Chem. Soc.*, 102, 939, 1980.
73. Leach, A.R., *Molecular Modelling Principles and Applications*, 2nd ed., Prentice Hall, New York, 2001.
74. Leenheer, J.A., Organic substance structures that facilitate contaminant transport and transformations in aquatic sediments, in *Humics and Soils*, vol. 1 of *Organic Substances and Sediments in Water* series, Lewis Publishers, Chelsea, CA, 1991.
75. Schlautman, M.A. and Morgan, J.J., Effects of aqueous chemistry on the binding of polycyclic aromatic hydrocarbons by dissolved humic materials, *Environ. Sci. Technol.*, 27, 961, 1993.
76. Nanny, M.A., Bortiatynski, J.M., and Hatcher, P.G., Noncovalent interactions between acenaphthenone and dissolved fulvic acid as determined by ^{13}C NMR T_1 relaxation measurements, *Environ. Sci. Technol.*, 31, 530, 1997.
77. van Duin, A.C.T. and Larter, S.R., Application of molecular dynamics calculations in the prediction of dynamical molecular properties, *Org. Geochem.*, 29, 1043, 1998.
78. van Duin, A.C.T. and Larter, S.R., A computational chemical study of penetration and displacement of water films near mineral surfaces, *Geochem. Trans.*, 2, 35, 2001.
79. Xing, B. and Pignatello, J.J., Dual-mode sorption of low-polarity compounds in glassy poly(vinyl chloride) and soil organic matter, *Environ. Sci. Technol.*, 31, 792, 1997.
80. Paul, E.A. and Clark, F.E., *Soil Microbiology and Biochemistry*, Academic Press, San Diego, 1996.
81. Desai, M.V.M., Matthew, E., and Ganguly, A.K., Differential interaction of marine humic and fulvic acids with alkaline earth and rare earth elements, *Curr. Sci.*, 39, 429, 1970.
82. Hair, M.E. and Bassett, C.R., Dissolved and particulate humic acids in an east-coast estuary, *Estuary and Coastal Mar. Sci.*, 1, 107, 1973.

83. Hedges, J.I. and Parker, P.L., Land-derived organic matter in surface sediments from the Gulf of Mexico, *Geochim. Cosmochim. Acta*, 40, 1019, 1976.
84. Sullivan, D.J. et al., The rates of water exchange in Al(III)-salicylate and Al(III)-sulfosalicylate complexes, *Geochim. Cosmochim. Acta*, 63, 1471, 1999.
85. Kubicki, J.D. and Heaney, P.J., Modeling Interactions of Aqueous Silica and Sorbitol: Complexation, Polymerization and Association, paper presented at the Goldschmidt Conference, Homestead, VA, 2001.
86. Mayer, L.M. et al., Bioavailability of sedimentary contaminants subject to deposit feeder digestion, *Environ. Sci. Technol.*, 30, 2641, 1996.
87. Chen, Z. and Mayer, L.M., Sedimentary metal bioavailability determined by the digestive constraints of marine deposit feeders: Gut retention time and dissolved amino acids, *Mar. Ecol.-Prog. Ser.*, 176, 139, 1999.
88. Chen, Z. et al., High concentrations of complexed metals in the guts of deposit feeders, *Limnol. Oceanogr.*, 45, 1358, 2000.

6 Silicon-Organic Interactions in the Environment and in Organisms

Nita Sahai

CONTENTS

6.1 Introduction 146
6.2 Phospholipid–Oxide Surface Interactions 146
  6.2.1 Differential Effect of Oxides on Cell Membrane Structure 146
  6.2.2 Molecular Composition of Cell Membranes 147
  6.2.3 Phospholipid Adhesion to Oxides 147
6.3 Thermodynamic Basis for the Membranolytic Potential of Oxides: Crystal Chemistry and Interfacial Solvation 148
  6.3.1 Effect of Exchange-Free Energy 148
  6.3.2 Effect of Adsorption Entropy 149
  6.3.3 Model Interpretation 151
  6.3.4 Model Prediction for Behavior of Stishovite 151
  6.3.5 Relevance to Cell Attachment 151
6.4 Role of Dissolved Si-Organic Complexes in Biology 152
  6.4.1 Approach to Studying Putative Si-Organic Complexes: Rationale for Using MO Calculations 153
  6.4.2 Computational Details 154
  6.4.3 Computational Results and Interpretation 154
    6.4.3.1 Si-Polyalcohol Complexes 154
    6.4.3.2 Si-Amino Acid Complexes 159
  6.4.4 Relevance to Biological Silicon Uptake 162
6.5 Summary 162
References 163

1-56670-623-8/03/$0.00+$1.50

6.1 INTRODUCTION

Currently, there is in the geochemical and soil science literature a great emphasis on the environmental significance of interactions between inorganic species, mineral surfaces, organic molecules, and microbes. Much of the emphasis has been on the chelation of heavy metal nutrients from minerals by organically mediated mineral dissolution, and on bacterial attachment to mineral surfaces. In this chapter, I focus on a metalloid, silicon, which is the most abundant element, after oxygen, in the Earth's crust. I summarize results of theoretical studies for silicon-organic interactions where the silicon exists in the solid phase as quartz or amorphous silica, and where the silicon is dissolved in aqueous solution. The studies are motivated by a desire to understand silicon-organic interactions in the natural environment and in organisms.

6.2 PHOSPHOLIPID–OXIDE SURFACE INTERACTIONS

Interactions of amphiphilic molecules with oxide surfaces can serve as a model to understand cell attachment at mineral surfaces, an early step in some types of cell-mediated mineral dissolution. Attachment of cell membranes at oxide surfaces is also significant to medical geology for determining the fate of inhaled dust particles in the lung, which can cause respiratory disorders such as silicosis, and for understanding the attachment of living tissue to bioceramic prosthetic implants in designing biocompatible devices. Industrial processes of relevance include biofouling of wastewater treatment plants, and biobeneficiation of ores by froth floatation.

6.2.1 Differential Effect of Oxides on Cell Membrane Structure

Interactions of entire cells, and of the molecular constituents of cell membranes with oxide surfaces have been studied previously with respect to their ability to rupture the cell membrane. Membranolysis refers to the rupture of cell membranes when placed in contact with fine particles of oxide, a few microns in size. The membranolytic ability of oxides was taken as a proxy for the ability of those oxides to produce silicosis, a respiratory disorder caused by prolonged exposure to fine mineral dusts.[1–3] Oxides are lytic towards liposomes, lysosomes, and eukaryotic blood cells, such as the erythrocyte (the red blood cell) and the macrophage (a type of white blood cell).[4–7] The sequence for decreasing membranolytic ability of oxides is tridymite ~ cristobalite > quartz > coesite ~ amorphous silica > corundum >stishovite > anatase.[8–14] It is interesting that the octahedrally coordinated silica polymorph (stishovite) behaves less like quartz and more like the other oxides, and that amorphous silica is less lytic than quartz and the other tetrahedral crystalline silica polymorphs. Furthermore, amorphous SiO_2 bioceramics apparently do not share the membranolytic properties of the tetrahedral crystalline SiO_2 polymorphs. Rather, silica bioceramics are biocompatible, and are used in constructing dental and orthopedic prosthetic implants. Clearly, both the chemistry and the structure of the oxide influence its membranolytic potential.

Despite more than four decades of research, there is, as yet, no consensus in the silicosis literature for the fundamental reasons behind the different membranolytic activity of the various silica polymorphs and the other oxides. One possible reason for this slow progress is the neglect of mineralogical differences between the various phases, in terms of both chemistry and crystal structure, as reflected in the usage of terms such as "quartz glass" (*sic*) in the literature.[12]

Along another line of investigation, it has been found that the strength of bacterial adhesion on oxides decreases as hematite > corundum > quartz.[15–18] This is intriguing, because it is exactly the reverse of the membranolytic ability sequence. The stronger and more prolific attachment of some bacterial strains to oxides such as Fe_2O_3 and Al_2O_3, compared to quartz may be related, in part, to the fact that the bacteria can extract and utilize the iron for metabolic processes. At least one additional, plausible factor is that quartz is more deleterious to cell membrane structure than the other oxides. Thus, various lines of evidence suggest that different oxides have different effects on cell membrane structure.

6.2.2 Molecular Composition of Cell Membranes

Prokaryotic cell walls and cell membranes are composed primarily of lipopolysaccharides, teichoic acids, and phospholipids (PLs).[19–20] Eukaryotic cell membranes consist of phospholipids and proteins.[21] Phospholipids are amphiphilic organic molecules, where the "head group" part of the molecule is polar or charged, and, hence, is hydrophilic. The polar head group contains phosphoryl and often secondary, tertiary, or quaternary amine groups such as tetramethyl ammonium (TMA^+).[21] The remaining hydrocarbon "tail" of the molecule is hydrophobic. One might anticipate that the interfacial properties of such molecules would be quite different from those of the usual suspects in natural water, that is, small dissolved ions such as the alkali metals, alkaline earth metals, transition and heavy metals, oxyanions, and low-molecular-weight organic acids. Still, the interaction of the small head group moeity with oxide surfaces may be tractable using electric double-layer theory developed for small ions. This is the approach taken below, but first, a brief description of the entire PL attachment process is provided.

6.2.3 Phospholipid Adhesion to Oxides

The attachment of PLs to oxide surfaces occurs in several distinct steps characterized by the main contributing physical factors at each step. The earliest step involves electrostatic attraction between the polar head group and the charged oxide surface. This stage is followed by adsorption driven mainly by hydrophobic forces where tail-to-tail attraction occurs, resulting in the formation of PL aggregates at the surfaces. The next stage involves an increase in the number of PLs per aggregate and, ultimately, surface micelles are formed.[22–24]

The electrostatic attraction step can be examined using the Solution and Electrostatic (SE) model for ion adsorption.[25] The model uses a thermodynamic approach based on crystal chemistry and interfacial solvation. The SE model was developed originally for oxide surface-small ion interactions,[26–30] and cannot pretend to represent entire PL–oxide surface interactions. PL–oxide surface attachment has been modeled previously in various ways such as the DLVO model, the extended DLVO model, and the hydrophobic interfacial model.[31–33] These are all excellent approaches to modeling PL and bacterial-mineral surface interactions, but the specific nature of the solid is usually given minimal attention.

The aim of the present work is to take a first stab at representing the specificity of the oxide in the initial stages of PL–oxide surface interactions. The hypothesis is that disruption to cell membrane structure inherited from the early electrostatic adsorption stage affects membrane rupture in the overall cell attachment process.

6.3 THERMODYNAMIC BASIS FOR THE MEMBRANOLYTIC POTENTIAL OF OXIDES: CRYSTAL CHEMISTRY AND INTERFACIAL SOLVATION

6.3.1 Effect of Exchange-Free Energy

The SE model applies to electrostatic adsorption of an ion at an oppositely charged site on the metal oxide surface. The model assumes that the total free energy of solvation of the i-th ion ($\Delta G^0_{ads,i}$) can be resolved into two contributions.[26,30] The first part is the solvation energy change in moving a solvated ion from bulk aqueous solution to the oxide surface ($\Delta G^0_{solv,i}$). The second is an ion-intrinsic contribution related to the physical and chemical properties of the ion ($\Delta G^0_{ii,i}$). In terms of model parameters, the solvation energy depends on the inverse of the average static dielectric constant of the k-th oxide ($1/\varepsilon_k$) and the charge-to-effective radius ratio of the sorbed ion ($z/R_{e,i}$). The intrinsic contribution depends only on the charge-to-effective radius ratio of the ion in aqueous solution ($z/r_{e,i}$).

The population of charged surface sites is determined by the first and second surface protonation constants of the oxide ($\Delta G^0_{ads,H^+,1}$ and $\Delta G^0_{ads,H^+,2}$). These terms depend, in turn, on the crystal-chemical parameter, s/r, which is the Pauling bond strength[34] per metal-OH bond length, and on $1/\varepsilon_k$.[35–37]

At pH > ~3, the quartz and amorphous silica surface are negatively charged.[38–43] The negative charge favors electrostatic attraction of the TMA^+ moiety of the PL head group. The surfaces of the other oxides are positively charged at ambient pHs. For example, TiO_2 is positively charged up to pH ~ 7,[44–45] Fe_2O_3 (up to pH ~ 8), and Al_2O_3 to pH ~ 9.[40,46–50] Thus, cation adsorption is more favorable on the quartz and amorphous silica surfaces in the pH range of normal surface waters (pH ~ 4 to 8.5), human blood plasma (pH = 7.2), and the lysosome in the macrophage (pH = 4.8) wherein digestion of foreign particles inhaled and retained in the lungs occurs.

The ambient aqueous solution in membranolysis experiments or in the natural environment usually also contains some other ion such as Na^+ or Ca^{2+} as the dominant cation. TMA^+, Na^+, and Ca^{2+}, thus compete for negatively charged surface sites. The relevant thermodynamic quantity to consider, then, is the free energy of the *exchange* reaction at each oxide surface, $\Delta G_{exc,TMA^+/Na^+} = \Delta G_{ads,TMA^+} - \Delta G_{ads,Na^+}$.[25] This quantity is negative for quartz and amorphous silica, but is positive for the other oxides at

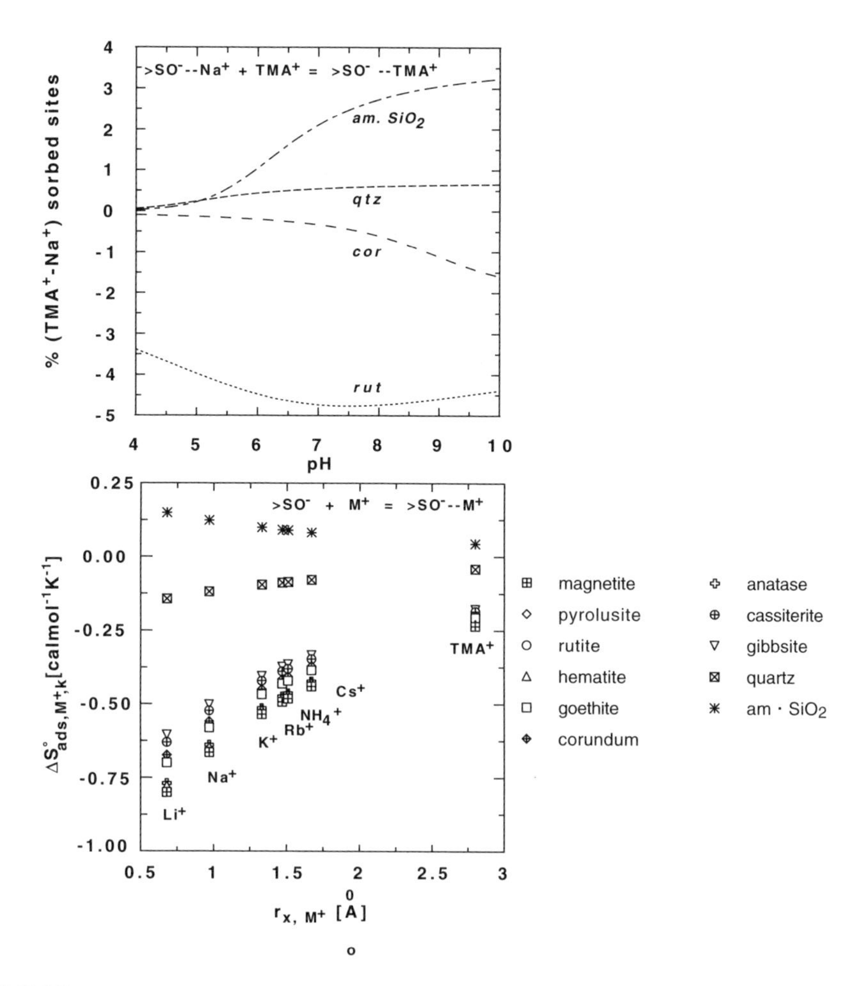

FIGURE 6.1 Factors that influence the nature of electrostatic cation adsorption at oxide surfaces, and thus, PL membrane structure. (a) Percentage of surface sites occupied by TMA^+ ions compared to Na^+ ions. This number is another way of representing $\Delta G_{exc,TMA^+/Na^+}$. (b) Standard state entropy of cation adsorption. (Modified and reprinted from Sahai, N., *J. Colloid Interface Sci.*, 252, 309, copyright 2002, and Sahai, N., *Geochim. Cosmochim. Acta*, 64, 3629, copyright 2000. With permission from Elsevier Science.)

ambient pHs. Hence, TMA^+ can out-compete Na^+ at the quartz and amorphous silica surface but not on other oxides (Figure 6.1a). In this manner, ΔG_{exc}, $\Delta G_{exc,TMA^+/Na^+}$ distinguishes quartz and amorphous silica from the other oxides.

6.3.2 Effect of Adsorption Entropy

The entropy of TMA^+ adsorption is another important factor controlling lysis.[25] The relevant quantity now is not $\Delta S_{ads,TMA^+}-\Delta S_{ads,Na^+}$, but rather, $\Delta S_{ads,TMA^+}$. This is because regardless of the entropy change in replacing Na^+ by TMA^+, an unfavorable change in entropy of the head group upon adsorption will cause conformational change of the PL and thus, membrane rupture. Lacking an easy way to calculate $\Delta S_{ads,TMA^+}$ at ambient conditions, I will make do with standard state $\Delta S^0_{ads,TMA^+}$ to provide conceptual understanding.

The entropy change predicted for the adsorption reaction is unfavorable ($\Delta S^0_{ads,TMA^+} < 0$) for all oxides except for amorphous silica (Figure 6.1b). Thus, $\Delta S^0_{ads,TMA^+}$ distinguishes quartz from amorphous silica. The negative adsorption entropy on quartz suggests that PL head-group conformation is changed, with greater disruption to membrane structure. A positive value of $\Delta S^0_{ads,TMA^+}$ for amorphous silica suggests that the change in head group conformation will be smaller and disrupt the membrane less than for oxides that have $\Delta S^0_{ads,TMA^+} < 0$. The effect of entropy and the extent of TMA^+/Na^+ adsorption balance each other out for amorphous silica, so that the net membranolytic ability observed is similar to or slightly less than for quartz. The change in adsorption entropy of TMA^+ is predicted to be negative for oxides such as TiO_2, Fe_2O_3, and Al_2O_3, suggesting some degree of membrane disruption. But these oxides cannot cause substantial damage because $\Delta G_{exc,TMA^+/Na^+}$ is small or negative at the low pHs of the ambient solutions, because of positive surface charge.

In summary, both factors ($\Delta G_{exc,TMA^+/Na^+} < 0$ and $\Delta S_{ads,TMA^+} < 0$) need to be fulfilled in order to disrupt the membrane sufficiently to cause lysis (Figure 6.2). This happens at the surface of quartz and other crystalline polymorphs of silica containing four-fold coordinated silicon. The $\Delta G_{exc,TMA^+/Na^+}$ term distinguishes tetrahedrally coordinated crystalline and amorphous silica polymorphs from other oxides. The $\Delta S_{ads,TMA^+}$ contribution distinguishes tetrahedral silica polymorphs from amorphous silica.[25]

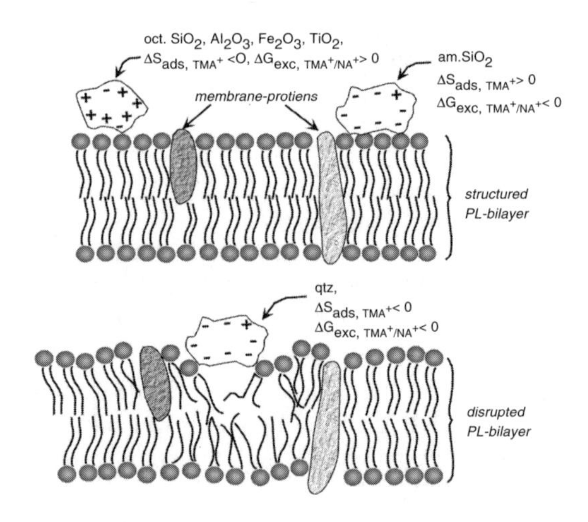

FIGURE 6.2 Cartoon representation of the thermodynamic effect of oxide chemistry and structure on PL membrane structure. (Reprinted from Sahai, N., *J. Colloid Interface Sci.*, 252, 309, copyright 2002. With permission from Elsevier Science.)

6.3.3 Model Interpretation

The differences among the various oxides can be explained by the following interpretation of the SE model.[25] Oxides such as quartz and amorphous silica, which have smaller values of the average static dielectric constant, are interpreted as more polarizable, more covalent, chemically "softer,"[51] and less hydrophilic compared to oxides that have larger values of the dielectric constant. When a "harder"[51] hydrated solute such as Na^+ approaches the SiO_2 surface, it would have the more unfavorable interaction, with a more positive $\Delta G^0_{solv,i}$ value. Conversely, the softer TMA^+ ion would have the less unfavorable interaction with a smaller positive $\Delta G^0_{solv,i}$ value. The opposite would be true for $\Delta G^0_{solv,i}$ and total $\Delta G^0_{ads,i}$ at the surfaces of Al_2O_3, Fe_2O_3, or TiO_2, because they are less polarizable, "harder," more hydrophilic solids. Quartz has a static dielectric constant of value intermediate between amorphous SiO_2 and the other oxides, so it should be chemically harder than amorphous silica but less so than the other oxides. This interpretation is consistent with the conclusion that cristobalite and quartz are more hydrophilic than amorphous silica, based on water vapor, NH_3, pyridine, and alcohol adsorption experiments on silica polymorph surfaces.[52–57]

6.3.4 Model Prediction for Behavior of Stishovite

The preceding discussion allows us to predict how an oxide such as stishovite would behave. Stishovite has a smaller value of s/r than quartz, so the former should have a positively charged surface at low pH, like Al_2O_3 or TiO_2. The average static dielectric constant of stishovite can be estimated to equal ~ 7 to 11.[25] This can be compared to the dielectric constant values of ~ 3 to 4 for amorphous silica and quartz, and ~ 10 and 121, respectively, for corundum and rutile. The value of $1/\varepsilon_k$ and the corresponding SE model contribution of $\Delta G^0_{solv,i}$ to total $\Delta G^0_{ads,i}$, then, decreases as amorphous SiO_2 > quartz > corundum > stishovite > anatase > hematite > rutile. Combining the surface charge and surface solvation behavior, the membranolytic activity of stishovite should resemble Al_2O_3 or TiO_2 rather than the other SiO_2 polymorphs. With the exception of amorphous SiO_2, the predicted and observed hemolytic sequence is consistent with increasing value of the average static dielectric constant. Since ε_k depends on both polarizability and molar volume, the model is also consistent with the observation that hemolytic ability is inversely related to molar volume and density of the oxide.[58,59]

6.3.5 Relevance to Cell Attachment

Thus far, the interactions of phospholipid head groups have been considered, because the model was applied toward rationalizing the membranolysis of eukaryotic cells such as erythrocytes, and PLs are the primary constituents of eukaryotic cell membranes. A reasonable question to ask at this time is whether the above results are relevant to prokaryotic membranes. Although PLs constitute a smaller proportion of the prokaryotic cell wall and cell membrane, the other constituent molecules such as liopolysaccharides and teichoic acids, are also amphiphilic. The general structure of a hydrophilic portion attached to a hydrophobic tail is common to most of these amphiphilic molecules. The fundamental concepts of the model discussed above, and the thesis that early changes in membrane structure ultimately affect membranolysis should, therefore, be broadly relevant to both prokaryotic and eukaryotic cell membranes.

Extending this idea one step further, bacteria may have evolved to produce extracellular polymeric substances (EPSs) in order to make mineral surfaces more favorable for attachment. This would be an important evolutionary step, especially if the earliest bacteria utilized minerals for respiration and nutrition.[25] According to the present model, oxides other than quartz also have unfavorable entropic interactions with the head group PL ($\Delta S^0_{ads,i} < 0$). EPSs should then be exuded on the surfaces of many oxide (and silicate) minerals. As discussed above, quartz is the most harmful, so greater production of EPSs should be expected on quartz, all other factors being equal. Consistent with this hypothesis, the nature of the substrate and of the bacterial surfaces does, in fact, affect the amount of EPS produced.[60–62] The idea that surfaces become more hydrophilic by bacterial attachment also underlies the biobeneficiation of ores during mineral separation by floatation.

An additional piece of evidence in support of these ideas is the observation that quartz is not known to be precipitated as a biomineral by any organism. The only reported case of quartz found within an organism was interpreted to be a case of a whole quartz particle being entrained and retained by a cactus plant.[63] Biogenic amorphous silica, on the other hand, is produced voluminously even though quartz is thermodynamically the more stable phase. Quartz precipitation at environmental temperatures is, of course, kinetically inhibited. But if this were the only reason, organisms could have evolved enzymes to accelerate quartz precipitation. The deleterious effect of quartz on lipid membranes may explain the lack of quartz biomineralization.[25] In contrast, the favorable entropic effect at the amorphous silica surface indicates that cells can attach without significant damage. This may also explain why prosthetic implants made of silica bioceramics are biocompatible.

In summary, an early step in cell attachment to oxide surfaces involves electrostatic attraction of the polar head group of amphiphilic compounds constituting the membrane. This early stage can be modeled using classical electrostatic and solvation theory along with crystal chemistry to explain the specific influence of the oxide on the process. It is reiterated that amphiphile attachment depends on electrostatics only in the early stages, and involves many complicated hydrophobic interactions for full attachment to occur. It is, nevertheless, an intriguing thesis that entropically determined disruption of membrane structure inherited from this stage can have a significant influence on the entire process.

6.4 ROLE OF DISSOLVED SI-ORGANIC COMPLEXES IN BIOLOGY

An important part of the global biogeochemical silicon cycle involves biogenic silicon removal from ocean water by eukaryotic microorganisms such as diatoms, sponges, and radiolaria. It has been suggested in the literature that organic-silicon molecules may be involved in the uptake of silicon by these microorganisms and in the nucleation of biogenic silica. Si-organic interactions could, thus, have a tremendous impact on the global biogeochemical cycle of Si.

Furthermore, diatoms and radiolaria exert strict species-specific control over the patterns and pore sizes of the silica, which is precipitated at ambient conditions. Controlled industrial synthesis of silica, on the other hand, requires extreme pHs and/or temperatures. Biomimetic silica synthesis has, therefore, attracted much interest among materials scientists and phycologists for potential industrial applications, such as components of industrial separators, catalysts, and electronic devices.[64–67]

Organic-silicon starting compounds have been proposed in biogenic silica precipitation, because such compounds need first to be hydrolyzed and then polymerized. This process slows down the precipitation process, allowing time for greater structural control. Many studies have been aimed at identifying which, if any, organic molecules may be involved in biological silicon uptake and in biogenic silica precipitation. Results show that the molecules are mainly polysaccharides, and glycoproteins enriched in glycine, histidine, and lysine, and in OH- or COOH-terminated amino acids, such as serine, threonine, aspartic acid, and glutamic acid.[68–71] Silicon-organic complexes have been proposed in the literature, with H-bonds between the silanol and the –OH or –COOH groups. Complexes with direct C-O-Si covalent bonds where Si is quadra-, penta-, and hexa-coordinated have also been suggested.[68,69,72–78] In contrast to these suggestions, based on the ^{29}Si NMR shifts measured for natural biogenic silica samples, one research group had concluded previously that there are no direct C-O-Si bonds in solid biogenic silica.[79] Thus, there is considerable ambiguity in the type of silicon-organic complexes, with respect to the different types of bonds, the coordination numbers for Si, and whether such complexes exist at all in natural systems. Other unresolved questions in silica biomineralization have been reviewed in detail elsewhere.[80]

6.4.1 Approach to Studying Putative Si-Organic Complexes: Rationale for Using MO Calculations

A new approach was initiated recently to test some of the existing ideas in the literature on biomineralization.[81–83] *Ab initio* Hartree–Fock (HF) molecular orbital calculations were used to determine the stable geometry, formation energy, and ^{29}Si NMR chemical shifts of the putative silicon-organic complexes. The results can be used in at least two ways. First, the theoretically calculated NMR shifts and formation energies can help to eliminate putative complexes, which have been proposed in the literature in order to explain experimentally observed ^{29}Si NMR shifts. The calculations can also be used to predict the shifts of new Si-organic structures previously not considered in the literature. If the newly proposed complexes exist in natural systems, they should be identifiable experimentally based on their ^{29}Si NMR shifts. Thus, the ^{29}Si NMR shifts serve as diagnostic fingerprints. The ultimate goal is to determine whether organic-silicic acid complexes are likely for silicon uptake.

Only one group has reported another computational approach, based on the semi-empirical AM1 and PM3 methods, to understanding organic-silicon complexation from a biomineralization perspective.[75,84,85] Compared to the semi-empirical approach, the *ab initio* HF method is completely theoretical without the need for parameterization on experimental data, but the trade-off is in the size of molecular clusters that can be treated.

6.4.2 Computational Details

The computational method is discussed in detail elsewhere.[82,83] Briefly, energies and geometries were obtained at the HF/6–31G* level using the programs GAMESS, Gaussian 94 and Gaussian 98.[86–90] Corrections for basis set superposition error[91] and thermal effects were made based on calculations for a smaller model consisting of CH_3OH and $Si(OH)_4$ and their H-bonded and covalently bonded complexes.[82] The self-consistent reaction field model was used for solvation, where the solvent is treated as a dielectric continuum with a spherical cavity for the solute.[89] The radius (a_0) of each molecule was determined for the optimized gas-phase geometry by selecting the Volume option in Gaussian. These radii were then used for the size of the cavity in the dielectric continuum. The dielectric constant was set equal to that of water, 78.5. A value of –76.0265 Hartree was used for the energy for liquid water, which is calculated as the gas-phase value for a single H_2O molecule plus the experimental heat of vaporization equal to –0.0158 Hartree.[92] The theoretically calculated value for the hydration energy of H^+ is –0.4259 Hartree.[93] Because of these approximations and the neglect of electron correlation in the HF formalism, relative reaction energies are more reliable than absolute values.

^{29}Si NMR chemical shifts were calculated for each molecule relative to the theoretical shielding for tetramethylsilane (TMS), at the HF/6–311+G(2d,p)[86] level using the GIAO method,[94] as implemented in Gaussian 94 and Gaussian 98. Shifts for gas-phase molecules are reported because the inclusion of solvation via the SCRF method was found to have little effect on the predicted shifts.[83] Comparison of calculated shifts with experimental values for compounds with well-known structures yielded an error estimate of about 1 to 8% for quadra-coordinated silicon and 2 to 9% for penta-coordinated silicon.

All Si-organic molecules shown were drawn using the program MacMolplt.[95] Quadra-, penta- and hexa-coordinated Si are called ^{Q}Si, ^{P}Si, and ^{H}Si below. serOH refers to the serine molecule consisting of the amino acid backbone plus the OH side chain.

6.4.3 Computational Results and Interpretation

6.4.3.1 Si-Polyalcohol Complexes

Si complexes with polyalcohols were examined from two up to five carbon atoms, that is, ethylene glycol ($C_2H_4O_2$), glycerol ($C_3H_6O_6$), threitol ($C_4H_8O_4$); and arabitol ($C_5H_{10}O_5$), a six-carbon polyalcohol ($C_6H_{12}O_6$) and the corresponding sugar-acid, gluconic acid ($C_6H_{12}O_7$).[83] Spirocyclic complexes are those where the ligand:Si ratio is >1, and the ratio is equal to one for monocyclic complexes.

The calculations are summarized in Tables 6.1 and 6.2, and structures for selected molecules are shown in Figures 6.3 and 6.4. The salient features are discussed in detail elsewhere.[83] One important result for the ^{Q}Si complexes is that as ring size increases (i.e., size of the polyalcohol increases), the shifts initially become more

TABLE 6.1
^{29}Si NMR Isotropic Shifts (δ) and Anisotropic Shifts ($|\sigma_{33}-\sigma_{11}|$) Reported in ppm with Respect to TMS Standard, for Silicon–Polyalcohol Spirocyclic Complexes

Model Complex	Ligand: Si Ratio	δ, Experimental	Calculated δ, HF/6–311+G (2d,p)	Calculated δ, B3LYP/6–311 +G(2d,p)	Calculated $\|J_{33}-J_{11}\|$,[a] HF/6–311+ G(2d,p)
		Effect of Ring Size, ^{Q}Si			
$(C_2H_4O_2)_2{}^QSi$	2:1	–43.6[b]	–44.0	–42.2	46.9
$(C_3H_6O_3)_2{}^QSi$	2:1	–81.7[b,c,d]	–83.4	–83.4	31.7
$(C_4H_8O_4)_2{}^QSi$ (Figure 6.3a)	2:1	–81.6[b,d]	–75.0	–75.0	8.8
		Effect of Ring Size and Hypercoordination, ^{P}Si			
$[(C_2H_4O_2)_2{}^PSiOH]^-$ (Figure 6.3b)	2:1	–102.7[d]	–108.9	–100.3	102.7
$[(C_3H_6O_3)_2{}^PSiOH]^-$	2:1	–130.4[d]	–140.3	–144.1	133.9
$[(C_4H_8O_4)_2{}^PSiOH]^-$ (Figure 6.3(c))	2:1	–130.8, –131.1[d]	–133.3	–133.6	127.9
		Effect of Decreasing Ligand:^{P}Si Ratio at Fixed Ring Size			
$[(C_2H_4O_2)_3{}^PSiH]^-$ (Figure 6.3(d))	3:1	–103, –104, –105[e]	–110.2		81.7
$[(C_2H_4O_2)_5{}^PSi_2]^{2-}$ (Figure 6.3e)	5:2	–103, –104, –105[e]	–114.4, –114.4		76.8, 90.9
		Effect of Protonating an O_{br}[f]			
$[(C_2H_4O_2)_3{}^PSiH_2]$ bidentate → $[(C_2H_4O_2)_3{}^QSiH_2]$ monodentate	3:1		–64.7		33.2
		Standard			
$^QSi(CH_3)_4$ (TMS)			0.0 (σ = 386.0)	0.0 (σ = 327.9)	0.0

[a] Anisotropic shifts reported are the absolute value of the difference between the maximum and minimum eigenvalves of the shielding tensor.
[b] See Kemmitt and Milestone, 1995.
[c] Assigned to polymeric Si spirocyclic complex in Kemmitt and Milestone, 1995.
[d] Values for the analogous *dialato* complexes.
[e] See Herreros, Carr, and Klinowski, 1994.
[f] O_{br} represents a bridging oxygen.

Source: Reprinted with permission from Sahai, N. and Tossell, J.A., *Inorg. Chem.*, 41, 748, copyright 2002. American Chemical Society.

TABLE 6.2
^{29}Si HF/6–311+G(2d,p) NMR Isotropic Shifts (δ) and Anisotropic Shifts ($|\sigma_{33}-\sigma_{11}|$) Reported with Respect to TMS Standard, for Silicon–Polyalcohol Monocyclic Complexes

Model Complex	d (ppm)	$\|\sigma_{33}-\sigma_{11}\|$ (ppm)[a]
	Effect of Ring Size, ^{Q}Si	
$C_2H_4O_2{}^{Q}Si(OH)_2$	–57.2	24.5
$C_3H_6O_3{}^{Q}Si(OH)_2$	–77.0	42.5
$C_4H_8O_4{}^{Q}Si(OH)_2$	–75.2	35.6
$C_4H_{10}O_4...{}^{Q}Si(OH)_4$	–71.2	22.0
	Effect of Ring Size and Hypercoordination	
$[C_2H_4O_2{}^{P}Si(OH)_3]^-$	–116.9	130.6
$[C_3H_6O_3{}^{P}Si(OH)_3]^-$	–130.4	139.0
$[C_4H_8O_4{}^{P}Si(OH)_3]^-$ *I* Si bonded to O on C_1, C_4 (Figure 6.4a)	–128.3[b]	140.8
$[C_4H_8O_4{}^{P}Si(OH)_3]^-$ *II* Si bonded to O on C_1, C_2 (Figure 6.4b)	–117.3	124.4
$[C_5H_{10}O_5{}^{P}Si(OH)_3]^-$	–128.0	137.5
$[C_4H_8O_4{}^{H}Si(OH)_4]^{2-}$ hexacoordinate Si	–179.9[c]	8.4
	Effect of Protonating an O_{br}	
$C_2H_4O(OH){}^{P}Si(OH)_3$ bidentate → $C_2H_4O(OH){}^{Q}Si(OH)_3$ monodentate	–75.0	14.3
$[C_4H_8O_3(OH){}^{P}Si(OH)_3]$ *I* bidentate → $[C_4H_8O_3(OH){}^{Q}Si(OH)_3]$ *I* monodentate (Figure 6.4c)	–75.2	21.7
$[C_4H_8O_3(OH){}^{P}Si(OH)_3]$ *II* bidentate → $[C_4H_8O_3(OH){}^{Q}Si(OH)_3]$ *II* monodentate (Figure 6.4d)	–73.8	17.3
	Effect of Chelate Denticity	
$[C_4H_7O_4{}^{P}Si(OH)_2]^-$ Si bonded to O on C_1, C_2, C_4 (tridentate ligand)	–122.4	101.7
	Effect of Polymerization, ^{Q}Si and ^{P}Si	
$(C_3H_6O_2)_3{}^{Q}Si_3O_3$ (Figure 6.4e)	–96.9	32.8
$[(C_3H_6O_2)_3{}^{Q}Si_2O_2{}^{P}SiO(OH)]^-$ (Figure 6.4f)	–87.5, –90.3, –143.1 Mean = –107.0	61.1, 97.2, 42.6
	Standard	
$^{Q}Si(CH_3)_4$ (TMS)	0.0 (σ = 386.0)	0.0

[a] Anisotropic shifts reported refer to the absolute value of the difference between the maximum and the minimum eigenvalues of the shielding tensor.

[b] This structure was proposed to explain a ^{29}Si resonance seen at –102 to –103 ppm in an experimental study of aqueous Si-polyalcohol solutions. (See Kinrade, S.D. et al., *Science*, 285, 1542, 1999).

[c] This structure was proposed to explain a ^{29}Si resonance seen at –145 to –147 ppm. (See Kinrade, S.D. et al., *Science*, 285, 1542, 1999).

Source: Reprinted with permission from Sahai, N. and Tossell, J.A., *Inorg. Chem.*, 41, 748, copyright 2002. American Chemical Society.

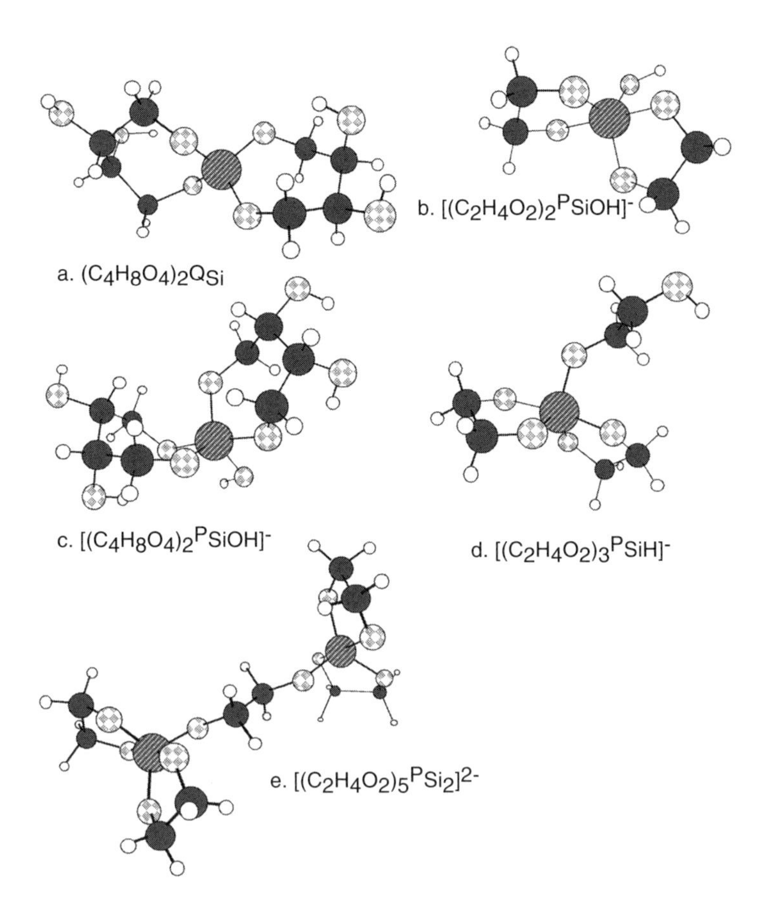

FIGURE 6.3 Spirocyclic Si-polyalcohol complexes with Si in quadra- and penta-coordination. Corresponding ^{29}Si NMR shifts are reported in Table 6.1. Key to circles representing atoms: solid white, H; solid grey, C; square hatch, O; diagonal hatch, Si. (Modified and reprinted with permission from Sahai, N. and Tossell, J.A., *Inorg. Chem.*, 41, 748, copyright 2002. American Chemical Society.)

negative but then the effect seems to level off. The ^PSi complexes have large, negative calculated shifts from −116 to −140 ppm depending on polyalcohol size. For the ethylene glycol spirocyclic compound, the shift is much smaller than the acyclic compound but the values are comparable for the glycerol and threitol analogs. The monocyclic complex containing hexa-coordinated Si and threitol has a large negative shift of −180 ppm. The monocyclic complexes of ^PSi with a six-carbon polyalcohol and with gluconic acid produced calculated shifts of −128 to −130 ppm, and a ^QSi-gluconic acid complex yielded a value of −75 ppm (Sahai, unpublished results).

The ^{29}Si NMR spectra recorded experimentally in threitol solutions showed a peak at −102 to −103 ppm, which was assigned to a ^PSi-threitol complex. A peak at ~ −145 to −147 ppm found in threitol and in gluconic acid solutions peaks was assigned to mono- and spirocyclic ^HSi-ligand complexes.[76,77] Comparison of our calculated results for the structures proposed in the literature with the actually measured values allows

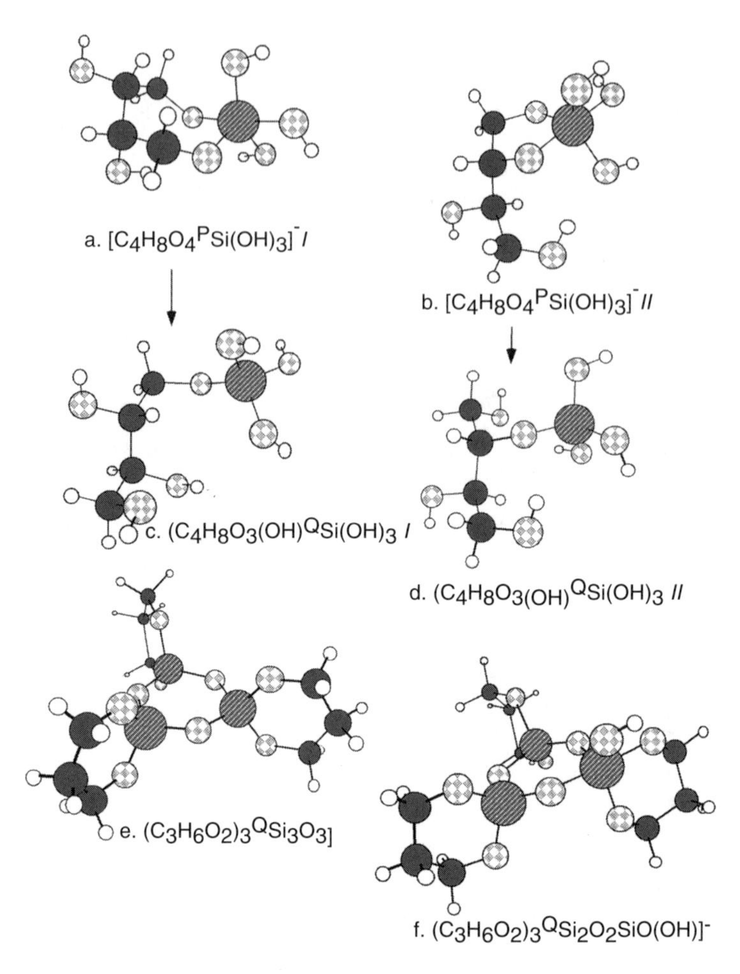

FIGURE 6.4 Monocyclic Si-polyalcohol complexes with Si in quadra- and penta-coordination (a) to (d); polymeric ring-complexes (e), (f). Corresponding ^{29}Si NMR shifts are reported in Table 6.2. Key to circles representing atoms: solid white, H; solid grey, C; square hatch, O; diagonal hatch, Si. (Modified and reprinted with permission from Sahai, N. and Tossell, J.A., *Inorg. Chem.*, 41, 748, copyright 2002. American Chemical Society.)

us to draw the following conclusions. First, the ^{Q}Si monocyclic complexes containing ligands larger than ethylene glycol have shifts in the –71 to –77 ppm range, which are close to the –70 to –80 ppm range seen for inorganic silicic acid (H_4SiO_4) and for the dimer of silicic acid. Thus, ^{29}Si NMR isotropic chemical shifts are not necessarily diagnostic of the presence of ^{Q}Si-organic complexes. Further, none of the structures examined theoretically yielded shifts in the –102 ppm range for ligands larger than ethylene glycol. All of these structures examined thus far contain Si-organic seven-membered rings. In order to explain the observed –102 ppm peak, we examined some alternative structures. The complex $[C_4H_8O_4{}^{P}Si(OH)_3]^-$ *II* (Figure 6.4b) is similar to the structure $[C_4H_8O_4{}^{P}Si(OH)_3]^-$ *I* (Figure 6.4a) originally proposed in the literature, but differs from *I* in that Si is now bonded to the oxygen of the C_1 and C_2 carbons.[83]

Structure II is a five-membered ring and I is a seven-membered ring. The calculated shift for *I* is −128 ppm, quite different from the observation at −102 ppm, whereas the calculated shift for *II* is −117.3 ppm. The latter value is closer to the observed resonance, but even so, it is beyond the range of computational error.

The next two alternative structures are polymeric silicon-threitol complexes in the cyclic trimer form. The Si-organic structure forms six-membered rings. In both these complexes, each silicon atom is bonded through a bridging oxygen (O_{br}) to two other silicon atoms and to two tetrahedral carbon atoms. In the first case, $(C_3H_6O_2)_3{}^QSi_3O_3$ (Figure 6.4e), all three silicon atoms are quadra-coordinated and the shift is −96.9 ppm, which, within computational error, is very close to the observed −102 ppm peak. For the next complex, $[(C_3H_6O_2)_3{}^QSi_2O_2{}^PSiO(OH)]^{1-}$ (Figure 6.4e), one Si is penta-coordinated and the remaining two are ^QSi (Figure 6.4f). The calculated resonances are −143.1 ppm, −87.5 ppm, and −90.3 ppm, respectively. If rapid exchange occurs between these silicon sites, then the measured value would reflect the arithmetic mean equal to −107 ppm. If computational error is considered, this value is also very close to the −102 ppm peak seen in the experiment. Among the structures considered thus far, the cyclic trimers are the most likely candidates for explaining the peaks seen at −102 ppm.

Finally, the lack of a peak at −180 to −190 ppm suggests that ^HSi was not observed at all in the experiment. The calculations show that the −145 ppm peak cannot be due to seven-membered ring ^HSi complexes. Most likely, this peak is due to some ^PSi complexes, which is consistent with the high pHs at which the experimental study was conducted, and even this value is slightly beyond the range of calculated values including computational error.

The conclusions above, based on NMR shifts, are supported by the calculated formation energies for the covalently bonded ^QSi complexes, H-bonded ^QSi complexes and covalently bonded ^PSi complexes.[83] The relative stability of the Si-threitol complexes decreases as C-O-^QSi > COH...HO^QSi >> C-O-^PSi. Formation of the covalent five-fold coordinated complex, $[C_4H_8O_4{}^PSi(OH)_3]^{1-}$, is highly endothermic when H_4SiO_4 is the reactant. The formation reaction only becomes more favorable when the reactant is changed to $H_3SiO_4^-$, which would occur at high pHs.

6.4.3.2 Si-Amino Acid Complexes

An enzyme called silicatein, containing serine and histidine residues at the active site, has been isolated recently from the organic matter occluded within the silica of a sponge spicule. At neutral pH and room temperature, the enzyme promotes *in vitro* hydrolysis and polymerization of tetreethoxyorthosiliate or TEOS, $Si(OC_2H_5)_4$. Based on homology with a known hydrolytic enzyme, cathepsin, Shimizu et al.,[70] Zhou et al.,[98] and Cha et al.[99] proposed a base-catalyzed nucleophilic mechanism, where histidine acts as a nucleophilic base towards Si in the starting silicon compound. The mechanism assumes that the organism utilizes an organic-Si compound, TEOS, as the starting silicon substrate. In this mechanism, a penta-coordinated Si intermediate is formed, [ser-O-$^PSi(OC_2H_5)_3$Nhis]$^-$, with four Si-OC bonds and one Si-N bond. The Si-OC bonds would be longer in the ^PSi complex than in a normal ^QSi complex, and would be more susceptible to hydrolysis than in a ^QSi complex.

Complexes of Si with serine and with histidine have been studied.[82] In the first case, silicic acid (H_4SiO_4) is combined with serine to form H-bonded(Figure 6.5a) and covalently bonded complexes where Si is quadra- and penta-coordinated (Figures 6.5b and c). In the second case, the starting silicon compound is already an organic compound, tetramethoxysilicate, $Si(OCH_3)_4$. $Si(OCH_3)_4$ was used as a model for the starting silicon compound that has been implicitly suggested as the form in which sponges utilize silicon.[98,99]

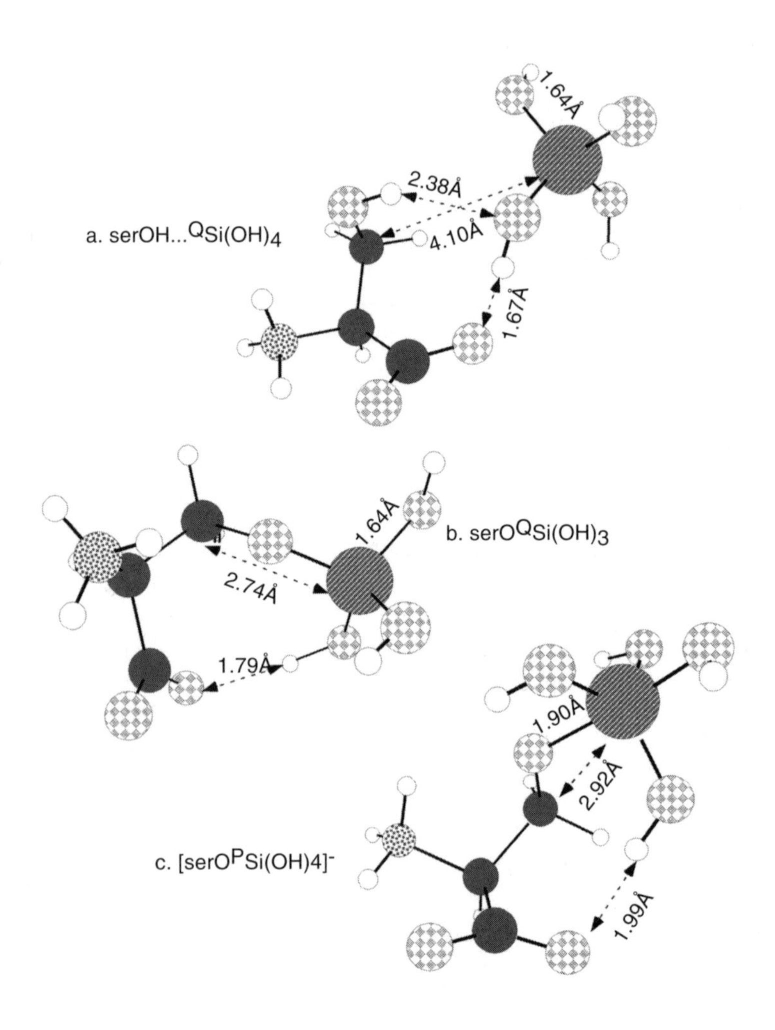

FIGURE 6.5 Si-serine complexes containing H-bonds and direct Si-O-C bonds, with Si in quadra- and penta-coordination. Key to circles representing atoms: solid white, H; solid grey, C; stipple, N; square hatch, O; diagonal hatch, Si. (Modified and reprinted from Sahai, N. and Tossell, J.A., *Geochim. Cosmochim. Acta*, 65, 2043, copyright 2001. With permission from Elsevier Science.) (*continued*)

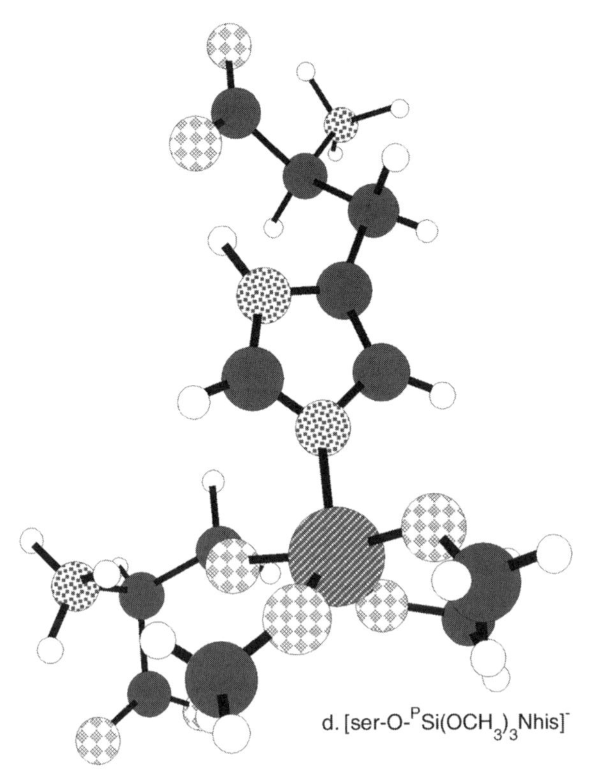

FIGURE 6.5 (*continued*) Si-serine complexes containing H-bonds and direct Si-O-C bonds, with Si in quadra- and penta-coordination. Key to circles representing atoms: solid white, H; solid grey, C; stipple, N; square hatch, O; diagonal hatch, Si. (Modified and reprinted from Sahai, N. and Tossell, J.A., *Geochim. Cosmochim. Acta*, 65, 2043, copyright 2001. With permission from Elsevier Science.)

For the silicic acid-serine system, our results show that if H-bonded and/or covalently bonded complexes containing ^{Q}Si did exist in biogenic silicification, they would not be detected by ^{29}Si NMR because their predicted isotropic shifts are similar to inorganic ^{Q}Si. A similar result was obtained for the polyalcohol complexes as discussed above. The penta-coordinated complex, $[serO^{P}Si(OH)_4]^-$, would be detectable because of large ^{29}Si isotropic (−121 to −140.1 ppm) and anisotropic shifts (110 to 142 ppm), and diagnostically long ^{P}Si-O bond lengths (1.7 to 1.9 Å).[13] C shifts were found to be insensitive to the type of bonding.[17] O shifts were the most sensitive to bonding, as might be expected, since it is the most directly involved nucleus in the C-O-Si bonding. The formation reaction of the covalent quadra-coordinated complex, $serO^{Q}Si(OH)_3$ is 10 kcalmol^{-1} more exothermic than the H-bonded complex, $serOH...^{Q}Si(OH)_4$. The penta-coordinated complex, $[serO^{P}Si(OH)_4]^-$, is thermodynamically unfavorable in the acidic pH of the silica deposition vesicle in diatoms when compared with $Si(OCH_3)_4$ as a starting compound. Silicic acid is energetically favored over silicon-organic complexes as the form of dissolved silicon taken up by the organism at the basic pH of seawater.

Preliminary calculations for the molecule, [ser-O-$^{P}Si(OCH_3)_3$Nhis]$^-$, result in a shift of −132 ppm.[82] A transient resonance at −145 ppm lasting about 6 hours has been reported in diatom cultures that are actively synthesizing silica. This was a very weak peak indeed, but if it is accepted as an identifiable peak, one can say that the predicted value for the N-bearing ^{P}Si complex is within computational error of the observed value.

6.4.4 Relevance to Biological Silicon Uptake

Seawater has a mildly basic pH of 8.5, and the pK_a of silicic acid is 9.5. H_4SiO_4 is, therefore, the dominant form of silicon expected in seawater and $H_3SiO_4^-$ would only become dominant at higher pHs. The calculated relative reaction energies show that the hypercoordinated penta-coordinated complex is favored only when $H_3SiO_4^-$ is the reactant. These observations suggest that the concentration of hypercoordinated complexes formed in reaction with seawater would be very low. Diatoms deposit silica within a vesicle called the silica deposition vesicle. It was recently determined that within the SDV, the solution has a pH ~5 to 6.[100] Significant concentrations of hypercoordinated silicon-serine and silicon-polyalcohol complexes are even less likely in this acidic environment. Thus, it is improbable that such complexes are involved in the transport and utilization of dissolved silicon by marine organisms.

It is, however, entirely possible that hypercoordinated Si species, organic and inorganic, are involved as transient species in the silica polymerization process. Such species are known to exist briefly, from a few hours up to 24 h, in the mother liquor and in the solid phase during sol-gel synthesis of silica at extreme pHs and temperatures around 150°C.[97,101] The formation of these species at ambient pHs and temperatures may be catalyzed enzymatically by the organism during silica polymerization. A ^{29}Si resonance at −131.5 ppm has been reported in a diatom, and was assigned to a hexa-coordinated Si-organic complex.[78] That signal, however, was exceedingly weak and does not correspond to the theoretically calculated ^{29}Si shifts of −180 ppm for ^{H}Si.

The weak yet thermodynamically feasible complexation of four-fold coordinated silicon with serine and polyalcohols at near-neutral pHs is consistent with experimental and theoretical results for other organics such as carboxylic acids, phenol, catecholamines, glycine, and serine.[102–104] The organism may utilize four-fold coordinated SiO-organic complexes as intracellular storage pools. Silicic acid, the dominant form of dissolved silicon in most natural waters, is the most likely form of silicon for biological silicon transport from seawater into cell. These results are consistent with recent studies showing that diatoms take up H_4SiO_4, rather than $H_3SiO_4^-$.[105]

6.5 SUMMARY

Silicon-organic interactions, where silicon is present as solid silica and as a dissolved species in aqueous solution, are examined using classical electrostatic and solvation theory, crystal chemistry, thermodynamics, and *ab initio* molecular orbital calculations. Specifically, I address why various silica polymorphs and other oxides interact differently with cell membrane phospholipids, and examine the likelihood of the existence of hypercoordinated silicon-organic, ligand-silicon complexes in ambient aqueous solutions. The results are relevant to cell adhesion on mineral substrates and to biological silicon uptake in the environment.

Cell-membrane PLs are amphiphilic, with hydrophobic tails, and a hydrophilic, polar head group often containing quaternary ammonium moieties, such as TMA^+. The head group adsorbs initially by electrostatic forces on charged oxide surfaces, where the quaternary ammonium must displace Na^+, the dominant cation in cell fluids and in seawater. The quartz surface has a favorable free energy for the exchange of TMA^+ with Na^+ ($\Delta G_{exc,TMA^+/Na^+} = \Delta G_{ads,TMA^+} - \Delta G_{ads,Na^+} < 0$), and an unfavorable entropy of adsorption of TMA^+ ($\Delta S_{ads,TMA^+} < 0$). As a result, membrane structure is disrupted when the PL adsorbs. For amorphous silica, membrane structure is less disrupted because $\Delta S_{ads,TMA^+} > 0$. Other oxides such as Al_2O_3, Fe_2O_3, TiO_2, and stishovite (the octahedral silica polymorph) are characterized by $\Delta S_{ads,TMA^+} < 0$, but now, $\Delta G_{exc,TMA^+/Na^+} > 0$. Limited PL adsorption occurs, so these oxides are more benign. Thus, the model explains the observed sequence for decreasing ability to rupture cell membranes, quartz > amorphous silica > Al_2O_3 > Fe_2O_3 > TiO_2, and the sequence for increasing strength of bacterial adhesion, quartz < amorphous silica < Al_2O_3 < Fe_2O_3 < TiO_2.

Organisms that produce exoskeletons of amorphous silica transport dissolved silicon across the cell membrane and process the silicon intracellularly. The chemical form in which silicon is utilized has been examined using *ab initio* HF molecular orbital theory by calculating relative formation energies and diagnostic ^{29}Si NMR shifts of putative Si-organic compounds proposed in the literature. The organic compounds considered are polyalcohols, polysaccharides (sugar acids), and the amino acids, serine and histidine. Results indicate that covalently bonded hypercoordinated Si-organic aliphatic compounds are thermodynamically unfavorable at ambient conditions. Four-fold coordinated Si-organic complexes may be involved in intracellular silicon storage and transport but silicic acid is the most likely form of silicon transport from seawater into the cell. Hypercoordinated Si species may occur, however, as transient species during the polymerization of silica.

REFERENCES

1. Nash, T., Allison, A.C., and Harington, J.S., Physico-chemical properties of silica in relation to its toxicity, *Nature*, 210, 259, 1966.
2. Nolan, R.P et al., Quartz hemolysis as related to its surface functionalities, *Environ. Res.*, 26, 503, 1981.
3. Fubini, B., Health effects of silica, in *The Surface Properties of Silicas*, Legrand, A.P., Ed., John Wiley & Sons, Chichester, 1998, p. 415.
4. Harley, J.D. and Margolis, J., Haemolytic activity of colloidal silica, *Nature*, 189, 1010, 1961.
5. Stalder, K. and Stöber, W., Haemolytic activity of suspensions of different silica modifications and inert dusts, *Nature*, 207, 874, 1965.
6. Parazzi, E., et al., Studies on the cytotoxic action of silica dusts on macrophages *in vitro*, *Arch. Environ. Health*, 17, 850, 1968.

7. Weissman, G. and Rita, G.A., Molecular basis of gouty inflammation: Interaction of monosodium urate crystals with lysosomes and liposomes, *Nature New Biol.*, 240, 167, 1972.
8. King, E. J. et al., The action of different forms of pure silica on the lungs of rats, *Br. J. Ind. Med.*, 10, 9, 1953.
9. King, E. J. et al., Effect of modifications on the surface of quartz on its fibrogenic properties in the lungs of rats, *A. M. A. Arch. Ind. Hyg.*, 7, 455, Research Report No. 80, Sheffield, U.K., 1953.
10. Zaidi, S.H. et al., Fibrogenic activity of different forms of free silica, quartz, cristobalite and tridymite on the livers of mice, *A. M. A. Arch. Ind. Hyg.*, 13, 112, 1956.
11. Brieger, H. and Gross, P., On the theory of silicosis. I. Coesite, *Arch. Environ. Health*, 13, 38, 1966.
12. Brieger, H. and Gross, P., On the theory of silicosis. II. Quartz Glass, *Arch. Environ. Health*, 14, 299, 1967.
13. Brieger, H. and Gross, P., On the theory of silicosis. III. Stishovite, *Arch. Environ. Health*, 15, 751, 1967
14. Stöber, W. and Brieger, H., On the theory of silicosis. IV. The topochemical interactions, *Arch. Environ. Health*, 16, 706, 1968
15. Mills, A.L. et al., Effect of solution ionic strength and iron coatings on mineral grains on the sorption of bacterial cells to quartz sand, *Appl. Environ. Microbiol.*, 60, 3300, 1994.
16. Deo, N. and Natarajan, K.A., Surface modification and biobeneficiation of some oxide minerals using Bacillus polymyxa, *Miner. Metall. Process.*, 14, 32, 1997.
17. Deo, N., Natarajan, K.A., and Somasundaran, P., Mechanisms of adhesion of *Paenibacillus polymyxa* onto hematite, corundum and quartz, *Int. J. Miner. Process.*, 62, 27, 2001.
18. Yee, N., Fein, J., and Daughney, C.J., Experimental study of the pH, ionic strength, and reversibility behavior of bacteria-mineral adsorption, *Geochim. Cosmochim. Acta*, 64, 609, 2000.
19. Wicken, A.J. and Knox, K.W., Bacterial cell surface amphiphiles, *Biochim. Biophys. Acta*, 604, 1, 1980.
20. Beveridge, T.J., Ultrastructure, chemistry and function of the bacterial wall, *Int. Rev. Cytol.*, 72, 229, 1981.
21. Lodish, H. et al., *Molecular Cell Biology*, 3rd ed., Scientific American Books, New York, 1995, p. 596.
22. Somasundaran, P. and Furstenau, D.W., Mechanisms of alkyl sulfonate adsorption at the alumina-water interface, *J. Phys. Chem.*, 70, 91, 1966.
23. Chandar, P., Somasundaran, P., and Turro, N.J., Fluorescence probe studies on the structure of the adsorbed layer of dodecyl sulfate at the alumina-water interface, *J. Colloid Interface Sci.*, 117, 31, 1987.
24. Rapuano, R. and Carmona-Ribeiro, A.M., Physical adsorption of bilayer membranes on silica, *J. Colloid Interface Sci.*, 193, 104, 1997.
25. Sahai, N., Biomembrane phospholipid–oxide surface interactions: Crystal chemical and thermodynamic basis, *J. Colloid Interface Sci.*, 252, 309, 2002.
26. Sahai, N. and Sverjensky, D.A., Solvation and electrostatic model for specific electrolyte adsorption, *Geochim. Cosmochim. Acta*, 61, 2827, 1997.
27. Sahai, N. and Sverjensky, D.A., Evaluation of internally-consistent parameters for the triple-layer model by the systematic analysis of oxide surface titration data, *Geochim. Cosmochim. Acta*, 61, 2801, 1997.

28. Sahai, N. and Sverjensky, D.A., GEOSURF: A computer program for modeling adsorption on mineral surfaces from aqueous solution, *Comput. Geosci.*, 24, 853, 1998.
29. Sahai, N., Estimating adsorption enthalpies and affinity sequences of monovalent electrolyte ions on oxide surfaces in aqueous solution, *Geochim. Cosmochim. Acta*, 64, 3629, 2000.
30. Sahai, N., Is silica really an anomalous oxide? Surface acidity and aqueous hydrolysis revisited, *Environ. Sci. Technol.*, 36, 445, 2002.
31. Lyklema, J., Proteins at solid liquid interfaces-a colloid-chemical review, *Colloids Surfaces*, 10, 33, 1984.
32. van Oss, C.J., Hydrophobicity of biosurfaces-origin, quantitative determination and interaction energies, *Colloids Surfaces B Biointerfaces*, 5, 91, 1995.
33. Hermansson, M., The DLVO theory in microbial adhesion, *Colloids Surfaces B Biointerfaces*, 14, 105, 1999.
34. Pauling, L., *The Nature of the Chemical Bond*, 3rd ed., Cornell University Press, Ithaca, NY, 1960.
35. Sverjensky, D.A., Zero-point-of-charge prediction from crystal chemistry and solvation theory, *Geochim. Cosmochim. Acta*, 58, 3123, 1994.
36. Sverjensky, D.A. and Sahai, N., Theoretical prediction of single-site surface protonation equilibrium constants for oxides and silicates in water, *Geochim. Cosmochim. Acta*, 60, 3773, 1996.
37. Sverjensky, D.A. and Sahai, N., Theoretical prediction of single-site enthalpies of surface protonation for oxides and silicates in water, *Geochim. Cosmochim. Acta*, 62, 3703, 1998.
38. Bolt, G.H., Determination of the charge density of silica sols, *J. Phys. Chem.*, 61, 1166, 1957.
39. Abendroth, R.P., Behavior of a pyrogenic silica in simple electrolytes, *J. Colloid Interface Sci.*, 34, 591, 1970.
40. Yates, D.E., The Structure of the Oxide/Aqueous Electrolyte Interface, Ph.D. thesis, University of Melbourne, 1975.
41. Riese, A.C., Adsorption of Radium and Thorium onto Quartz and Kaolinite: A Comparison of Solution/Surface Equilibria Models, Ph.D. thesis, Colorado School of Mines, 1982.
42. Brady, P.V., Silica surface chemistry at elevated temperatures, *Geochim. Cosmochim. Acta*, 56, 2941, 1992.
43. Casey, W.H., Enthalpy changes for bronsted acid-base reactions on silica, *J. Colloid Interface Sci.*, 163, 407, 1994.
44. Berube, Y.G. and De Bruyn, P.L., Adsorption at the Rutile-solution interface: I. Thermodynamic and experimental study, *J. Colloid Interface Sci.*, 27, 305, 1968.
45. Fokkink, L.G.J., Surface Charge Formation and Cadmium Binding on Rutile and Hematite, Ph.D. thesis, Agricultural University, Wageningen, The Netherlands, 1987.
46. Huang, C. and Stumm, W., Specific adsorption of cations on hydrous gamma-Al_2O_3, *J. Colloid Interface Sci.*, 43, 409, 1973, in *Characterization of Aqueous Colloids by Their Electrical Double-Layer and Intrinsic Surface Chemical Properties*, James, R.O. and Parks, G.A., Eds., vol. 12 of *Surface and Colloid Science*, Matijevic, E., Plenum Press, New York, 1982, p. 119.
47. Breeusma, A., Adsorption of Ions on Hematite (α-Fe_2O_3), Ph.D. thesis, Agricultural University, Wageningen, The Netherlands, 1973.

48. Liang, L., Effects of Surface Chemistry on Kinetics of Coagulation of Submicron Iron Oxide Particles (α-Fe_2O_3) in Water, Ph.D. thesis, California Institute of Technology, 1988.
49. Sprycha, R., Electrical double layer at alumina/electrolyte interface I. Surface charge and zeta potential, *J. Colloid Interface Sci.*, 127, 1, 1989.
50. Hayes, K.F. et al., Surface complexation models: An evaluation of model parameter estimation using FITEQL and oxide mineral titration data, *J. Colloid Interface Sci.*, 142, 448, 1991.
51. Pearson, R.G., Absolute electronegativity and hardness: Applications to inorganic chemistry, *Inorg. Chem.*, 27, 734, 1988.
52. Bolis, V. et al., Hydrophilic and hydrophobic sites on dehydrated crystalline and amorphous silicas, *J. Chem. Soc. Faraday Trans.*, 87, 497, 1991.
53. Bolis, V., Cavenago, A., and Fubin, B., Surface heterogeneity on hydrophilic and hydrophobic silicas: Water and alcohols as probes for H-bonding and dispersion forces, *Langmuir*, 13, 895, 1997.
54. Fubini, B. et al., Structural and induced heterogeneity at the surface of some SiO_2 polymorphs from the enthalpy of adsorption of various molecules, *Langmuir*, 9, 2712, 1993.
55. Fubini B. et al., Relationship between surface properties and cellular responses to crystalline silica: Studies with heat-treated cristobalite, *Chem. Res. Toxicol.*, 12, 737, 1999.
56. Cauvel, A., Brunel, D., and Di Renzo, F., Hydrophobic and hydrophilic behavior of micelle-templated mesoporous silica, *Langmuir*, 13, 2773, 1997.
57. Zoungrana, T. et al., Competitive interactions between water and organic solvents onto mineral solid surfaces studied by calorimetry, *Langmuir*, 11, 1760, 1995.
58. Wiessner, J. H. et al., The effect of crystal structure on mouse lung inflammation and fibrosis, *Am. Rev. Resp. Dis.*, 138, 445, 1990.
59. Mandel, G. and Mandel, N., The structure of crystalline SiO_2, In *Silica and Silica-Induced Lung Diseases*, Castranova, V., Vallyathan, V., and Wallace, W.E., Eds., CRC Press, Boca Raton, FL, 1996, p. 63.
60. Eginton, P.J. et al., The influence of substratum properties on the attachment of bacterial cells, *Colloids Surfaces B Biointerfaces*, 5, 153, 1995.
61. van Loosdrecht, M.C.M. et al., Influence of interfaces on microbial activity, *Microbiol. Rev.*, 54, 75, 1990.
62. van Loosdrecht M.C.M. et al., The role of bacterial-cell wall hydrophobicity in adhesion, *Appl. Environ. Microbiol.*, 53, 1893, 1987.
63. Monje, P.V. and Baran, E.J., First evidences of the bioaccumulation of alpha-quartz in cactaceae, *J. Plant Physiol.*, 157, 457, 2000.
64. Birchall, J.D., The importance of the study of biominerals to materials technology, in *Biomineralization: Chemical and Biochemical Perspectives*, Mann, S., Webb, S., and Williams, R.J.P., Eds., VCH Publishers, Weinheim, Germany, 1989, p. 223.
65. Morse, D.E., Silicon biotechnology: Harnessing biological silica production to construct new materials, *Trends Biotechnol.*, 17, 230, 1999.
66. Tacke, R., Milestones in the biochemistry of silicon: From basic research to biotechnological applications, *Angew. Chem. Int. Ed.*, 38, 3015, 1999.
67. Vrieling, E.G. et al., Diatom silicon biomineralization as an inspirational source of new approaches to silica production, *J. Biotechnol.*, 70, 39, 1999.
68. Hecky, R., et al., The amino-acid and sugar composition of diatom cell-walls, *Mar. Biol.*, 19, 323, 1973.

69. Swift, D. and Wheeler, A.P., Evidence of an organic matrix from diatom silica, *J. Phycol.*, 28, 202, 1992.
70. Shimizu, K. et al., Silicatein α: Cathepsin L-like protein in sponge biosilica, *Proc. Natl. Acad. Sci.*, 95, 6234, 1998.
71. Kroger, N., Deutzmann, R., and Sumper, M., Polycationic peptides from diatom biosilica that direct silica nanosphere formation, *Science*, 286, 1129, 1999.
72. Sullivan, C.W., Silicification by diatoms, in *Silicon Biochemistry*, Ciba Foundation Symposium 121, John Wiley & Sons, Chichester, 1986, p. 59.
73. Pickett-Heaps, J., Schmidt, A.M., and Edgar, L.A., The cell biology of diatom valve formation, in *Prog. Phycol. Res.*, Round, F.E. and Chapman, D.J., Eds., 1990, p. 1.
74. Frausto da Silva, J.J.R. and Williams, R.J.P., *The Biological Chemistry of the Elements: The Inorganic Chemistry of Life*, Clarendon Press, Oxford, 1991.
75. Lobel, K.D. and Hench, L.L., In vitro adsorption and activity of enzymes on reaction layers of bioactive glass substrates, *J. Biomed. Mater. Res.*, 39, 575, 1998.
76. Kinrade, S.D. et al., Stable five- and six-coordinated silicate anions in aqueous solution, *Science*, 285, 1542, 1999.
77. Kinrade, S.D. et al., Aqueous hypervalent silicon complexes with aliphatic sugar acids, *J. Chem. Soc. Dalton Trans.*, 961, 2001.
78. Kinrade S.D. et al., Silicon-29 NMR evidence of a transient hexavalent silicon complex in the diatom *Navicula pelliculosa*, *J. Chem. Soc. Dalton Trans.*, 307, 2002.
79. Perry, C.C. and Mann, S., Aspects of biological silicification, in *Origin, Evolution, and Modern Aspects of Biomineralization in Plants and Animals*, Crick, R.E., Ed., Plenum Press, New York, 1989, p. 419.
80. Gordon, R. and Drum, R.W., The chemical basis of diatom morphogenesis, *Int. Rev. Cytol.*, 150, 243, 1994.
81. Sahai, N. and Tossell, J.A., Molecular orbital study of apatite ($Ca_5(PO_4)_3OH$) nucleation at silica bioceramic surfaces, *J. Phys. Chem. B*, 104, 4322, 2000.
82. Sahai, N. and Tossell, J.A., Formation energies and NMR chemical shifts calculated for putative serine-silicate complexes in silica biomineralization, *Geochim. Cosmochim. Acta*, 65, 2043, 2001.
83. Sahai, N. and Tossell, J.A., ^{29}Si NMR shifts and relative stabilities calculated for hypercoordinated silicon-polyalcohol complexes: Role in sol-gel and biogenic silica synthesis, *Inorg. Chem.*, 41, 748, 2002.
84. West, J.K., Latour, R., and Hench, L.L., Molecular modeling study of poly-L-lysine onto silica glass, *J. Biomed. Mat. Res.*, 37, 585, 1997.
85. Lobel, K.D. and Hench, L.L., In vitro adsorption and activity of enzymes on reaction layers of bioactive glass substrates, *J. Biomed. Mater. Res.*, 39, 575, 1998.
86. Hehre, W.J. et al., *Ab Initio Molecular Orbital Theory*, John Wiley & Sons, New York, 1986.
87. Schmidt, M.W. et al., General atomic and molecular electronic structure system, *J. Comput. Chem.*, 1347, 1993.
88. Frisch, M J. et al., *Gaussian 94, Rev. B.3*, Gaussian, Pittsburgh, 1995.
89. Frisch, M.J. et al., *Gaussian 98, Rev. B.3*, Gaussian, Pittsburgh, 1999.
90. Frisch, A. and Frisch, M.J., *Gaussian 98 User's Reference*, 2nd ed., Gaussian, Pittsburgh, 1999.
91. Boys, S.F. and Bernardi, F., The calculation of small molecular interactions by the difference of separate total energies: Some procedures with reduced errors, *Molec. Phys.*, 19, 553, 1970.
92. Ben-Naim, A. and Marcus, Y., Solvation thermodynamics of nonionic solutes, *J. Chem. Phys.*, 81, 2016, 1984.

93. Tawa, G.J. et al., Calculation of the aqueous solvation free energy of the proton, *J. Chem. Phys.*, 109, 4852, 1998.
94. Wolinski, K., Hinton, J.F., and Pulay, P., Efficient implementation of the guage-dependent atomic orbital method for NMR chemical shift calculations, *J. Am. Chem. Soc.*, 112, 8251, 1982.
95. Bode, B.M. and Gordon, M.S., MacMolPlt: A graphical user interface for GAMESS, *J. Mol. Graphics Modeling*, 16, 133, 1998.
96. Kemmitt, T. and Milestone, N.B., The ring size influence on ^{29}Si N. M. R. chemical shifts of some spirocyclic tetra- and penta-coordinate dialato silicates, *Aust. J. Chem.*, 48, 93, 1995.
97. Herreros, B., Carr, S.W., and Klinowski, J., 5-coordinate Si compounds a intermediates in the synthesis of silicates in nonaqueous media, *Science*, 263, 1585, 1994.
98. Zhou, Y. et al., Efficient catalysis of polysiloxane synthesis by silicatein α requires specific hydroxy and imidazole functionalities, *Angew. Chem. Int. Ed.*, 38, 780, 1999.
99. Cha, J.N. et al., Silicatein filaments and subunits from a marine sponge direct the polymerization of silica and silicones *in vitro*, *Proc. Natl. Acad. Sci. USA*, 96, 361, 1999.
100. Vrieling, E.G., Gieskes, W.W.C., and Beelen, T.P.M., Silicon deposition in diatoms: Control by the pH inside the silicon deposition vesicle, *J. Phycol.*, 35, 548, 1999.
101. Belot, V. et al., Sol-gel chemistry of hydrogenosiliconates: The role of hypervalent silicon species, in *Better Ceramics Through Chemistry IV*, Zelinski, B.J.J. et al., Eds., Materials Research Society Symposium Proceedings, Warrendale, PA, 1990, p. 3.
102. Öhman, L.-O. et al., Equilibrium and structural studies of silicon (IV) and aluminum (III) in aqueous solution. 28. Formation of soluble silicic acid ligand complexes as studied by potentiometric and solubility measurements, *Acta Chem. Scand.*, 45, 335, 1991.
103. Sedeh, I.F., Sjöberg, S., and Öhman, L.-O., Equilibrium and structural studies of silicon (IV) and aluminum (III) in aqueous solution. 31. Aqueous complexation between silicic acid and the catecholamines dopamine and L-DOPA, *J. Inorg. Biochem.*, 50, 119, 1993.
104. Pokrovski, G.S. and Schott, J., Experimental study of the complexation of silicon and germanium with aqueous organic species: Implications for germanium and silicon transport and Ge/Si ratio in natural waters, *Geochim. Cosmochim. Acta*, 62, 3413, 1998.
105. Amo, Y.D. and Brzezinski, M., The chemical form of dissolved Si taken up by marine diatoms, *J. Phycol.*, 35, 1162, 1999.

7 Formation of Biogenic Manganese Oxides and Their Influence on the Scavenging of Toxic Trace Elements

Yarrow M. Nelson and Leonard W. Lion

CONTENTS

7.1 Introduction ... 169
7.2 Kinetics of Biological Mn Oxidation ... 170
7.3 Trace Metal Adsorption to Biogenic Mn Oxides ... 175
7.4 Trace Metal Adsorption to Mixtures of Biogenic Mn Oxides and Fe Oxides ... 179
7.5 Role of Mn Oxides in Controlling Trace Metal Adsorption to Natural Biofilms ... 180
7.6 Conclusions ... 183
Acknowledgments ... 184
References ... 184

7.1 INTRODUCTION

Biological Mn oxidation is an important process in the environment because it not only controls the cycling and bioavailability of Mn itself, but also is likely to exert controls on the cycling and bioavailability of other trace metals, either toxic (Nelson et al., 1999a) or nutrients (Bartlett, 1988), that strongly bind to sparingly soluble biogenic Mn oxides. Biogenic Mn oxides may also play important roles in the abiotic oxidation of complex organic compounds (Sunda and Kieber, 1994) and as terminal electron acceptors in biologically mediated degradation reactions (Nealson and Myers, 1992). Because of the importance of Mn cycling in the environment, a significant amount of research has been devoted to determining the enzymatic pathways responsible for biological oxidation of Mn(II) (Tebo et al., 1997). Also, a

1-56670-623-8/03/$0.00+$1.50

kinetic model for biological Mn oxidation has recently been developed (Zhang et al., 2002), and the implications of strong trace metal binding by biogenic Mn oxides have been explored through sampling and analysis in aquatic environments (Nelson et al., 1999a; Dong et al., 2000; Wilson et al., 2001).

The abiotic oxidation of Mn(II) is kinetically inhibited below pH 9 (Morgan and Stumm, 1964; Langen et al., 1997), and therefore Mn oxidation in natural aquatic environments is expected to be predominantly driven by biological processes, either by direct enzymatically catalyzed oxidation (Ghiorse, 1984) or indirectly by biologically induced microenvironment changes such as a localized pH increase caused by algae (Aguilar and Nealson, 1994). Biological Mn oxidation has been demonstrated in the field in both freshwater (Nealson et al., 1988) and marine (Tebo and Emerson, 1986; Moffett, 1997) environments. Several microorganisms with a demonstrated ability to catalyze Mn oxidation have been isolated in pure culture, including *Leptothrix discophora* (Ghiorse, 1984), a bacterium isolated from freshwater wetlands; the marine bacterium *Bacillus subtilis* which forms spores that catalyze Mn oxidation on their surfaces (Tebo et al., 1988); and the freshwater bacterium *Pseudomonas putida* MnB1 (Douka, 1977, 1980). The enzymology and genetics for the biological oxidation of Mn(II) by these model organisms have been reviewed by Tebo et al. (1997).

We focus here on the kinetics of biological Mn oxidation and the reactivity of biogenic Mn oxides. Although rates of Mn oxidation have been reported from field and laboratory studies, only recently has a rate law for biologically catalyzed Mn oxidation been derived. Similarly, it has long been suspected that biogenic Mn oxides might exhibit greater reactivity than abiotic Mn oxides, but only recently have laboratory studies with pure biogenic Mn oxides demonstrated this high reactivity. Finally, we explore the implications of the high surface reactivity of biogenic Mn oxides by describing recent field studies of trace metal adsorption by natural biofilms containing biogenic Mn oxides.

7.2 KINETICS OF BIOLOGICAL Mn OXIDATION

When the kinetics of Mn(II) oxidation were first described mathematically, the model was restricted to pH > 9 and ignored biological activity (Morgan and Stumm, 1964; Stumm and Morgan, 1996). Because the abiotic rate of Mn(II) oxidation is extremely slow at circumneutral pH (Langen et al., 1997), the reaction is unlikely to proceed without biological catalysis, and it is therefore important to develop a kinetic model applicable to the biologically mediated reaction. In natural waters at circumneutral pH, biological Mn oxidation can be orders of magnitude faster than abiotic Mn oxidation (Nealson et al., 1988; Tebo, 1991; Wehrli et al., 1992). Reported rates of biological catalysis of Mn oxidation vary from 65 nM/h in marine environments (Tebo, 1991) to 350 nM/h in freshwater (Tipping, 1984). As expected, the rate of biological Mn oxidation varies with solution conditions as required by the organism and enzyme systems involved. In natural environments a strong dependence of biological Mn oxidation rate on both temperature and pH has been observed (Tebo and Emerson, 1985; Sunda and Huntsman, 1987). Similarly, pure cultures of *Leptothrix discophora* SS1 exhibit a strong dependence on environmental conditions, with a maximum rate of Mn oxidation at pH 7.5 and an optimum temperature of 30°C (Adams and Ghiorse, 1987). The mechanisms of both abiotic and biological Mn oxidation have been reviewed by Tebo et al. (1997).

Given the importance of biological Mn oxidation, we have measured Mn oxidation rates by the model bacterium *Leptothrix discophora* SS1 under a range of controlled laboratory conditions and have developed a rate law for biologically mediated Mn(II) oxidation as a function of environmental conditions including temperature, pH, and the concentrations of cells, Mn(II), O_2, and Cu. *L. discophora* SS1 (ATCC 43821), a heterotrophic, freshwater proteobacterium, was used to produce biogenic Mn oxides in controlled laboratory bioreactors with a defined growth medium (Zhang et al., 2002). The use of a defined medium allowed for the determination of metal speciation without interference from buffers that could complex Mn (Table 7.1). The observed Mn oxidation rate was found to be directly proportional to cell and O_2 concentrations and exhibited a pH optimum of 7.5 and temperature optimum of 30°C. Mn oxidation kinetics by *L. discophora* SS1 obeyed Michaelis–Menten enzyme kinetics with respect to Mn(II) concentration (Figure 7.1). This result agrees with earlier field experiments that observed Michaelis–Menten kinetics for Mn oxidation in freshwater environments (Tebo and Emerson, 1986; Sunda and

TABLE 7.1
Manganese and Copper Speciation in MMS Medium ([Mn(II)] = 50 μM, [Cu(II)] = 0.1 μM, $P_{CO_2} = 10^{-3.5}$ atm, T = 25°C

	% of Total Metal as Each Species					
	pH					
	6.0	6.5	7.0	7.5	8.0	8.5
Mn Species						
Mn^{2+}	78.5	78.5	78.4	78.2	35.6	3.6
$MnSO_4$	10.9	10.9	10.9	10.0	5	0
$MnHPO_4$	9.9	10	10	10	10	0
Rhodochrosite[$MnCO_3$(s)]	0	0	0	0	48.8	95.8
Cu Species						
Cu^{2+}	83.6	69.9	28.4	4.2	N/A	N/A
$Cu(OH)^+$	0	2.2	2.8	1.3	N/A	N/A
$Cu(OH)_2$ aq	1.7	14.6	59.4	88.4	N/A	N/A
$CuSO_4$	13.1	10.9	3.3	0	N/A	N/A
$CuCO_3$ aq	0	0	4.4	5	N/A	N/A

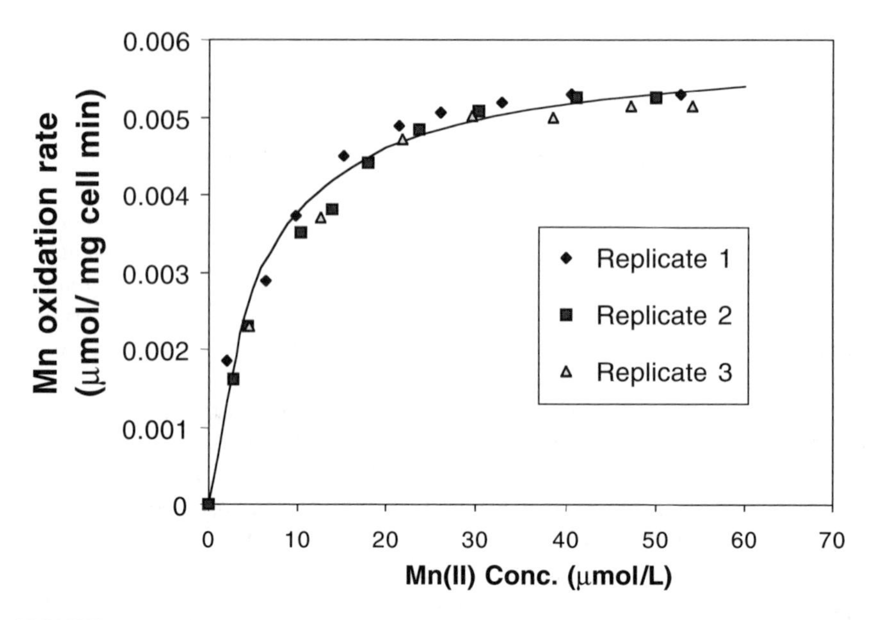

FIGURE 7.1 Michaelis–Menten oxidation kinetics for Mn(II) at T = 25°C, pH = 7.5, O_2 = 8.05 mg/l and zero added Cu. (Reprinted from Zhang, J. et al., *Geochim. Cosmochim. Acta*, 65, 773, copyright 2002. With permission from Elsevier Science.)

Huntsman, 1987; Moffett, 1994). At the optimum pH and temperature, the maximum oxidation rate (v_{max}) was 0.0059 μmol Mn(II)/min-mg cell (at 25°C, pH = 7.5, and a dissolved oxygen concentration of 8.05 mg/l) (Zhang et al., 2002). The half-velocity coefficient (K_s) for Mn oxidation by *L. discophora* in the controlled bioreactors was 5.7 μmol Mn(II)/l. This value of K_s is similar to that determined previously under less-controlled conditions using buffers to regulate pH (Adams and Ghiorse, 1987).

Recent investigations of the molecular biology of Mn-oxidizing bacteria have implicated copper-containing enzymes in Mn(II) oxidation. Multi-copper oxidase enzymes have been reported to mediate bacterial Mn oxidation for *Pseudomonas putida, Bacillus* SG1 spores and *L. discophora* (Brouwers et al., 2000a,b). Copper addition increased Mn(II) oxidation rates of *P. putida* by a factor of five (Brouwers et al., 2000a,b), and also increased Mn oxidation rates by spores of *Bacillus* SG1 (VanWaasbergen et al., 1996). Copper stimulated the activity of supernatant obtained from stationary phase suspensions of *L. discophora* SS1 when the cells were grown in the presence of Cu; however, Cu did not stimulate Mn(II) oxidation when added directly to the spent medium supernatant subsequent to growth of the bacterium (Brouwers et al., 2000a,b). Our research examined the effect of copper concentration on Mn oxidation rates by *L. discophora* SS1 using the controlled bioreactor and defined medium described above. In the bioreactor experiments copper inhibited cell growth rate and yield at Cu concentrations as low as 0.02 μM (Figure 7.2), but enhanced Mn(II) oxidation rates (Table 7.2) (Zhang et al., 2002).

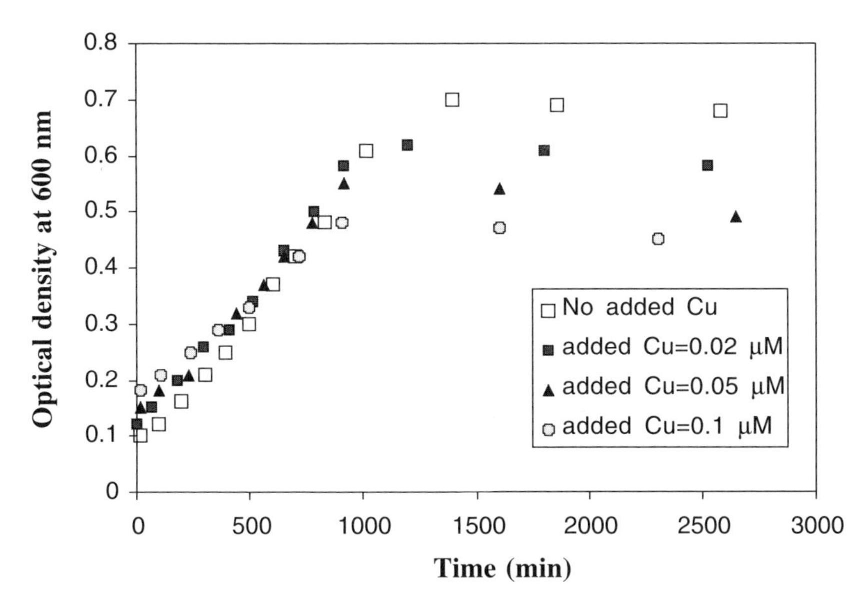

FIGURE 7.2 Effect of added Cu on *L. discophora* SS1 growth curves at pH = 7.5, T = 25°C, and O_2 = 8.05 mg/l. (Reprinted from Zhang, J. et al., *Geochim. Cosmochim. Acta*, 65, 773, copyright 2002. With permission from Elsevier Science.)

TABLE 7.2
Effect of Cu on *L. discophora* SS-1 Mn Oxidizing Activity[a]

[Cu(II)] μM	Specific Mn(II) Oxidizing Activity (per Equal Cell Weight), %
0	100
0.02	125
0.05	158
0.1	189

[a] The activity of a culture without added Cu(II) are defined as 100%.

Source: Reprinted from Zhang, J. et al., *Geochim. Cosmochim. Acta,* 65, 773, copyright 2002. With permission from Elsevier Science.)

Using the controlled bioreactor experiments described above, we developed a general rate law for biological Mn oxidation by *L. discophora* that shows the mathematical dependence on cell concentration, Mn concentration, pH, temperature, dissolved oxygen concentration, and copper concentration (Zhang et al., 2002).

$$-\frac{d[Mn(II)]}{dt}=\frac{k[X][Mn(II)]}{K_S+[Mn(II)]}(k_{o_2}[O_2])(Ae^{-Ea/RT})\left(\frac{k_{pH}}{1+[H^+]/K_1+K_2/[H^+]}\right)(1+k_c[Cu(II)])$$

where

$[X]$ is cell concentration, mg/l
$[O_2]$ is dissolved oxygen concentration, mg/l
$[Cu]$ is total dissolved copper concentration, μmol/l
k = 0.0059 μmol Mn(II)/(mg cell• min)
K_S = 5.7 μmol Mn(II)/l
k_{O_2} = 1/8.05 = 0.124/(mg/l) ([O_2] = 8.05 mg/l at 25°C
E_a = 22.9 kcal/(g cell• mole)
$A = 2.3\times10^{14}$
$K_1 = 3.05\times10^{-8}$
$K_2 = 2.46\times10^{-8}$
k_{pH} = 4.52
k_C = 8.8/(μmol Cu/l).

At 25°C, pH = 7.5, [O_2] = 8.05 mg/l and zero added copper (i.e., Cu < 5nM), the above rate law for Mn(II) oxidation simplifies to the following Michaelis–Menten expression for biological Mn oxidation rate:

$$-\frac{d[Mn(II)]}{dt}=\frac{kX[Mn(II)]}{K_S+[Mn(II)]}$$

It is interesting to compare the expected rates of Mn oxidation via abiotic mechanisms with the rates expected from the biological kinetic rate law described above. Abiotic Mn oxidation rates at pH 8.03 were measured in seawater by von Langen et al. (1997) who reported a first-order rate constant of 1.1×10^{-6} (normalized for P_{O_2} = 1 atm and T = 25°C). At this pH and for similar conditions, the cell concentration of *L. discophora* required to obtain the same rate would be only 0.30 μg/l (Zhang et al., 2002) (i.e., approximately 3×10^5 cells/l). It is reasonable to assume that cell populations of Mn-oxidizing bacteria far greater than this would be possible in natural environments. Even smaller population sizes would be required to match abiotic rates (if they could be measured) at lower pH values.

It should be noted that the kinetic rate law described above is quantitatively applicable only to the strain of *L. discophora* used in the experiments described. Different Mn-oxidizing bacteria and even different stains of *L. discophora* would be expected to exhibit different rates of catalysis of Mn oxidation. For example, recent investigations in a wetland in New York State found many different genetic strains of *Leptothrix*, each exhibiting different rates of catalysis of Mn oxidation (Verity, 2001). However, the general form of the rate law could be expected to be similar for different species of Mn oxidizers.

The rate law for biological Mn oxidation described above could potentially be incorporated into environmental models to describe the cycling of Mn in natural environments and the associated fate of other trace metals or of organic compounds that are oxidized by reaction with biogenic Mn oxides. However, successful use of these models will require a better understanding of the population size of Mn oxidizers in natural aquatic environments.

7.3 TRACE METAL ADSORPTION TO BIOGENIC Mn OXIDES

The environmental fate and behavior of toxic transition metals are governed by interactive biogeochemical processes, such as adsorption, complexation and multiple biological interactions (Krauskopf, 1956; Jenne, 1968; Turekian, 1977; Vuceta and Morgan, 1978; Westall et al., 1995; Nelson et al., 1999a). Microorganisms have the potential to adsorb significant concentrations of trace metals, and many bacteria produce extracellular polymers that have well-established metal binding properties (Lion et al., 1988; Pradhan and Levine, 1992; Herman et al., 1995; Nelson et al., 1995). However, depending on the trace metal of interest, adsorption by Fe and Mn oxides can be far greater than that by organic materials (Lion et al., 1982). While the role of iron oxides in metal scavenging has received considerable attention (Dzombak and Morel, 1990), and trace metal adsorption by Mn oxide minerals (presumably of abiotic origin) has been studied (Catts and Langmuir, 1986), far less is known about the properties of biogenic Mn oxides and their role as metal scavenging agents. Trace metal adsorption by biogenic Mn oxides is relevant because biological Mn oxidation is expected to dominate in circumneutral environments (see above). Here we describe some recent measurements of trace metal binding to biogenic Mn oxides prepared under controlled laboratory conditions, as well as to defined mixtures of Fe oxides and biogenic Mn oxides.

Biologically oxidized Mn oxides are generally believed to be amorphous or poorly crystalline and of mixed oxidation state (Hem and Lind, 1983; Murray et al., 1985; Adams and Ghiorse, 1988; Wehrli et al., 1992; Mandernack et al., 1995; Mandernack et al., 1995). Investigations using x-ray absorption fine structure (XAFS) have begun to elucidate the mineralogy of microcrystalline regions in Mn oxides and Mn oxyhydroxides (Manceau and Combes, 1988; Friedl et al., 1997). Co, Zn, Ce, and trivalent lanthanides have been reported to be incorporated into biogenic Mn oxides via the same enzymatic pathways as Mn oxidation (Moffett and Ho, 1995), but adsorption of these elements to already-formed Mn oxides was not described. He and Tebo (1998) measured the surface area of Mn-oxidizing *Bacillus* SG1 spores and Cu adsorption to these spores, but not to the Mn oxides formed by the spores. Trace metal adsorption to natural materials containing Mn oxides and Fe oxides has been measured (Tessier et al., 1996), but the specific role of Mn oxides in these experiments is difficult to isolate because the extracted Mn oxides were likely mixed with Fe oxides and residual organic material from the field.

The first measurement of trace metal adsorption to laboratory-prepared biogenic Mn oxides was reported by Nelson et al. (1999b). In these experiments both Mn oxidation and trace metal adsorption were carried out under controlled conditions with *L. discophora* grown in a defined medium. The use of pH controllers in these experiments eliminated the need for buffers that could potentially complex with either Mn ions or adsorbing cations. Also, competing trace metals were excluded during both production of Mn oxides and adsorption measurements (except for 0.1 μM Fe, which was necessary for Mn oxidation) (Nelson et al., 1999b). Under these controlled laboratory conditions at pH 6.0 and 25°C, Pb adsorption by *L. discophora* cells with biogenic Mn oxide coatings was two orders of magnitude greater than Pb adsorption by cells without Mn (Figure 7.3) (Nelson et al., 1999b). This result was expected because of the known affinity of Pb for metal oxides compared to organic material. Even more interesting was the comparison between Pb adsorption by the biogenic Mn oxide and abiotically prepared Mn oxides. Adsorption isotherms at pH 6.0, I = 0.05 M, and T = 25°C show that the biogenic Mn oxide exhibited five times the adsorption capacity of a freshly precipitated abiotic Mn oxide, and a much steeper adsorption isotherm (Figure 7.4) (Nelson et al., 1999b). The steep isotherm is important because it indicates that the difference between the adsorption of the biogenic Mn oxide and the abiotic Mn oxide is even more significant at low Pb concentrations as would be encountered in natural aquatic environments. For comparison, Pb adsorption of the biogenic Mn oxide is several orders of magnitude greater than that of abiotic pyrolusite Mn oxide minerals and more than an order of magnitude greater than that of colloidal Fe oxyhydroxide under the same conditions (Figure 7.5) (Nelson et al., 1999b). The global abundance of Mn is less than Fe and thus concentration of iron oxides is expected to exceed that of Mn oxides in many natural aquatic systems. However, the enhanced reactivity of Mn oxides with respect

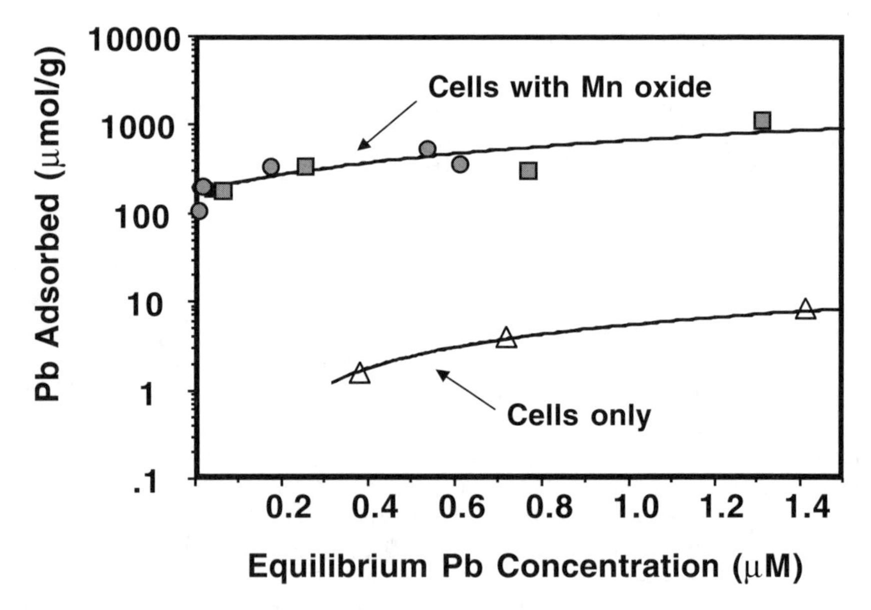

FIGURE 7.3 Effect of biogenic Mn oxide deposits on Pb adsorption by *L. discophora* cells at pH 6.0 and 25°C. Cell concentration = 63 mg/l, Mn loading = 0.8 mmol/g cells. (From Nelson, Y.M. et al., *Appl. Environ. Microbiol.*, 65, 175, 1999b. With permission.)

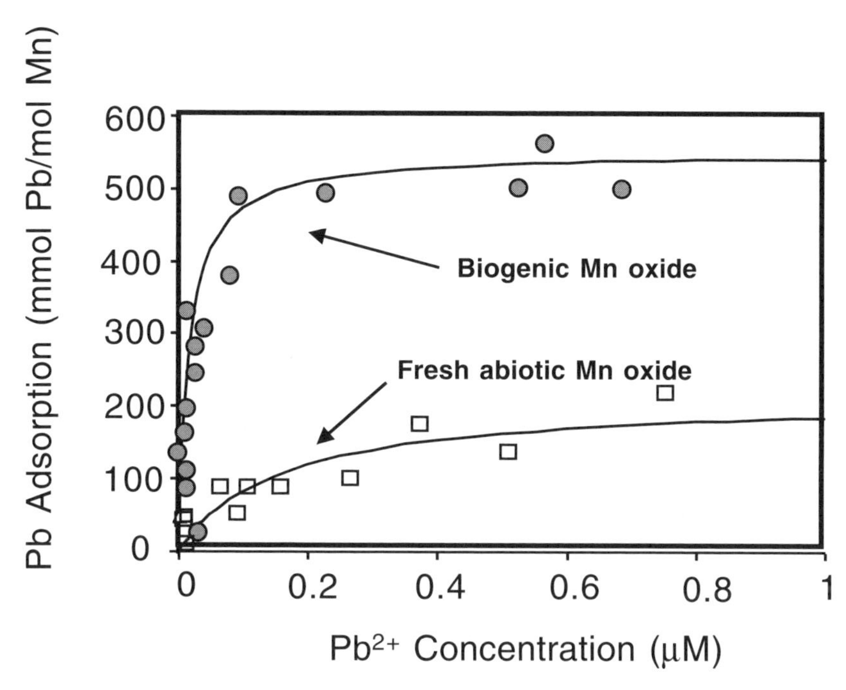

FIGURE 7.4 Pb adsorption of biogenic Mn oxide compared to that of a fresh abiotically prepared Mn oxide. (From Nelson, Y.M. et al., *Appl. Environ. Microbiol.*, 65, 175, 1999b. With permission.)

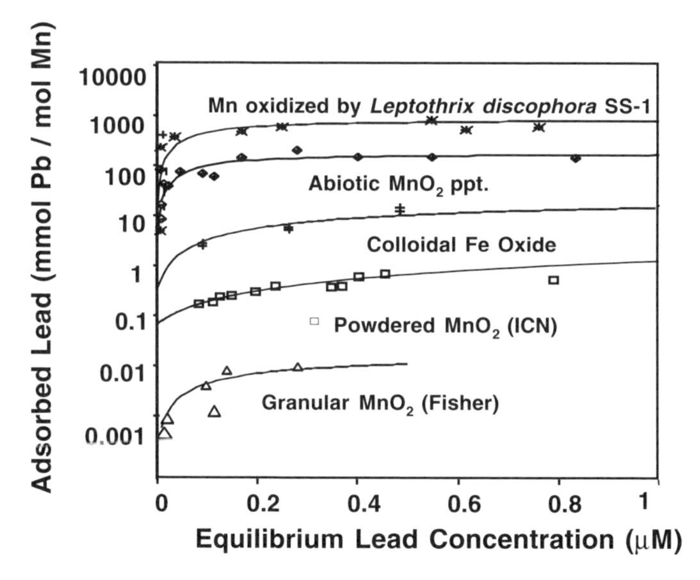

FIGURE 7.5 Pb adsorption of biogenic Mn oxide compared to that of colloidal Fe oxyhydroxide and abiotic Mn oxide minerals (pyrolusite). (From Nelson, Y.M. et al., *Appl. Environ. Microbiol.*, 65, 175, 1999b. With permission.)

to adsorption of Pb suggests that their importance may equal or exceed that of Fe oxides in the scavenging of some toxic metals.

The biogenic Mn oxides formed under the controlled conditions described above were determined to be amorphous by x-ray diffraction analysis, with very small peaks matching the spectra of ramsdellite, suggesting a partially orthorhombic structure (Nelson et al., 2001) (Figure 7.6). The specific surface area of the biogenic Mn oxide was 220 m^2/g, and was significantly greater than that of the other Mn oxides tested (Table 7.3). The observed Pb adsorption correlated with specific surface area, although the ratio of Pb adsorption Γ_{max} to surface area was significantly greater for the biogenic Mn oxide and the abiotic Mn oxide than for the pyrolusite Mn oxides (Table 7.3).

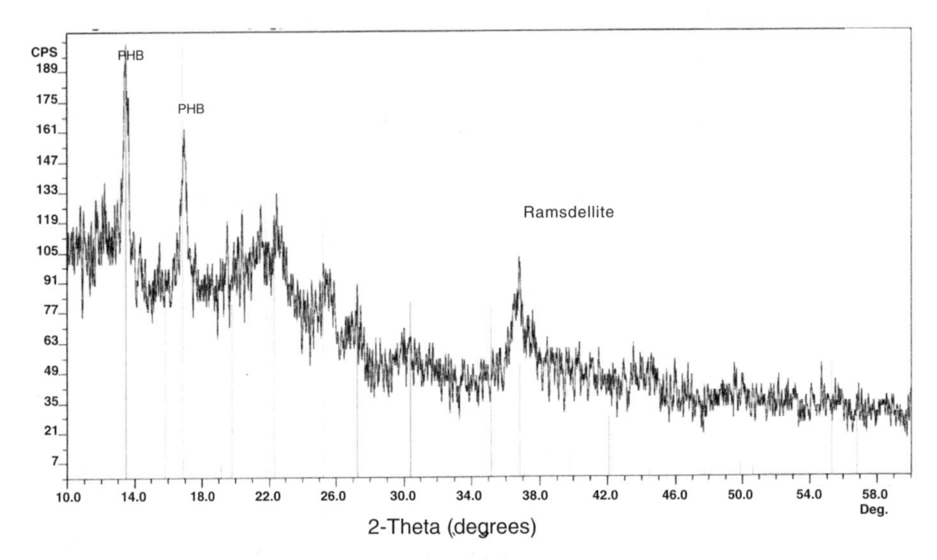

FIGURE 7.6 X-ray diffraction pattern of biogenic Mn oxide produced by *L. discophora* prepared at pH 7.5. Peaks corresponding to polyhydroxybutyrate are labeled PHB, and the peak corresponding to ramsdellite is labeled accordingly

TABLE 7.3
Specific surface area and Pb adsorption capacity of biogenic Mn oxide, fresh abiotic Mn oxide precipitate and pyrolusite minerals.

Mn Oxide	BET Surface Area, S.A. (m^2/g)	Max. Pb Ads., Γmax (μmol Pb/mmol Mn)	Ratio Γmax/SA, (μmol-m^2/mmol-g)
Biogenic Mn oxide	224	550	2.5
Fresh abiotic Mn oxide	58	220	3.8
Pyrolusite (powdered)	4.7	1.2	0.26
Pyrolusite (granular)	0.048	0.031	0.69

Source: From Nelson, Y.M. et al., *Appl. Environ. Microbiol.*, 65, 175, 1999b. With permission.

As expected, trace metal adsorption to biogenic Mn oxides is dependent on pH. Wilson et al. (2001) measured and modeled the pH dependence of Pb to biogenic Mn oxides and compared it to that of colloidal iron oxides. As seen in Figure 7.7, the adsorption of Pb to Mn oxide exceeds that of Fe oxide (on a per mole of Mn or Fe basis) at pH values below 8.5. The pH dependence of Pb adsorption to Fe was greater than that for Pb adsorption to biogenic Mn oxides and thus, adsorption to Fe oxides is expected to become relatively more important as pH increases.

7.4 TRACE METAL ADSORPTION TO MIXTURES OF BIOGENIC Mn OXIDES AND Fe OXIDES

In natural environments, Mn and Fe oxides often coexist as mixtures, making it important to determine if their trace metal adsorption properties are altered by interactions with each other. To test this hypothesis we measured Pb adsorption to mixtures of biogenic Mn oxide and colloidal Fe oxide. The mixtures were prepared in two ways. First, previously prepared biogenic Mn oxide was mixed in suspension with previously precipitated amorphous Fe(III) oxide. Second, Mn(II) was biologically oxidized in the presence of a previously prepared Fe(III) oxide suspension (Nelson et al., 2002). As above, all adsorption measurements were made in a chemically defined medium (Nelson et al., 1999b). In both cases, observed Pb adsorption by the Fe/Mn oxide mixtures was similar to that predicted using Langmuir

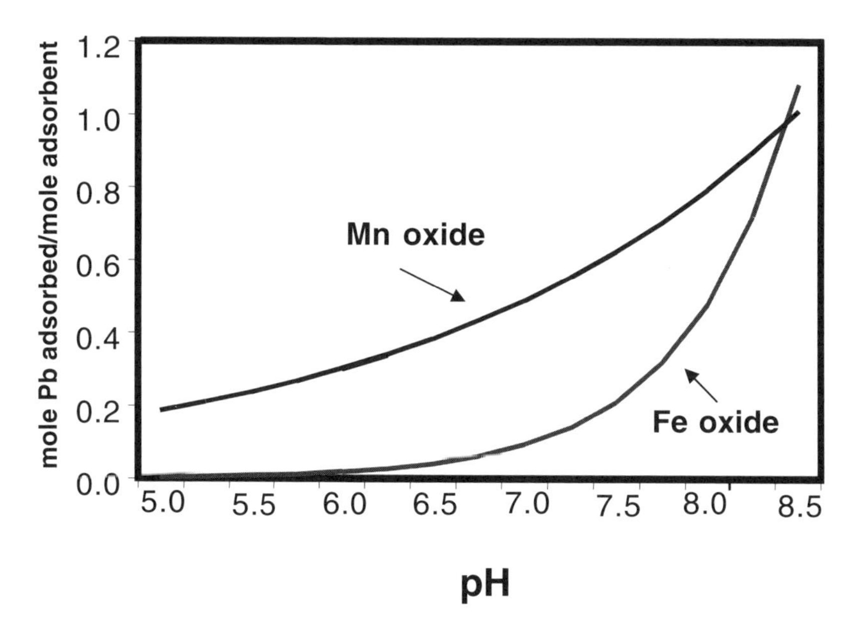

FIGURE 7.7 pH dependence of Pb adsorption to biogenic Mn oxide versus Fe oxide. (See Wilson, A.R. et al., 2001.)

TABLE 7.4
Pb Adsorption to Fe/Mn (Hydr)oxides: Fe(III) (Hydr)oxide Mixed with Biogenic Mn (Hydr)oxide after Biological Oxidation

Equilibrium Pb^{2+} Conc. (mM Pb)	Observed Pb Binding (μM Pb)	Particulate Mn Conc. (μM Mn)	Particulate Fe Conc. (μM Fe)	Fe/Mn Ratio	Pb Adsorption Calculated from Isotherms		
					On Fe (μM Pb)	On Mn (μM Pb)	Total (μM Pb)
0.025	0.142	0.75	22.18	29.4	0.080	0.130	0.210
0.069	0.451	0.77	22.93	29.6	0.204	0.227	0.432
0.157	0.823	0.87	23.27	26.9	0.382	0.326	0.708
0.880	1.214	0.73	21.19	29.0	0.775	0.338	1.113
1.638	0.873	0.42	13.13	31.5	0.548	0.197	0.745

Source: Reprinted with permission from Nelson, Y.M. et al. *Environ. Sci. Technol.*, 36, 421, copyright 2002. American Chemical Society.

TABLE 7.5
Pb Adsorption to Fe/Mn (Hydr)oxides: Fe(III) (Hydr)oxide Present During Biological Mn Oxidation

Equilibrium Pb^{2+} Conc. (μM Pb)	Observed Pb Binding (μM Pb)	Particulate Mn Conc. (μM Mn)	Particulate Fe Conc. (μM Fe)	Fe/Mn Ratio	Pb Adsorption Calculated from Isotherms		
					On Fe (μM Pb)	On Mn (μM Pb)	Total (μM Pb)
0.013	0.181	1.10	24.22	22.0	0.047	0.119	0.166
0.164	0.242	0.00	13.57	—	0.230	0.000	0.230
0.275	0.600	0.42	18.99	45.4	0.438	0.217	0.655
0.678	1.669	1.07	27.40	25.7	0.928	0.572	1.501
1.565	1.253	0.45	19.16	42.2	0.793	0.247	1.040

Source: Reprinted with permission from Nelson, Y.M. et al. *Environ. Sci. Technol.*, 36, 421, copyright 2002. American Chemical Society.

isotherm predictions for the individual oxides and assuming additivity of their Langmuir isotherms (compare columns 2 to 8 in Tables 7.4 and 7.5). These results suggest that interactions between the Mn and Fe oxides under these conditions did not alter the adsorption properties of the oxides.

7.5 ROLE OF Mn OXIDES IN CONTROLLING TRACE METAL ADSORPTION TO NATURAL BIOFILMS

Given the high adsorption capacity of biogenic Mn oxide, we conducted investigations to estimate the magnitude of the influence of biogenic Mn oxides on total trace metal adsorption to natural suspended particulate material and biofilms containing biogenic Mn oxides. Two different approaches were used to make these estimates.

One method used adsorption to laboratory surrogate materials in an adsorption additivity model to estimate the relative contributions of Mn and Fe oxides and organic materials to Pb adsorption to natural biofilms. The second method made use of a novel selective extraction approach in which trace metal adsorption to surface coatings was measured before and after the selective removal of constituents.

For the surrogate adsorption and additivity model, the Pb adsorption behavior of biogenic Mn oxide was used as a surrogate to represent naturally occurring Mn minerals. Surrogates were also selected for Fe oxides, Al oxides, and several organic materials and their Ph adsorption behaviors were similarly characterized. The adsorption behavior of these surrogates was then used in an additive model described by Nelson et al. (1999a) to predict the Pb adsorption by heterogeneous natural materials. Under the experimental conditions (at pH 6.0) with the observed composition of natural surface coatings from several freshwater lakes, biogenic Mn oxide was predicted to be the dominant adsorbent sink for Pb and the cumulative adsorption predicted for all selected surrogates provided a good match to the observed adsorption behavior (Figure 7.8) (Nelson et al., 1999a). We recently extended this work by examining Pb adsorption over a range of pH (Wilson et al., 2001). Mn oxide was revealed to be the dominant adsorbent for Pb (on the surface coating materials obtained from Cayuga Lake, Ithaca, New York, at the Fe and Mn concentrations sampled) below pH 6.5 and was second to Fe oxides above this pH (Figure 7.9) (Wilson et al., 2001).

For the selective extraction experiments, surface coatings were collected in a freshwater lake (Cayuga Lake, New York) on glass slides and Pb and Cd adsorption was measured under controlled conditions before and after extractions to determine by difference the adsorptive properties of the extracted component(s). 0.01 M

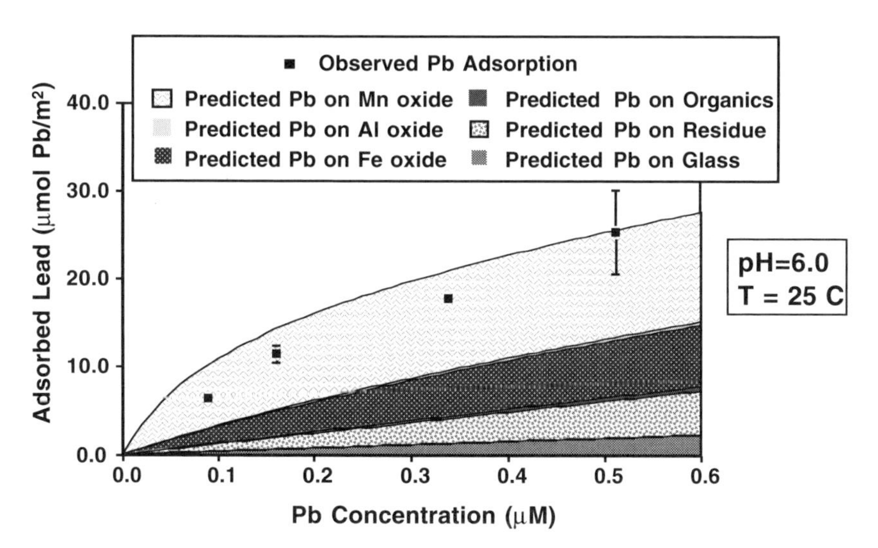

FIGURE 7.8 Predicted Pb adsorption to components of biofilms collected at Cayuga Lake, New York, showing dominant role of biogenic Mn oxides. (From Nelson, Y.M. et al., *Limnol. Oceanogr.*, 44, 1715, 1999a. With permission.)

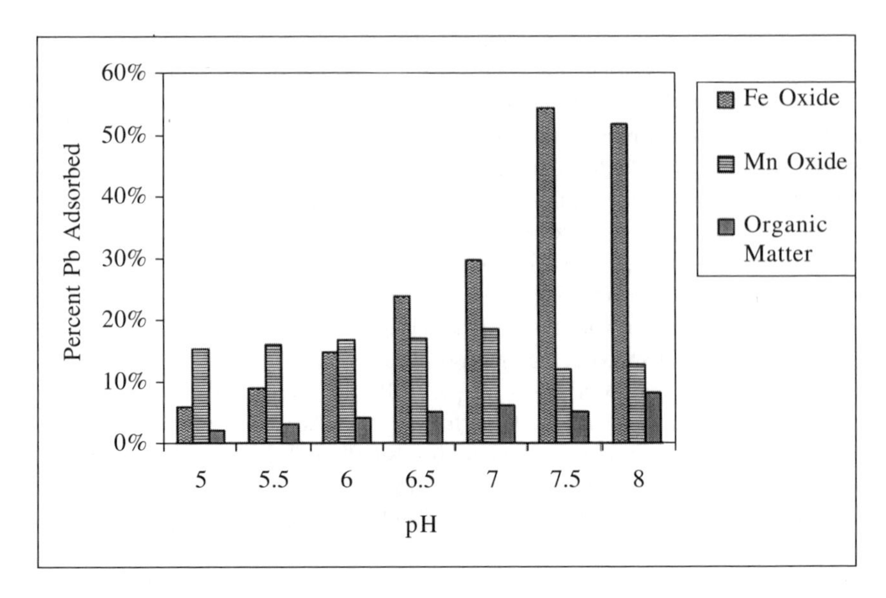

FIGURE 7.9 Relative contribution to Pb adsorption by biogenic Mn oxide. Fe oxide and natural organic matter in a natural surface coating as a function of pH. Below pH 6.5, biogenic Mn oxide dominates Pb adsortion, while above pH 6.5, Pb adsorption to biogenic Mn oxide is second to that of Fe oxide. (See Wilson, A.R. et al., 2001.)

TABLE 7.5
Pb Adsorption to Fe/Mn (Hydr)oxides: Fe(III) (Hydr)oxide Present During Biological Mn Oxidation

Equilibrium Pb^{2+} Conc. (μM Pb)	Observed Pb Binding (μM Pb)	Particulate Mn Conc. (μM Mn)	Particulate Fe Conc. (μM Fe)	Fe/Mn Ratio	Pb Adsorption Calculated from Isotherms		
					On Fe (μM Pb)	On Mn (μM Pb)	Total (μM Pb)
0.013	0.181	1.10	24.22	22.0	0.047	0.119	0.166
0.164	0.242	0.00	13.57	—	0.230	0.000	0.230
0.275	0.600	0.42	18.99	45.4	0.438	0.217	0.655
0.678	1.669	1.07	27.40	25.7	0.928	0.572	1.501
1.565	1.253	0.45	19.16	42.2	0.793	0.247	1.040

Source: Reprinted with permission from Nelson, Y.M. et al. *Environ. Sci. Technol.*, 36, 421, copyright 2002. American Chemical Society.

$NH_2OH{\cdot}HCl$ + 0.01 M HNO_3 was used to selectively remove Mn oxides; 0.3 M $Na_2S_2O_4$ was used to remove Mn and Fe oxides; and 10% oxalic acid was used to remove metal oxides and organic materials. Nonlinear regression analysis of the observed Pb and Cd adsorption, based on the assumption of additive Langmuir adsorption isotherms, was used to estimate the relative contributions of each surface coating constituent to total Pb and Cd binding of the biofilms. The results agreed with the experiments described above, in that Pb adsorption was dominated by Mn

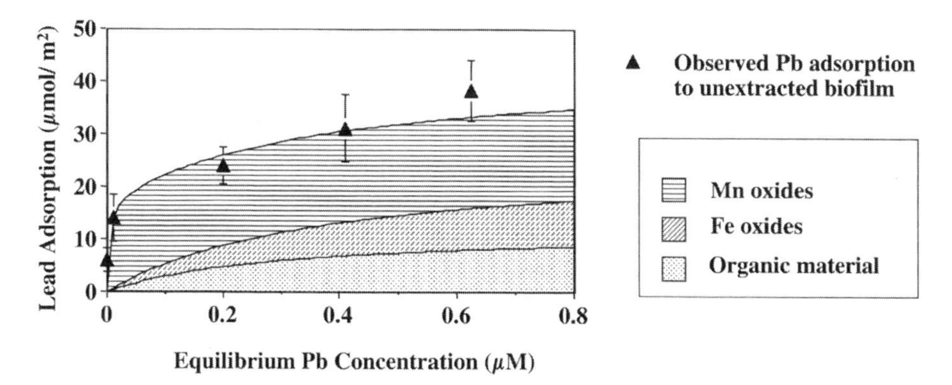

FIGURE 7.10 Relative roles of biogenic Mn oxide, Fe oxide, and organic material in controlling Pb adsorption to biofilms as determined using a novel selective extraction technique. (From Dong, D. et al., *Water Res.*, 34, 427, 2000. With permission.)

oxides, with lesser roles of Fe oxides and organic material, and the contribution of Al oxides to Pb adsorption was insignificant (Figure 7.10) (Dong et al., 2000). Interestingly, Cd adsorption to the lake biofilms was dominated by Fe oxides, with lesser roles of Mn and Al oxides and organic material (Dong et al., 2000); thus the relative importance of adsorptive scavenging by Mn versus Fe oxides depends upon the trace metal of interest.

The extraction experiments described above used Pb and Cd adsorption to assess the reactivity of components of the heterogeneous surface coating materials under laboratory conditions. However, Dong et al. (in press) recently extended the extraction results to actual lake conditions by measuring the ambient-adsorbed Pb released by each extraction, and these results confirmed surface Mn oxides to be the dominant sink for the trace levels of adsorbed Pb that occur in the lake.

7.6 CONCLUSIONS

These recent research results verify that biogenic Mn oxides are amorphous and exhibit greater specific surface area than typical abiotically prepared Mn oxides, and that the biogenic Mn oxides have a very high adsorption capacity for Pb. Further experiments should be done to determine the reactivity of the biogenic Mn oxides to other trace metals. Since biogenic Mn oxides are formed rapidly at circumneutral pH, they are likely to be the prevalent Mn solid phase present in many natural aquatic systems. This indicates a significant role for biogenic Mn oxides in the scavenging of toxic transition metals in aquatic environments. Additional experiments in natural systems will likely extend our understanding of the diverse microbial species involved in Mn oxidation. Similar work in soil environments should also be conducted to help establish the role of biogenic Mn oxides in controlling trace metal behavior (e.g., bioavailability) as well as the availability and mobility of Mn. Eventually, it is hoped that this work will lead to the development of useful models to describe the cycling of Mn in natural environments and coupled cycles of other trace metals.

ACKNOWLEDGMENTS

This research was supported, in part, by the following grants from the National Science Foundation: BES-9706715 and CHE-9708093. We acknowledge the assistance of William Ghiorse, Michael Shuler, Paul Koster van Groos, Elizabeth Costello, Barbara Eaglesham, Rebecca Verity, Cameron Willkens, Alyson Wilson, and Jinghao Zhang. We are also grateful for copyright permission from the editors of *Applied and Environmental Microbiology*, *Environmental Science and Technology*, *Geochimica et Cosmochimica Acta*, and *Limnology and Oceanography*.

REFERENCES

Adams, L.F. and Ghiorse, W.C., Characterization of extracellular Mn^{2+}-oxidizing activity and isolation of an Mn^{2+}-oxidizing protein from *Leptothrix discophora* SS-1, J. Bacteriol., 169, 1279, 1987.

Adams, L.F. and Ghiorse, W.C., Oxidation state of Mn in the Mn oxide produced by *Leptothrix discophora* SS-1, *Geochim. Cosmochim. Acta*, 52, 2073, 1988.

Aguilar, C. and Nealson, K.H., Manganese reduction in Oneida Lake, New York: Estimates of spatial and temporal manganese flux, *Can. J. Fisheries Aquatic Sci.*, 51, 185, 1994.

Bartlett, R.J., Manganese redox reactions and organic interactions in soils, in *Manganese in soils and plants*, Graham, R.D., Hannam, R.J., and Uren, N.C., Eds., Kluwer, Dordrecht, 1988, p. 59.

Brouwers, G.J. et al., Stimulation of Mn^{2+} oxidation in *Leptothrix discophora* SS-1 by Cu^{2+} and sequence analysis of the region flanking the gene encoding putative multicopper oxidase mofA, *Geomicrobiol. J.,* 17, 25, 2000a.

Brouwers, G.J., et al., Bacteria Mn^{2+} oxidation multicopper oxidases: An overview of mechanisms and functions, *Geomicrobiol. J.,* 17, 1, 2000b.

Catts, J.G. and Langmuir, D., Adsorption of Cu, Pb and Zn by $dMnO_2$: Applicability of the site binding–surface complexation model, *Appl. Geochem.*, 1, 255, 1986.

Dong, D. et al., Adsorption of Pb and Cd onto metal oxides and organic material in natural surface coatings as determined by selective extractions: New evidence for the importance of Mn and Fe oxides, *Water Res.*, 34, 427, 2000.

Dong, D., Derry, L., and Lion, L.W., Pb scavenging from a freshwater lake by Mn oxides in heterogeneous surface coating materials, *Water Res.*, in press.

Douka, C., Study of bacteria from manganese concretions, *Soil Biol. Biochem.*, 9, 89, 1977.

Douka, C.E., Kinetics of manganese oxidation by cell-free extracts of bacteria isolated from manganese concretions from soil, *Appl. Environ. Microbiol.*, 39, 74, 1980.

Dzombak, D.A. and Morel, F.M.M., *Surface Complexation Modeling. Hydrous Ferric Oxide*, John Wiley & Sons, New York, 1990.

Friedl, G., Wehrli, B., and Manceau, A., Solid phases in the cycling of manganese in eutrophic lakes: New insights from EXAFS spectroscopy, *Geochim. Cosmochim. Acta*, 61, 275, 1997.

Ghiorse, W.C., Biology of iron- and manganese-depositing bacteria, *Ann. Rev. Microbiol.,* 38, 515, 1984.

He, L.M. and Tebo, B.M., Surface charge properties of and Cu(II) adsorption by spores of the marine *Bacillus* sp. strain SG-1, *Appl. Environ. Microbiol.*, 64, 1123, 1998.

Hem, J.D. and Lind, C.J., Nonequilibrium models for predicting forms of precipitated manganese oxides, *Geochim. Cosmochim. Acta,* 47, 2037, 1983.

Herman, D.C., Artiola, J.F., and Miller, R.M., Removal of cadmium, lead, and zinc from soil by a rhamnolipid biosurfactant, *Environ. Sci. Technol.*, 29, 2280, 1995.

Jenne, E.A., Controls on Mn, Co, Ni, Cu and Zn concentrations in soil and water: The significant role of hydrous Mn and Fe oxides, in *Trace Inorganics in Water,* Gould, R.F., Ed., American Chemical Society, Washington, D.C., 1968, p. 337.

Krauskopf, K.B., Factors controlling the concentration of thirteen rare metals in seawater, *Geochim. Cosmochim. Acta,* 9, 1, 1956.

Lion, L.W., Altmann, R.S., and Leckie, J.O., Trace-metal adsorption characteristics of estuarine particulate matter: Evaluation of contributions of Fe/Mn oxide and organic surface coatings, *Environ. Sci. Technol.*, 16, 660, 1982.

Lion, L.W. et al., Trace metal interactions with microbial biofilms in natural and engineered systems, CRC Crit. *Rev. Environ. Control,* 17, 273, 1988.

Manceau, A. and Combes, J.M., Structure of Mn and Fe oxides and oxyhydroxides: A topological approach by EXAFS, *Phys. Chem. Miner.*, 15, 283, 1988.

Mandernack, K.W. et al., Oxygen isotope analyses of chemically and microbially produced manganese oxides and manganates, *Geochim. Cosmochim. Acta,* 59, 4409, 1995.

Mandernack, K.W., Post, J., and Tebo, B.M., Manganese mineral formation by bacterial spores of the marine Bacillus, strain SG-1: Evidence for the direct oxidation of Mn(II) to Mn(IV), *Geochim. Cosmochim. Acta*, 59, 4393, 1995.

Moffett, J.W., The relationship between cerium and manganese oxidation in the marine environment, *Limnol. Oceanogr.,* 39, 1309, 1994.

Moffett, J.W., The importance of microbial Mn oxidation in the upper ocean: A comparison of the Sargasso Sea and equatorial Pacific, *Deep Sea Res. Part I Oceanogr. Res. Pap.*, 44, 1277, 1997.

Moffett, J.W. and Ho, J., *Microbially mediated incorporation of trace-elements into manganese oxides in seawater,* Abstr. Pap. Am. Chem. Soc., 209, 103-Geoc, 1995.

Morgan, J.J. and Stumm, W., Colloid-chemical properties of manganese dioxide, *J. Colloid Sci.*, 19, 347, 1964.

Murray, J.W. et al., Oxidation of Mn(II): Initial mineralogy, oxidation state and aging, *Geochim. Cosmochim. Acta*, 49, 463, 1985.

Nealson, K.H. and Myers, C.R., Microbial reduction of manganese and iron: New approaches to carbon cycling, *Appl. Environ. Microbiol.*, 58, 439, 1992.

Nealson, K.H., Tebo, B.B., and Rosson, R.A., Occurrence and mechanisms of microbial oxidation of manganese, *Adv. Appl. Microbiol.*, 33, 299, 1988.

Nelson, Y.M. et al., Lead distribution in a simulated aquatic environment: Effects of bacterial biofilms and iron oxide, *Water Res.*, 29, 1934, 1995.

Nelson, Y.M. et al., Lead binding to metal oxides and organic phases of natural aquatic biofilms, *Limnol. Oceanogr.*, 44, 1715, 1999a.

Nelson, Y.M. et al., Production of biogenic Mn oxides by *Leptothrix discophora* SS-1 in a chemically defined growth medium and evaluation of their Pb adsorption characteristics, *Appl. Environ. Microbiol.*, 65, 175, 1999b.

Nelson, Y.M. et al., Adsorption Properties of Mixed Ferric and Manganese Oxides, paper presented at American Chemical Society, 221st national meeting, San Diego, CA, 2001.

Nelson, Y.M. et al., Effect of oxide formation mechanisms on lead adsorption by biogenic manganese (hydr)oxides, iron (hydr)oxides, and their mixtures, *Environ. Sci. Technol.,* 36, 421, 2002.

Pradhan, A.A. and Levine, A.D., Experimental evaluation of microbial metal uptake by individual components of a microbial biosorption system, *Water Sci. Technol.*, 26, 2145, 1992.

Stumm, W. and Morgan, J.J., *Aquatic Chemistry: Chemical Equilibria and Rates in Natural Waters*, John Wiley & Sons, New York, 1996.

Sunda, W.G. and Huntsman, S.A., Microbial oxidation of manganese in a North Carolina estuary, *Limnol. Oceanogr.*, 32, 552, 1987.

Sunda, W.G. and Kieber, D.J., Oxidation of humic substances by manganese oxides yields low-molecular-weight organic substrates, *Nature*, 367, 62, 1994.

Tebo, B.M., Manganese(II) oxidation in the suboxic zone of the Black Sea, *Deep Sea Res.*, 38, S883, 1991.

Tebo, B.M. and Emerson, S., Effect of oxygen tension, manganese-II concentration and temperature on the microbially catalyzed manganese-II oxidation rate in a marine fjord, *Appl. Environ. Microbiol.*, 50, 1268, 1985.

Tebo, B.M. and Emerson, S., Microbial manganese (II) oxidation in the marine environment: A quantitative study, *Biogeochemistry*, 2, 149, 1986.

Tebo, B.M. et al., Bacterially mediated mineral formation: Insights into manganese(II) oxidation from molecular genetic and biochemical studies, in *Geomicrobiology: Interactions Between Microbes and Minerals*, Banfield, J.F. and Nealson, K.H., Eds., Mineralogical Society of America, Washington, D.C., 1997, p. 225.

Tebo, B.M., Mandernack, K., and Rosson, R.A., Manganese oxidation by spore coat or expression protein from spores of a manganese (II) oxidizing marine Bacillus, *Abstr. Annu. Meet. Am. Soc. Microbiol.*, I-121, 201, 1988.

Tessier, A. et al., Metal sorption to diagenetic iron and manganese oxyhydroxides and associated organic matter: Narrowing the gap between field and laboratory measurements, *Geochim. Cosmochim. Acta,* 60, 387, 1996.

Tipping, E., Temperature dependence of Mn(II) oxidation in lakewaters: A test of biological involvement, *Geochim. Cosmochim. Acta*, 48, 1353, 1984.

Turekian, K.K., The fate of metals in the oceans, *Geochim. Cosmochim. Acta,* 41, 1139, 1977.

VanWaasbergen, L.G., Hildebrand, M., and Tebo, B.M., Identification and characterization of a gene cluster involved in manganese oxidation by spores of the marine *Bacillus* sp. strain SG-1, *J. Bacteriol.,* 178, 3517, 1996.

Verity, R., Investigations of Manganese Oxidizing Bacteria, Master's thesis, Cornell University, 2001.

von Langen, P.J.V. et al., Oxidation kinetics of manganese (II) in seawater at nanomolar concentrations, *Geochim. Cosmochim. Acta*, 61, 4945, 1997.

Vuceta, J. and Morgan, J.J., Chemical modeling of trace metals in fresh waters: Role of complexation and adsorption, *Environ. Sci. Technol.*, 12, 1302, 1978.

Wehrli, B., Friedl, G., and Manceau, A., Reaction rates and products of manganese oxidation at the sediment–water interface, in *Advances in Chemistry Series 244, Aquatic Chemistry: Interfacial and Interspecies Processes*, Huang, C.P., O'Melia, C.R., and Morgan, J.J., Eds., American Chemical Society, Washington, D.C., 1992, p. 111.

Westall, J.C. et al., Models for association of metal ions with heterogeneous environmental sorbents. 1. Complexation of Co(II) by leonardite humic acid as a function of pH and $NaClO_4$ concentration, *Environ. Sci. Technol.*, 29, 951, 1995.

Wilson, A.R. et al., The effects of pH and surface composition on Pb adsorption to natural freshwater biofilms, *Environ. Sci. Technol.*, 35, 3182, 2001.

Zhang, J. et al., Kinetics of Mn oxidation by *Leptothrix discophora* SS1, *Geochim. Cosmochim. Acta*, 65, 773, 2002.

8 Zinc Speciation in Contaminated Soils Combining Direct and Indirect Characterization Methods

Darryl Roberts, Andreas C. Scheinost, and Donald L. Sparks

CONTENTS

8.1 Introduction 188
8.2 Approaches to Determining Metal Speciation in Soils 191
 8.2.1 Single Extraction Methods 191
 8.2.2 Selective Sequential Extraction Methods 191
 8.2.3 Analytical Techniques 192
  8.2.3.1 Synchrotron-Based Methods 193
  8.2.3.2 Microspectroscopic Approaches 194
 8.2.4 Advantage of Combining Techniques 195
8.3 Case Study: Zn-Contaminated Soil in the Vicinity of a Smelter 195
 8.3.1 Site Description, Sampling, and Soil Characteristics 195
 8.3.2 XRD and EMPA Analysis 197
 8.3.3 Sequential Extractions 198
 8.3.4 Bulk EXAFS Spectroscopy 202
  8.3.4.1 EXAFS Data Analysis 202
 8.3.5 EXAFS of Soil Samples 205
  8.3.5.1 Surface Soil 205
  8.3.5.2 Subsurface Soils 207
 8.3.6 EXAFS Combined with Sequential Extractions 209
 8.3.7 Synchotron-μ-XRF 215
 8.3.8 μ-EXAFS 217
  8.3.8.1 Surface Soil 217
  8.3.8.2 Subsurface Soil 218
 8.3.9 Desorption Studies 219

1-56670-623-8/03/$0.00+$1.50

8.4 Conclusions and Environmental Significance 221
8.4.1 Fate of Zn in Soils 221
8.4.2 Summary of Speciation Techniques 222
References 223

8.1 INTRODUCTION

The contamination of surface and subsurface environments via the anthropogenic and natural input of heavy metals has established the need to investigate and comprehend metal–soil interactions. The pathways for heavy metal introduction into soil and aquatic environments are numerous, and include the land application of sewage sludge and municipal composts, mine wastes, dredged materials, fly ash, and atmospheric deposits.[1] In addition to these anthropogenic sources, heavy metals can be introduced to soils naturally as reaction products via the dissolution of metal-bearing minerals that are found in concentrated deposits. Of a thousand Superfund sites named in the U.S. Environmental Protection Agency's National Priority List of 1986, 40% were reported to have elevated levels of heavy metals relative to background levels.[2] The fate and mobility of these metals in soils and sediments are of concern because of potential bioaccumulation, food chain magnification, degradation of vegetation, and human exposure.[3]

The effective toxicity of heavy metals to soil ecosystems depends not only on total metal concentrations, but also, and perhaps more importantly, on the chemical nature of the most mobile species. The long-term bioavailability to humans and other organisms is determined by the resupply of the metal to the mobile pool from more stable phases. Thus, quantitative speciation of metal species as well as their variation with time is a prerequisite for long-term risk assessments. The complex and heterogeneous array of mineral sorption sites, organic materials, metal oxides, macro- and micro-pores, and microorganisms in soils provide a matrix that may strongly sequester metal ions. Noncrystalline aluminosilicates (allophanes), oxides, and hydroxides of Fe, Al, and Mn, and even the edges of layer silicate clays, to a lesser extent, provide surface sites for the specific adsorption and interaction of transition and heavy metals.[4] Before any remediation strategy is attempted, it is wise to determine and understand the nature of the interactions of metal ions with these reactive sites. These interactions can be considered one portion of the overall concept of metal speciation in soils. However, the determination of metal speciation in complex and heterogeneous systems such as soils and sediments is far from a trivial task.

Speciation encompasses both the chemical and physical form an element takes in a geochemical setting. A detailed definition of speciation includes the following components: (1) the identity of the contaminant of concern or interest; (2) the oxidation state of the contaminant; (3) associations and complexes to solids and dissolved species (surface complexes, metal-ligand bonds, surface precipitates); and (4) the molecular geometry and coordination environment of the metal.[5] The more of these parameters that can be identified the better one can predict the potential risk of toxicity to organisms by heavy metal contaminants. Prior to the application of sequential extraction techniques and analytical tools, researchers often relied on total metal concentration as an indication of the degree of bioavailability of a heavy metal. However, several studies have shown that the form the metal takes in soils is of much greater importance than the total concentration of the metal with regards to the bioavailability to the organism.[6,7] Metal speciation in soil and aquatic systems continues to be a dynamic topic and of interest to soil scientists, engineers, toxicologists, and geochemists alike, as there remains no sufficient method to characterize metal contaminants in all natural settings.

The lack of a universal method of determining heavy metal speciation in natural settings comes as a result of the complexity of soil, sediment, and aquatic environments. The multiple solid phases in soils include primary minerals, phyllosilicates, hydrous metal oxides, and organic debris. Metals can potentially bind to these sorbents by a number of sorption processes, including both chemical and physical mechanisms. The mechanism(s) of metal binding strongly influences the fate and bioavailability of metals in the environment. In addition to solid phases, the soil solution is also heterogeneous in nature, containing dissolved organic matter and other metal-binding ligands over a range of concentrations. This leads to metal-ligand complexes in the soil solution and ternary complexes at the solid–solution interface. The presence of ligands in an ion-sorbent complex has been shown to influence the atomic coordination environment of the ion and, therefore, may lead to differences in the stability of metal sorption complexes.[8] The partitioning of metal contaminants between solid and solution phases is a dynamic process and an accurate description of this process is important in constructing models capable of predicting heavy metal behavior in surface and subsurface environments.

A metal that has received a fair amount of attention due to its ubiquitous nature in soils and sediments and role as a plant essential nutrient, is Zn. Zinc is mined in 50 countries and smelted in 21 countries.[9] At background levels it poses no serious threat to biota and vegetation, while in areas that have elevated levels of Zn as a result of smelting, land application of biosolids, or other anthropogenic processes, it is often a detriment to the environment.[10] At acidic pH values, Zn toxicity to plants is the third most common metal toxicity behind Al and Mn.[10] Under acidic oxidizing conditions, Zn is one of the most soluble and mobile of the trace metal cations. It does not complex tightly with organic matter at low pH; therefore, acid-leached soils often have Zn deficiencies because of depletion of this element in the surface layer. The degree of Zn bioavailability and, therefore Zn toxicity, is by and large determined by the nature of its complexation to surfaces found in soils, such as phyllosilicates, metal oxides, and organic matter. Research investigating Zn sorption using laboratory-based macroscopic sorption experiments using oxide and clay minerals as sorbents suggests Zn has variable reactivity and speciation in soils. Sorption studies have shown that Zn can adsorb onto Mn oxides, Fe (hydr)oxides and Al (hydr)oxides, and aluminosilicates.[11–18] At alkaline pH values and at high initial Zn concentrations, the precipitation of $Zn(OH)_2$, $Zn(CO)_3$, and $ZnFe_2O_4$ may control Zn solubility.[19,20] In these studies, however, direct determination of Zn sorption mechanisms and speciation using spectroscopic and/or microscopic approaches was not employed, allowing room for further interpretation of the results.

With the advent of more sophisticated analytical techniques and their application to soils and sediments, further information on the nature of Zn sorption complexes in clay mineral and metal oxide systems has been gleaned. Waychunas et al.[21] studied Zn sorption to ferrihydrite using x-ray absorption fine structure (XAFS) spectroscopy and found that Zn forms inner-sphere adsorption complexes at low Zn sorption densities, changing to the formation of Zn hydroxide polymers with increasing Zn sorption densities, and finally transforming to a brucite-like solid phase at the highest sorption densities in the study. In a study of Zn sorption on goethite, inner-sphere surface complexes were observed using XAFS.[22] In investigations using Al-bearing mineral phases as sorbents and at neutral to basic pH values, researchers have demonstrated that Zn can form both inner-sphere surface complexes and Zn hydrotalcite-like phases upon sorption to Al-bearing minerals.[23,24] Zn sorption on manganite resulted in both inner-sphere and multinuclear hydroxo-complexes.[25] Perhaps the most significant finding in many of these studies is the fact that Zn-bearing precipitate phases often formed under reaction conditions well below the solubility limit of known Zn solid phases, suggesting that their formation in soils and sediments may have been overlooked using conventional approaches. For example, the sorption kinetics of Zn on hydroxyapatite surfaces had an initial rapid sorption step followed by a much slower rate of Zn removal from solution.[26] It was conceded that x-ray diffraction (XRD) and scanning electron microscopy (SEM) were not sensitive enough to determine if precipitation was a major mechanism at high pH values (>7.0).

With a substantial amount of Zn sorption studies performed using a combination of sophisticated analytical tools such as XAFS in mineral and metal oxide systems, there is a natural progression to investigate Zn speciation in actual soils and sediments. By applying XAFS and electron microscopy to Zn-contaminated soils and sediments, Zn has been demonstrated to occur as ZnS in reduced environments, often followed by repartitioning into Zn hydroxide and/or ZnFe hydroxide phases, adsorption to Fe(oxyhydr)oxides, or incorporation into phyllosilicates upon oxidation.[27–30] Manceau et al.[31] employed a variety of techniques, including XRD, XAFS, and micro-focused XAFS to demonstrate that upon weathering of Zn-mineral phases in soils, Zn was taken up by the formation of Zn-containing phyllosilicates and, to a lesser extent, by adsorption to Fe and Mn (oxyhydr)oxides. In addition to adsorption and precipitation as the primary mechanisms for Zn removal from solution, Zn may be effectively removed from solution via diffusion of Zn ions into the micropores of Fe oxides.[32,33] These studies demonstrate that in any given system, Zn may be present in one of several forms making direct identification of each species difficult using traditional approaches. The majority of studies employed to characterize the reactivity in Zn has dealt with relatively simplistic systems, with one or two sorbent phases in question. Clearly, natural environments are much more complex and only after extensive studies in the above systems can one focus on natural samples. To better illustrate this point, we now turn our attention to the various approaches that have been used to identify metal species in soils and sediments, followed by a specific scenario of applying these techniques to Zn-contaminated soils.

8.2 APPROACHES TO DETERMINING METAL SPECIATION IN SOILS

8.2.1 Single Extraction Methods

For most contaminated sites where practical considerations of limited money and resources are operational, the most efficient and cost-effective method of determining heavy metal speciation is often desired. One of the most commonly used approaches has been to measure total metal concentration and correlate this to the amount of metal that may be bioavailable, based on thermodynamic considerations. However, total concentration approaches overlook the fact that not all of the metal may be labile or available for uptake.[7] Slightly more discriminating in the amount of metal extracted is the approach of single extractions using chemicals such as EDTA and DTPA. This approach has been successfully applied to soils for both fertility assessments and for estimating the degree of contamination for heavy-metal impacted sites.[34,35] These approaches generally cannot estimate the amount of slowly available metal that is released over time since extractions are carried out over a period of several hours. Moreover, the exact speciation of the metal is not gleaned using these types of approaches. However, these approaches continue to be developed and are of great benefit given their relatively low cost and availability.

8.2.2 Selective Sequential Extraction Methods

A more rigorous and complete alternative to determining metal speciation via total metal concentration and one-step extractions is the use of sequential extractions. Sequential extraction methods for heavy metals in soils and sediments have been developed and employed in an effort to provide detailed information on metal origin, biological and physicochemical availability, mobilization, and transport.[36,37] After many studies and refinements, the chemical extractions steps are designed to selectively extract physically and chemically sorbed metal ions, as well as metals occluded in carbonates, Mn (hydr)oxides, crystalline and amorphous Fe (hydr)oxides, and metal sulfides. The resulting extract is operationally defined based on the proposed chemical association between the extracted species and solid phases in which it is associated. Given that the extraction is operationally defined, the extracted metal may or may not truly represent the defined chemical species, so care must be taken to report the step in which it was removed rather than the phases it is associated with. Many studies investigating the impact of mining and metallurgic activities on soils have utilized various sequential extraction techniques in an effort to speciate heavy metals.[38–40]

The use of sequential extractions for metal speciation has other limitations and pitfalls as well. These include (1) the incomplete dissolution of a target phases; (2) the removal of a nontarget species; (3) the incomplete removal of a dissolved species due to re-adsorption on remaining soil components or due to re-precipitation with the added reagent; and (4) change of the valence of redox-sensitive elements.[41–45] These limitations are becoming more evident with the progress in research coupling sequential extractions with analytical techniques capable of directly determining metal speciation in soils and sediments.[39,41,43,44,46] These studies, and future studies, will certainly aid in explaining why selectively extracted metal fractions are often not or only weakly correlated to bioavailable metals.[45] In doing so, the process of sequential extractions will become more complete and universal, significantly improving our understanding of metal partitioning and mobility in soils. Despite the limitations of these approaches, however, sequential extractions continue to be valuable for relative comparisons between contaminated sites, and due to their widespread availability and relative ease.

8.2.3 Analytical Techniques

Several analytical tools prevalent in characterization of materials in the surface sciences, chemistry, physics, and geology have been applied to direct speciation of heavy metals in soils and sediments for many years. The clear advantage in using direct techniques over chemical extractions is the lower risk of sample alteration and transformations of metal species from using extracting solutions. When selecting an analytical technique to speciate and quantify the form of metals in complex heterogeneous materials such as soils and sediments, a selective and nondestructive one is favorable.[47] One of the most widely used analytical techniques is XRD. For characterization of crystalline phases and minerals, XRD is extremely useful. However, metal-contaminated soils and sediments often contain the metal in a form such that it is a minority phase below the detection limit of the instrument, or the important reactive phase is amorphous and only produces a large background in the diffractogram. Other x-ray–based techniques include x-ray fluorescence (XRF) spectroscopy and x-ray photoelectron spectroscopy (XPS). XRF has been used for decades to determine the concentration of trace metals in soils and sediments, with lower detection limits becoming more common with technological advances.[48] However, this technique only provides elemental concentrations with no insight into metal speciation. XPS, however, is a surface-sensitive analytical technique that provides elemental chemical state and semi-quantitative information.[49] The pitfall to this technique is that it is *ex situ*, and requires samples be dried and placed under ultra high vacuum that may lead to experimental artifacts.[50] Other techniques that provide useful information on elemental speciation in soils and sediments but also are *ex situ* include auger electron spectroscopy (AES) and secondary mass spectroscopy (SIMS).[50]

Given the myriad of reactive phases in soils and their complex distribution in the soil matrix, a technique capable of providing spatial and morphological information on heavy metal speciation is desired. Microscopic techniques may resolve the different reactive sites in soil at the micron level, thus allowing for a more selective approach to speciation. Examples of these techniques include SEM, electron microprobe analysis (EMPA), and transmission electron microscopy (TEM). In order to glean elemental information and ratios, all the above techniques are often coupled with an energy dispersive spectrometer (EDS). While the above techniques have given insight into elemental associations and metal distributions in contaminated soils and sediments, they do have a few drawbacks. The most notable are that EDS is only sensitive to greater than 0.1% elemental concentration, it is insensitive to oxidation states of target elements, and it does not provide crystallographic data.[29,41,43] A study investigating Zn speciation in contaminated sediments found that SEM coupled with x-ray EDS only provided elemental concentrations, but discerning between Zn sulfate and Zn sulfide was not possible.[29] Similarly, EMPA was unable to locate Hg grains within a Hg-contaminated sample and was unable to distinguish between polymorphs of Hg-bearing phases (cinnabar and metacinnabar).[51] Several other studies have pointed out similar shortcomings of these techniques in speciating metal phases in soils and sediments.[41,43] In all of these studies, the authors mention and/or use XAFS as a more robust technique to characterize the metal phases and complexes found in their samples. Indeed, given its sensitivity to amorphous species, minority phases, and adsorbed complexes, XAFS is one of the few *in situ* techniques capable of discerning between the myriad of possible surface species occurring on the submicrometer scale in soils and sediments. We now turn our attention to the use of this technique in determining metal speciation in natural environments.

8.2.3.1 Synchrotron-Based Methods

The application of synchrotron light sources to address environmental issues has provided insight into the reaction mechanisms of heavy metals at interfaces between sorbent phases found in soils and the soil solution. The most widely used technique for this has been XAFS. The term XAFS is a general term encompassing several energies around an absorption edge for a specific element, namely the pre-edge, near-edge (XANES), and extended portion (EXAFS). Each region provides specific information on an element depending on the selected energy range, making XAFS an element-specific technique. Several articles provide excellent overviews on the use of this technique in environmental samples.[52,53] Briefly, in the XANES region, electron transitions lead to an absorption edge from which chemical information of the target element, such as oxidation state, can be deduced. EXAFS can provide the identity of the ligands surrounding the target element, specific bond distances, and coordination numbers of first- and second-shell ligands.[53] This information is extremely useful in speciation of metals in soils and sediments as it provides quantitative information on the geometry, composition, and mode of attachment of a metal ion at a sorbent interface.[5] Given the intensity of synchrotron facilities, this technique has a detection limit down to 50 ppm and can target a specific element, potentially with little interference from other elements in the complex matrix in which it is located. Gleaning this type of information *in situ* is not possible with any other technique. Features that have dramatically increased the use of XAFS in environmental studies include more available synchrotron facilities, more routine data analysis due to computer-based packages, and word of mouth via professional meetings and journal articles. Many studies can be found in the literature detailing the use of this technique in order to speciate metals in soils and sediments.[3,31,41,43,45,47,51,54–57] Nonetheless, XAFS does have limitations and is by no means the only technique one should use for speciation of heavy metals in environmental samples.

In soils and other natural samples, metal ions may partition to more than one reactive site, with each sorbent–sorbate complex providing a unique spectroscopic signal. In addition, the x-ray beam hitting the sample will inevitably bombard the sorbent phase or other minerals in the matrix which may cause fluorescence, resulting in an interference with the spectrum of the central element of interest (e.g., for Co and Ni in samples containing Fe oxides in a significant amount).[58] The EXAFS spectra obtained in doing these types of measurements represents the sum of all the geometric configurations of the sorbing ion, weighted by the abundance of each.[58] Therefore, the determination of all metal species is only as good as the ability to analyze the data successfully. In order to discriminate between species and quantify them in a multispecies system, the target species must have different oxidation states, or vary in atomic distances by ≥0.1 Å and/or coordination numbers by ≥1.[59] Using a nonlinear least-squares fit of the raw data or a shell-fitting approach of Fourier-transformed data, typically only two species may be detected within a given sample and there is a tendency to overlook soluble species with weak or missing second-shell backscattering in the presence of minerals with strong second-shell backscattering.[31] This latter point often leads to an inability to successfully detect minor metal-bearing phases, even though they may be the most reactive or significant in the metal speciation. Discrimination among species has also been achieved using the linear combination fit (LCF) technique, where spectra of known reference species are fitted to the spectrum of the unknown sample. LCF has been successfully employed to identify and quantify up to three major species, including minerals and sorption complexes.[43,55] The success of the speciation depends critically on a spectral database containing all the major species coexisting in the unknown sample, underscoring the need to have a thorough database of reference spectra. One way to determine single species in a multispecies system separated by space is to use micro-focused XAFS (μ-XAFS), which will be discussed below.

Logistical drawbacks to using XAFS include the availability of synchrotron light sources, the increased demand for beam time at these facilities, and the difficulty in analyzing data. Clearly, the number of metal-impacted sites requiring metal speciation information far exceeds the amount of time available at synchrotron facilities. The combination of XAFS with more routine speciation techniques, such as sequential extractions, is important, as the former technique has been able to detect artifacts and other shortcomings of the latter technique and may eventually lead to more specific and defined extraction procedures.[41,45] By combining sequential extraction techniques with XAFS, the number of species may be reduced by chemical separation prior to attempting their identification by XAFS. Moreover, the use of two independent methods for determining metal speciation in soils may provide a more reliable result than either of the methods alone.

8.2.3.2 Microspectroscopic Approaches

To date, standard bulk XAFS has been the most widely used synchrotron-based technique used to characterize heavy metals in environmental samples. However, in soils and sediments, microenvironments exist that have isolated phases in higher concentrations relative to the average of the total matrix.[53] For example, the microenvironment of oxides, minerals, and microorganisms in the rhizosphere has been shown to have a quite different chemical environment compared to the bulk soil.[60] Often these phases may be very reactive and of significance in the partitioning of heavy metals, but may be overlooked using other analytical techniques that measure an area constituting the average of all phases. With focusing mirrors and other devices, the x-ray beam bombarding a sample may go down to a few square microns in area, nearing the size of the most reactive species in soils, enabling one to distinguish between individual species in a heterogeneous system. In order to determine the exact location to place the focused x-ray beam on the sample, μ-XAFS is often combined with microsynchrotron-based XRF (μ-SXRF), allowing elemental maps to be obtained prior to analysis. While EMPA is often not sensitive enough to detect trace metals in soil, μ-SXRF offers sufficient sensitivity to investigate the spatial distribution of trace metals and their spatial correlation with other elements. Until recently, most studies have employed μ-XANES to determine the oxidation state of target elements in environmentally relevant samples since first- and second-generation light sources were not bright enough to achieve decent results for μ-EXAFS.[31,61–63] With the advent of brighter, third-generation sources, μ-EXAFS has been used to speciate metals in soils and sediments.[30,31,64,65]

8.2.4 Advantage of Combining Techniques

In this brief overview of the approaches to speciation of metals in soils, sediments, and other environmentally relevant settings, it is clear that no single technique enables one to get an accurate and precise determination of metal speciation. In fact, several of the aforementioned studies that used a combination of chemical extraction and analytical techniques such as XRD, microscopy, and x-ray absorption techniques arrived at the conclusion that the most thorough results were achieved in combining techniques.[38,41,54] Since no single characterization method gives a complete description of surface structure or the geometric details of sorption complexes, it is important to employ a variety of methods that provide complementary information.[66] To further illustrate this point, the remainder of the chapter focuses on the combination of several analytical techniques in determining and quantifying Zn speciation in a soil contaminated as a result of smelting operations. In addition, results from a leaching experiment will serve to link metal speciation to metal bioavailability. Each technique is presented in its own section, with a summary comparing and contrasting the usefulness of each result. This has been the focus of two separate papers, and many of the figures and discussion can be found therein.[64,67] We hope that the advantages of combining techniques will become clear, particularly when it comes to determining the shortcomings of each technique.

8.3 CASE STUDY: Zn-CONTAMINATED SOIL IN THE VICINITY OF A SMELTER

8.3.1 Site Description, Sampling, and Soil Characteristics

Emissions from the Palmerton smelting plant in Palmerton, Pennsylvania have contaminated over 2000 acres of land on the north-facing slope of nearby Blue Mountain in the Appalachians (Figure 8.1). The Zn smelting facilities (Smelters I and II) are located in east-central Pennsylvania near the confluence of Aquashicola Creek and

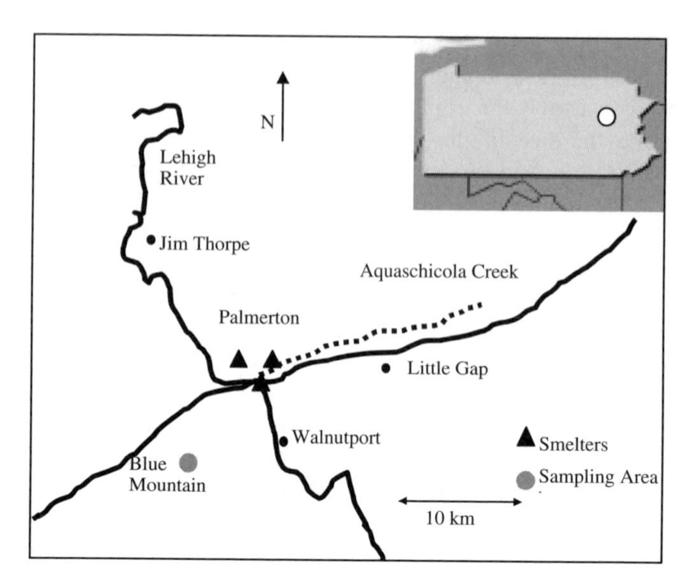

FIGURE 8.1 Location of Blue Mountain sampling site in the vicinity of the Palmerton Smelter, Palmerton, Pennsylvania.

the Lehigh River in the town of Palmerton.[68] The first of two smelting plants was opened in 1898 by the New Jersey Zinc Company in order to process zinc sulfide (sphalerite) from New Jersey ore. In 1980, the plants stopped Zn smelting and in 1982 the U.S. Environmental Protection Agency placed the facilities on its national priorities list as a Superfund site. The sphalerite ores contained approximately 55% zinc, 31% sulfur, 0.15% cadmium, 0.30% lead, and 0.40% copper.[69] For 82 years the facilities had an average annual output of metals measuring 47 Mg of Cd, 95 Mg of Pb, and 3,575 Mg of Zn. Daily metal emissions since 1960 ranged from 6000 to 9000 kg of Zn, from 70 to 90 kg of Cd, and less than 90 kg of Pb and Cu.[69,70] Sulfuric acid produced by smelting processes was also deposited in the surrounding areas, contributing to strongly acidic soil pH values. As a consequence, the dense forest vegetation of Blue Mountain was completely lost and soils on hill slopes almost completely eroded, exposing the underlying bedrock. Several attempts have been made to remediate the site and some revegetation has been successful, but exposed soil surfaces and bedrock are still prevalent.[69]

The most heavily contaminated soil collected from a profile directly above Smelter II was selected for detailed experiments. The soil was collected from a pit between exposed bedrocks, where a shallow soil profile <15 cm in depth persisted. The topsoil consisted of a 3- to 6-cm thick layer of dark, hydrophobic organic debris consisting of only partially decomposed plant residues and soil organic matter. The accumulation of this amount of organic matter, which does not exist in surrounding forest soils, is an indication of drastically reduced biodegradation. The consolidated subsoil about 20 cm in thickness is most likely the remainder of the original Dekalb and Laidig stony loam soils derived from shale, sandstone, and conglomerate.[68] Undisturbed and bulk samples were collected from both topsoil and subsoil. In addition, a sediment sample was collected from an artificial pond nearby, which was dry at the time of sampling.

Soil samples were air dried, sieved to collect the <2-mm size fraction, and then ground in a mortar for sample homogenization. For EMPA and μ-XAFS measurements, the undisturbed samples, aggregates of several centimeters in diameter, were air dried, embedded in acrylic resin (LR-White), cut, and polished into thin sections of various thicknesses (30 to 350 μm). For XRD analysis, soils were dispersed in DDI water for 24 h, sonified to break up aggregates, and wet sieved to collect the <250-μm fraction. From the subsoil sample, dark concretions 0.5 to 2 mm in diameter were hand collected and ground in a mortar and pestle for XAFS analysis. Soil pH was determined in 0.01 M $CaCl_2$. Total metal concentrations were measured with an X-Lab 2000 energy-dispersive x-ray fluorescence spectrometer (Spectro) equipped with a sequence of secondary targets (Mo, Al_2O_3, B_4C/Pd, Co, and HOPG) to generate polarized x-rays. The lower detection limit was 0.5 mg/kg for most metals. Results of these analyses are presented in Table 8.1.

8.3.2 XRD and EMPA Analysis

Bulk mineralogy of the <250-μm fraction of the surface and subsurface soils was determined by powder XRD using a Philips Norelco 1720 instrument equipped with a Cu tube (40 kV, 40 mA). Diffractograms were collected between 3° and 70° 2θ, with 0.04° steps and a counting time of 5 sec per step. Results indicated that quartz is the most abundant mineral in both the topsoil and subsoil. In addition to quartz, the subsoil contained gibbsite, an Al-interlayered clay mineral (determined by ion saturation and heating), and evidence of amorphous Fe and Mn oxides. Diffractograms of the topsoil showed peaks from franklinite ($ZnFe_2O_4$), a spinel-type mineral. XRD analysis of the subsoil did not reveal the presence of any Zn-bearing minerals. Similar studies done on soils with increased levels of heavy metals also had difficulty identifying metal-bearing species using XRD, even when these species were readily identified using XAFS or other techniques.[40,51] However, while XRD may not be

TABLE 8.1
Palmerton Soil Sample Characteristics

		Topsoil	Subsoil	Sediment
pH		3.2	3.9	4.5
C	g/kg	320	50	25
S	g/kg	6.4	0.8	1.0
Mn	g/kg	1.5	0.5	0.1
Fe	g/kg	33	25	22
Zn	mg/kg	6200	900	2500
Pb	mg/kg	7000	62	406
OM content	%	12	.8	3

Source: Reprinted with permission from Scheinost, A.C. et al., *Environ. Sci. Technol.* 36, 5021, copyright 2002. American Chemical Society.

the best way to identify metal-bearing phases in soils, it does provide information on possible sorbent phases present in the soil that may be capable of complexing metal ions.

Electron microprobe analysis was performed on the resin-embedded thin sections (30 to 100 μm thick) mounted on pure quartz slides, using a JEOL JXA-8600 microprobe equipped with wavelength-dispersive spectrometers (WDSs). Several elements (Si, Al, S, P, K, Ca, Zn, Mn, Fe, and Pb) were mapped, and then the compositions of selected sample areas were determined with higher precision. The images were taken using a backscattered electron detector so that low Z elements appear dark and high Z elements are bright. Backscattered electron images (BSEs) and selected elemental distributions collected by EMPA analysis are shown for the topsoil (Figure 8.2) and the subsoil (Figure 8.3). The main spherical entity in the topsoil image is an organic aggregate with moieties of metal-bearing phases distributed throughout, indicated by the bright white spots in the BSE. Several Zn grains were found using this technique, measuring 1 to 4 μm in diameter. Qualitatively, most of these are associated with Fe and S. Detailed quantitative WDSs of such spots gave Fe/Zn ratios of 1 to 2 in agreement with those in franklinite, and S/Zn ratios of about 1 indicating either Zn sulfide or sulfate. Regions of enriched Si and K were also present, most likely representing quartz and K-feldspars, respectively. EMPA was less successful in identifying Zn-bearing phases in the subsoil, indicating that Zn was not found in abundant portions in any one phase. Aluminum, Si, and Fe in the maps in the subsurface soil indicated the presence of clay minerals and/or metal oxides. While not considered in this study, Pb was identified in both the surface and subsurface samples using EMPA. In the surface soil, Pb was more evenly distributed throughout the organic aggregates, and concentrated in two Pb-rich particles in the subsoil sample. This suggests, qualitatively, that Zn and Pb behave differently in these soils.

8.3.3 Sequential Extractions

Sequential extractions (SSEs) were performed on the surface and subsurface soils, as summarized in Table 8.2. The first six steps follow the method of Zeien and Brümmer,[71] and Brümmer and Herms.[72] The residual phase, that is, the phase that remained after step 6, was digested with a two-step microwave procedure. For the first step, 250 mg of sample were added to 4 ml of 65% HNO_3, 2.5 ml of 32% HCl and 1 ml of 48% HF, and then heated for 5 min at 150 W and for 15 min at 350 W. For the second step the sample was cooled, and 10 ml of a 6% H_3BO_3 were added and heated for 25 min at 250 W until a pressure of 15 bars was reached. No visible residues were left after this procedure. Extracted Zn was measured by atomic absorption spectrometry (Varian Spectra 220 Fast Sequential). The percentage of total Zn removed in each of the extraction steps is presented in Figure 8.4. The sum of all SSE steps was in good agreement with the total amount determined by XRF (±5%).

In the topsoil, 86% of Zn was extracted in steps 6 and 7, indicating that Zn was predominantly bound by iron oxides and other, more stable minerals/oxides (Table 8.3). The amount of readily exchangeable Zn (step 1) accounted for only 7%, and

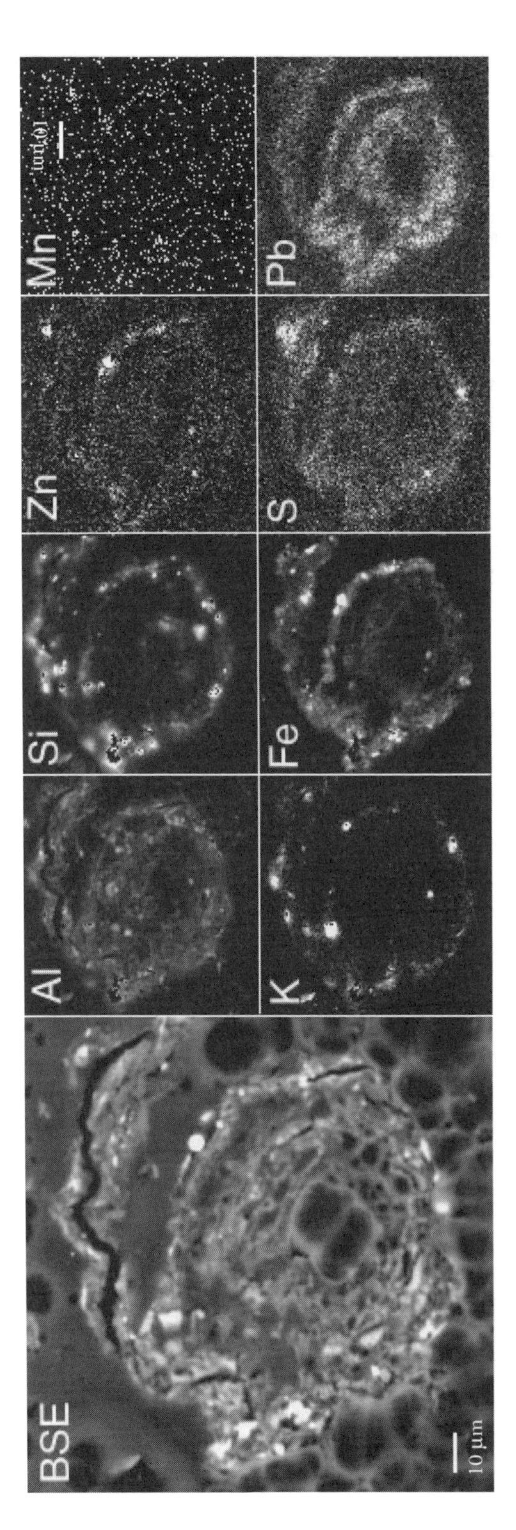

FIGURE 8.2 Backscattered electron image of surface soil and corresponding x-ray elemental dot maps. White colors indicate highest concentration of target elements, and dark spots indicate low concentration.

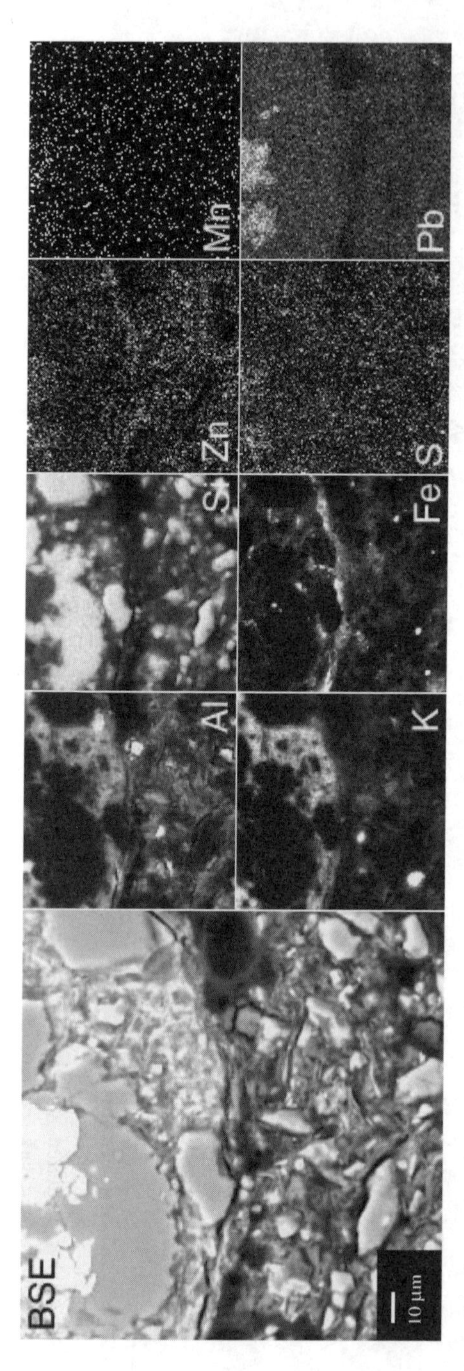

FIGURE 8.3 Backscattered electron image of subsurface soil and corresponding x-ray elemental dot maps. White colors indicate highest concentration of target elements, and dark spots indicate low concentration.

TABLE 8.2
Summary of Sequential Extraction Procedure

Extraction Step	Extracting Solution	Operational Definition	Reaction Time and Temperature
		Untreated sample	
1	1 M NH_4NO_3	Exchangeable metal ions, water soluble metal salts	24 h, 20°C
2	1 M NH_4OAc (pH 6)	Weakly complexed metals and metals bound by carbonates	24 h, 20°C
3	0.1 M NH_3OHCl + 1 M NH_4OAc (pH 6)	Metals bound by Mn (hydr)oxides	0.5 h, 20°C
4	0.025 M NH_4-EDTA (pH 4.6)	Metals bound by organic matter	1.5 h, 20°C
5	0.2 M NH_4-oxalate (pH 3.25)	Metals bound by Fe (hydr)oxides of low crystallinity	4 h, 20°C
6	0.1 M ascorbic acid + 0.2 M NH_4-oxalate (pH 3.25)	Metals bound by crystalline Fe (hydr)oxides	0.5 h, 97°C
7	Conc. HNO_3, HCl, HF	Metals bound by residual fraction	Microwave

Source: Reprinted with permission from Scheinost, A.C. et al., *Environ. Sci. Technol.*, 36, 5021, copyright 2002. American Chemical Society.

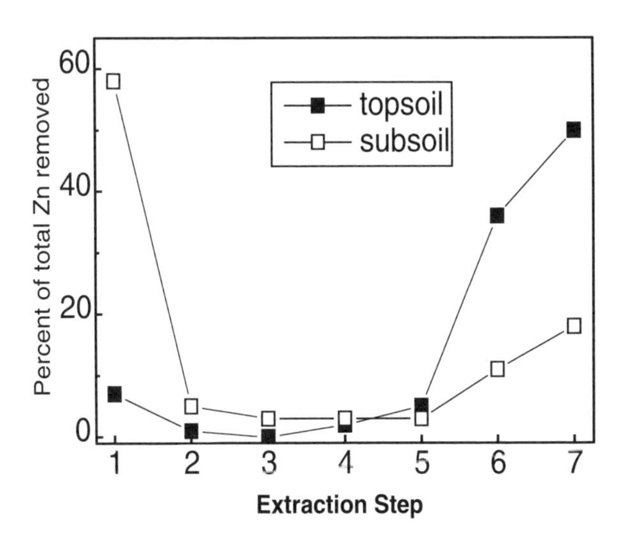

FIGURE 8.4 Amount of Zn removed from surface and subsurface soils at each step of the selective sequential extraction procedure. Error bars indicate the standard deviation of replicates. (Reprinted with permission from Scheinost, A.C. et al., *Environ. Sci. Technol.*, 36, 5021, copyright 2002. American Chemical Society.)

the remaining 7% of Zn was released during steps 2 to 5. In addition, substantial amounts of Mn were released during all steps, with a maximum of 36% for step 6, suggesting that Mn was distributed among different species (Table 8.3). While most Fe was released during step 6, large amounts remained in the residual fraction, indicating the presence of a fairly stable Fe phase (Table 8.3). In the subsoil, which was less acidic (pH 3.9) and contained less Zn (900 mg/kg compared to 6200 mg/kg) than the topsoil, 58% of the total Zn was the readily exchangeable form (Figure 8.4, Table 8.3). The remaining Zn species were almost exclusively extracted during steps 6 and 7, suggesting that a similarly stable phase(s) as in the topsoil were present. Based on this first set of results, it was clear that the quantities of Zn species, and perhaps the identity of species, are different in the surface and subsurface soils.

8.3.4 Bulk EXAFS Spectroscopy

Zinc K-edge (9659 eV) EXAFS spectra of soil samples and Zn reference compounds were collected at beam line X-11A at the National Synchrotron Light Source (NSLS), Upton, New York. Details on the experimental setup and sample preparation can be found elsewhere.[64] Briefly, unaltered soil samples were placed in Teflon sample holders and sealed with Kapton tape. For the surface soil, samples were dry sieved and the <2-mm fraction was collected. For the subsurface soil, the <2-μm and <250-μm fractions were collected. In addition, dark nodules measuring 0.5 to 2 mm in diameter were collected from the subsurface soil and ground in an agate mortar and pestle. All samples were measured at room temperature in fluorescence-yield mode using a Stern-Heald-type (Lytle) detector filled with Kr gas. Data scans were measured in at least triplicate and up to ten scans, depending on Zn concentration in the sample, and then averaged to improve the signal-to-noise ratio. Once raw XAFS data were collected, they were converted to wave vector (k) units by assigning the origin of the abscissa to the first inflection point of the edge. EXAFS chi(k) functions were derived from the spectra by modeling the post-edge region with a spline function. The chi functions were k^3-weighted and then Fourier-transformed using a Hanning window, resulting in radial structure functions (RSFs). The interatomic distances shown in the RSF graphs are uncorrected for phase shift so that the true distance is not represented.

8.3.4.1 EXAFS Data Analysis

Two approaches were used in analysis of EXAFS data in order to most accurately determine Zn speciation in the samples: a multishell fitting approach of the RSF data and LCF of the chi(k)*k^3 data combined with principal component analysis (PCA). Both methods rely on a linear least-squares fitting procedure. For the multishell fitting, structural parameters of the first- and second-coordination shells were determined for model compounds and soil samples using theoretical paths generated by FEFF7.[73] The estimated error using this approach is as follows: for bond distance (R), the first shell was accurate to R ±0.02 Å and the second shell to R ±0.05 Å. For coordination number (N), the first shell was accurate to N ±20% and the second shell to N ±40%.[64] Using WinXAS97 2.1, EXAFS parameters, including R, N, and

TABLE 8.3
Percent Metals Removed from Palmerton Soil Profile by SSE

Sample	Step	Mn	Fe	Zn
Topsoil	1	13	0	7
	2	8	0	1
	3	7	1	0
	4	8	23	2
	5	18	10	5
	6	36	39	36
	7	10	27	50
Subsoil	1	27	0	58
	2	32	0	5
	3	27	1	3
	4	1	3	3
	5	2	21	3
	6	4	45	11
	7	7	30	18

the Debye-Waller factor (σ^2), were determined for soil samples using the above approach.[74] By comparing the resulting bond distances and coordination numbers for the soil samples to the same parameters for reference materials (Figure 8.5, after Fourier transformation to R space), one is able to arrive at conclusions as to the possible Zn species present in the sample. However, soil samples are heterogeneous in nature and the speciation represents an average of all species present. In multiphase systems such as soils, multishell fitting may lead to misjudging the number of structural parameters.[31] Nonetheless, this method can be quite useful if care is taken and if combined with other fitting approaches.

The second approach to fitting EXAFS data relied on linear least-squares fitting of reference chi(k)*k^3 to the experimental chi data. In order to select the number and identity of Zn-bearing reference spectra to be used for LCF, PCA was employed. With PCA consideration is given to the statistical variance within an experimental dataset and breaks it down into principal components. The statistically meaningful number of components to regenerate the original input spectra and whether these components correspond to specific species is possible with PCA.[30,75] Both the number and identity of species in a set of samples can be estimated without requiring *a priori* assumptions. In order to make a large database of Zn-bearing references available, a large number of samples were collected or synthesized and analyzed using EXAFS and are outlined in the following paragraph. The selection of the reference minerals and sorption samples made was based on the mineralogy of the soil, reports of other researches in the literature, and common phases encountered in laboratory studies.

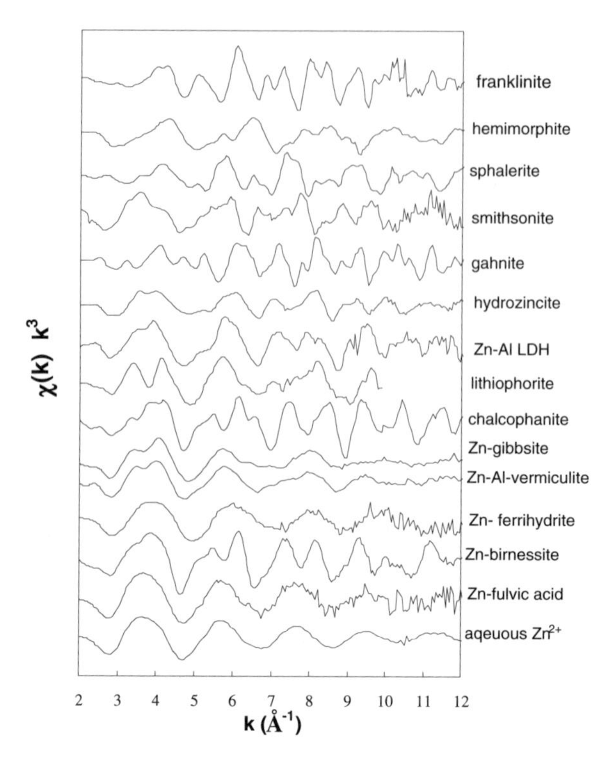

FIGURE 8.5 Normalized Zn-EXAFS k^3-weighted chi spectra of reference materials and sorption samples used as empirical models for linear combination fitting. (Reprinted with permission from Roberts, D.R. et al., *Environ. Sci. Technol.*, 36, 1742, copyright 2002. American Chemical Society.)

Minerals provided by the Museum of Natural History, Washington, D.C., include: franklinite ($ZnFe_2O_4$), hydrozincite ($Zn_5(OH)_6(CO_3)_2$), smithsonite ($ZnCO_3$), hemimorphite ($Zn4Si_2)_7(OH)_2 \cdot H_2O$), and chalcophanite ($(Zn,Fe,Mn)Mn_3O_7.3H_2O$); and sphalerite (ZnS) (Aldrich, 99.9+% purity). The Mineral Collection of the Swiss Federal Institute (ETH), Zurich, provided gahnite ($ZnAl_2O_4$) (MPS-ETH V.S. #7355) from Bodenmais, Bayrischer Wald, Germany. A natural lithiophorite sample $(Al,Li,Zn)MnO_2(OH)_2$ was provided by the Museum of Natural History, Bern, Switzerland. Aqueous Zn^{2+} was prepared by dissolving 10 mmol/l of $Zn(NO_3)_2$ (Zn nitrate, Aldrich, 99.9+% purity) in DDI H_2O and adjusting the pH to 6. A Zn-Al layered, double-hydroxide phase was synthesized in the laboratory following the method of Ford and Sparks.[23] Sorption samples were prepared by reacting Zn with ferrihydrite (two-line, freshly precipitated); high-surface area gibbsite (synthesized and aged 30 days, 90 m2 g^{-1}); birnessite (45 m^2 g^{-1}); hydroxy-Al interlayered vermiculite (Al-verm) (University of Missouri Source Clays Repository, Sanford vermiculite, cleaned, 90 m^2 g^{-1}); and fulvic acid (Aldrich, 99% purity).[76,77] For sorption samples, in an N_2 atmosphere, 10 g/l of solids were titrated with a 0.1 M $Zn(NO_3)_2$ stock solution to achieve Zn loadings of ca. ±0.5 μmol/m^2 for ferrihydrite and 1.5 μmol/m^2 for the remaining sorbents. The pH was adjusted to 6.0 ±0.3 and maintained during a 24-h

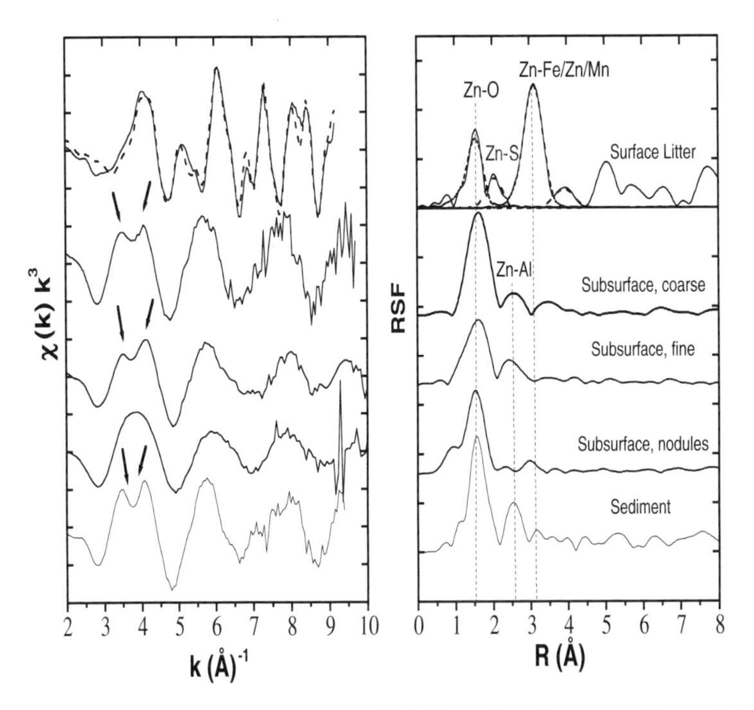

FIGURE 8.6 Bulk Zn-EXAFS k^3-weighted chi (left panel) and corresponding radial structure functions (right panel) resulting from Fourier analysis of chi data for surface and subsurface soil samples. The dotted line in the top spectrum of the left panel results from LCF fitting, and in the right panel from a shell-fitting approach. (Reprinted with permission from Roberts, D.R. et al., *Environ. Sci. Technol.*, 36, 1742, copyright 2002. American Chemical Society.)

period. Solids were separated by centrifuging at 10,000 rpm for 10 min and stored in a refrigerator as wet pastes until analysis. The raw chi(k) × k^3 EXAFS data for the reference mineral and sorption samples are presented in Figure 8.5. Visually, one can easily identify characteristic backscattering features of heavier elements that can assist in the initial identification of mineral samples, while adsorbed samples show spectra dominated by backscattering from the first O shell. Fit results for these reference samples are presented in Table 8.4.

8.3.5 EXAFS of Soil Samples

8.3.5.1 Surface Soil

The k^3-weighted chi spectra and corresponding radial structure functions for the surface, subsurface, and sediment samples are presented in Figure 8.6. The top spectrum in both the left and right panels show representative fits for LCF and multishell fitting, respectively (dashed lines = fits and solid lines = raw data). For the topsoil sample, multishell fitting revealed that Zn was tetrahedrally coordinated to both O and S in the first coordination shell (N≈4). The second-shell contribution could be fit with either a Zn or Fe atom at a distance of 3.49 Å. Coordination numbers and distances of the O shell and the Zn shell are in line with those of franklinite.

TABLE 8.4
EXAFS Parameters for Zn Reference Minerals and Sorption Samples

Sample	Formula/Conditions	First Shell			Second/Third Shells		
		CN[a] and element	R [Å][b]	σ^2 [Å^2][c]	CN and element	R[Å]	σ^2[Å^2]
Franklinite	$(Zn,Fe,Mn)^{II}(Fe,Mn)^{III}{}_2O_4$	4.0 O	1.97	0.003	12.0 Fe	3.51	0.007
Hemimorphite	$Zn_4Si_2O_7(OH)_2\ H_2O$	4.1 O	1.94	0.006	21.4 Zn	3.33	0.028
Sphalerite	ZnS	3.9 S	2.34	0.004	9.0 Zn, 9.1 S	3.81, 4.46	0.009, 0.010
Smithsonite	$ZnCO_3$	5.9 O	2.10	0.0028	7.8 Zn	3.71	0.0072
Gahnite	$ZnAl_2O_4$	4.4 O	1.97	0.005	14.5 Al	3.41	0.005
Hydrozincite	$Zn_5(CO_3)_2(OH)_6$	4.1 O	2.02	0.0041	2.2 Zn	3.22	0.0061
Zn-Al LDH	Synthesized (Ford et al, 2000)	6.3 O	2.07	0.009	3.9 Zn, 2.4 Al	3.10, 3.06	0.008, 0.009
Lithiophorite	$(Li,Al,Zn)(Mn)O_2(OH)_2$	6.4 O	2.02	0.009	4.7 Al	2.95	0.004
Chalcophanite	$(Zn,Fe,Mn)^{II}Mn3^{IV}O_7\ 3H_2O$	6.5 O	2.01	0.013	6.3 Mn	3.42	0.007
Zn-gibbsite	pH 6.0; 1.5 mmol/m^2 Zn	5.1 O	2.01	0.0097	4.4 Al	3.02	0.008
Zn-Al verm.	pH 6.0; 1.5 mmol/m^2 Zn	5.8 O	1.97	0.0090	2.5 Al	3.05	0.0051
Zn-HFO	pH 6.0; 0.5 mmol/m^2 Zn	3.9 O	1.94	0.0050	1.9 Fe	3.34	0.0170
Zn-birnessite	pH 6.0; 3.5 mmol/m^2 Zn	5.6 O	2.07	0.0070	7.7 Mn	3.49	0.0085
Zn-fulvic acid	pH 6.0; 2.0 mmol/m^2 Zn (Aldrich F.A.)	6.8 O	2.06	0.0100			
Aqueous Zn^{2+}	10 mmol/l $Zn(NO_3)_2$ in DDI H_2O	5.7 O	2.07	0.010			

[a] Coordination number.
[b] Interatomic distance.
[c] Debye–Waller factor.

Coordination number and distance of the S shell are indicative of sphalerite (Tables 8.4 and 8.5). The peaks in the RSF from the range of 4.5 Å to 6 Å are most likely a result of backscattering from another set of O or S and no attempts were made to fit these contributions. Using the shell-fitting approach, the main Zn species appear to be franklinite and sphalerite, confirming the EMPA results. The LCF approach estimated 59% franklinite and 28% sphalerite, and 14% aqueous Zn (a sum of 1.01 and residual of 22.5) (Table 8.6). The aqueous component for the topsoil sample most likely is present as a result of Zn complexed to organic material in the sample. Metals complexed to organic compounds generally yield a weak EXAFS signal, which can be masked by the more intense signal from inorganic compounds.[31] This is why the Zn-fulvic acid spectrum strongly resembles the aqueous Zn^{2+} spectrum (see Figure 8.5). The fact that aqueous Zn was the best candidate for the LCF may be due to the better signal-to-noise ratio of this sample.

8.3.5.2 Subsurface Soils

In the subsurface soil samples, the chi(k) × k^3 spectra have fewer, weaker beats, relative to the surface soil indicating a lack of significant higher-shell backscattering from a high Z element such as Zn, Mn, or Fe (Figure 8.6, left panel). The chi spectra for the coarse (<2 mm) and fine (<250 μm) fractions of the soil have similar structural features, with a larger shoulder at 5.5 $Å^{-1}$ being a major distinction between them. With the exception of the nodule sample, all spectra have a split in the first peak of chi spectra at 3.8 $Å^{-1}$ (see arrows, Figure 8.6). The splitting feature has previously been attributed to the presence of "light" Al atoms in the coordination shell of Zn.[78] Incidentally, many of the reference chi spectra of Zn sorbed to Al-bearing minerals show the same oscillation (Figure 8.5). In addition, the RSF data for all spectra with the split oscillation at 3.8 $Å^{-1}$ in their chi spectrum all have an Al atom in the second shell around Zn, with fitting resulting in approximately 1 to 2 Al atoms at distances of approximately 3.05 Å. This is an indication that Zn is in a solid phase that bears Al or is complexed to Al atoms in a mineral as an inner-sphere sorption complex. The possible phases that Zn may be part of will be discussed in a later section. For the first-shell coordination of Zn in these samples, Zn-O distances have values between those expected for octahedral and tetrahedral coordination (approximately 2.03 Å), suggesting Zn is in both coordination environments, either in the same or different Zn-bearing phases. According to crystallographic data, the distance between O and Zn should be approximately 1.96 to 1.98 Å for tetrahedral coordination, but should increase from 2.06 to 2.08 Å for octahedral coordination.[24] It is common for Zn to be in both tetrahedral and octahedral coordination due to the lack of crystal-field stabilization energy, allowing Zn(II) to easily switch between both types of coordination.[79] The second shell for the subsurface bulk and fine soils can be fit with one to two Al atoms at approximately 3.5 Å. Prior results have shown that Zn complexed to gibbsite has similar first shell distances for Zn-O and Zn-Al, but this does not definitively prove Zn bound to Al is the main form of Zn in these samples.[80] The sediment sample has a chi(k) × k^3 and RSF spectra very similar to the subsurface soils (Figure 8.6). Zinc is found in octahedral coordination and can be fit with approximately 2.6 Al atoms at 3.00 Å (Table 8.5).

TABLE 8.5
EXAFS Parameters for Shell Fitting of Soil Sample Spectra

Sample	First Shell				Second Shell			
	Atom	CN[a]	R (Å)[b]	σ^2(Å^2)[c]	Atom	CN	R (Å)	σ^2(Å^2)
Bulk XAFS								
Surface soil, <2mm	Zn-O	3.7	1.98	0.0071	Zn-Fe/Zn	11.2	3.49	0.0089
	Zn-S	3.8	2.35	0.0080				
Subsurface soil, <2mm (coarse)	Zn-O	5.7	2.08	0.0051	Zn-Al	1.2	3.04	0.0030
Subsurface soil, <250 μm (fine)	Zn-O	6.2	2.03	0.0065	Zn-Al	1.9	3.06	0.005[f]
Subsurface soil, nodules	Zn-O	6.3	2.01	0.0070	Zn-Fe/Mn	1.1	3.46	0.0100
Sediment sample	Zn-O	7.0	2.00	0.0090	Zn-Al	2.6	3.00	0.003
Micro XAFS								
Surface soil, spot 1[d]	Zn-O	4.0	1.97	0.0079	Zn-Fe/Zn	13.2	3.52	0.0091
Surface soil, spot 2[d]	Zn-O	4.1	1.98	0.0070	Zn-Fe/Zn	8.1	3.51	0.0100
Subsurface soil, spot 1[e]	Zn-O	5.6	2.04	0.005[f]	Zn-Al	1.5	3.01	0.005[f]
Subsurface soil, spot 2[e]	Zn-O	4.1	2.00	0.005[f]	Zn-Fe	1.9	3.25	0.005[f]
Subsurface soil, spot 3[e]	Zn-O	3.7	1.98	0.005[f]	Zn-Fe/Mn	1.4	3.45	0.005[f]

[a] Coordination number.
[b] Interatomic distance.
[c] Debye-Waller factor.
[d] Spot number in Figure 8.9.
[e] Spot number in Figure 8.10.
[f] Value fixed during fitting.

To glean more information on Zn speciation, the LCF fitting approach was used for the subsurface samples. In contrast to the topsoil, the PCA performed with the subsoil spectra failed to determine the number of components. This may be due to the smaller number of spectra available for this sample. The best LCF for the coarse fraction of the soil was achieved with three components, 60% aqueous Zn, 30% Zn-gibbsite, and 10% Zn-ferrihydrite. For the subsurface fine fraction, the amount of aqueous Zn decreased to 35%, Zn-ferrihydrite dropped to 5%, and Zn-gibbsite increased up to 60%. This indicates that Zn is probably bound via outer-sphere complexation in the subsurface soil and this complex was altered upon fractionation to the fine particle size via wet sieving. The formation of Zn-Al-O,OH complexes is unlikely at the low pH value of this soil (3.9), and we hypothesize that this reference spectrum may represent Zn near the surface of a trioctahedral Al hydroxide layer. Candidate phases include Zn occurring in the trioctahedral Al(Li) hydroxide layers sandwiched between $Mn^{III,IV}$ oxide layers in lithiophorite or Zn bound in Al-hydroxide layers intercalated between negatively charged phyllosilicate layers, such as Al-hydroxy interlayered vermiculite or montmorillonite.[81,82] The acidic nature of the soil would favor either scenario, but at this point both shell fitting and LCF do not lead to definitive species identification. As we shall see, employing sequential extractions along with EXAFS measurements may help decipher the nature of the Zn-Al association in the subsurface soils. Whatever the case, Zn clearly has a different speciation in the subsurface soil as compared to the surface soil, suggesting dissolution of primary Zn-bearing phases followed by re-precipitation or partitioning to new phases.

The EXAFS spectra of the blackish, Mn- and Fe-rich nodules separated from the subsoil sample could be fit with about one Mn or Fe atoms at a distance of 3.46 Å (Table 8.5), indicating that Zn is sorbed to a either a Mn or Fe (hydr)oxide, or both simultaneously. With the LCF, Zn sorbed to ferrihydrite and Zn sorbed to birnessite were present in near equal portions of about 34% (Table 8.6). Aqueous Zn contributed approximately 25% and Zn sorbed to gibbsite only 10% of the total. Comparing the Zn-Mn distance to that of Zn-adsorbed birnessite, the values are quite close (Table 8.4). The comparison between both fitting approaches clearly shows that the shell fitting tends to reveal only the species with the strongest second-shell backscattering, while the linear combination fit reveals several other species including the aqueous one. In combination, the results suggest that Zn resides in three species in the subsoil: outer-sphere or organic complexes; sorbed as inner-sphere complexes by both Al and Fe/Mn (hydr)oxides; or as a minor constituent in a neo-formed mineral. Using the LCF approach, the sediment spectrum was best fit with 70% Zn-gibbsite, 20% aqueous Zn^{2+}, and 13% Zn-ferrihydrite, similar to the subsurface samples (Table 8.6).

8.3.6 EXAFS Combined with Sequential Extractions

Zn-edge EXAFS spectra (chi data), and Fourier transforms (RSFs) of the untreated topsoil sample and after each extraction step (1 to 6) are shown in Figure 8.7. All spectra appear nearly unchanged from steps 0 to 4, coinciding with the small losses of Zn during extraction steps 1 to 4 (Figure 8.4), suggesting that little alteration to

TABLE 8.6
Results from LCF of EXAFS Chi(k) k^3 Spectra

Sample	Franklinite (%)	Sphalerite (%)	Zn-Birnessite (%)	$Zn^{2+}aq$ (%)	Zn-Gibbsite (%)	Zn-Ferrihydrite (%)	Sum (%)	Residual Rp (%)
Bulk EXAFS								
Surface soil, <2mm	59	28		14			101	22.5
Subsurface soil, <2 mm				60	33	10	103	27.1
Subsurface soil, <2 µm				35	62	6	103	26.7
Subsurface soil, nodules			34	26	11	33	104	27.0
Sediment sample				22	72	12	106	26.6
Micro EXAFS								
Surface soil, spot 1[d]	102						102	23.5
Surface soil, spot 2[d]		86				16	102	22.6
Subsurface soil, spot 1[e]				35	52	16	103	30.5
Subsurface soil, spot 2[e]				46	16	41	103	33.6
Subsurface soil, spot 3[e]			25	26	17	36	104	25.3

Note: Fit region — 3.0 to 10.0 $Å^{-1}$ for surface soil, 1.5 to 8.0 $Å^{-1}$ for subsurface soil.

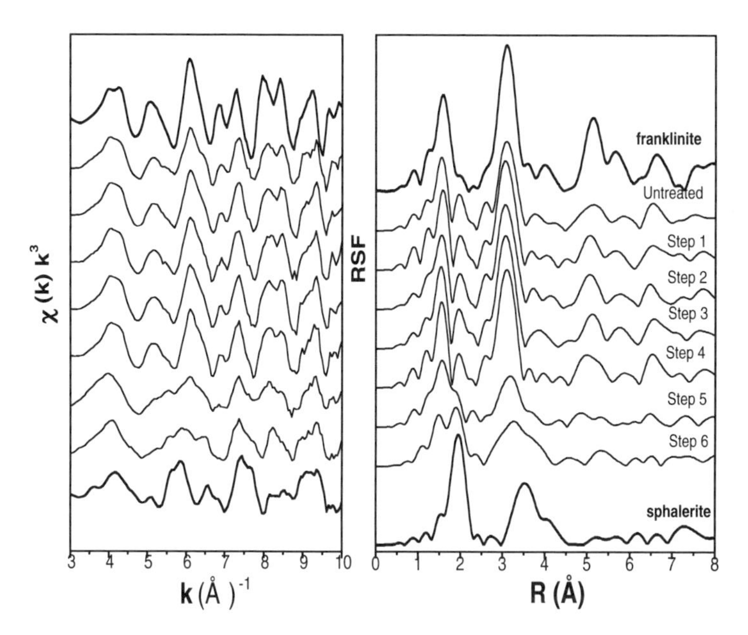

FIGURE 8.7 Bulk Zn-EXAFS k^3-weighted chi (left panel) and corresponding radial structure functions (right panel) resulting from Fourier analysis of chi data for surface soil sample after each step in the sequential extraction. (Reprinted with permission from Scheinost, A.C. et al., *Environ. Sci. Technol*, 36, 5021, copyright 2002. American Chemical Society.)

the sample is occurring and the Zn species composition remains unchanged. With step 5, however, the EXAFS spectral change is noticeable, yet the extraction step coinciding with this change extracts little Zn from the sample. Moreover, extraction step 6 removes 36% of total Zn from the sample, but the observed spectral changes between EXAFS spectra 5 to 6 are relatively small (Figure 8.7). The RSFs of samples at steps 5 and 6 show two additional shells at 1.9 and 3.5 Å, which match shell positions in the spectrum of sphalerite. The two peaks indicative of franklinite (1.6 and 3.1 Å) are still present in spectra 5 and 6, but with lower intensity compared to spectra 0 to 4. Since strongly reducing reagents like dithionite were not used in steps 5 or 6, the formation of sphalerite during these two extraction steps can be excluded and it can be concluded that sphalerite was present in the untreated sample. While the existence of sphalerite was suggested in the untreated sample due to the presence of a Zn-S bond, more definitive proof for the presence of sphalerite was derived after sequential extraction, since this phase was isolated and not masked by the franklinite EXAFS signal.

While the observed spectral changes from steps 4 to 6 were consistent with the partial dissolution of franklinite in extraction step 6, the spectral changes with step 5 were unexpected. This change of spectra without a significant release of Zn from the sample suggests alteration of the main Zn species and formation of a new Zn species with little loss of Zn. In an effort to establish the potential new speciation Zn takes upon this extraction step, we utilized information from EXAFS spectra. After extraction step 5, a substantial part of Zn was in mixed octahedral-tetrahedral coordination, as revealed by the Zn-O distance of 2.01 Å. This in contrast to the purely tetrahedral coordination for Zn in franklinite and sphalerite, which are the predominant phases before step 5 and after step 6, respectively. Moreover, there was evidence for the presence of franklinite after step 5 as indicated by the Zn-Fe distance in the second coordination shell. These results suggest the formation of a new species with octahedral Zn-O coordination during extraction step 5. The decrease of the RSF peaks of franklinite suggests that the new species formed after partial dissolution of franklinite. The small amount of Zn removed during step 5 could indicate re-adsorption of dissolved Zn to the remaining minerals or organic matter.[43,44] Alternatively, the use of oxalate in step 5 may have induced the precipitation of Zn oxalate, analogous to the precipitation of Pb-oxalate observed before.45 Unfortunately, Zn adsorption complexes and Zn-oxalate dihydrate have identical first-shell coordination and may be distinguished only by their second shell, which is very weak relative to the strong backscatter peaks from franklinite and sphalerite. While the formation of new adsorption complexes or precipitation of Zn-bearing phases induced by the extracting solution have not been proven definitively, their existence or coexistence is likely.

The results of LCF for the untreated surface sample and after each extraction step are compiled in Table 8.6. Depending on the extraction step, either aqueous Zn^{2+} (untreated sample), Zn oxalate (steps 5 and 6), or no third species in addition to franklinite and sphalerite (steps 1 to 4) gave the best fits. The sum of all fractions varied between 0.98 and 1.04 only, confirming that the number of species included in the fit was reasonable. A franklinite to sphalerite ratio of about 2 was maintained for samples from the untreated sample up to step 4, and then decreased to 1.7 (after step 5) and 0.5 (after step 6). The EXAFS results can be used to summarize the processes that occur during sequential extraction of the soil sample: step 1 removes aqueous species or species sorbed as outer-sphere complexes, which cannot be discriminated by EXAFS. Steps 1 to 4 had no effect on franklinite and sphalerite, while step 5 dissolves about 50% of both minerals, most likely components with low crystallinity. The dissolved Zn reprecipitates as Zn-oxalate, explaining why only 5% of Zn was removed during step 5 while the EXAFS data dramatically changed. Finally, step 6 removed 36% of total Zn by dissolving part of the neo-formed Zn oxalate (due to a high temperature step) and more franklinite than sphalerite. Sphalerite dominated the residual phase with 50% of total Zn. In summary, the topsoil contained about 10% of aqueous and exchangeable Zn species, 30% of sphalerite, and 60% of franklinite. For the aqueous fraction, one can estimate the error from the difference between sequential extraction and the LCF as to ≤5% absolute or ≤50% relative. For the mineral phases, variations of the LCF from extraction step 0 to 4 suggest an absolute error of ≤2% and relative errors between 3 and 7%. Therefore, LCF of EXAFS spectra allows one to quantify the mineral phases with a relatively high precision, while the quantification of the aqueous Zn^{2+} seems to be more biased.

The EXAFS spectra for the untreated and extracted subsoil samples are shown in Figure 8.8. For simplicity, only extraction steps 1 and 6 are shown, as there is little difference in the spectra between each step. The chi spectra (left panel) have a characteristic double beat at 3.5 and 4.2 $Å^{-1}$ in all three samples, which has been

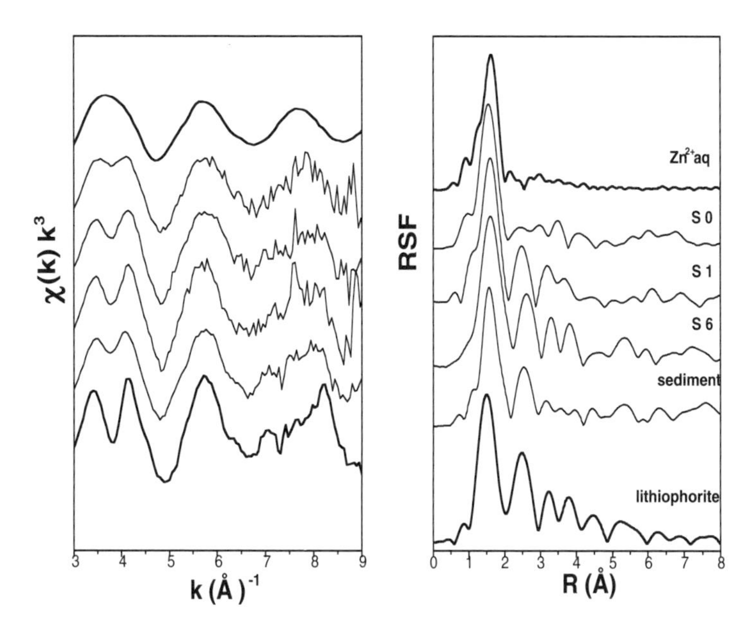

FIGURE 8.8 Bulk Zn-EXAFS k^3-weighted chi (left panel) and corresponding radial structure functions (right panel) resulting from Fourier analysis of chi data for untreated subsurface soil sample, as well as after extraction step 1 and 6. (Reprinted with permission from Scheinost, A.C. et al., *Environ. Sci. Technol*, 36, 5021, copyright 2002. American Chemical Society.)

ascribed to Al atoms in the second coordination shell as previously noted. In agreement with visual observation of the chi spectra, the small backscattering peak at about 2.5 Å(uncorrected for phase shift) in the RSF could be fitted with 1.9 Al atoms at 3.00 to 3.05 Å, gradually increasing up to 6.3 Al atoms after the final extraction step (Table 8.5). The increase of the Al coordination numbers from 1.9 to 6.3 with increasing removal of Zn by the extraction steps can be explained by a removal of Zn that has substituted for Al in a trioctahedral Al hydroxide layer, since Zn may be coordinated to two, three, four, and six Al atoms at a distance of about 3 Å in such a layer. The increase in coordination number is due to preferential dissolution of Zn-containing hydroxide edges, while Zn enclosed by six Al atoms remains up to extraction step 7.

The EXAFS results reported for the untreated samples (see Section 8.3.4) led to the conclusion that Zn may form highly ordered inner-sphere sorption complexes with gibbsite surfaces or substitute into an octahedral Al-hydroxide layer of some sort. The use of sequential extraction enabled more concrete conclusions to be made. For the nonextracted soil samples (bulk and coarse), second-shell Al coordination numbers did not exceed four, in line with the dioctahedral structure of gibbsite sheets (only two out of three metal positions are occupied). Elsewhere, a gradual increase was observed in Al coordination up to six with each extraction step, indicating that Zn is part of a fully occupied, trioctahedral Al-Zn^{2+} layer and not part of gibbsite or another dioctahedral Al compound.[67] While dioctahedral Al-hydroxide layers are essentially charge balanced and thus energetically very stable, fully occupied trioctahedral-hydroxide layers composed of a relatively small amount of divalent metals in addition to Al^{3+} possess a large net positive charge.[82] Therefore, they are stable only when this excess charge is counterbalanced by alternating layers of net negative charge, as can be found in the mineral lithiophorite (trioctahedral Al (Ni, Cu, Zn) hydroxide layers between $Mn^{III,IV}$ oxide layers), and in Al-hydroxo-interlayered smectite or vermiculite. While the chi spectra do resemble lithiophorite (Figure 8.8), shell fitting revealed significant differences in the first- and second-shell bond distances between all soil samples and the lithiophorite reference material. Moreover, the amount of Mn in the sample is too small to form lithiophorite with all of the remaining Zn (160 mg/kg). In contrast, formation of Zn-substituted, Al-hydroxide interlayered vermiculite or montmorillonite seems more likely due to the acidic pH of the subsoil sample, where sufficient free Al^{3+} is available to convert vermiculite and montmorillonite into their interlayered forms. However, formation of Zn-substituted, trioctahedral 1:1 or 2:1 phyllosilicates seems to be less likely due to the structural charge imbalance and their high solubility at low pH.[31] A more thorough overview of this process, including a proposed model, can be found elsewhere.[67]

LCF of the subsoil proved to be more difficult compared to the surface soil. The PCA performed with the subsoil spectra failed to determine the number of components, as indicated by indicator values increasing from the first component instead of approaching a minimum.[74] The best LCF fit to the untreated sample was achieved with three components: 30% of Zn-gibbsite, 10% of Zn-ferrihydrite, and 60% of aqueous Zn^{2+} (Table 8.6). The last value is close to the 58% determined by the sequential extraction (Table 8.3). However, an aqueous Zn species could not be fit for EXAFS samples after steps 1 and 6, which confirms a complete removal of this species by step 1. The Zn-gibbsite reference remained a significant fraction even for after step 6. In this model compound, Zn is bound to 0.6 Al atoms at a distance of 2.99 Å indicative of an inner-sphere sorption complex (Table 8.4, Figure 8.5). While formation of such a sorption complex is unlikely at the low subsoil pH, this reference may represent Zn near the substituted in the surface of trioctahedral Al-hydroxide layers. A substantial part of this species is dissolved by step 2 through 6, indicated by the drop from 59% in step 1 to 26% in step 6 (Table 8.3) and an increase of the Al coordination numbers from 3.4 to 6.3.[67] The two-fold increase of the residual R_p indicates a large fit error for sample 6, which may be caused primarily by the structural differences between the solid phase in the residual fraction and the available Zn-gibbsite reference compound (Table 8.4). Due to the lack of the proper references, quantitative speciation of the subsoil sample is most likely highly biased in comparison to that of the topsoil and is not 100% conclusive. Recently, more reference samples were prepared and collected with EXAFS in order to elucidate the Zn-bearing phase responsible for the fairly nonlabile Zn.[67] The findings are not discussed here; instead we focus on using microspectroscopic studies to further elucidate Zn speciation in the subsurface soil. This approach can be applied in order to isolate and spatially resolve the various Zn species in the soils. Similar approaches have been successful in isolating important species in heterogeneous soil samples.[30,31,64,65]

8.3.7 Synchotron-μ-XRF

For the topsoil sample, μ-SXRF was measured on GSECARS beam line 13-ID-C of the Advanced Photon Source (APS), Argonne National Lab, Chicago, Illinois. The higher brightness of APS enables one to characterize samples with lower metal concentrations relative to other synchrotron light sources. It also makes it possible to focus the beam to a small size (down to a micron diameter) without the loss of substantial flux, allowing one to successfully measure μ-SXRF and μ-EXAFS. A pair of Si(111) channel-cut crystals were employed as monochromators, and the beam was focused down to approximately 2-5 μm using a set of grazing incidence, platinum-coated, elliptically bent, Kirkpatrick–Baez focusing mirrors.[83] Resin-embedded thin sections of the surface and subsurface soil samples (10 to 50 μm thick) were placed on a digital x-y-z stage and set at an angle of 45° to the incident beam. The use of thin sections is necessary in order to get a sample of uniform thickness, eliminating differences in concentration due to heterogeneous sample thickness. Fluorescence x-rays were detected with a Ge nine-element detector positioned approximately 1 to 2 cm from the sample, depending on the Zn concentration in the sample. The μ-SXRF elemental maps were collected over an area of 800 × 800 μm^2 (surface sample) or 300 × 300 μm^2 (subsurface sample) with a 5-μm step size and 500 msec dwell time. Only elements with $Z > 20$ are detected using this technique, so correlations to soft elements such as S, O, N, and C cannot be determined.

The μ-SXRF maps collected on the surface and subsurface soils are presented in Figures 8.9 and 8.10, respectively. The colored scale bar at the right of each panel represents the relative concentrations of the elements. The absolute concentrations of the elements cannot be directly compared, but one is able to get a sense of the relative concentrations of an element within its own map. The surface soil displays a strong correlation between Zn and Fe in concentrated spots throughout the sample, similar to the EMPA results (Figure 8.2). Correlations between elements were confirmed by plotting the elements versus one another (data not shown). Little

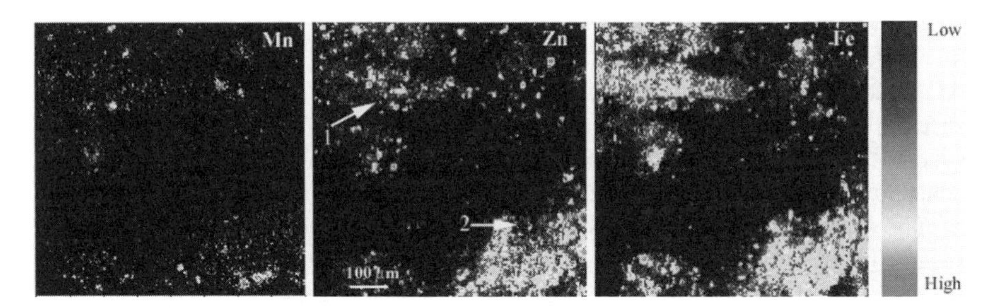

FIGURE 8.9 (See color insert following page 112) Micro-SXRF elemental maps for the surface soil sample. The numbered arrows indicate the micro-EXAFS spectra in Figure 8.11. The color bar gives relative amounts of each element in arbitrary units. (Reprinted with permission from Roberts, D.R. et al., *Environ. Sci. Technol.*, 36, 1742-1750. Copyright 2002. American Chemical Society.)

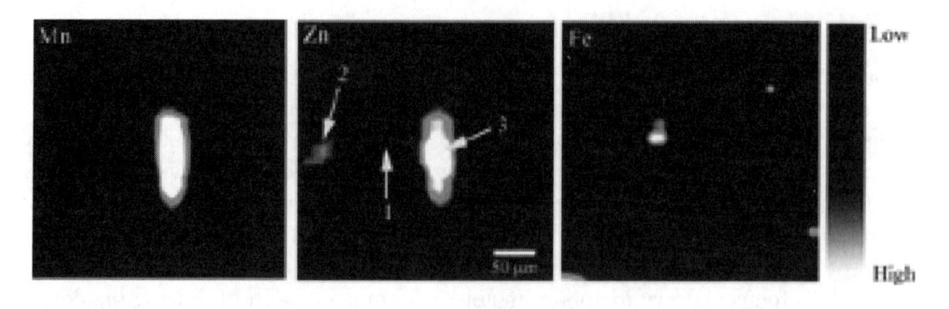

FIGURE 8.10 (See color insert) Micro-SXRF elemental maps for the subsurface soil sample. The numbered arrows indicate the micro-EXAFS spectra in Figure 8.11. The color bar gives relative amounts of each element in arbitrary units. (Reprinted with permission from Roberts, D.R. et al., *Environ. Sci. Technol.*, 36, 1742-1750. Copyright 2002. American Chemical Society.)

correlation between Zn and Mn was evident in the surface soil, and no significant correlations existed between Zn and other elements using this technique. If it were possible to collect carbon data, we speculate there would be a correlation to Zn since this soil is OM-rich and Zn-bearing particles are dispersed throughout the organic matrix. However, such a correlation would not necessarily imply that Zn is bound to the element that it is spatially correlated to. Such is the case with Zn and Fe in the surface sample: At this point at least a physical correlation can be assigned and one must be careful not to assign a chemical relationship between two elements based on a physical correlation alone. This is why combining a spatial technique with one that provides direct chemical information is necessary in order to achieve proper speciation information.

The subsurface sample has a different spatial arrangement of Zn compared to the surface sample (Figure 8.10). First, the Zn is not in small Zn-rich spots, as in the surface soil, but instead is more diffusely arranged in the sample, which implies that Zn is not in the same Zn-bearing grains so abundant in the surface sample. Moreover, Zn has a stronger correlation to Mn than it does to Fe, and the highest concentrations of Zn and Mn in their respective maps are in the same region. As mentioned previously, the correlation does not necessarily imply that Zn is directly bound to Mn. As shown above, little information was gleaned on Zn and Mn speciation using EMPA for the subsurface sample, pointing to the utility of using synchrotron-based μ-SXRF when dealing with a sample with relatively low metal concentrations. The finding that Zn is associated with Mn in the subsurface sample is quite an interesting discovery since it was previously shown that the Zn-EXAFS spectra in the subsurface soil resembled that of lithiophorite, which may have Zn substituting into trioctahedral Al-hydroxide layers situated between $Mn^{III,IV}$ oxide layers. However, EXAFS results for the soil sample and a reference lithiophorite had enough discrepancies such that it was not possible to reasonably conclude Zn was in this mineral phase. The reference lithiophorite used for comparison is only one of many that could possibly form, which may result in different Zn coordination environments, and a conclusion that the EXAFS did not reveal the presence of lithiophorite. Moreover, the Zn coordination environment was an average of values for all Zn species in the sample, which may account for the differences in the Zn atomic-coordination environment between references and the soil. To gain more insight into Zn speciation in this sample and the surface sample, μ-EXAFS was employed, potentially allowing one to spatially segregate individual Zn species that would normally be averaged together using bulk EXAFS.

8.3.8 μ-EXAFS

8.3.8.1 Surface Soil

The μ-EXAFS spectra were collected using the same settings as for the bulk-EXAFS spectra collected at NSLS. To assess systematic deviations in EXAFS data collection at the two different beam lines, selected reference samples were run at both locations. Independent of the beam line employed, μ-EXAFS and bulk-EXAFS spectra were highly reproducible (similar EXAFS parameters on reference materials; data not shown). Using the μ-SXRF maps, Zn μ-EXAFS was collected at various locations in the sample (the numbered arrows in Figures 8.9 and 8.10). For the surface soil, two Zn-rich spots separated by several microns were selected for EXAFS analysis and the results are shown in Figure 8.11 (top two spectra). The k^3-weighted chi

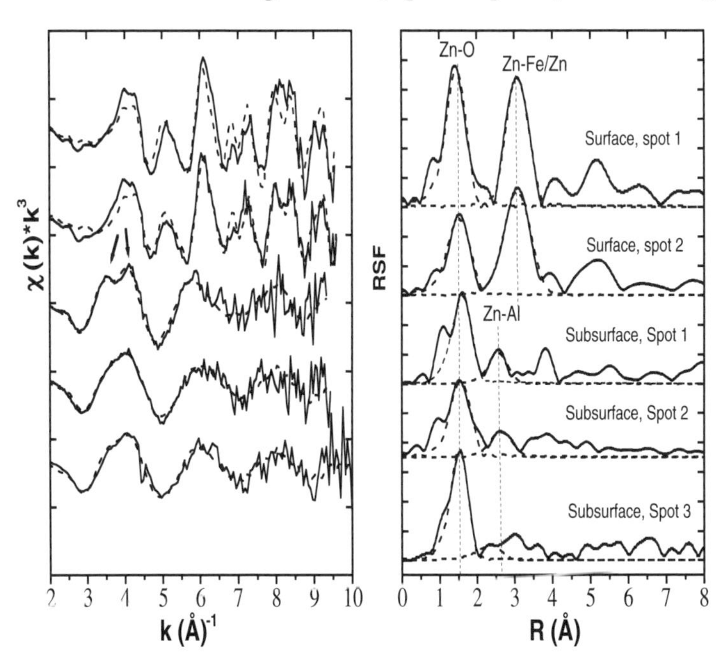

FIGURE 8.11 Micro Zn-EXAFS k^3-weighted chi (left panel) and corresponding radial structure functions (right panel) resulting from Fourier analysis of chi data for surface and subsurface soil samples. The dotted line in the top spectrum of the left panel results from LCF fitting, and in the right panel from a shell-fitting approach. (Reprinted with permission from Roberts, D.R. et al., *Environ. Sci. Technol.*, 36, 1742, copyright 2002. American Chemical Society.)

spectra are in the left panel (solid line = raw data and dashed line = LCF result) and the corresponding radial structure functions (solid line = raw data and dashed line = shell fitting result) in the right panel. LCF revealed that the spectrum of spot 1 was best fit by 100% franklinite and spot 2 by 85% franklinite and 15% Zn sorbed to ferrihydrite (Table 8.6). Using a shell fitting approach, 4 O atoms at distances of 1.98 Å were fit in the first shell, and a second shell was fit with 8 to 13 Zn or Fe atoms (not able to distinguish) at a distance of 3.51 Å (Table 8.5). Comparing the results to reference franklinite (4 O at 3.7 Å; 11 Zn or Fe at 3.50 Å), it is possible to conclude that both Zn-rich grains in the μ-SXRF map were mostly franklinite. The fact that the second spot had a better fit by including Zn-ferrihydrite reference spectra may be due to the partial dissolution of the franklinite grain and subsequent adsorption or co-precipitation with a ferrihydrite-like phase, also a product of franklinite dissolution.[30,84] The sphalerite contribution detected using bulk EXAFS was not observed using this approach, suggesting that bulk EXAFS is a better approach for determining the overall Zn speciation in the sample, whereas μ-EXAFS is suitable for spatially resolving individual contributions to the overall speciation. Therefore the latter technique is especially helpful in the case when over a small sample area one expects different speciation, such as for example with the subsurface soil.

8.3.8.2 Subsurface Soil

The k^3-weighted chi spectra corresponding to spots 1 to 3 in the subsoil (Figure 8.10) are shown in the left panel of Figure 8.11 (bottom three spectra). Immediately evident as one visually examines the three spectra are the contrasts with the two surface spectra, and the similarity to the bulk EXAFS spectra for the subsurface soil (Figure 8.6). However, two of the three spectra do not have the split oscillation at 3.8 $Å^{-1}$ indicative of the presence of Al atoms in the Zn coordination environment. Unfortunately, one is not able to observe Al or Si using μ-SXRF mapping, so showing the correlation between Zn and Al proves difficult. Beginning with spot 1 (area rich in Zn, Fe, and Mn), the chi spectrum shows the distinct splitting at 3.8 $Å^1$. The results from multishell fitting (Table 8.5) indicate that Zn is octahedrally coordinated (5.6 O atoms at 2.04 Å) and bound to approximately 1.5 Al atoms at 3.01 Å. The first- and second-shell data around the central Zn atom strongly resemble the Zn bound to Al oxides[80] (Table 8.4). In addition, the LCF results (Table 8.6) show that the best-fit result was obtained with 50% Zn-gibbsite, 35% aqueous Zn, and 15% Zn-ferrihydrite. While the adsorption of Zn to Al-bearing oxides is not favored at the low pH found in this soil, there is clearly an association between Zn and Al. Moreover, there is evidence Zn is also associated with both Fe and Mn based on the μ-SXRF maps, and this association may be the reason for the large shell at around 3.9 Å in the RSF. Conclusive speciation of Zn in this spot of the μ-SXRF map cannot be established, although it appears Zn may be bound as an inner-sphere complex to Al and Fe-bearing oxides, or present in a neo-formed mineral or Al-hydroxy interlayer. The other consideration is that even in this small area (approximately 5 μm), Zn may be in more than one species, leading to complex μ-EXAFS data.

For Zn spots 2 and 3, the Zn-O distance was more indicative of tetrahedral coordination (R_{Zn-O} = 2.00 and 1.98 Å, respectively). Spot two had two Fe atoms at 3.25 Å while spot 3 could be fit with either 1.4 Fe or Mn atoms at 3.45 Å (Table 8.5). LCF revealed that spots 2 and 3 were best fit with contributions from Zn-ferrihydrite, aqueous Zn^{2+}, Zn-birnessite, and Zn-gibbsite (Table 8.6). The most significant difference between these two spots was in the Zn-birnessite contribution, which was not present in spot 2 but accounted for 25% of the Zn species in spot 3, agreeing with the observations from μ-SXRF (Figure 8.10) and from shell fitting (Table 8.5). The results suggest that in these two spots, a portion of Zn is tetrahedrally bound to Fe and/or Mn oxides. The difference in Zn speciation among all three spots over such a small sample clearly demonstrate the microheterogeneity of metal speciation in soils and the fact that bulk techniques like EXAFS truly do detect an average of many species. This type of microscale approach to metal speciation gives the unique perspective of what a plant root, microorganism, or aqueous ion may encounter in the soil matrix. Moreover, the capability of this technique to segregate species has been demonstrated, even if concrete speciation of the actual species themselves remains elusive.

8.3.9 Desorption Studies

A speciation technique that may closely approximate the availability of a metal to plants and microorganisms is leaching with a neutral salt solution.[85,86] In essence, this type of technique may simulate a rainfall event or some other introduction of a solvent capable of perturbing the system and subsequently making the metal species of interest available for plant and microbial uptake, transport through pores, and/or immobilization. The experimental setup for the Zn desorption studies from the contaminated surface and subsurface soils are described elsewhere.[86] Briefly, a stirred-flow reaction chamber was connected to an HPLC pump at one end and to a fraction collector at the other end. A 0.2-M $CaCl_2$ solution adjusted to the pH of the soil was pumped through the chamber at a flow rate of 0.5 ml/min^{-1}. The chamber was equipped with a 25-mm filter membrane with 0.2-μm pore size to separate the solid from the solution. The chamber volume of ca. 9 ml contained a suspension of 28 g/l^{-1} of soil, which was stirred at 400 rpm with a magnetic stir bar. Fractions of 5 ml were sampled at the chamber outlet and analyzed by atomic absorption spectrometry (AAS).

Figure 8.12 displays the results from the stirred-flow desorption experiments. After leaching with 70 chamber volumes, nearly 65% of total Zn was removed from the subsoil, but only 11% from the topsoil sample. Moreover, the subsurface sample released Zn more rapidly, with over 60% of the total Zn removed from the soil after only five chamber volumes as compared to a more gradual release from the surface material. In terms of total amounts, 620 mg/kg Zn were released from the subsoil, and 800 mg/kg from the topsoil. Thus, almost similar amounts were released from both horizons, although the topsoil Zn concentration was about an order of magnitude higher than the subsoil Zn concentration. In the surface soil, we suspect that Zn would be preferentially released from sphalerite, rather than franklinite, since sphalerite is less stable in oxidizing environments.[87] In addition, Zn was most likely

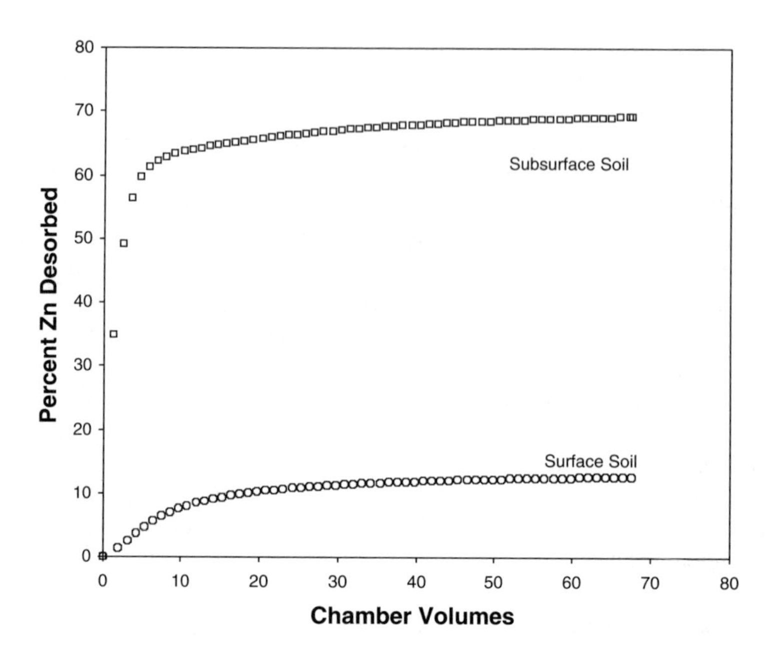

FIGURE 8.12 Zn released from surface and subsurface soil samples using stirred-flow desorption. Curves represent the amount of Zn released relative to the total amount of Zn in the sample. (Reprinted with permission from Roberts, D.R. et al., *Environ. Sci. Technol.*, 36, 1742, copyright 2002. American Chemcial Society.)

released from organic-matter complexes, as this species of Zn was previously noted in both the sequential extraction section process and from LCF of EXAFS data. The gradual release of Zn from this sample with time suggests the Zn source is slowly dissolving, or is diffusion limiting. In the subsurface soil, the Zn is rapidly released initially followed by a slower desorption step. In this case, we postulate that labile Zn that is adsorbed as an outer-sphere complex to organic matter or oxide/mineral surfaces is initially removed, followed by a slower release due to Zn bound to Fe and Al minerals/oxides and/or Zn in the structure of a neo-formed mineral phase. The sequential extraction suggested Zn was bound in a relatively stable phase, and this may account for the Zn still remaining in the soil after 70 chamber volumes. The subsurface soil curve has not yet reached a plateau, suggesting that more Zn will be released over time. While this method of speciation reveals little on the actual Zn species causing the observations, the results are consistent with the speciation of Zn determined using the other approaches and may give the most realistic information on the behavior of Zn in the soils. Combining the different results clearly indicates that Zn speciation, rather than total Zn concentration is controlling Zn desorption from these soils. This is a crucial finding, considering total metal concentration has often been one of the main criteria in assessing the risk associated with a metal-contaminated site.

8.4 CONCLUSIONS AND ENVIRONMENTAL SIGNIFICANCE

8.4.1 Fate of Zn in Soils

In the topsoil samples, the presence of franklinite and sphalerite can be explained if one considers the history of the smelting facility. The main Zn ore used in the smelting process at Palmerton was sphalerite. During smelting, sphalerite is exothermally converted at 900°C to zinc oxide, a more soluble Zn mineral phase.[88] Due to the presence of significant amounts of iron in the sphalerite ores, zinc ferrite ($ZnFe_2O_4$, franklinite) also forms during the roasting process. Given smelting inefficiencies, portions of sphalerite may not be oxidized and can be released in the smelter emissions.[89] In addition, franklinite can be released by the same mechanism or be introduced from the slag piles.[90] The rounded particles of sphalerite and franklinite (Figure 8.2) indicate they are stack emitted and have undergone the filtering process, rather than formed authigenically.[90] Since the topsoil is exposed to the atmosphere, one would expect ZnS to oxidize and release Zn and SO_4^{2-} and therefore no longer be present after several decades. While some of the ZnS has presumably dissolved and repartitioned or mobilized, decreased biological activity in the surface material is probably minimal due to deposition of sulfuric acid, leading to reduced microbial oxidation of metal-bearing particles. The presence of Pb and Cd in the sphalerite may decrease its solubility, as well as its particle size. The fact that the introduced franklinite is the main Zn-bearing phase in the topsoil and remains for decades is in agreement with thermodynamic data, which states that it is one of the most stable Zn minerals found in soil environments.[19] However, the acidic environment will promote franklinite dissolution and as it dissolves, ferrihydrite would be the first Fe oxyhydroxide to form, explaining its presence in the surface soil. Other soils contaminated as a result of metallurgical activities also revealed the persistence of sphalerite and franklinite, often with the readsorbing to Fe-bearing oxides and clay minerals.[28,91,92]

The Zn speciation in the subsoil is not as easily explained by the deposition of primary material. Rather, sphalerite and franklinite have undergone dissolution in the topsoil and a portion of the Zn has migrated into the subsoil, since there are few secondary minerals capable of sorbing Zn in the surface as revealed by XRD (data not shown). In contrast, the subsoil has more in terms of minerals capable of sorbing Zn: gibbsite, Al-hydroxy interlayered montmorillonite, and Fe and Mn oxides. Once introduced into the subsurface soil, the speciation of Zn is dependent on both reaction conditions and the sorbent phases present. In the soil studied here, the pH is low enough to limit the neo-formation of Zn minerals and also limits complexation to reactive sites on minerals and oxides. However, EXAFS revealed the presence of Zn complexed to Fe, Mn, and Al oxides, as well as incorporated into hydroxy-Al interlayers between clay minerals.[67] In addition to those relatively stable phases, most Zn was found in aqueous complexes, most likely complexed as outer-sphere species to minerals or organic material. Clearly, the Zn speciation in the subsurface soil is much more complex and more difficult to determine relative to the surface soil. As demonstrated with the stirred-flow desorption experiments, much of the Zn in the subsoil is still relatively labile and the species of Zn are most likely not stable in the long term. However, some of the Zn is stable and further mineral transformations with time may lead to more stable Zn-bearing phases. At such a low pH, dissolved Al and Si would be prevalent, and if pH gradually increases over time, there is a good chance Zn will coprecipitate in phyllosilicates, as has been demonstrated in neutral pH soils.[30] Another approach is to increase the soil pH by liming, thereby promoting precipitation of more stable Zn-bearing phases. This approach, combined with biosolids and composts, has been successful in the lower-lying regions in the Palmerton area.[93]

8.4.2 Summary of Speciation Techniques

The elevated concentration of Zn in the surface soil along with the occurrence of Zn-bearing minerals made for relatively easy species determination. Indeed, XRD identified franklinite and EM showed the elemental association between Zn and Fe and Zn and S and provided insight into particle size, morphology, and spatial distribution. The sequential extraction approach for the surface soil also revealed that Zn was heavily associated with Fe, but beyond that it was difficult to determine the exact species. None of these techniques was able to show the presence of aqueous or outer-sphere Zn^{2+} as revealed by EXAFS. This technique also directly identified and allowed for quantification of the phases franklinite, sphalerite, and aqueous Zn^{2+}. The use of μ-EXAFS allowed for the identity of the individual Zn-bearing phases in the soil and they were revealed to be mostly franklinite, in agreement with the findings from bulk EXAFS. However, the sphalerite was not identified with the microfocused approach, indicating the utility of bulk XAFS in identifying all species even in a heterogeneous system. One of the most important findings was the discovery of artifacts from the used of sequential extractions as determined by EXAFS. Since the neo-formed phase was amorphous, most traditional analytical techniques would have been unable to detect it.

In contrast to the surface soil, each individual analytical technique was less successful in determining Zn speciation when employed separately. Sequential extraction did indicate that a large portion (60%) of Zn was readily exchangeable, which may be attributed to one of many Zn species. As expected, XRD and EMPA revealed little in terms of the Zn species in this soil, as the majority of Zn was not in a detectable form or below the detection limit of the instrument. However, XRD was useful in identifying minerals present in the soil, specifically the occurrence of Al-hydroxy interlayered montmorillonite, which was useful in confirming the Zn species as identified using EXAFS.[67] This is a good example of when a method may not be able to directly identify a species, but it provides insight into potential sorbent phases and other metal-bearing phases, thereby providing indirect speciation information. Bulk EXAFS was much more useful in providing direct species information, indicating that Zn was in an outer-sphere and/or aqueous state, in addition to more stable phases. In addition, bulk EXAFS suggested that Zn was in different species within the soil, as indicated by the spectral changes with the size fraction of the soil and by hand picking dark nodules suspected of bearing Zn. However, identifying the exact Zn species was difficult, and several could be proposed as candidate phases. By combining the sequential extraction with bulk EXAFS, more information was gleaned, specifically the identity of the more stable phase that was extracted in step

6 of the sequential extractions and remained after the stirred-flow experiment. The μ-SXRF detected the Zn in the sample that EM was not able to do, showing the correlation of Zn and Mn, and Zn and Fe to a lesser extent. The collection of μ-EXAFS from various spots in the sample provided detailed atomic coordination on Zn, showing Zn coordinated to Al and Mn/Fe in different parts of the small sample area. Fitting the μ-EXAFS and bulk EXAFS data using the LCF approach was difficult since not all the possible phases in the soil were represented with the references. This is a major consideration when using this approach to fitting data and all steps should be taken to have as large as a reference library of spectra as possible. Even the extensive amount of reference spectra collected for this study was insufficient. Combining all the synchrotron-based techniques combined with sequential extraction proved to be the most valuable approach to determining and quantifying the Zn species in the subsurface sample.

REFERENCES

1. Förstner, U., Land contamination by metals — global scope and magnitude of problem, in *Metal Speciation and Contamination of Soil*, Allen, H.E., Huang, C.P., and Bailey, G.W., Eds., CRC Press, Boca Raton, FL, 1995, p. 1.
2. U.S. Environmental Protection Agency, *National Priorities List Fact Book*, U.S. Environmental Protection Agency, Washington, D.C., 1986.
3. Hering, J.G., Implication of complexation, sorption and dissolution kinetics for metal transport in soils, in *Metal Speciation and Contamination of Soil*, Allen, H.E., Huang, C.P., and Bailey, G.W., Eds., CRC Press, Boca Raton, FL, 1995, p. 59.
4. McBride, M., Reactions controlling heavy metal solubility in soils, in *Advances in Soil Science*, vol. 10, Springer-Verlag, New York, 1989.
5. Brown, G.E.J., Foster, A.L., and Ostergren, J.D., Mineral surfaces and bioavailability of heavy metals: A molecular scale perspective, *Proc. Natl. Acad. Sci. USA*, 96, 3388, 1999.
6. Gasser, U.G. et al., Lead release from smelter and mine waste impacted materials under simulated gastric conditions and relation to speciation, *Environ. Sci. Technol.*, 30, 761, 1996.
7. Davis, A. et al., Mass balance on surface-bound, mineralogical, and total lead concentrations as related to aggregate bioaccessibility, *Environ. Sci. Technol.*, 31, 37, 1997.
8. Bargar, J.R. et al., Characterization of U(VI)-carbonate ternary complexes on hematite: EXAFS and electrophoretic mobility measurements, *Geochim. Cosmochim. Acta*, 64, 2737, 2000.
9. Dudka, S. and Adriano, D.C., Environmental impacts of metal ore mining and processing: A review, *J. Environ. Qual.*, 26, 590, 1997.
10. Chaney, R.L., Zinc phytoxicity, in *Zinc in Soils and Plants*, Robson, A.D., Ed., Kluwer, Dordrecht1993, p. 135.
11. Zasoski, R.J. and Burau, R.G., Sorption and sorptive interaction of cadmium and zinc on hydrous manganese oxide, *Soil Sci. Soc. Am. J.*, 52, 81, 1988.
12. Stahl, R.S. and James, B.R., Zinc sorption by manganese-oxide-coated sand as a function of pH, *Soil Sci. Soc. Am. J.*, 55, 1291, 1991.
13. Murray, J.W., The interaction of metal ions at the manganese dioxide-solution interface, *Geochim. Cosmochim. Acta*, 39, 505, 1975.

14. Melis, P. et al., Sulfate-zinc interaction on aluminum hydroxide surfaces, *Z. Pflanzenernahr. Bodenk.*, 150, 99, 1987.
15. Kinniburgh, D.G., Jackson, M.L., and Syers, J.K., Adsorption of alkaline earth, transition, and heavy metal cations by hydrous oxide gels of iron and aluminum, *Soil Sci. Soc. Am. J.*, 40, 796, 1976.
16. Huang, C.P. and Rhoads, A., Adsorption of Zn(II) onto hydrous aluminosilicates, *J. Colloid Interface Sci.*, 131, 289, 1989.
17. Spark, K.M., Johnson, B.B., and Wells, J.D., Characterizing heavy-metal adsorption on oxides and oxyhydroxides, *Eur. J. Soil Sci.*, 46, 621, 1995.
18. Ladonin, D.V., Specific adsorption of copper and zinc by some soil minerals, *Eurasian Soil Sci.*, 30, 1478, 1997.
19. Sadiq, M., Solubility and speciation of zinc in calcareous soils, *Water, Air, Soil Pollut.*, 57–58, 411, 1991.
20. Metwally, A.I. et al., Effect of pH on zinc adsorption and solubility in suspensions of different clays and soils, *Z. Pflanzenernähr. Bodenk.*, 156, 131, 1993.
21. Waychunas, G.A. et al., Surface chemistry of ferrihydrite: Part 1. EXAFS studies of the geometry of coprecipitated and adsorbed arsenate, *Geochim. Cosmochim. Acta*, 57, 2251, 1993.
22. Schlegel, M.L., Manceau, A., and Charlet, L., EXAFS study of Zn and ZnEDTA sorption at the goethite (a-FeOOH)/water interface, *J. Phys. IV*, 7, 823, 1997.
23. Ford, R.G. and Sparks, D.L., The nature of Zn precipitates formed in the presence of pyrophyllite, *Environ. Sci. Technol.*, 34, 2479, 2000.
24. Trainor, T.P., Brown Jr., G.E., and Parks, G.A., Adsorption and precipitation of aqueous Zn(II) on alumina powders, *J. Colloid Interface Sci.*, 231, 359, 2000.
25. Bochatay, L. and Persson, P., Metal ion coordination at the water-manganite (g-MnOOH) interface II. An EXAFS study of zinc(II), *J. Colloid Interface Sci.*, 229, 593, 2000.
26. Xu, Y., Schwartz, F.W., and Traina, S.J., Sorption of Zn^{2+} and Cd^{2+} on hydroxyapatite surfaces, *Environ. Sci. Technol.*, 28, 1472, 1994.
27. Hesterberg, D. et al., X-ray absorption spectroscopy of lead and zinc speciation in a contaminated groundwater aquifer, *Environ. Sci. Technol.*, 31, 2840, 1997.
28. O'Day, P.A., Carroll, S.A., and Waychunas, G.A., Rock-water interactions controlling zinc, cadmium, and lead concentration in surface waters and sediments, U.S. tri-state mining district. 1. Molecular identification using x-ray absorption spectroscopy, *Environ. Sci. Technol.*, 32, 943, 1998.
29. Webb, S.M., Leppard, G.G., and Gaillard, J.-F., Zinc speciation in a contaminated aquatic environment: Characterization of environmental particles by analytical electron microscopy, *Environ. Sci. Technol.*, 34, 1926, 2000.
30. Isaure, M.-P. et al., Quantitative Zn speciation in a contaminated dredged sediment by m-PIXE, m-SXRF, EXAFS spectroscopy and principal component analysis, *Geochim. Cosmochim. Acta*, 66, 1549, 2002.
31. Manceau, A. et al., Quantitative Zn speciation in smelter-contaminated soils by EXAFS spectroscopy, *Am. J. Sci.*, 300, 289, 2000.
32. Gerth, J., Brümmer, G.W., and Tiller, K.G., Retention of Ni, Zn, and Cd by Si-associated goethite, *Z. Pflanzenernähr. Bodenk.*, 156, 123, 1992.
33. Brümmer, G.W., Gerth, J., and Tiller, K.G., Reaction kinetics of the adsorption and desorption of nickel, zinc and cadmium by goethite. I. Adsorption and diffusion of metals, *J. Soil Sci.*, 39, 37, 1988.
34. Lindsay, W.L. and Norvell, W.A., Development of a DTPA soil test for zinc, iron, manganese, and copper, *Soil Sci. Soc. Am. J.*, 42, 421, 1978.

35. McGrath, D., Application of single and sequential extraction procedures to polluted and unpolluted soils, *Sci. Total Environ.*, 178, 37, 1996.
36. Tessier, A., Campbell, P.G.C., and Bisson, M., Sequential extraction procedure for the speciation of particulate trace metals, *Anal. Chem.*, 51, 844, 1979.
37. Sposito, G., Lund, L.J., and Chang, A.C., *Soil Sci. Soc. Am. J.*, 46, 260, 1981.
38. Venditti, D., Durécu, S., and Berthelin, J., A multidisciplinary approach to assess history, environmental risks, and remediation feasibility of soils contaminated by metallurgical activities. Part A: chemical and physical properties of metals and leaching ability, *Arch. Environ. Contam. Toxicol.*, 38, 411, 2000.
39. Adamo, P. et al., Chemical and mineralogical forms of Cu and Ni in contaminated soils from the Sudbury mining region and smelting region, Canada, *Environ. Pollut.*, 91, 11, 1996.
40. Song, Y. et al., Chemical and mineralogical forms of lead, zinc and cadmium in particle size fraction of some wastes, sediments and soils in Korea, *Appl. Geochem.*, 14, 621, 1999.
41. La Force, M.J. and Fendorf, S., Solid-phase iron characterization during common selective sequential extractions, *Soil Sci. Soc. Am. J.*, 64, 1608, 2000.
42. Brümmer, G.W. et al., Adsorption-desorption and/or precipitation-dissolution processes of Zn in soils, *Geoderma*, 31, 337, 1983.
43. Ostergren, J.D. et al., Quantitative speciation of lead in selected mine tailings from Leadville, CO, *Environ. Sci. Technol.*, 33, 1627, 1999.
44. Gruebel, K.A., Davis, J.A., and Leckie, J.O., The feasibility of using sequential extraction techniques for arsenic and selenium in soils and sediments, *Soil Sci. Soc. Am. J.*, 52, 390, 1988.
45. Calmano, W., Mangold, S., and Welter, E.F., An XAFS investigation of the artifacts caused by sequential extraction analyses of Pb-contaminated soils, *J. Anal. Chem.*, 371, 823, 2001.
46. Henderson, P.J. et al., The chemical and physical characteristics of heavy metals in humus and till in the vicinity of the base metal smelter at Flin Flon, Manitoba, Canada, *Environ. Geol.*, 34, 39, 1998.
47. Manceau, A. et al., Direct determination of lead speciation in contaminated soils by EXAFS spectroscopy, *Environ. Sci. Technol.*, 30, 1540, 1996.
48. Karathanasis, A.D. and Hajek, B.F., Elemental analysis by X-ray fluorescence spectroscopy, in *Methods of Soil Analysis. Part 3. Chemical Methods*, Sparks, D.L., Ed., Soils Science Society of America and American Society of Agronomy, Madison, WI, 1996, p. 161.
49. Vempati, R.K. and Cocke, D.L., X-ray photoelectron spectroscopy, in *Methods of Soil Analysis. Part 3. Chemical Methods*, Sparks, D.L., Ed., Soils Science Society of America and American Society of Agronomy, Madison, WI, 1996, p. 377.
50. Sparks, D.L. et al., Kinetics and mechanisms of metal sorption at the mineral–water interface, in *Mineral–Water Interfacial Reactions: Kinetics and Mechanisms*, in Sparks, D.L. and Grundl, T.J., Eds., American Chemical Society, Washington, D.C., 1998, p. 108.
51. Kim, C.S., Brown Jr., G.E., and Rytuba, J.J., Characterization and speciation of mercury-bearing mine wastes using x-ray absorption spectroscopy, *Sci. Total Environ.*, 261, 157, 2000.
52. Fendorf, S.E. et al., Applications of x-ray absorption fine structure spectroscopy to soils, *Soil Sci. Soc. Am.*, 58, 1583, 1994.

53. Schulze, D.G. and Bertsch, P.M., Synchrotron x-ray techniques in soil, plant, and environmental research, in *Advances in Agronomy*, vol. 55, Sparks, D.L., Ed., Academic Press, New York, 1995, p. 1.
54. O'Day, P.A. et al., Metal speciation and bioavailability in contaminated estuary sediments, Alameda Naval Air Station, California, *Environ. Sci. Technol.*, 34, 3665, 2000.
55. Morin, G. et al., XAFS determination of the chemical form of lead in smelter-contaminated soils and mine tailings: Importance of adsorption processes, *Am. Mineral.*, 84, 420, 1999.
56. Pickering, I.J., Brown Jr., G.E., and Tokunaga, T.K., Quantitative speciation of selenium in soils using x-ray absorption spectroscopy, *Environ. Sci. Technol.*, 29, 2457, 1995.
57. Roberts, D.R., Scheidegger, A.M., and Sparks, D.L., Kinetics of mixed Ni-Al precipitate formation on a soil clay fraction, *Environ. Sci. Technol.*, 33, 3749, 1999.
58. Brown Jr., G.E., Parks, G.A., Bargar, J.R., and Towle, S.E., Use of X-ray absorption spectroscopy to study reaction mechanisms at metal oxide-water interfaces, in *Mineral–Water Interfacial Reactions: Kinetics and Mechanisms*, in Sparks, D.L. and Grundl, T.J., Eds., American Chemical Society, Columbus, OH, 1998, p.14.
59. O'Day, P.A. et al., Extended x-ray absorption fine structure (EXAFS) analysis of disorder and multiple-scattering in complex crystalline solids, *J. Am. Chem. Soc.*, 116, 2938, 1994.
60. Wang, Z.W., Shan, X.Q., and Zhang, S.Z., Comparison between fractionation and bioavailability of trace elements in rhizosphere and bulk soils, *Chemosphere*, 46, 1163, 2002.
61. Duff, M.C. et al., Mineral associations and average oxidation states of sorbed Pu on tuff, *Environ. Sci. Technol.*, 33, 2169, 1999.
62. Duff, M. et al., Comparison of two micro-analytical methods for detecting the spatial distribution of sorbed Pu on geological materials, *J. Contam. Hydrol.*, 47, 211, 2001.
63. Hunter, D.B. and Bertsch, P.M., In situ examination of uranium contaminated soil particles by micro-x-ray absorption and micro-fluorescence spectroscopies, *J. Radioanal. Nucl. Chem.*, 234, 237, 1998.
64. Roberts, D.R., Scheinost, A.C., and Sparks, D.L., Zinc speciation in a smelter-contaminated soil profile using bulk and microspectroscopic techniques, *Environ. Sci. Technol.*, 36, 1742, 2002.
65. Strawn, D. et al., Microscale investigation into the geochemistry of arsenic, selenium, and iron in soil developed in pyritic shale materials, *Geoderma*, 108, 237, 2002.
66. Brown Jr., G.E., Spectroscopic studies of chemisorption reaction mechanisms at oxide–water interfaces, in *Mineral-Water Interface Geochemistry*, Hochella, M.F. and White, A.F., Eds., Mineralogical Society of America, Washington, D.C., 1990, p. 309.
67. Scheinost, A.C. et al., Combining selective sequential extractions, x-ray absorption spectroscopy and principal component analysis for quantitative zinc speciation in soil, *Environ. Sci. Technol.*, 36, 5021, 2002.
68. Storm, G.L., Yahner, R.H., and Bellis, E.D., Vertebrate abundance and wildlife habitat suitability near the Palmerton zinc smelters, Pennsylvania, *Arch. Environ. Contam. Toxicol.*, 25, 428, 1993.
69. M. J. Buchauer, Contamination of soil and vegetation near a zinc smelter by zinc, cadmium, copper, and lead, *Environ. Sci. Technol.*, 7, 131, 1973.
70. Lalo, J., Pennsylvania's dead mountain, *Am. Forests*, March/April, 55, 1988.
71. Zeien, H.B. and Brümmer, G.W., *Mitt. Dtsch. Bodenkundl. Ges.*, 66, 439, 1991.
72. Brümmer, G.W.G. and Herms, U., *Z. Pflanzenernaehr. Bodenkd*, 149, 382, 1986.

73. Zabinsky, S.L. et al., Multiple-scattering calculations of x-ray absorption spectra, *Phys. Rev. B Condens. Matter*, 52, 2995, 1995.
74. Ressler, T., WinXAS: a program for x-ray absorption spectroscopy data analysis under MS-Windows, *J. Synchrotron Rad.*, 5, 118, 1998.
75. Ressler, T. et al., Quantitative speciation of Mn-bearing particulates emitted from autos burning (methylcyclopentadienyl) manganese tricarbonyl-added gasolines using XANES spectroscopy, *Environ. Sci. Technol.*, 34, 950, 2000.
76. Schwertmann, U. and Cornell, R.M., *Iron Oxides in the Laboratory: Preparation and Characterization*, Weinheim, New York, 1991.
77. McKenzie, R.M., The synthesis of birnessite, cryptomelane, and some other oxides and hydroxides of manganese, *Miner. Mag.*, 38, 493, 1971.
78. Manceau, A. et al., Crystal chemistry of trace elements in natural and synthetic goethite, *Geochim. Cosmochim. Acta*, 64, 3643, 2000.
79. Barak, P. and Helmke, P.A., The chemistry of zinc, in *Zinc in Plants and Soils*, Robson, A.D., Ed., Kluwer, Dordrecht, 1993, p. 1.
80. Roberts, D.R., Ford, R.G., and Sparks, D.L., Kinetics and mechanisms of Zn sorption on metal oxides using EXAFS, *J. Colloid Interface Sci.*, in revision.
81. Manceau, A. and Calas, G., Absence of evidence for Ni/Si substitution in phyllosilicates, *Clay Miner.*, 22, 357, 1987.
82. Barnhisel, D.C. and Bertsch, P.M., Chlorites and hydroxy interlayered vermiculite and smectite, in *Minerals in Soil Environ.*, 2nd ed., Dixon, J.B. and Weed, S.B., Eds., Soil Science Society of America, Madison, WI, 1989, p. 729.
83. MacDowell, A.A. et al., Progress towards sub-micron hard x-ray imaging using elliptically bent mirrors, in *X-ray Microfocusing: Applications and Techniques*, McNulty, I., Ed., vol. 3449, 1998, p. 137, SPIE Proceedings, Bellingham, WA.
84. Waychunas, G.A., Fuller, C.C., and Davis, J.A., Surface complexation and precipitate geometry for aqueous Zn(II) sorption on ferrihydrite I: X-ray absorption extended fine structure spectroscopy analysis, *Geochim. Cosmochim. Acta*, 66, 1119, 2002.
85. Yin, Y. et al., Kinetics of mercury(II) adsorption and desorption on soil, *Environ. Sci. Technol.*, 31, 496, 1997.
86. Strawn, D.G. and Sparks, D.L., Effects of soil organic matter on the kinetics and mechanisms of Pb(II) sorption and desorption in soil, *Soil Sci. Soc. Am. J.*, 64, 144, 1998.
87. Lindsay, W.L., *Chemical Equilibria in Soils*, John Wiley & Sons, New York, 1979.
88. Elgersma, F., Schinkel, J.N., and Weijnen, M.P.C., Improving environmental performance of a primary lead and zinc smelter, in *Heavy Metals: Problems and Solutions*, Salomons, W., Förstner, U., and Mader, P., Eds., Springer-Verlag, Berlin, 1995, p. 193.
89. Sobanska, S. et al., Microchemical investigations of dust emitted by a lead smelter, *Environ. Sci. Technol.*, 33, 1334, 1999.
90. Goodarzi, F. et al., Sources of lead and zinc associated with metal smelting activities in the Trail area, British Columbia, Canada, *J. Environ. Monitoring*, 4, 400, 2002.
91. Venditti, D., Berthelin, J., and Durécu, S., A multidisciplinary approach to assess history, environmental risks, and remediation feasibility of soils contaminated by metallurgical activities. Part B: Direct metal speciation in the solid phase, *Arch. Environ. Contam. Toxicol.*, 38, 421, 2000.
92. Buatier, M.D., Sobanska, S., and Elsass, F., TEM-EDX investigation on Zn- and Pb-contaminated soils, *Appl. Geochem.*, 16, 1165, 2001.
93. Li, Y.M. et al., Response of four turfgrass cultivars to limestone and biosolids-compost amendment of a zinc and cadmium contaminated soil at Palmerton, Pennsylvania, *J. Environ. Qual.*, 29, 1440, 2000.

9 Reduction/Cation Exchange Model of the Coincident Release of Manganese and Trace Metals following Soil Reduction

Dean M. Heil, Grant E. Cardon, and Colleen H. Green

CONTENTS

9.1 Introduction 230
9.1.1 Association of Trace Metals with Mn Oxides 230
9.1.2 Previous Studies on the Effect of Soil Reduction on Metal Solubility 230
9.1.3 Influence of Electrolyte Concentration and Cation Exchange Reactions 231
9.1.4 Processes and Reactions Controlling the Solubility of Mn and Trace Metals Following Reduction 231
9.2 Case Study 232
9.3 Reduction/Cation Exchange Model 233
9.3.1 Reduction Model 233
9.3.2 Cation Exchange Model 235
9.3.3 Calculation of Cation Exchange Coefficients 235
9.3.4 Comparison of Model Predictions to Experimental Data 236
9.3.4.1 Prediction of Ca, Mg, Mn, and Sr 236
9.3.4.2 Prediction of Ni and Zn 239
9.3.5 Model Limitations 242
9.3.6 Applications to Chemical Transport Modeling 242
9.4 Summary 242
References 243

1-56670-623-8/03/$0.00+$1.50

9.1 INTRODUCTION

The effect of soil reduction on trace metal solubility has important implications to both plant availability and toxicity, and chemical transport. The release of metals associated with Mn and Fe oxides following reductive dissolution is an important mechanism that can potentially increase the soluble concentrations of metals.[1,2] The potential for the release of trace metals following soil reduction appears to be the greatest for slightly to moderately reduced soils, with redox potentials between 100 and 400 mV. Under these conditions, redox potentials are sufficiently low to dissolve Mn or Fe oxides, but not low enough to precipitate metal sulfides. In highly reduced soils with redox potentials less than approximately 0 mV, the precipitation of metal sulfides limits the soluble concentration of trace metals.[1,3] Dissolution of Mn oxides precedes Fe oxide dissolution because of the lower redox potential required to dissolve Fe oxides.[4] Manganese (IV) oxides become unstable at a redox potential (E_H) of approximately 300 mV, whereas Fe (III) oxides are stable until E_H decreases to less than 100 mV, with the exact values of E_H required to initiate reductive dissolution dependent on pH. Consequently, the dissolution of Mn oxides may play a more important role in metal solubilization in the early stages of soil reduction when redox potential is low enough to dissolve Mn oxides, but Fe oxides may still be stable.

9.1.1 Association of Trace Metals with Mn Oxides

Manganese oxides have a high affinity for many of the trace metals.[5,6] In addition to surface adsorption, trace metals accumulate in Mn oxides by substitution and co-precipitation.[7] The adsorptive properties of Mn oxides for metals observed in the laboratory are verified in soils, as Mn oxide nodules separated from soils contain concentrations of trace metals that are considerably greater than the metal concentrations in the bulk soil.[7,8] The potential for association of trace metals with Mn oxides via co-precipitation or substitution is high when soils are subject to alternate wetting and drying cycles,[9] and Mn oxide crystals are forming.

9.1.2 Previous Studies on the Effect of Soil Reduction on Metal Solubility

Several researchers have reported an increase in the soluble concentrations of trace metals under reducing conditions. Chuan et al.[10] found that the release of soluble Pb, Cd, and Zn from a soil increased as E_H was decreased from 325 to −100 mV at a constant pH. Davranche and Bollinger[11] observed that Pb and Cd adsorbed to synthetic Mn or Fe oxide was released into solution as the solid phases were progressively dissolved by increasing concentrations of a reducing agent. The destabilization of Fe and Mn oxides following the addition of a reducing agent to a contaminated soil caused an increase in the soluble concentrations of both Cd and Pb.[11] Soil adsorbents not dissolved by reductive dissolution were considered to have a large effect on the solubility of the metals, as Cd concentrations did not increase substantially until pH was less than approximately 6, and Pb did not increase until pH was less than 4. The authors noted that this difference in the behavior of Cd and

Pb is consistent with a 2-pH unit difference in the adsorption edges of these two metals for natural colloids. Charlatchka and Cambier[2] reported that soluble concentrations of Pb and Cd increased with time in flooded soil cores, and concluded that a decrease in pH caused by reduction processes played a critical role in elevating soluble metal concentrations. Incubation of the soil in a pH stat-redox cell revealed that at fixed pH, soluble concentrations of Pb, Cd, and Zn increased with incubation time, coinciding with a decrease in redox potential.[2] Destabilization of Mn and Fe oxides was considered to be an important mechanism for the release of trace metals under steady pH. In cases where pH increases following reduction, trace metal solubilities have been observed to decrease. Kashem and Singh[12] reported that for all three soils that were studied, the soluble concentrations of Cd and Zn decreased following saturation, and Ni decreased in two of the three soils. These decreases in metal solubility coincided with a pH increase in all three soils, and were attributed to enhanced sorption and possibly greater stability of metal oxides or other minerals at higher pH.

9.1.3 Influence of Electrolyte Concentration and Cation Exchange Reactions

Another mechanism that could influence the solubilization of trace metals under reducing conditions could be the displacement of exchangeable metals by high concentrations of dissolved Mn released following the dissolution of Mn oxides. Soluble Ca and Mg have been observed to increase following soil reduction, and this has been attributed to the displacement of those cations from exchange sites by dissolved Mn and Fe.[13,14] This process may be described by the reaction:

$$CaX_2 + Mn^{2+} = MnX_2 + Ca^{2+}. \tag{9.1}$$

The increased concentration of divalent cations could also be expected to displace trace metals from cation exchange sites as well as Ca and Mg. Although the exchangeable metal concentrations in many soils are low, exchangeable metal concentrations are generally much greater than soluble metals concentrations, and could act as a source to the solution phase under reducing conditions when electrolyte concentration (EC) is increased.

9.1.4 Processes and Reactions Controlling the Solubility of Mn and Trace Metals Following Reduction

The soluble concentration of Mn under reducing conditions will depend on the amount of Mn oxide dissolved and the extent of re-precipitation or adsorption of the released Mn(II) to soil colloids. Xiang and Banin[15] found that a significant fraction of the Mn released by Mn oxide dissolution within 3 days of saturation was redistributed to cation exchange sites. Manganese can also be retained by specific adsorption to Fe oxides, organic matter, and layer silicate clay minerals.[5,16,17] A review of the mechanisms controlling adsorption of Mn to soil constituents is provided by Khattack and Page.[18]

The solubility of trace metals under reducing conditions will depend on the amount of Mn oxide dissolved, the concentration of the trace metals initially associated with the Mn oxide fraction, and the retention of the released metals to soil solids following dissolution of the Mn oxide. Trace metals initially associated with the Mn oxide fraction may also be retained by specific adsorption reactions,[19] involving surface hydroxyl sites on mineral and organic soil colloids. The partitioning of metals between these sites and the solution phase is highly dependent on pH, as predicted by the following reaction[20]:

$$=SOH + M^{z+} = =SOM^{(z-1)+} + H^{+}, \quad (9.2)$$

where =SOH is a surface hydroxyl functional group. In cases where pH changes significantly during soil reduction, we can expect that it will be necessary to include specific adsorption reactions to model the changes in solubility of both Mn and trace metals.

9.2 CASE STUDY

The solubility of Mn, Zn, Ni, and Sr following saturation of soil columns was studied for two soils collected from the Alamosa River Basin, Colorado. These two soils are classified as the LaJara (coarse-loamy, mixed (calcareous), frigid typic haplaquolls) and Mogote (fine-loamy, mixed (calcareous), frigid aquic ustorthents) series. Soils in this region have a history of irrigation with water impacted by acid mine drainage. Basic soil chemical and physical properties are shown in Table 9.1. Samples were collected from the base of the soil columns at 12-h intervals up to 84 h. This time frame was chosen to simulate the period of saturation following flood irrigation of soils in this region. Details of procedures are described by Green.[21] Reduction experiments are often performed with the addition of a carbon source to accelerate a decrease in E_H. The data used in the present model are from treatments that did not receive an amendment with an additional carbon source.

TABLE 9.1
Chemical and Physical Properties of LaJara and Mogote Soils

Soil	pH[a]	CEC[b] (cmol kg^{-1})	CCE[c] (g kg^{-1})	Fe_{ox}[d] (g kg^{-1})	OC[e] (g kg^{-1})	Sand (%)	Silt (%)	Clay (%)
LaJara	5.95	15.5	0.9	2.45	8.1	57	16	27
Mogote	6.80	17.8	1.6	2.14	8.8	44	19	37

[a] 24-h 1:1 soil:water pH.
[b] Summation method.
[c] Calcium carbonate equivalent.
[d] Iron oxide content.
[e] Organic carbon.

The results of these studies are described by Green et al.[22] The major findings are summarized here. The soluble concentrations of Mn, Zn, Ni, and Sr increased in both soils following reduction. Total electrolyte concentration also increased following reduction, and this change was due mainly to increases in soluble Ca and Mg concentrations. Redox potentials decreased to values that were sufficient to initiate the dissolution of Mn oxides within 24 h after saturation, and remained nearly constant through 84 h. Iron oxides were apparently stable under the redox conditions and time frame of our experiments, as increases in soluble Fe were not observed. Total electrolyte concentration (EC) also increased continuously throughout the 84-h saturation period, with most of the change in EC associated with increased concentrations of soluble Ca and Mg. Soluble concentrations of Pb and Cd were also measured, but were below the instrument detection limits for many samples. For these reasons, we chose to test the fit of the data from these experiments to a cation exchange model including Mn, Ca, Mg, Sr, Ni, and Zn. The data used to test the cation exchange model were taken from the average of duplicate columns for each of the two soils.

9.3 REDUCTION/CATION EXCHANGE MODEL

9.3.1 Reduction Model

The amount of Mn oxide dissolved over an 84-h time period for each experiment was calculated based on the assumption that the observed increase in EC was due to displacement of exchangeable cations by dissolved Mn^{2+}. Electrolyte concentration was calculated by summing the contributions from the divalent cations:

$$EC = 2\,([Ca^{2+}] + [Mg^{2+}] + [Mn^{2+}] + [Sr^{2+}] + [Ni^{2+}] + [Zn^{2+}]), \qquad (9.3)$$

where EC is in $mol_c\ l^{-1}$. Although soluble concentrations of K and Na changed slightly between 24 and 84 h, these cations were not included in the calculation of EC as they were not included in the cation exchange model. The exclusion of Na and K from the cation exchange model was based on the observation that Na and K accounted for only 9% and 7%, respectively, of the total increase in electrolyte concentration in the LaJara and Mogote soils. The total concentration of Mn dissolved between 24 and 84 h was calculated as

$$[Mn]_d = (EC_{84} - EC_{24})/2 \qquad (9.4)$$

with $[Mn]_d$ expressed as mol l^{-1}. A constant dissolution rate of Mn oxides was also assumed by dividing the total Mn dissolved into five 12-h intervals, beginning with 24 h at which time dissolution of Mn(IV) oxides began. This yielded a concentration of Mn of 1.89 E-4 M for the LaJara soil and 4.10 E-4 M for the Mogote soil for each 12-h interval. In terms of the concentration of Mn in the soil, the amount of Mn dissolved after 84 h was 5.41 E-4 mol kg^{-1} for the LaJara soil, and 1.17 E-3 mol kg^{-1} for the Mogote soil. Compared to the total concentration of reducible Mn (Table 9.2), this represents 30% of the reducible Mn oxide for the LaJara soil, and

TABLE 9.2
Metal Concentrations in Exchangeable, Mn-Oxide, and Total Fractions of LaJara and Mogote Soils[a]

	LaJara (mol kg^{-1})			Mogote (mol kg^{-1})		
	Exchangeable	Mn-oxide	Total	Exchangeable	Mn Oxide	Total
Ca	5.82 E-2	—	—	6.39 E-2	—	—
Mg	1.83 E-2	—	—	2.47 E-2	—	—
Mn	5.46 E-4	1.79 E-3	856	1.15 E-4	2.28 E-3	868
Sr	2.39 E-4	1.28 E-5	74.9	2.77 E-4	1.03 E-5	98.1
Zn	2.37 E-5	4.56 E-5	155	2.80 E-5	6.68 E-5	143
Ni	3.17 E-6	1.42 E-5	12.8	1.17 E-6	8.09 E-6	13.7

[a] Exchangeable cation concentrations are from 1-h 1 M extraction with KCl.

51% for the Mogote soil. For each time period, the total Mn concentration available for cation exchange reactions, in units of molarity, was calculated as

$$[Mn]_{T(t)} = [Mn]_i + [Mn]_{d(t)}, \quad (9.5)$$

where $[Mn]_i$ is the initial total concentration of Mn available for cation exchange reactions at the beginning of the experiment, and $[Mn]_{d(t)}$ is the concentration of Mn(II) released by dissolution of Mn oxide after each time period. The initial concentration of Mn was calculated as

$$[Mn]_i = [Mn]_{exch} + [Mn]_s \quad (9.6)$$

with $[Mn]_{exch}$ the initial concentration of exchangeable Mn, and $[Mn]_s$ the concentration of soluble Mn from the saturated columns at the 24-h time period. The concentrations of exchangeable metals were converted from mol kg^{-1} to mol l^{-1} by multiplying by the solid:solution ratio of the soil at saturation, which was 1.7 kg l^{-1} for both soils. The total concentrations of Ca and Mg were fixed based on the initial exchangeable and soluble concentrations, as in equation 9.6.

For modeling the release of Sr, Ni, and Zn, two approaches were taken. In the first model, the total concentration of Sr, Ni, and Zn available for cation exchange reactions was fixed as the sum of the initial concentrations of soluble and exchangeable concentrations as in equation 9.6. In the second model, the concentrations of Zn, Ni, and Sr released by dissolution of Mn oxide were included and were considered to be available for cation exchange reactions. The corresponding equation for the total concentration of each metal for the second model was

$$[M]_{T(t)} = [M]_{exch} + [M]_s + f_{M,Mn\text{-}ox} [Mn]_{d(t)} \quad (9.7)$$

where $f_{M,Mn\text{-}ox}$ is the number of moles of metal M per mole of Mn in the Mn oxide fraction. The quantity of metals associated with the Mn oxide fraction used to calculate $f_{M,Mn\text{-}ox}$ for each metal was measured by a sequential extraction procedure[23] (Table 9.2).

9.3.2 Cation Exchange Model

Cation exchange reactions involving Ca, Mg, Mn, Sr, Ni, and Zn were modeled based on the following reaction[24]:

$$CaX_2 + M^{2+} = MX_2 + Ca^{2+} \quad (9.8)$$

The cation exchange equation corresponding to this reaction is

$$K_{ex} = \frac{MX_2[Ca^{2+}]}{CaX_2[M^{2+}]} \quad (9.9)$$

where K_{ex} is the cation exchange coefficient. The cation exchange reaction may be separated into two component half-reactions to facilitate computer modeling[25,26]:

$$M^{2+} + 2\ X^- = MX_2, \quad (9.10)$$

with the equilibrium constant for the formation of MX_2 represented by K_f. The corresponding mass balance equation for cation exchange sites as applied to this problem was

$$X_T = 2\ (CaX_2 + MgX_2 + MnX_2 + SrX_2 + NiX_2 + ZnX_2) \quad (9.11)$$

where X_T is the total concentration of cation exchange sites in $mol_c\ kg^{-1}$. In order to represent fixed charge sites where the concentration of uncomplexed X^- is essentially zero, the convention used by Stadler and Schindler[26] was followed, with the log K_f for the formation of CaX_2 in equation 9.10 set equal to 20.0. We verified that following this convention resulted in less than 0.1% of exchange sites unoccupied by cations, and modeling results were not dependent on the value of log K_f for CaX_2 for values between 10 and 20. The log K_f values for equation 9.10 for cations other than Ca were obtained by adding the value of log K_{ex} to 20.0. The MINTEQA2 computer speciation program[27] was used for modeling.

9.3.3 Calculation of Cation Exchange Coefficients

Values for cation exchange coefficients were calculated using exchangeable cation concentrations from 1-h 1-M KCl extraction (Table 9.2). The quantity of exchangeable cations were calculated based on the surface excess of each cation[24]:

$$q_i = n_i - M_w\ m_i \quad (9.12)$$

TABLE 9.3
Cation Exchange Coefficients and Equilibrium Constants for LaJara and Mogote Soils

	LaJara		Mogote	
	K_{ex}	log K_f	K_{ex}	log K_f
Ca		20.000		20.000
Mg	0.864	19.937	1.000	20.00
Mn	0.588	19.769	0.605	19.782
Sr	1.091	20.0379	1.138	20.0563
Zn	0.212	19.325	0.207	19.317
Ni	0.582	19.700	0.340	19.532

where q_i is the surface excess in mol kg^{-1}, n_i is the total number of moles of the cation extracted per kilogram of dry soil, M_w is the gravimetric water content of the slurry, and m_i is the molality of the cation in the supernatant solution. We obtained values of m_i by performing 1-h extractions with water at the same solid:solution ration as for the 1-M KCl extractions. We found that correction for the concentration of soluble cations significantly reduced the calculated exchangeable concentrations of Zn and Ni, and this would have a substantial effect on the model predictions for these elements if uncorrected values were used. Data for soluble metal and cation concentrations were taken from the 24-h time period of the reduction experiments. The soluble cation and metal concentrations at 24 h in the soil columns were similar to the 1-h batch, water-soluble concentrations. Values for K_{ex} for the overall cation exchange reactions, based on Ca as the initial cation occupying exchange sites corresponding to equation 9.8 were first calculated from experimental data (Table 9.3). For modeling purposes, values for log K_f for the half-reactions for each metal corresponding to equation 9.10 were then determined (Table 9.3).

9.3.4 Comparison of Model Predictions to Experimental Data

9.3.4.1 Prediction of Ca, Mg, Mn, and Sr

The concentrations of soluble Ca, Mg, and Sr were consistent with the cation exchange model for both soils studied (Figures 9.1 to 9.4). Experimental Ca concentrations were greater than model predictions between 36 and 60 h; this could be a result of deviation from linear dissolution of the Mn oxide as was assumed in the model. Soluble Mn concentrations increased by a factor of 5 times between 24 and 84 h for the LaJara soil (Figure 9.2) and 28 times for the Mogote soil (Figure 9.4). The greater relative increase in soluble Mn concentration in the Mogote soil is a result of the increased amount of Mn oxide dissolved for that soil, as noted above. This is consistent with the higher amount of reducible Mn in the Mogote versus LaJara soils (Table 9.2), as well as a slightly lower E_H in the Mogote soil during reduction.[21] The model predicted these changes in Mn solubility with fair accuracy, with the soluble Mn at 84 h underestimated by the model by 17% for the LaJara soil, and overestimated by the model by 20% for the Mogote soil. Comparison of the soluble Mn concentrations

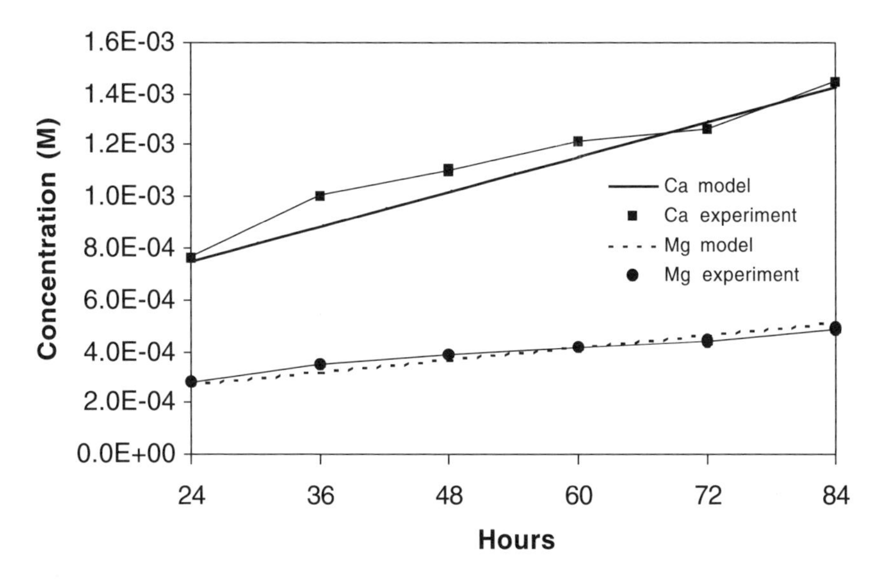

FIGURE 9.1 Soluble concentrations of Ca and Mg from LaJara soil.

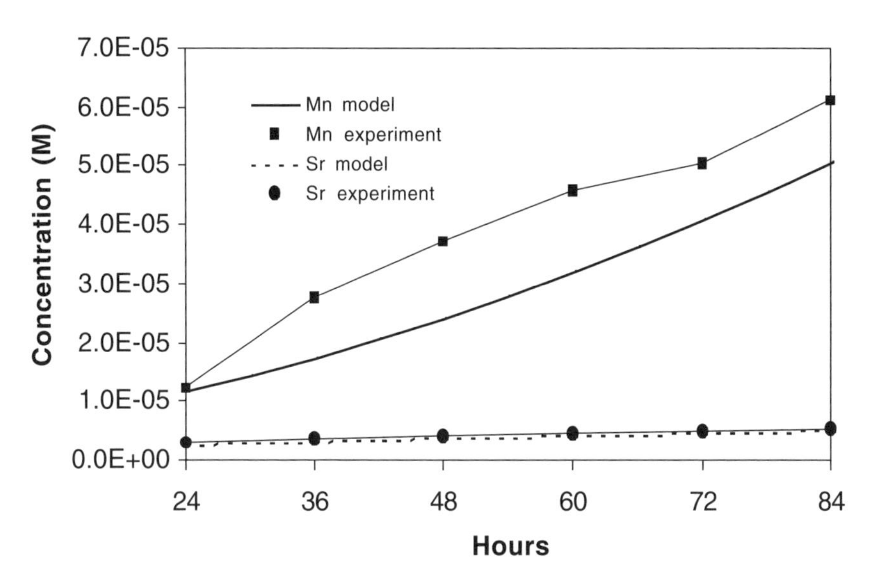

FIGURE 9.2 Soluble concentrations of Mn and Sr from LaJara soil.

at 84 h to the total amount of Mn released by Mn-oxide dissolution reveals that approximately 6% and 3%, respectively, of the released Mn remained in solution for the LaJara and Mogote soils, with the balance retained by cation exchange sites. The cation exchange capacity (CEC) of these two soils is very similar (Table 9.2). Therefore, the tendency for a smaller fraction of the dissolved Mn to be proportioned to exchange sites in the LaJara versus the Mogote soils is probably due to the higher amount of initial exchangeable Mn in the LaJara soil (Table 9.2). The addition of Zn,

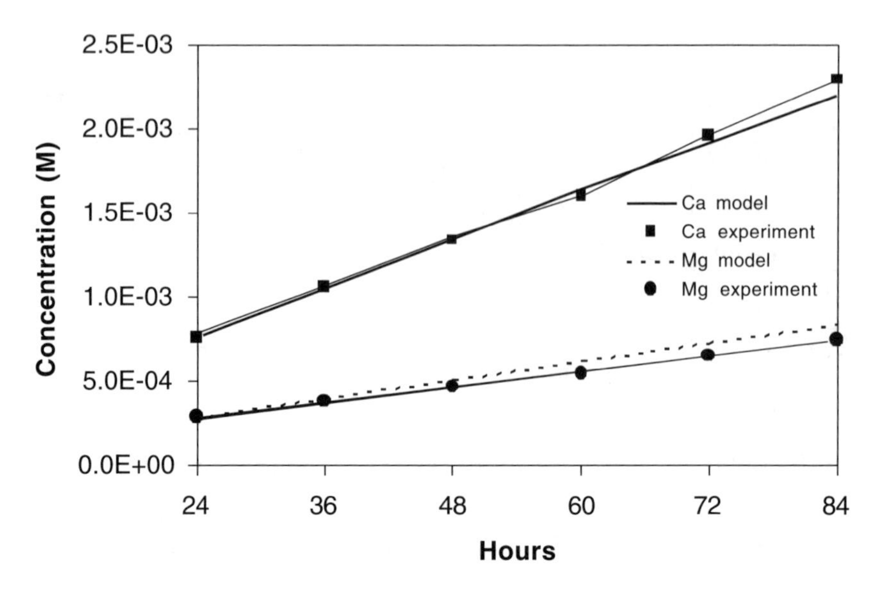

FIGURE 9.3 Soluble concentrations of Ca and Mg from Mogote soil.

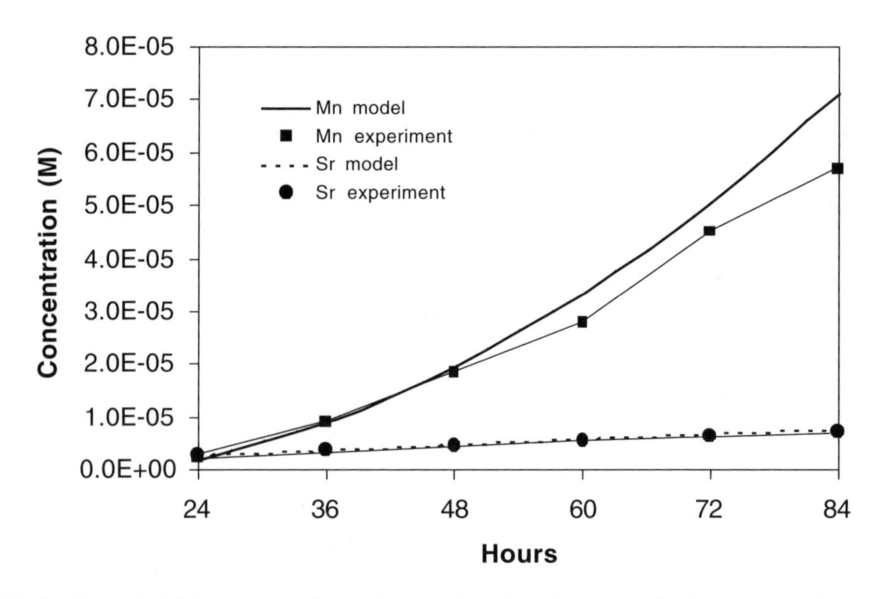

FIGURE 9.4 Soluble concentrations of Mn and Sr from Mogote soil.

Ni, and Sr from the Mn oxide fraction caused only slight changes in the predicted concentrations of Ca, Mg, Mn, and Sr. The model results shown in Figures 9.1 to 9.4 were plotted using the results of the model without the addition of Ni, Zn, and Sr initially associated with Mn oxides. However, these results are representative of both models for those elements. Although Sr is present in the Mn oxide fraction, the amount of Sr released as a result of Mn oxide dissolution is small compared to the initial exchangeable Sr concentration, which explains the small effect of adding Sr from the Mn oxide fraction on the modeled Sr solubility. The close fit of the cation exchange model to Sr solubility for both soils (Figures 9.2 and 4) indicates that Sr is behaving as an exchangeable cation in the soil, very similar to Ca and Mg. The relative increase in the concentrations of Sr, Ca, and Mg between 24 and 84 h was very similar to the relative increase in EC over this time period for both soils.

9.3.4.2 Prediction of Ni and Zn

For both soils, the increase in soluble Ni from the beginning to the end of the reduction experiments was accounted for by considering the initial exchangeable Ni as the only source available for cation exchange reactions (Figures 9.5 and 9.6). Addition of the Ni released by Mn oxide dissolution created a considerable over-prediction of soluble Ni concentration by the model in both soils. This result indicates that Ni initially associated with Mn oxide was retained by specific adsorption mechanisms as opposed to redistribution to exchange sites following soil reduction. The soluble Zn concentration at 84 h for the LaJara soil was underestimated without the inclusion of Zn initially associated with Mn oxide (Figure 9.7). Addition of the Zn initially associated with Mn oxide to the amount of Zn available for cation exchange reactions improved the accuracy of the model (Figure 9.7). It appears that for the LaJara soil, the Zn released by Mn oxide dissolution was converted to exchangeable plus soluble Zn. For the Mogote soil, Zn solubility was underestimated without including Zn initially associated with Mn oxide, but overestimated when it was included (Figure 9.8).This suggests that for the Mogote soil, some fraction of the Zn released by Mn oxide dissolution was retained by exchange sites. The

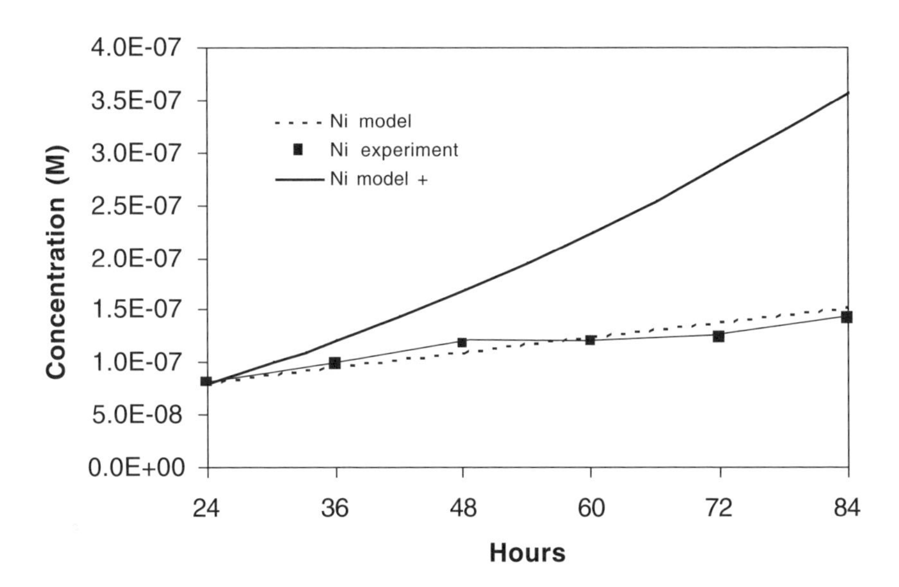

FIGURE 9.5 Soluble concentrations of Ni from LaJara soil. Ni model represents the model without addition of Ni from the Mn oxide fraction; Ni model + represents the model with addition of Ni from the Mn oxide fraction.

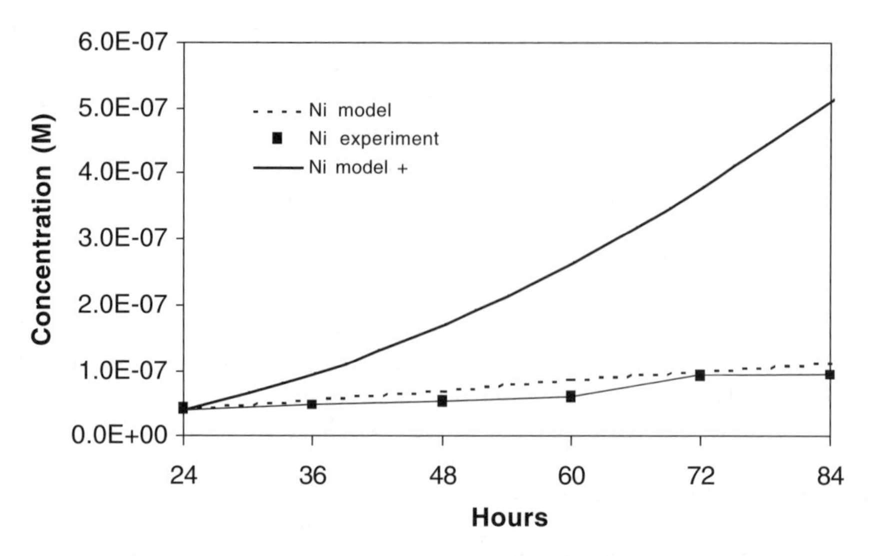

FIGURE 9.6 Soluble concentrations of Ni from the Mogote soil. Ni model represents the model without addition of Ni from the Mn oxide fraction; Ni model + represents the model with addition of Ni from the Mn oxide fraction.

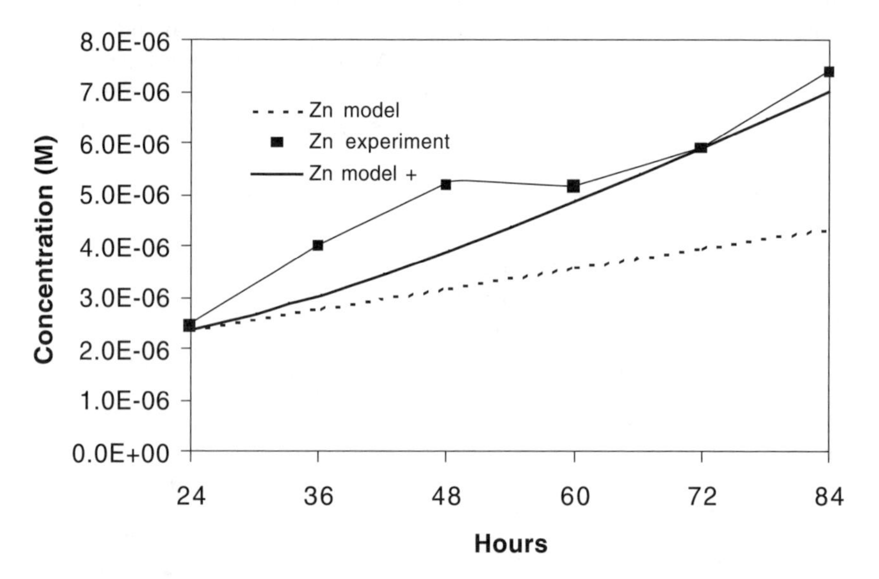

FIGURE 9.7 Soluble concentrations of Zn from the LaJara soil. Zn model represents the model without addition of Zn from the Mn oxide fraction; Zn model + represents the model with addition of Zn from the Mn oxide fraction.

tendency of Zn to redistribute to exchange sites in the LaJara soil as compared to redistribution to both exchange and specific adsorption sites in the Mogote soil is consistent with the lower pH of the LaJara soil during reduction. The average pH of the LaJara soil effluents between 24 and 84 h was 6.2, compared to an average

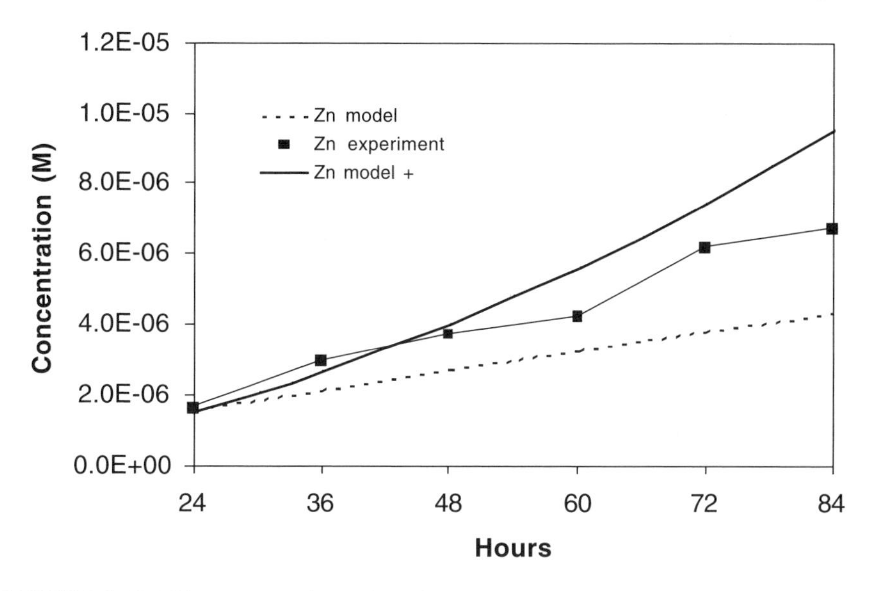

FIGURE 9.8 Soluble concentrations of Zn from the Mogote soil. Zn model represents the model without addition of Zn from the Mn oxide fraction; Zn model + represents the model with addition of Zn from the Mn oxide fraction.

pH of 6.6 for the Mogote soil. Cowan et al.[28] illustrated that the relative importance of cation exchange versus specific adsorption of metals is increased at lower pH. The apparent greater specific adsorption of Ni by the Mogote soil cannot be explained in terms of differences in the capacity for specific adsorption, as the contents of organic matter and Fe oxides are very similar for the two soils (Table 9.1). The apparent differences in the chemical behavior of Ni versus Zn in both soils suggests that Zn tends to be redistributed to cation exchange sites while Ni tends to be retained by specific adsorption. This result is consistent with the greater affinity of soil organic matter for Ni versus Zn.[29] Adsorption of Ni and Zn to organic matter may be more important than adsorption to Fe oxides in these soils, since Fe oxides consistently exhibit a greater affinity for Zn over Ni.[7,30]

For Ni and Zn, although the model fit the experimental data fairly well at the 84-h time period (depending on the assumption regarding the fate of released metals), the nonuniform changes in soluble metal concentrations over short time segments (Figure 9.7) were not predicted by the model. An assumption inherent to the application of equation 9.2 is that each trace metal was distributed uniformly throughout the Mn oxide particles. This may not be accurate if the dominant mechanism of metal association was surface adsorption as opposed to substitution within the Mn oxide crystal structures. If the dominant mechanism was surface adsorption, we might expect a high concentration of the adsorbed metals to be released in the early stages of reduction. Furthermore, if the particle size of the Mn oxide particles was not uniform, then smaller Mn oxide particles with a greater surface area and a corresponding metal adsorption capacity would also be expected to exhibit the highest rate of dissolution, again leading to a relatively high release of metals in the early stages.

9.3.5 Model Limitations

The cation exchange model presented cannot be expected to describe changes in the soluble concentrations of Mn or trace metals following reduction for all soils. Cation exchange models do not account for the effects of pH on metal release. This was not a limitation for applying the model to our data, because both the pH and E_H of our soil columns remained at a nearly constant value between 24 and 84 h for both soils. Changes in pH that are often observed following soil reduction can be expected to cause changes in the partitioning of adsorbed or precipitated metals. In addition, the rate of dissolution of Mn oxides is expected to change with both pH and E_H.[31] In alkaline soils, Mn(II) is known to precipitate as manganese carbonate (rhodochrosite) following soil reduction.[32,33] Furthermore, the dissolution of calcite by acid generated from reduction processes is believed to be largely responsible for the increases in soluble Ca observed following the reduction of calcareous soils.[2] Clearly, the dissolution of Mn oxide cannot be considered as the sole process responsible for the changes in EC and subsequent effects on cation exchange equilibria in alkaline soils. Changes in both the concentrations and chemical nature of dissolved organic compounds which complex trace metals following soil reduction may also play a critical role in trace metal solubility.[2] This mechanism may increase in importance for soils containing amendment with an organic carbon source and when large changes in redox potential are observed.

9.3.6 Applications to Chemical Transport Modeling

The incorporation of cation exchange reactions into chemical transport models has been described by Selim et al.[34] and Kretzschmar and Voegelin.[35] In our model, we determined the dissolution rate of Mn oxides based on the difference in EC between the beginning and end of the experiment. In order to apply this model with a strictly predictive approach, it would be necessary to have an independent measurement or estimate of Mn oxide dissolution rate for the soil of interest. It would also be necessary to be able to predict the dynamic changes in both E_H and pH as reduction proceeds. If pH changes significantly during reduction, then the model must account for both the change in the dissolution rate of Mn oxides and the repartitioning of metals to specific adsorption sites.

9.4 SUMMARY

The changes in the concentrations of divalent cations (Ca, Mg) and trace metals (Sr, Zn, Ni) were consistent with a cation exchange model. Sr solubility was modeled accurately by including only the initial exchangeable and soluble Sr as a source because the amount of Sr released by Mn oxide dissolution was small compared to the initial exchangeable Sr concentration. The increase in soluble Ni concentration in both soils following reduction can be explained based on the displacement of exchangeable Ni initially present. The addition of Ni released by Mn oxide dissolution to the cation exchange system resulted in substantial over-prediction by the model of soluble Ni concentrations at the end of the reduction experiments for both soils. This suggests that the Ni released from the Mn oxide fraction as a result of reduction was retained by specific adsorption or precipitation mechanisms. The behavior of Zn was intermediate between that of Ni and Sr. Exclusion of the Zn released by Mn oxide dissolution caused an underestimation by the model of the soluble Zn concentrations in both soils. Inclusion of the Zn from the Mn oxide fraction provided improved fit of the model to experimental data for the LaJara soil; however, Zn solubility was then overestimated for the Mogote soil. Cation exchange reactions appear to have an important influence on the changes in the solubility of Mn and other trace metals under reducing conditions for these soils. In certain situations, it will be necessary to include specific adsorption in the model, especially when attempting to model metals with a high affinity for specific adsorption sites such as Cu and Pb, and also when conditions are such that pH changes significantly during reduction.

REFERENCES

1. Reddy, C.N. and Patrick, W.H., Effect of redox potential and pH on the uptake of Cd and Pb by rice plants, *J. Environ. Qual.*, 6, 259, 1977.
2. Charlatchka, R. and Cambier, P., Influence of reducing conditions on solubility of trace metals in contaminated soils, *Water Air Soil Pollut.*, 118, 143, 2000.
3. Brennan, E.W. and Lindsay, W.L., The role of pyrite in controlling metal ion activities in highly reduced soils, *Geochim. Cosmochim. Acta,* 60, 3609, 1996.
4. Patrick, W.H. and Jugsujinda, A., Sequential reduction and oxidation of inorganic nitrogen, manganese, and iron in a flooded soil, *Soil Sci. Soc. Am. J.*, 56, 1071, 1992.
5. McKenzie, R.M., The adsorption of lead and other heavy metals on oxides of manganese and iron, *Aust. J. Soil Res.*, 18, 61, 1980.
6. Trivedi, P. and Axe, L., Modeling Cd and Zn sorption to hydrous metal oxides, *Environ. Sci. Technol.*, 34, 2215, 2000.
7. McKenzie, R.M., Manganese oxides and hydroxides, in *Minerals in Soil Environments*, 2nd ed., Dixon, J.B., and S.B. Weed, Eds., Soil Science Society of America, Madison, WI, 1989, chap. 9.
8. Palumbo, B. et al., Trace metal partitioning in Fe-Mn nodules from Sicilian soils, Italy, *Chem. Geol.*, 173, 257, 2001.
9. McBride, M.B., *Environmental Soil Chemistry*, Oxford University Press, Oxford, 1994.
10. Chuan, M.C., Shu, G.Y., and Liu, J.C., Solubility of heavy metals in a contaminated soil: Effects of redox potential and pH, *Water Air Soil Pollut.*, 90, 543, 1996.
11. Davranche, M. and Bollinger, J., Heavy metals desorption from synthesized and natural iron and manganese oxyhydroxides: Effect of reductive conditions, *J. Colloid Interface Sci.*, 227, 531, 2000.
12. Kashem, M.A. and Singh, B.R., Metal availability in contaminated soils: I. Effects of flooding and organic matter on changes in Eh, pH, and solubility of Cd, Ni, and Zn, *Nutrient Cycling in Agroecosystems*, 61, 247, 2001.
13. Phillips, I.R. and Greenway, M., Changes in water-soluble exchangeable ions, cation exchange capacity, and phosphorus_{max} in soil under alternating waterlogged and drying conditions, *Commun. Soil Sci. Plant Anal.*, 29, 51, 1998.
14. Narteh, L.T. and Sahrawat, K.L., Influence of flooding on electrochemical and chemical properties of West Africa soils, *Geoderma*, 87, 179, 1999.

15. Xiang, H.F. and Banin, A., Solid-phase manganese fractionation changes in saturated arid-zone soils: Pathways and kinetics, *Soil Sci. Soc. Am. J.,* 60, 1072, 1996.
16. Khan, S.U., Interaction between the humic acid fraction of soils and certain metallic cations, *Soil Sci. Soc. Am. Proc.*, 33, 851, 1969.
17. Ikhsan, J., Johnson, B.B., and Wells, J.D., A comparative study of the adsorption of transition metals on kaolinite, *J. Colloid Interface Sci.,* 217, 403, 1999.
18. Khattack, R.A. and Page, A.L., Mechanism of manganese adsorption on soil constituents, in *Biogeochemistry of Trace Metals*, Adriano, D.C., Ed., Lewis Publishers, Boca Raton, FL., chap. 14, 1992
19. McBride, M.B., Reactions controlling heavy metal solubility in soils, in *Advances in Soil Science,* Stewart, B.A., Ed., Springer-Verlag, New York, 1989, chap. 10.
20. Stumm, W., *Chemistry of the Solid–Water Interface: Processes at the Mineral–Water and Particle–Water Interface in Natural Systems*, John Wiley & Sons, New York, 1992.
21. Green, C.H., The Solubility of Manganese and Coincident Release of Metals Based on the Reduction of Alamosa River Basin Soils, Colorado, Master's thesis, Colorado State University, 1992.
22. Green, C.H. et al., Solubilization of manganese and trace metals in soils impacted by acid mine runoff, *J. Environ. Qual.*, July/August.
23. Miller, W.P., Martens, D.C., and Zelazny, L.W., Effect of sequence in extraction of trace metals from soils, *Soil Sci. Soc. Am. J.,* 50, 598, 1986.
24. Sposito, G., *The Chemistry of Soils*, Oxford University Press, Oxford, 1989.
25. Fletcher, P. and Sposito, G.S., The chemical modeling of clay/electrolyte interactions for montmorillonite, *Clay Miner.,* 24, 375, 1989.
26. Stadler, M. and Schindler, P.W., Modeling of H^+ and Cu^{2+} adsorption on calcium-montmorillonite, *Clays Clay Miner.,* 41, 288, 1993.
27. Allison, J.D., Brown, D.S., and Novo-Gradac, K.J., MINTEQA2/PRODEFA2, A Geochemical Assessment Model for Environmental Systems: Version 3.00 User's Manual, EPA-600/3-91-021, U.S. Environmental Protection Agency, Athens, GA, 1990.
28. Cowan, C.E. et al., Individual sorbent contributions to cadmium sorption on ultisols of mixed mineralogy, *Soil Sci. Soc. Am. J.*, 56, 1084, 1992.
29. Schnitzer, M. and Hansen, E.H., Organo-metallic interactions in soils: 8, An evaluation of methods for the determination of stability constants of metal-fulvic acid complexes, *Soil Sci.,* 109, 333, 1970.
30. Dzombak, D.A. and Morel, F.M.M., *Surface Complex Modeling. Hydrous Ferric Oxide*, Wiley & Sons, New York, 1990.
31. Banerjee, D. and Nesbitt, H.W., XPS study of reductive dissolution of birnessite by oxalate: Rates and mechanistic acpects of dissolution and redox processes, *Geochim. Cosmochim. Acta.,* 63, 3025, 1999.
32. Schwab, A.P. and Lindsay, W.L., The effect of redox on the solubility and availability of manganese in a calcareous soil, *Soil Sci. Soc. Am. J.,* 47, 217, 1983.
33. Lebron, I. and Suarez, D.L., Mechanisms and precipitation rate of rhodochrosite at 25 C as affected by P_{CO2} and organic ligands, *Soil Sci. Soc. Am. J.*, 63, 561, 1999.
34. Selim, H.M. et al., Modeling the transport and retention of cadmium in soils: Multireaction and multicomponent approaches, *Soil Sci. Soc. Am. J.,* 56, 1004, 1992.
35. Kretzschmar, R. and Voegelin, A., Modeling competitive sorption and release of heavy metals in soils, in *Heavy Metals Release in Soils,* Selim, H.M. and Sparks, D.L., Eds., CRC Press, Boca Raton, FL, 2001, chap. 3.

10 Behavior of Heavy Metals in Soil: Effect of Dissolved Organic Matter

Lixiang X. Zhou and J.W.C. Wong

CONTENTS

10.1 Introduction 245
10.2 Fractionation and Characterization of DOM 247
10.3 DOM Sorption in Soil 251
10.4 DOM Biodegradability 253
10.5 DOM Effect on Heavy Metal Sorption in Soils 254
10.6 Metal Dissolution as Affected by the Origin and Concentrations of DOM 261
10.7 Metals Bio-Availability as Affected by DOM 262
10.8 Summary 264
10.9 Conclusions 265
10.9.1 Future Research Needs 266
References 266

10.1 INTRODUCTION

Heavy metal contamination of soils has received much attention with regard to plant uptake, deterioration of soil microbial ecology, and contamination of groundwater or surface waters (Cunningham et al., 1975; Riekerk and Zasoski, 1979). The increased application of pesticides, urban wastes such as municipal refuse and sewage sludge, and animal wastes on farmland or orchards led to heavy metal accumulation in soils. Cu concentrations of as high as 1000 mg/kg in poultry litter and pig manure were not uncommon due to the supplement of Cu in animal feed as a common practice for many years (Van der Watt et al., 1994; Giusquiani et al., 1998). In many orchard soils, especially for the well-aged orchard, the Cu level has exceeded more than 300 mg/kg due to the application of Bordeaux mixture as

1-56670-623-8/03/$0.00+$1.50

pesticide for decades (Aoyama, 1998). The application of organic wastes such as animal manure, crop residues, green manure, and forest residues is very common practice to provide nutrients and to improve soil physical properties in many countries. In China, the practice of land application of farmyard manure can be traced back 2000 years, which effectively maintains high soil fertility and productivity. It is generally considered that these materials can immobilize metals by sorption of metal in particulate organic matter, which reduces the metal bioavailability in the contaminated soil. However, the effectiveness of *in situ* immobilization of metals by organic wastes depends on the origins and properties of the waste types used.

In general, the mobility of heavy metals in soil is severely limited by virtue of the strong sorption reactions between metal ions and the surface of soil particles. In numerous long-term sludge application experiments, however, evidence for metal translocation has been reported, especially in C-rich material-amended soils (Li and Shuman, 1996; Streck and Richter, 1997). Downward migration was observed 7 years after sludge application where soluble Cu, Zn, and Cd were greater at a depth of 40 to 60 cm in sludge-treated soil than in untreated soil (Campbell and Beckett, 1988). It has been well documented that dissolved organic matter (DOM) plays an important role in the mobility and translocation of many soil elements (such as N, P, Fe, Al and other trace metals) and organic and inorganic pollutants in soils (Qualls and Haines, 1991; McCarthy and Zachara, 1989; Kaiser and Zech, 1998; Berggren et al., 1990; Maxin and Kögel-Knabner, 1995). DOM can facilitate metal transport in soil and groundwater by acting as a "carrier" through formation of soluble metal-organic complexes (McCarthy and Zachara, 1989; Temminghoff et al., 1997). The drained groundwater of a field plot receiving the highest application of sludge DOM contained about twice the Cd concentration of the control plot during the first few weeks following sludge disposal (Lamy et al., 1993). Darmody et al. (1983) also noted that many metals were mobile in a silt loam receiving heavy sludge application, and Cu had greater downward movement than the other metals 3 years after the initial application.

Land application of organic manure, crop residue, and biosolids, which is an important means for disposal and recycling of wastes, has been shown to greatly increase the amount of DOM in soil (Zsolnay and Gorlitz, 1994; Han and Thompson, 1999), especially during the first few weeks following their application (Baham and Sposito, 1983; Lamy et al., 1993). Soil solution itself contains varying amounts of DOM, which originate from plant litter, soil organic matter, microbial biomass, and bacterial extracellular polymers or root exudates. DOM is defined operationally as a continuum of organic molecules of different sizes and structures that pass through a filter of 0.45-μm pore size (Kalbitz et al., 2000). It consists of low molecular substances such as organic acids, sugars, amino acids, and complex molecules of high molecular weight, such as humic substances. Similar to soil organic matter, a general chemical definition of DOM is impossible. However, it is feasible to fractionate and characterize DOM according to molecular weight and its "polarity" as hydrophilic/hydrophobic fractions by macroreticular exchange resins and other spectrum methods such as FT-IR, ^{13}C- and ^{1}H-NMR (Liang et al., 1996; Zhou et al., 2001; Keefer et al., 1984). Detailed information on fractionation and characterization of DOM has been reviewed by Herbert and Bertsch (1995).

Many reports have revealed that the DOM-associated transport of metal might be enhanced or inhibited depending on the nature of the DOM and its mobility in soils. Newman et al. (1993) and Jordan et al. (1997) observed the enhanced mobility of Cd, Cu, Cr, and Pb in the presence of DOM. However, Igloria and Hathhorn (1994) found opposite results: The mobility of the contaminants was limited in a pilot-scale lysimeter, which was attributed to the possibility of significant sorption of DOM and DOM metal on media (Jardine et al., 1989; McCarthy et al., 1993; David and Zech, 1990).

Despite intensive research in the past decade, most of the studies done in the laboratory have not yet been investigated in the field. In fact, many researches show that organic C and contaminants in aquatic ecosystems are partly from terrestrial ecosystems through runoff and percolation. However, it is impossible to predict how much DOM and DOM-facilitated solutes are transferred to aquatic environments without better understanding of the behavior of DOM itself and interaction of DOM and metals in soils. The aim of this chapter, based on a series of trials that we conducted, is to give a brief summary on the behavior of DOM derived from organic wastes in soils and its effect on heavy metal mobility, and to propose areas of future research.

10.2 FRACTIONATION AND CHARACTERIZATION OF DOM

The physical and chemical properties of DOM are difficult to define precisely because of the complexity of structure and components. In order to facilitate the study of DOM, a variety of techniques have been developed to fractionate samples into distinctive and hopefully less complex parts. Fractionation of a DOM sample does not result in pure homogeneous compounds but rather fractions in which one or more of the physical or chemical properties have a narrower range of values than the original sample. Commonly used fractionation procedures are based on "polarity" or molecular size of DOM.

DOM can be fractionated into six fractions in terms of "polarity" by macroreticular exchange resins as described by Leenheer (1981): hydrophilic acid (HiA), base (HiB), and neutral (HiN), and hydrophobic acid (HoA), base (HoB), and neutral (HoN). The distribution of various fractions of DOM in the selected organic wastes was given in Table 10.1.

The green manure (above-ground portions of field-grown broad bean) contained the highest amount of hydrophilic fractions while sludge compost and peat had the highest hydrophobic fractions. Although rice residue contained a lower amount of hydrophilic fractions than that of green manure, it had the highest percentage fraction of HiA among all organic wastes. Hydrophilic acid was the more dominant component of the hydrophilic fractions for all the organic wastes except for green manure. There was no significant difference between the amount of HiA and HiN in green manure. Hydrophobic acid represented the major component of the hydrophobic fractions of DOM from pig manure, sewage sludge, and sludge compost while hydrophobic neutral was the major component for peat,

TABLE 10.1
Distribution of Hydrophobic and Hydrophilic Fractions of DOM Derived from Organic Wastes (% of Total DOM)

DOM Sources	Hydrophobic Fractions			Total	Hydrophobic Fractions			Total
	HiA	HiB	HiN		HoA	HoB	HoN	
Green manure	32.84	7.71	37.86	78.41	4.03	1.62	15.93	21.58
Rice residue	41.78	3.32	10.74	58.55	6.89	4.80	29.75	41.45
Pig manure	25.22	12.71	7.42	45.35	44.17	3.91	6.57	54.65
Peat	25.64	0.65	12.23	38.52	20.42	8.63	32.43	61.48
Sewage sludge (1)[a]	22.66	17.72	8.24	48.62	34.21	1.06	16.11	51.38
Sewage sludge (2)[b]	39.4	16.2	4.18	59.84	38.5	0.81	0.85	40.16
Sludge (2) compost	21.2	2.57	1.86	25.63	52.0	0.43	22.0	74.37

[a] Sewage sludge (1) refers to dewatered anaerobically digested sludge collected from Wuxi Sewage Treatment plant, Jiangsu province, China.

[b] Sewage sludge (2) refers to dewatered sewage sludge collected from Taipo Wastewater Treatment plant, Hong Kong, China.

green manure, and rice residue. Hydrophobic acid and base each comprised of less than 7% of the total DOM of green manure and rice residue. According to Keefer et al. (1984), HiA mainly consists of simple organic acids; HiN, carbohydrates and polysaccharides; HiB, mostly amino acids; HoA, aromatic phenols; HoN, hydrocarbons; and HoB, complex aromatic amines. Hence, DOM from green manure should consist of more simple organic acids, polysaccharides, and amino acids, which would attribute to green manure a better complexing ability with metals than DOM from other sources.

Composting can drastically alter the amount and the fraction distribution of DOM in organic wastes. Following sludge composting, there was a decrease in DOM content due to the decomposition of the easily degradable organic compound by microbial activities (Liang et al., 1996; Raber and Kögel-Knabner, 1997). The acid fraction, HiA+HoA, was the major class of DOM in the fresh sludge and sludge compost. However, DOM of fresh sludge origin constituted of 50 to 60% of hydrophilic fractions and 40 to 50% of hydrophobic fractions. In contrast, the hydrophobic fraction (74%) of the compost DOM was much higher than the hydrophilic fraction (26%). Compared to the hydrophilic fraction, the hydrophobic fraction usually contains more large molecules such as acidic humic substances, which can be operationally defined as the fraction of DOM interacting with XAD-8 at pH 2 (Leenheer, 1981; Raber and Kögel-Knabner, 1997). Similar results were reported by Raber and Kögel-Knabner (1997) and Chefetz et al. (1998), who found that sewage sludge contained higher amounts of hydrophilic fraction but less hydrophobic fraction than sludge compost. Liang et al. (1996) reported that composting increased polymerization and cross-linking, which led to the formation of macromolecular hydrophobic fractions. The hydrophobic fractions have a stronger affinity to the soils or organic pollutants but weaker affinity to heavy metals than the hydrophilic fractions (Maxin and Kögel-Knabner, 1995; Totsche et al., 1997).

Gel permeation chromatography by using Sephadex G-15, G-75, or G-100 and ultrafiltration by using a range of polymer-based membrane filters, with nominal molecular weight cut-off values from 500 to 1,000,000 Da are commonly used to fractionate DOM in term of molecular weight. The distribution of various molecular-size fractions in DOM of the various organic wastes is listed in Table 10.2. In general, most of the DOM existed in the molecular-size fraction of <1000 Da or >25000 Da, whereas the intermediate molecular-size fractions comprised of less than 10% of total DOM except for pig manure, which contained 15% of total DOM in the 1000 to 3500 Da fraction. The distribution pattern of the various molecular-size fractions of DOM from the various organic materials was similar to that obtained for composts, leachate of waste disposal sites, and sewage sludge-amended soils in other studies (Han and Thompson, 1999; Homann and Grigal, 1992; Raber and Kögel-Knabner, 1997).

The fractions of DOM derived from the various wastes with a molecular-size fraction of <1000 Da followed the sequence: green manure (90%) > rice residue (79%) ≈ pig manure (76%) > peat (60%) > sewage sludge (45%). Ohno and Crannell (1996) found that the estimated molecular weight of DOM extracted from green manure ranged from 710 to 850 Da and 2000 to 2800 Da for animal manure DOM. Baham and Sposito (1983) also noted that approximately one-half of the organic compounds in the DOM from anaerobically digested sewage sludge had relative molecular mass of <1500 Da. Peat and sewage sludge contained a higher fraction of DOM with a molecular size >25,000 Da which could be explained by the degradation of compounds of low molecular weight during the formation of peat and the anaerobic digestion of sewage sludge. Liang et al. (1996) reported that following composting there was a decrease in DOM content from 4.2 to 2.5% of dissolved organic carbon (DOC) owing to the decomposition of the easily degradable organic compounds by microbial activities. Hence, fresh organic materials often contain a higher portion of DOM with small molecular size. Generally, hydrophilic fraction of DOM often contains higher amounts of lower-molecular weight fractions than hydrophobic fractions of DOM.

The three general characteristics of a chemical compound are the elemental composition, the arrangement of these elements in the chemical structure, and the types and locations of the functional groups in the structure (Swift, 1996). Spectroscopic methods that have already successfully used in general organic chemistry have been applied for DOM characterization to determine general structure of the component macromolecules of DOM. The spectra of FT-IR for the DOM derived from the five different organic wastes are depicted in Figure 10.1. The main absorption bands for all the samples were a broad band at 3300 to 3400 cm^{-1} (–OH and N-H stretch); a sharp peak at 2900 to 2960 cm^{-1} (symmetric and asymmetric C-H stretch of $-CH_2$); a shoulder peak at 1550 to 1660 (N-H deformation, COO– asymmetrical stretch and H-bonded C = O stretch); a peak around the 1400 cm^{-1} region (C-H deformation of aliphatic group and the $>C=C<$ stretch of the aromatic ring);

TABLE 10.2
Distribution of Molecular-Size Fractions (Dalton) of DOM Derived from Various Organic Wastes (% of Total DOM)

DOM Sources	<1000	1000–3500	3500–8000	8000–15000	15000–25000	>25000
Green manure	90.1	0.68	0.13	1.59	1.43	6.07
Rice residue	78.6	5.80	1.08	0.49	1.43	12.60
Pig manure	75.6	14.96	0.14	0.36	0.40	8.54
Peat	59.9	1.32	6.31	0.81	0.12	31.54
Sewage sludge	45.1	3.15	3.06	2.31	1.07	45.31

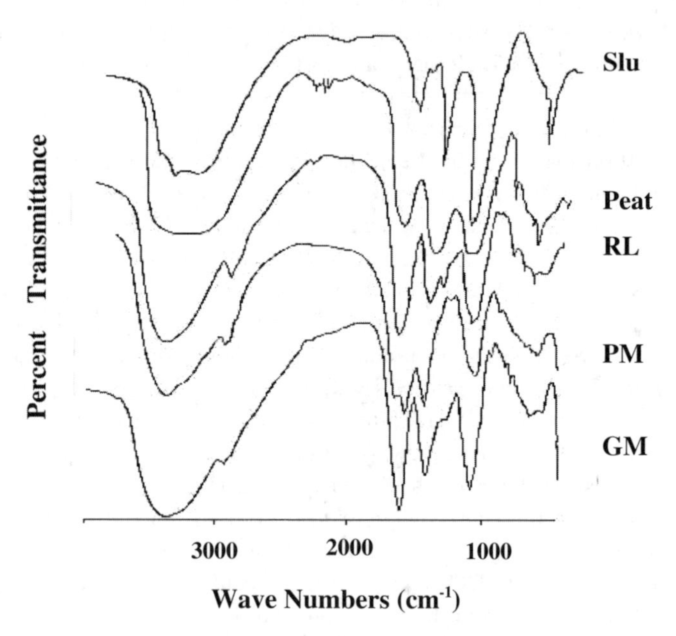

FIGURE 10.1 FT-IR spectra of DOM derived from green manure (GM), pig manure (PM), rice litter (RL), peat, and sewage sludge (Slu).

a peak at 1050 to 1130 cm^{-1} (C-N stretch in amino acid and C-O stretch in polysaccharides and carboxyl); and a sharp peak at 620 to 700 cm^{-1} (C-H bending in aromatic ring or C-H deformation of carbohydrates) (Morrison and Boyd, 1983).

The strong absorption bands at 1572 and 1603 cm^{-1} for DOM from pig manure and green manure, respectively, together with the relative strong bands at 1050 and 1110 cm^{-1} suggested that DOM of green manure and pig manure had a considerably higher amount of aliphatic organic acids or amino acids than that of peat and sewage sludge. The FT-IR spectrum showed that DOM of pig manure had more carboxylate instead of the protonated carboxyl groups in DOM of green manure (Figure 10.1). Discernible strong sharp bands at 628 and 675 cm^{-1} indicated the more aromatic feature of the DOM of peat. Based on the peaks displayed by aliphatic –NH and C = O or COOH, their relative amounts in DOM could be described by the following order: pig manure ≈ green manure>rice residue>peat ≈ sewage sludge.

The organic compounds containing carboxyl groups are important in mobilizing metals adsorbed onto soils through ligand exchange or their complexation reaction. Among the various waste materials, DOM from green manure and pig manure contained higher amounts of carboxyl-containing compounds (data not shown), which might result from the presence of more amino acids in these materials. This was supported by the strong absorption peaks of δ_{N-H}, ν_{C-N}, and ν_{-coo^-} existed in the FT-IR spectrum of pig and green-manure DOM. The level of carboxyl-containing compounds in the DOM of the different organic wastes followed the same sequence as that of the molecular-size fraction of <1000 Da. The findings suggested that most of the carboxyl-containing compounds in DOM might consist of low-molecular-weight aliphatic acid.

10.3 DOM SORPTION IN SOIL

The behavior of DOM itself in soil is an important factor affecting the mobility of metals. DOM in its mobile form is believed to enhance the transport of the associated contaminants through porous media (Newman et al., 1993). If DOM is immobilized during the transportation process, it will provide an adsorption site for pollutants. As a result, the mobility of the associated pollutants will be impeded (Jardine et al., 1989). Among the various chemical components of DOM, low-molecular-weight or hydrophilic fractions of DOM had stronger metal binding capability. Hence, DOM of these fractions was less retarded by soil (Liang et al., 1996; Gu et al., 1995; Kaiser and Zech, 1997). Figure 10.2 depicts the sorption isotherms for the DOM of

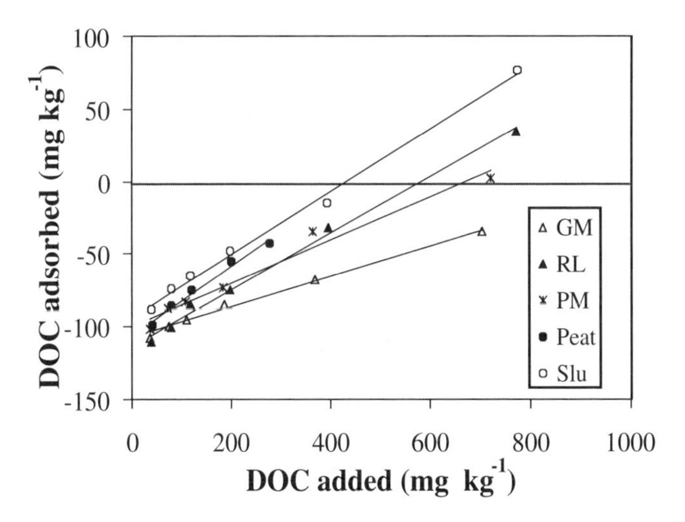

FIGURE 10.2 The initial mass isotherms of DOM from green manure (GM), pig manure (PM), rice litter (RL), peat, and sewage sludge (Slu) at 22°C with an equilibrium time of 2 h.

TABLE 10.3
Sorption Parameters and Distribution Coefficients from Initial Mass Isotherms

DOM Source	Slope (m)	Intercept (b)	K_d	R^2
Green manure	0.1059	107.69	0.474	0.991
Rice residue	0.1962	113.57	0.976	0.994
Pig manure	0.1493	99.82	0.702	0.974
Peat	0.2364	105.52	1.238	0.984
Sewage sludge	0.2168	93.62	1.107	0.996

the five organic wastes onto the calcareous sandy loam. The initial mass (IM) isotherm given by Nodvin et al. (1986) can be used to describe a linear regression of sorption against DOC concentration (Table 10.3):

$$RE = mXi - b$$

where RE = the release of DOC (mg/kg); m = the regression coefficient; Xi = the initial concentration of DOC in soil suspension, expressed as mg/kg soil; and b = the intercept (mg/kg). The distribution coefficient (K_d), an index of the affinity of DOM for soil, can be calculated according to Nodvin et al. (1986):

$$K_d = \frac{m}{1-m} X \frac{(volume\ of\ solution)}{(mass\ of\ soil)}$$

The IM approach has been shown to be a useful tool for describing the sorption of DOM in soils because it takes into consideration the release of indigenous DOC from soil (Kaiser and Zech, 1998; Donald et al., 1993). A significant net release of DOC was observed for soils receiving DOM from various organic materials at a concentration of <400 mg C/kg (i.e., 100 mg C/l) owing to the organic matter exhibited in the soil (Figure 10.2). As the amount of added DOM increased, the release of DOM from the soil decreased. A net sorption was observed at a DOM concentration of >700 mg C/kg for all organic materials except green manure. The slope (m) and distribution coefficient (K_d) obtained from the IM isotherm differed greatly for DOM from various sources. Green manure and pig manure DOM had lower m and K_d values, which indicated the relatively lower affinity of these DOM with the soil as compared to DOM derived from other sources. A significant negative correlation was found between the DOM affinity with soil in terms of the m value and the amount of the low-molecular-weight fraction or hydrophilic fractions of DOM. Hence, DOM fractions having higher amounts of larger-molecular-weight or hydrophobic fractions would be preferentially adsorbed by soil, which was in agreement with the results obtained in other studies (Jardine et al., 1989; Gu et al., 1995; Kaiser and Zech, 1998). Dissolved organic matter of peat exhibited the highest affinity with soil, which might be partly attributed to its higher aromatic nature as evidenced from the FT-IR spectrum. The preferential sorption of the DOM fraction of large molecular weight is likely due to the favorable chemical structures of the organic compounds in these DOM, depending on the nature of the organic materials. Oden et al. (1993) and McKnight et al. (1992) also found that DOM of a greater aromatic nature would favor its partitioning to the mineral surfaces. Therefore, composting organic wastes would increase the adsorption of DOM on soils because of the increase in aromatic carbon-containing compounds after composting (Liang et al., 1996; Chefetz et al., 1998; Inbar et al., 1989).

The affinity of DOM with soil was very low with an average DOM sorption percentage of about 22.4 ±4.8% to 31.2 ±5.2% only at an initial DOC concentration of 100 mg/l and 200 mg/l, respectively, for the five selected DOMs (Table 10.3). This result was supported by the small slope m of 0.11 to 0.24 and K_d of 0.47 to 1.23 ml/g obtained from the IM isotherms. Liang et al. (1996), who worked on a variety of soils with clay contents ranging from 3 to 54%, showed that the adsorption of the DOC by soils increased as the clay, organic matter contents, and the surface areas of the soils increased. The coarse texture of the selected calcareous soil and the characteristics of the selected DOM itself can explain the lower affinity of DOM with soil observed in the present study. In addition, the acidic soil with higher Fe-oxide and Mn-oxide content exhibited much higher DOC adsorption ability than calcareous soil rich in 2:1 minerals.

10.4 DOM BIODEGRADABILITY

Biodegradation of DOM in soil is another important factor affecting the interaction between DOM and metals. Low biodegradability can make DOM persist sufficiently long to permit transport and removal of DOM-bound metals. DOM contains polysaccharides, simple organic acids, amino acids, amino sugar, and proteinaceous material, which are important nutrients (C and N) for microbial growth (Holtzclaw and Sposito, 1978; Boyd et al., 1980). Soil incubation studies showed that DOM added to soil was readily decomposed under an optimum ambient temperature regardless of the origin of the DOM and incubation conditions (Figure 10.3). However, DOM derived from green manure was more susceptible to microbial decomposition compared to that from sewage sludge due to its small molecular size and relatively simple chemical components. Almost 90% of green manure DOM and 25% of sewage sludge DOM were decomposed within 1 day, and nearly 100% and 55% after 1 week following the addition of DOM, respectively, in the aerobic incubation trial. Similar results were also found in waterlogged incubation conditions. However, in incubation under waterlogged conditions, the biodegradable rate of DOM is 20 to 50% lower than under aerobic conditions, indicating that DOM can persist longer under waterlogged conditions.

In another DOM adsorption study, it was found that among the three selected organic wastes, DOM of green manure origin was most susceptible to microbial decomposition with a decrease in DOM as high as 84% after 24 h of shaking the soil suspension containing DOM, compared to only 19% and 18% reduction for pig manure and sewage sludge, respectively (Zhou and Wong, 2000). A marked decrease in DOM occurred mainly after 12 h of shaking for the different organic wastes, which accounted for 77%, 71%, and 66% of the total DOM decomposed within a 24-h experiment for green manure, pig manure, and sewage sludge, respectively.

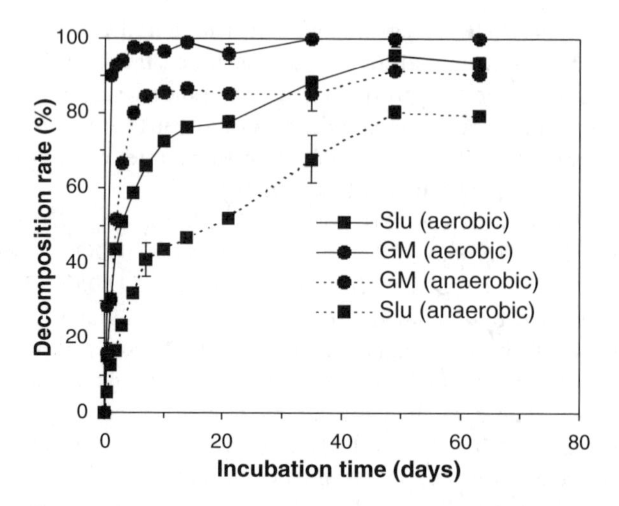

FIGURE 10.3 The kinetics of biodegradation of DOM from green manure (GM) and sewage sludge (Slu) in the contaminated sandy loam under aerobic and waterlogged incubation at 22±1°C.

This further revealed that the origin of DOM would be a major factor determining the susceptibility of DOM to microbial attack. Generally, DOM derived from green manure is considered to be most susceptible to microbial decomposition as compared to DOM of other origins, such as peat, animal manure, biosolids, forest litter, or crop residue, due to its small molecular size and relatively simple components (Ohon and Crannel, 1996).

10.5 DOM EFFECT ON HEAVY METAL SORPTION IN SOILS

Many studies indicated that in the presence of DOM, the metal sorption capacity decreased markedly for most soils, and the effect on the calcareous soil was greater than on the acidic sandy loam. Figure 10.4 shows the metal sorption equilibrium isotherms onto soils with or without the addition of 400 mg C/l of DOM. The equilibrium isotherms could be better depicted according to the linear Freundlich equation with the high value for the correlation coefficient of determination (r^2):

$$\text{Log}\,(x/m) = \text{Log}\,K + 1/n\,\text{Log}\,C$$

where x/m is the amount of metal adsorbed (mg/kg); C is the equilibrium metal concentration (mg/l); K is the equilibrium partition coefficient, and 1/n is the sorption intensity.

The calculated parameters of the Freundlich sorption isotherms are listed in Table 10.4. Theoretically, the higher the sorption intensity parameter (1/n), the lower the binding affinity of soil with metals. The equilibrium partition coefficient (k) is positively related to metal sorption capacity of soils. The sorption capacities and

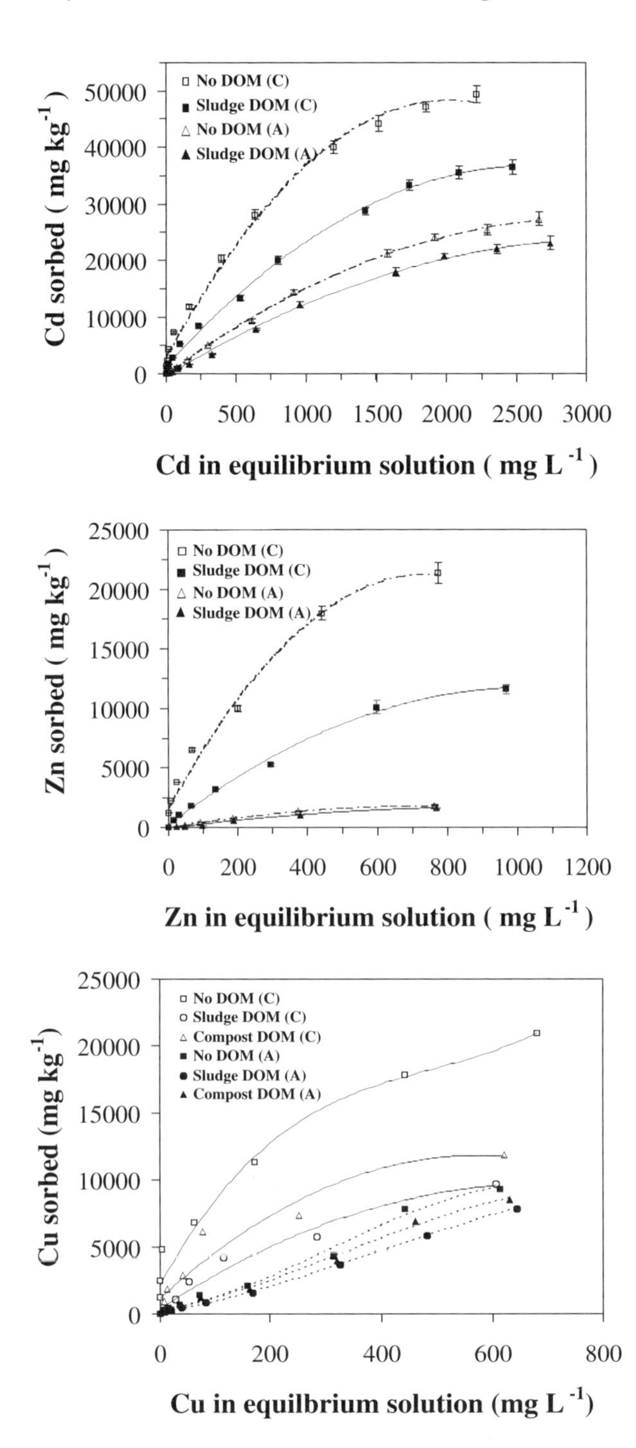

FIGURE 10.4 The sorption isotherm of Cu, Zn, and Cd onto the acidic (A) and calcareous (C) soils with and without the addition of 400 mg C/l of the sludge or sludge compost DOM. (From Zhou, L.X. and Wong, J.W.C., *J. Environ. Qual.,* 30(3) 2001. With permission.)

TABLE 10.4
Parameters of Freundlich Equation for Metal Sorption by Soils in Presence and Absence of 400 mg C/l of DOM

	Acidic Sandy Loam			Calcareous Clay Loam		
	No DOM	Sludge DOM	Compost DOM	No DOM	Sludge DOM	Compost DOM
Cu						
k	52.1	10.7	32.1	2776	47.9	379
1/n	0.786	1.003	0.850	0.301	0.873	0.552
R^2	0.976	0.995	0.991	0.942	0.930	0.958
Zn						
k	3.86	1.84		1227	89.7	
1/n	0.98	1.05		0.42	0.72	
R^2	0.950	0.978		0.984	0.997	
Cd						
k	22.2	9.54		1384	242	
1/n	0.93	1.01		0.46	0.65	
R^2	0.995	0.993		0.988	0.998	

the binding affinity of two soils for metals follows the order calcareous clay loam >> acidic lateritic sandy loam at the same equilibrium concentration of Cu, Zn, or Cd in the absence or presence of sludge DOM as indicated clearly by K and 1/n values listed in Table 10.4. Acidic soil demonstrated much less ability to retain the heavy metals than calcareous clay loam due to much lower pH in the former. The changes in surface negative-charge density and the formation of metal hydroxide precipitation might be responsible for the increased metal sorption at higher pH (Naidu et al., 1997). In addition, clay mineral types can explain the differences in metal sorption in various soils. Calcareous clay loam dominated by 2:1 minerals with higher surface negative-charge density adsorbed the largest amounts of Cu, Cd, or Zn. In contrast, the strongly weathered oxisols, such as the selected acidic sandy loam with low negative surface-charge densities and oxidic mineralogy adsorbed only small amounts of the metals. Similar results have been reported by other researchers (Zachara et al., 1993; Naidu et al., 1997). As shown in Figure 10.4 and Table 10.4, among the selected metals, Cu exhibited the strongest affinity and highest sorption capacity with soils compared to Cd and Zn without the addition of DOM. Cd and Zn were found to have similar binding affinity with each soil in terms of 1/n, but Cd sorption exhibited a higher k value than Zn sorption in each soil, especially in the acidic sandy loam, indicating that a higher sorption capacity for Cd^{2+} relative to Zn^{2+} occurred in the soils. Gerhard and Bruemmer (1999) also found similar results that, on the basis of Freundlich K values, Cd sorption (K = 71) was greater than Zn sorption (K = 26.7) in four soil samples of different compositions.

The addition of 400 mg C/l of DOM decreased markedly the Cu, Cd, and Zn sorption capacity by a factor of 4.8–58 for Cu, 2.3–5.7 for Cd, and 2.1–13.7 for Zn on two soils on the basis of the Freundlich K value, indicating that the inhibition effect of DOM on metal sorption follows the order Cu >> Zn > Cd. This may be attributed to the relatively strong affinity of Cu with the selected DOM. Boyd and Sommers (1990) used linear correlation analysis to predict the stability constants for complexes of sludge-derived soluble fulvic acid with the divalent cations of Mg, Ca, Mn, Fe, Ni, and Zn. At pH 5, the log K values for the fulvate-divalent metal ion complexes (in parenthesis) decreased in the following order: Pb (4.22) > Fe (3.96) > Mn (3.93) > Cu (3.88) > Ni (3.81) > Zn (3.54) > Ca (3.12) > Cd (3.04) > Mg (2.71). At pH 8, the log K values follow approximately the Irving and Williams (1984) stability sequence, that is, $Cu^{2+} > Ni^{2+} = Zn^{2+} > Co^{2+} > Mn^{2+} = Cd^{2+} > Ca^{2+} > Mg^{2+}$ (Bourg and Vedy, 1986; Bloom and McBride, 1979). Zinc forms slightly more stable complexes with organics than Cd, which explains the higher reduction in Zn sorption following the addition of DOC (Bunzl et al., 1976; Morley and Gadd, 1995). The role of DOM in reducing metal sorption could be attributed to the formation of soluble metal-organic complexes because metals can be strongly bound by organic matter, especially for Cu (Stevenson and Ardakani, 1972).

The inhibition effect of DOM on metal sorption depends on the various soil types. The addition of DOM caused a greater decrease in the K value for acidic sandy loam (4.8 for Cu, 2.1 for Cd, and 2.3 for Zn) than that of calcareous clay loam (58 for Cu, 5.72 for Cd, and 13.7 for Zn). Likewise, based on the Freundlich 1/n value, the binding affinity of metals with soils was also decreased in the presence of DOM especially for calcareous clay loam. Only slight reduction of the metal binding affinity was found for the acidic sandy soils. The role of DOM in reducing metal sorption, especially at higher pH soils with 2:1 minerals, could be attributed to the formation of soluble metal-organic complexes because sludge DOM contained many diverse metal-chelating groups responsible for the decreasing metal sorption in soils. In most agricultural soils with pH ranging from 5 to 8, such as soils containing larger amounts of 2:1 minerals, DOM, is mainly present in the mobile form instead of adsorbed form in these soils of negative charge surface. Only strongly weathered oxisols with low negative surface-charge densities and oxidic mineralogy could adsorb DOM partially, while the adsorbed DOM facilitated metal immobilization in soil solution through formation of a ternary metal-DOM-soil complex. Kalbitz and Wennrich (1998) found that DOM was of minor importance in the mobilization of heavy metals in soils with a low soil pH (<4.5).

The inhibition effect of DOM on metal sorption at different pH levels was shown in Figure 10.5. Addition of sludge DOM reduced Cu sorption at each respective pH for both soils. The reduction was especially obvious with an increase in pH, which implied that DOM could bind with Cu more readily and strongly at a higher pH. When pH was greater than 6.8, Cu sorption unexpectedly decreased with an increase in pH in the presence of sludge DOM for acidic and calcareous soils. Similar behavior was also observed by James and Barrow (1981). It was assumed that DOM might complex with Cu in different binding forms at various pH values as seen in the following illustration (McBride, 1994; Qin and Mao, 1993).

Hydroxyl groups bound with Cu could be easily ionized at high pH to yield a negative charge. Consequently, the Cu-DOM complex bearing a negative charge would be repelled by the soils of the same charge through which Cu sorption was

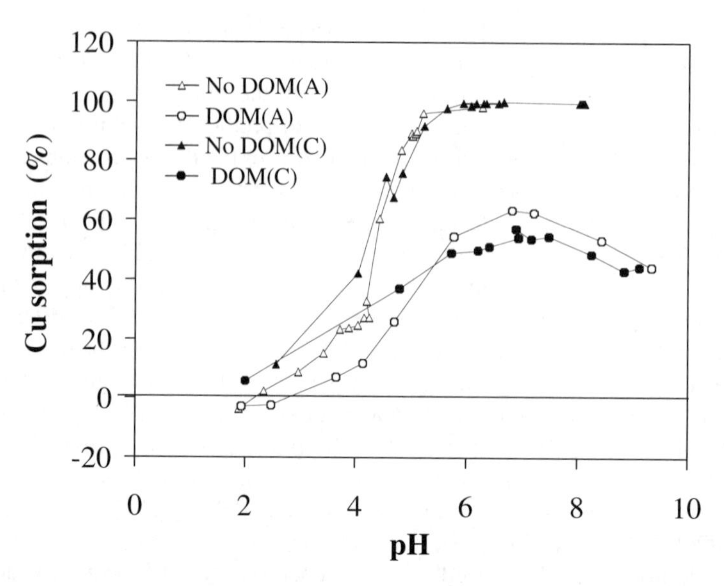

FIGURE 10.5 Effect of pH on Cu sorption onto the acidic (A) and calcareous (C) soils with or without the addition of 300 mg C/l of sludge DOM. (From Zhou, L.X. and Wong, J.W.C., *J. Environ. Qual.*, 30(3) 2001. With permission.)

$$\begin{matrix} R\backslash /COO^- \\ \| \\ R'/ \backslash OH \end{matrix} + Cu \xrightarrow{\text{Low pH}} \begin{matrix} R\backslash /COO\backslash \\ \| \quad\quad Cu\,(H_2O)_x \\ R'/ \backslash O/ \end{matrix} + H^+$$

$$\downarrow \text{High pH}$$

$$\begin{matrix} R\backslash /COO\backslash \quad /OH \\ \| \quad\quad Cu \\ R'/ \backslash O/ \quad \backslash (H_2O)_{x-1} \end{matrix}$$

reduced. Other researchers have also reported different binding forms of Cu with organic ligands at various pH levels (Messori et al., 1997). In addition, dissolved macromolecules exhibiting different structures in aqueous solution at various pH conditions might also be involved, which could modify the exposed surface area and alter the functional group chemistry of DOM. It has been observed that at high pH, organic molecules dispersed into aggregates of small size (<0.1 μm) and that such constituents exhibited a high affinity toward Cu (Myneni et al., 1999). Thus, it is clear that Cu sorption by soils was simultaneously affected by both pH and DOM concentration at a lower soil pH. At a pH condition of >6.8, however, Cu sorption was predominantly affected by DOM due to strong binding affinity of DOM with Cu.

It is noteworthy that a Cd or Zn sorption pattern over an extensive pH range and in the presence and absence of DOM differed from that of Cu (Figure 10.6). DOM produced a much stronger inhibition effect on Cu sorption than on Zn or Cd. Moreover, unlike Cd or Zn sorption patterns, the higher the pH, the greater

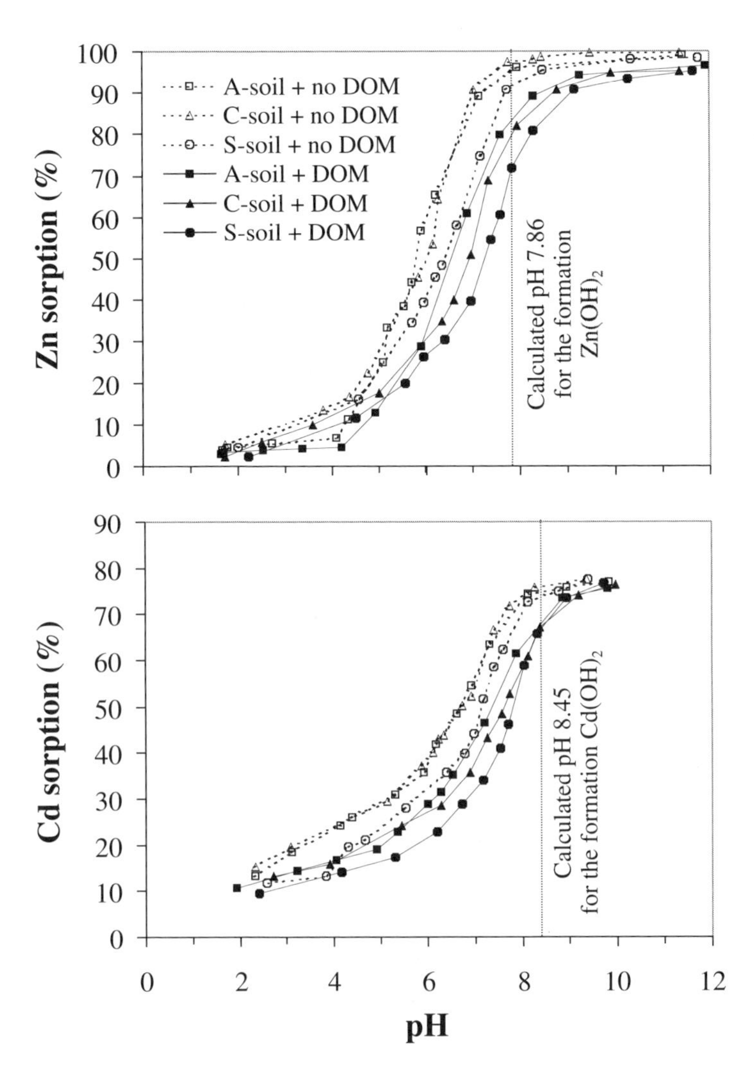

FIGURE 10.6 pH effect on Zn and Cd sorption on acidic sandy soil (A) and calcareous clay soil (C) in the presence and absence of 400 mg C/l^{-1} of sludge DOM.

the DOM effect on Cu sorption even if pH was higher than that at which $Cu(OH)_2$ would occur. The addition of DOM reduced the Zn and Cd sorption at each respective pH for the soils as indicated by the fact that the sorption curves with the presence of DOM shifted to the right compared to those without DOM added. In the pH range of 5 to 8, commonly found in agricultural soils, metal sorption reduction caused by DOM was very apparent with a maximum inhibition effect on metal sorption at pH 7 to 7.5, especially for Zn. However, when pH was raised to 7.8 for Zn and 8.4 for Cd at which $Zn(OH)_2$ or $Cd(OH)_2$ precipitations occurred, the discrepancy of Zn or Cd sorption resulting from the presence and absence of DOM tended to be minimized.

Increasing the sludge DOM concentration caused a significant reduction in the metal sorption onto soils. Taking Cu as an example, a significant negative linear correlation between DOM concentration and Cu sorption was observed for soil receiving a DOM concentration in the range of 0 to 400 mg C/l (r = 0.938 and 0.915 for sludge DOM and sludge compost DOM in the acidic sandy soil, respectively; r = 0.990 and 0.996 in the calcareous soil, respectively ($p<0.05$)) (Figure 10.7). Sorption decreased by 7.3% and 12.4% with a DOM increment of 100 mg C/l for sludge compost and sludge DOM, respectively, in calcareous soil, and correspondingly 1.5% and 6.8% in acidic sandy soil calculated using the linear regression equations obtained in Figure 10.5. When DOM concentration added was raised to 400 mg C/l, there was no Cu being sorbed by the acidic sandy soil with sludge DOM treatment, while in calcareous soil Cu sorption was reduced by 47% and 28% for sludge and sludge compost DOM treatments, respectively. Sludge DOM had a more significant effect on reducing Cu adsorption than that of compost DOM and the effect was more pronounced for calcareous soil than acidic soil.

Obviously, the DOM source also greatly affects metal sorption onto soils. As shown in Table 10.4 and Figure 10.7, the Cu sorption capacity and binding energy calculated for the two soils with two DOM treatments followed this order: no DOM > compost DOM > sludge DOM. The differences in metal sorption behavior caused by DOM of various original materials appeared to be closely related to the chemical components of DOM. Sludge compost DOM contained a relatively greater amount of high-molecular-weight hydrophobic fractions, especially hydrophobic acid (HoA) and hydrophobic neutral (HoN), but fewer hydrophilic fractions, especially hydro-

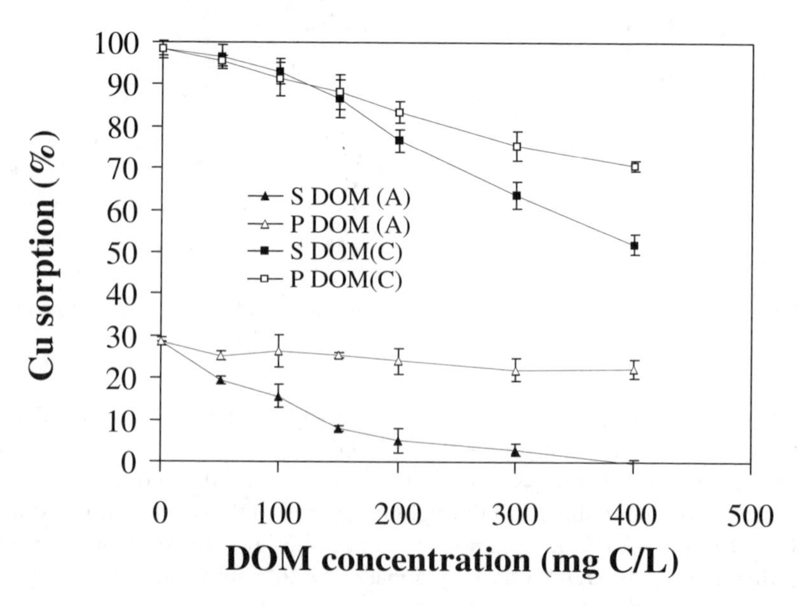

FIGURE 10.7 Effect of concentration of DOM derived from sludge (S) and sludge compost (P) on the Cu sorption onto the acidic (A) and calcareous (C) soils with initial Cu concentration of 40 mg/1. (From Zhou, L.X. and Wong, J.W.C., *J. Environ. Qual.*, 30(3) 2001. With permission.)

philic base (HiB) and hydrophilic acid (HiA), than sludge DOM (Table 10.1). FT-IR showed that sludge DOM appeared to have more C-N and C-O groups of chelating features, possibly from organic acids, amino acids, and amines than compost DOM, especially for the HiA, HiB, and HoA fractions (Zhou et al., 2000). Keefer et al. (1984) pointed out that the HiB fraction mainly consisted of the N-containing group, including most amino acids, amino sugar, low-molecular-weight amines, and pyridine, while the HiA fraction contained the component of the –COO functional group, such as uronic acids, simple organic acids, and polyfunctional acid, which resulted in the higher affinity of Cu with HiB. Sludge DOM contained more HiB fractions, which were not readily adsorbed by soils but could strongly associate with Cu. Thus, sludge DOM had a stronger capability to reduce Cu adsorption by soils than did compost DOM.

10.6 METAL DISSOLUTION AS AFFECTED BY THE ORIGIN AND CONCENTRATIONS OF DOM

In China and other countries, the application of C-rich organic materials to heavy metal–contaminated soils is a common practice. However, the potential risk of metal mobilization should not be overlooked when metal dissolution facilitated by DOM released from organic wastes exceeds metal immobilization caused by particular organic matter. Figure 10.8 demonstrates that addition of DOM derived from green manure, pig manure, peat, rice litter, and sewage sludge to a Cu-contaminated soil caused an increase in soil water-soluble Cu contents. Compared to the control (no

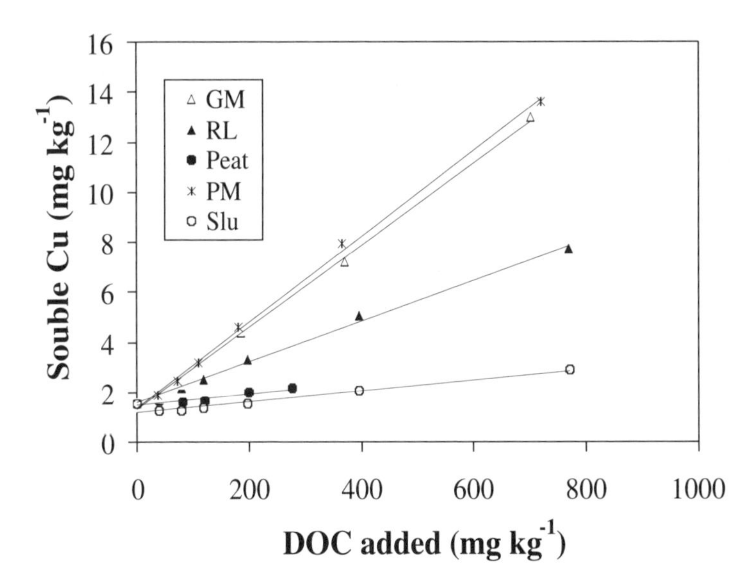

FIGURE 10.8 The release of Cu from the Cu-contaminated sandy loam after the addition of various concentrations of DOM from green manure (GM), pig manure (PM), rice litter (RL), peat, and sewage sludge (Slu).

addition of DOM), soluble Cu contents increased by a factor of 2 to 9 at a DOC concentration ranging from 100 to 200 mg/l. Hence, DOM concentration for a specific organic waste played an important role in mobilizing Cu sorbed on soil particles. While using a cupric electrode, speciation of Cu in the soil solution showed no detectable free Cu ions in the soil solution, especially at pH > 5.8, although the total Cu concentration in the soil solution was as high as 3.4 mg/l. Obviously, most of the Cu existed as Cu–DOM organic complexes in the soil solution. This can be explained by the stronger binding between DOM and Cu at high pH conditions, similar to those obtained by Oden et al. (1993) and Inskeep and Baham (1983). Even at a DOM concentration as low as 10 mg C/l derived from green manure, pig manure, and rice litter, there was a significant release of Cu from the soil. Apparent Cu release could only be observed at a DOM concentration of ≥50 mg/l for DOM derived from sludge and peat.

The dissolution of Cu from contaminated soil was positively and significantly correlated with the amount of carboxyl groups and the fractions of DOM with molecular size <3500 Da while only positively but insignificantly correlated with the percentage of hydrophilic fractions in the DOM. This explains why green and pig manure DOM had a higher ability to release Cu from the contaminated soil because of the higher amounts of small molecular-size fractions or the carboxyl-containing compounds in the DOM. Hydrophilic fractions of DOM exhibited a preferential role in mobilizing Cu as compared to hydrophobic fractions, although the correlation coefficient did not reach a significant level. Therefore, the percentage fraction of small molecular size, the amount of total carboxyl-containing compounds, and the affinity of DOM itself with soil are suitable parameters for assessing the capability of DOM in the dissolution of Cu. Obviously, the influence of DOM on Cu release from contaminated soil would be a function of DOM sources and concentrations.

The decomposition of DOM in soil will lead to the re-immobilization of DOM-associated metals. Under aerobic incubation conditions, DOM decomposition could be observed clearly with an increase in incubation time. As a result, the amount of Cu in soil solution decreased drastically, especially for green manure DOM (Figure 10.9). This may be due to the breakdown of the soluble Cu-DOM complex through microbial degradation, leading to the re-adsorption of Cu onto the soil matrix.

Although green manure DOM was about 7.8 times stronger in releasing Cu than that of sewage sludge DOM at the same level of DOM as calculated from the slope of the equations listed in Figure 10.8, soluble Cu content in the soil amended with green manure was lower than that with sewage sludge after 5 days of incubation. The drastic decrease in green manure DOM remaining in the soil was the major reason for the decrease in Cu. Nevertheless, the soil water-soluble Cu contents of this initial 5-day period for green manure might be high enough to produce a phytotoxic effect on plant growth.

10.7 METALS BIO-AVAILABILITY AS AFFECTED BY DOM

Mobilization of soil heavy metal is facilitated by DOM, which may result in an increase in metal phytotoxicity. However, direct evidence on the increase in metal

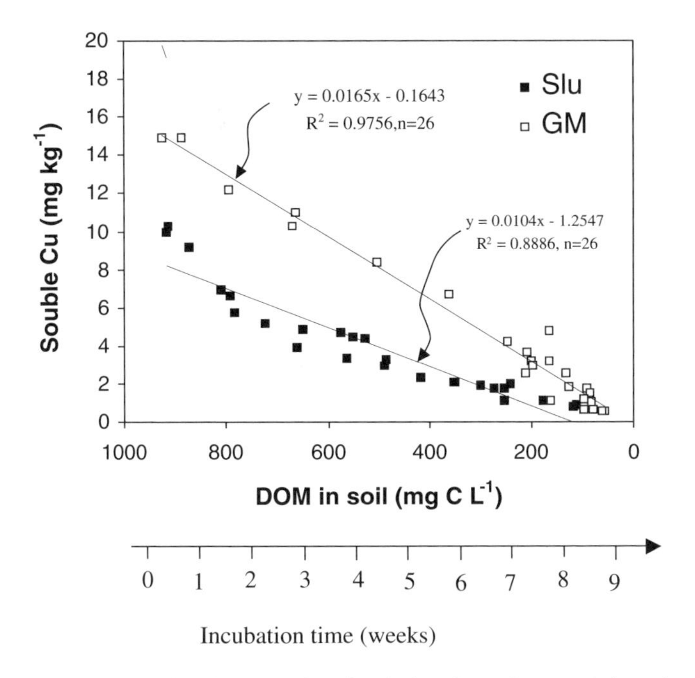

FIGURE 10.9 The correlation between Cu dissolution from Cu-contaminated sandy loam and remaining DOM derived from sludge (Slu) and green manure (GM) in the soil under aerobic incubation conditions.

uptake by plants following the addition of DOM into contaminated soils is scarce. In our greenhouse trial, a Cu-contaminated calcareous soil was amended separately with 2% each of the green manure, sewage sludge, and pig manure (w/w). Ryegrass seeds were sowed and allowed to grow in these amended soils for 7 weeks. The results showed that DOM concentrations in differently treated soils depended greatly on the organic waste types (Table 10.5). Green manure–treated soil exhibited the highest DOM concentration, varying from 780 mgc/kg soil in the first week to about 160 mgc/kg in the seventh week when the ryegrass was harvested. Soil amended with sewage sludge contained lower concentrations of DOM, which were only slightly higher than the control without the addition of any organic wastes. As mentioned above, the incorporation of organic wastes into metal-contaminated soil will produce two contrasting effects: metal immobilization subject to the particular types of organic waste and metal mobilization facilitated by the DOM released from the organic wastes. The final net mobilization or immobilization depended on the balance of these two contrasting effects. In the greenhouse trial, Cu concentration in the ryegrass grown in various organic waste–amended soils follows the order green manure–treated soil > pig manure–treated soil > control > sewage sludge–treated soil. The addition of green and pig manure into the Cu-contaminated soils led to 2.1 and 1.4 times higher Cu concentration in ryegrass than the control treatment. In contrast, sewage sludge could slightly reduce Cu bioavailability in Cu-contaminated soil. Similar results were found in terms of Cu uptake by ryegrass. Metal bioavailability in contaminated soils amended with various organic wastes

TABLE 10.5
Copper Uptake by Ryegrass Grown in Calcareous Soil Amended with Sewage Sludge, Pig Manure, and Green Manure in Greenhouse Experiment

	CK	Sewage Sludge		Pig Manure		Green Manure	
			Above CK %		Above CK %		Above CK %
Cu conc. (mg/kg)	67.43 (9.03)	65.40 (9.56)	−3.01	94.60 (1.94)	40.90	140.8 (1.03)	109
Cu uptake (μg/pot)	868 (9.03)	806 (9.56)	−7.14	1286 (1.94)	48.16	2094 (8.07)	141
Soil DOM conc. (mg/kg)	62-68	112–85		193–105		780–160	

Note: CK = Control treatment with adding the same amount of chemical N, P, and K fertilizers but without the addition of any organic wastes.

was in agreement with the DOM present in these soils (Table 10.5) and DOM properties as discussed previously, which further confirmed the important role of DOM in metal mobilization. This implies that remediation of heavy metal–contaminated soils through the addition of organic wastes should take into account of the types of organic wastes and should be assessed more cautiously.

10.8 SUMMARY

Dissolved organic matter (DOM) plays an important role in the mobilization, translocation, and toxicity of many inorganic and organic pollutants in soils. High concentration of DOM often occurs in the farmland amended with organic wastes. Dissolved organic matter could be fractionated and chemically characterized in terms of molecular size and their "polarity" as hydrophilic/hydrophobic fraction by macroreticular exchange resins. Dissolved organic matter from the readily decomposable organic wastes (green manure, rice litter, pig manure) are often dominated by hydrophilic fractions or low molecular size fraction. The percentage of total DOM with a molecular size fraction <3500 Da was 91% for green manure and pig manure, 84% for rice residue, 61% for peat, and 48% for sewage sludge. The sorption of DOM onto the soil was negatively correlated with the amount of small molecular size fraction or hydrophilic fractions in the respective DOM. Addition of DOM derived from these organic wastes caused an increase in soil water-soluble Cu especially for DOM from green manure and pig manure. The ability of DOM derived from different organic wastes in mobilizing Cu follows the following sequence: green manure ≈ pig manure > rice residue > peat ≈ sewage sludge. In addition, the effect of DOM on metal sorption or dissolution was related to soil pH and soil mineral types. There was a greater reduction in metal sorption in the presence of DOM in calcareous soil with higher pH and dominated by 2:1 minerals than that of acidic soils with low pH and 1:1 minerals. However, biodegradation of DOM in soils was very obvious especially for DOM with higher amounts of hydrophilic or low molecular-size fractions, which resulted in reimmobilization of DOM-associated soluble metal, although the decomposition of DOM was much slower under water-logged condition than that under aerobic condition. This was supported by the higher Cu uptake in ryegrass grown in Cu contaminated soil amended with green manure and pig manure, the more readily decomposable organic wastes, due to Cu mobilization resulted from the release of DOM from these organic wastes. It can be concluded that relatively "stable" organic wastes such as sewage sludge, compost and peat moss would likely be the more suitable amendment materials for the metal-contaminated soil.

10.9 CONCLUSIONS

Depending on organic wastes types, application of organic wastes on farmlands will, to a certain extent, increase soil DOM concentration, which will enhance soil metal availability and uptake by plants. However, the interaction of DOM-metal-soil complexes is affected by many factors, including the properties and behavior of DOM in soils, metal types, soil properties, and so on. DOM dominated by hydrophilic and lower molecular weight fractions exhibits a higher mobility in soil and hence a higher ability in mobilizing metals. Among the selected organic wastes, green manure contains a greater amount of DOC, accounting for as much as 15% of total dry sample weight followed by crop residue and animal wastes containing DOC of 2 to 5%. DOC in peat, sewage sludge, and mature compost often accounts for less than 1%, especially for peat with the lowest DOC content of 0.15%. Correspondingly, DOM extracted from organic wastes containing higher concentrations of DOC appeared to have a higher proportion of hydrophilic fractions. The addition of relatively readily decomposable organic wastes to metal-contaminated soils should be assessed carefully due to the possible metal mobilization facilitated by the released DOM with a high proportion of hydrophilic fractions or low molecular-size fractions. Metal sorption experiments in different soils or pH levels indicated that the reduction of metal sorption under the influence of DOM was much more obvious in calcareous clay loam than in acidic soil, especially for Cu. Unlike Cu sorption in soil without DOM addition, Cu sorption was unexpectedly decreased at pH >6.8 with an increase in pH in the presence of DOM. However, the discrepancy of Zn or Cd sorption resulting from the presence and absence of DOM tended to be minimal when pH was raised to 7.8 or above for Zn and 8.4 or above for Cd at which $Zn(OH)_2$ or $Cd(OH)_2$ precipitations occurred. In a DOM concentration of <200 mg C/l, metal sorption by soils decreased with an increase in DOM concentration. At a given DOM concentration increment, the increase in the inhibition of metal sorption is positively related to the amount of lower molecular-size fractions or hydrophilic fractions of DOM, and followed the order acidic sandy loam<calcareous sandy loam<calcareous clay loam.

The decomposition of DOM in soil will lead to reimmobilization of DOM-associated metal. Under aerobic incubation conditions, DOM decomposition could be observed clearly with an increase in incubation time especially within the first week after DOM was incorporated into soils. Nevertheless, in the ryegrass bioassessment trial, the increase in Cu uptake in the soils amended with the readily decomposable organic wastes, that is, green manure and pig manure, could be observed due to Cu mobilization facilitated by DOM released from these organic wastes, which exceeded the amount of Cu immobilized by particular organic matter.

10.9.1 Future Research Needs

Although intensive research in the last decade on the interaction between DOM and heavy metals has been carried out, quantitative prediction of metal behavior involving DOM at the field scale is not yet possible because there are many factors affecting the interaction of DOM and metals under field conditions. The DOM contribution to facilitate pollutants to move into aquatic environment from cropland is unclear. Also, the information concerning DOM dynamics or DOM fluxes under field conditions is limited. The mechanisms of pollutants migrating downward through preferential flow, pollutant uptake by plant, and pollutant microbial toxicity in the presence of DOM need to be further explored. The high sorption capacity of soil clay minerals and oxides for DOM as shown in laboratory studies may not control the transport of DOM in soils in the field if macropore fluxes are dominant under field conditions. The fluxes and translocation of dissolved organic nitrogen and phosphorus as related to DOM in the field are rarely studied. They are important factors that control eutrophication resulting from agricultural nonpoint pollution. Thus, future research should concentrate on (1) the importance of hydrological conditions for the release and fate of DOM; (2) the dynamics and biodegradability of DOM under various field conditions, including upland and waterlogged cultivation systems, fertilization, tilling systems, land management, and climate, among others; (3) quantification of the various DOM sources under field conditions, including crop litter, root exudates, microbial biomass, and soil humus, and so on, and their contribution to DOM release; (4) the mechanism to explain how DOM-metal complexes are absorbed and transferred in plant roots; (5) microbial molecular ecology as affected by DOM-metal complexes; (6) fractionation of DOM that occurs during soil percolation and its importance on heavy metal translocation in soil; and (6) DOM flux in the field and its effect on other pollutants, such as N, P, and herbicides.

REFERENCES

Aoyama, M., Effects of heavy metal accumulation in apple orchard soils in the mineralization of humified plant residues, *Soil Sci. Plant Nutr.*, 44, 209, 1998.

Baham, J. and Sposito, G., Chemistry of water-soluble, metal-complexing ligands extracted from an anaerobically digested sewage sludge, *J. Environ. Qual.*, 12, 96, 1983.

Berggren, D. et al., Metal solubility and pathways in acidified forest ecosystems of South Sweden, *Sci. Total Environ.*, 96, 103, 1990.

Bloom, P.G. and McBride, M.B., Metal ion binding and exchange with hydrogen and acid washed peat, *Soil Sci. Soc. Am. J.*, 43, 687, 1979.

Bourg, A.C.M. and Vedy, J.C., Expected speciation of dissolved trace metals in water of acid profiles, *Geoderma*, 38, 279, 1986.

Boyd, S.A. and Sommers, L.E., Humic and fulvic acid fractions from sewage sludges and sludge amended soils, in *Humic Substances in Soil and Crop Sciences,* Clapp, C.E., Malcolm, R.L., and Bloom, P.R., Eds., American Society of Agronomy, Inc. USA, 1990, pp. 203–220.

Boyd, S.A., Sommers, L.E., and Nelson, D.W., Changes in the humic acid fraction of soil resulting from biosolids application, *Soil Sci. Soc. Am. J.*, 44, 1179, 1980.

Bunzl, K., Schmidt, W., and Sansoni, B., Kinetics of ion exchange in soil organic matter. IV. Adsorption and desorption of Pb^{2+}, Cu^{2+}, Cd^{2+}, Zn^{2+}, and Ca^{2+} by peat, *J. Soil Sci.*, 118, 32, 1976.

Campbell, D.J. and Beckett, P.H.T., The soil solution in a soil treated with digested sewage sludge, *J. Soil Sci.*, 39, 283, 1988.

Chefetz, B. et al., Humic-acid transformation during composting of municipal solid waste, *J. Environ. Qual.*, 27, 794, 1998.

Cunningham, J.D., Ryan, J.A., and Keeney, D.R., Phytotoxicity and metal uptake from soil treated with metal-amended sewage sludge, *J. Environ. Qual.*, 4, 455, 1975.

Darmody, R.G. et al., Municipal sewage sludge compost-amended soils: Some spatiotemporal treatment effects, *J. Environ. Qual.*, 12, 231, 1983.

David, M.B. and Zech, W., Adsorption of dissolved organic carbon and sulfate by acid forest soils in the Fichtelgebirge, FRG, *Z. Pflanzenernaehr. Bodenkd.*, 153, 379, 1990.

Donald, R.G., Anderson, D.W., and Stewart, J.W.B., Potential role of dissolved organic carbon in phosphrous transport in forested soils, *Soil Sci. Soc. Am. J.*, 57, 1611, 1993.

Gerhard, W. and Bruemmer, G.W., Adsorption and solubility of ten metals in soil samples of different composition, *J. Plant Nutr. Soil Sci.*, 162, 155, 1999.

Giusquiani, P.L. et al., Fate of pig sludge liquid fraction in calcareous soil: Agricultural and environmental implications, *J. Environ. Qual.*, 27, 364, 1998.

Gu, B. et al., Adsorption and desorption of different organic matter fractions on iron oxide, *Geochim. Cosmochim. Acta*, 59, 219, 1995.

Han, N. and Thompson, L., Soluble organic carbon in a biosolids-amended mollisol, *J. Environ. Qual.*, 28, 652, 1999.

Herbert, B.E. and Bertsch, P.M., Characterization of dissolved and colloidal organic matter in soil solution: A review, in *Carbon Forms and Functions in Forest Soils*, Kelly, J.M. and McFee, W.W., Eds., Soil Science Society of America, Madison, WI, 1995, p. 63.

Holtzclaw, K.M. and Sposito, G., Analytical properties of the soluble, metal-complexing fractions in biosolids-soil mixtures: III. Unaltered anionic surfactants in fulvic acid, *Soil Sci. Soc. Am. J.*, 42, 607, 1978.

Homann, P.S. and Grigal, D.F., Molecular weight distribution of soluble orgaincs from laboratory-manipulated surface soils, *Soil Sci. Soc. Am. J.*, 56, 1305, 1992.

Inbar, Y., Chen, Y., and Hadar, Y., Solid-state carbon-13 nuclear magnetic resonance and infrared spectroscopy of composted organic matter, *Soil Sci. Soc. Am. J.*, 53, 1695, 1989.

Inskeep, W.P. and Baham, J., Adsorption of Cd (II) and Cu (II) by Na-Montmorillonite at low surface coverage, *Soil Sci. Soc. Am. J.*, 47, 660, 1983.

James, R.O. and Barrow, N.J., Copper reactions with inorganic components of soils including uptake by oxide and silicate minerals, in *Copper in Soils and Plants*, Loneragan, J.F., Robson, A.D., and Graham, R.D., Eds., Academic Press, Australia, 1981, p. 47.

Jardine, P.M., Weber, N.L., and McCarthy, J.F., Mechanisms of dissolved organic carbon adsorption on soil, *Soil Sci. Soc. Am. J.*, 53, 1378, 1989.

Jordan, T.E., Correll, D.L., and Weller, D.E., Effects of agriculture on discharges of nutrients from coastal plain watersheds of Chesapeake Bay, *J. Environ. Qual.*, 26, 836, 1997.

Kaiser, K. and Zech, W., Competitive sorption of dissolved organic matter fractions to the soils and related mineral phase, *Soil Sci. Soc. Am. J.*, 61, 64, 1997.

Kaiser, K. and Zech, W., Rates of dissolved organic matter release and sorption in forest soils, *Soil Sci.* 163, 714, 1998.

Kalbitz, K. and Wennrich, R., Mobilization of heavy metals and arsenic in polluted wetland soils and its dependence on dissolved organic matter, *Sci. Total Environ.*, 209, 27, 1998.

Kalbitz, K. et al., Controls on the dynamics of dissolved organic matter in soils: A review, *Soil Sci.*, 165, 277, 2000.

Keefer, R.F., Codling, E.E., and Singh, R.N., Fractionation of metal-organic components extracted from a sludge-amended soil, *Soil Sci. Soc. Am. J.*, 48, 1054, 1984.

Lamy, I., Bourgeois, S., and Bermond, A., Soil cadmium mobility as a consequence of sewage sludge disposal, *J. Environ. Qual.*, 22, 731, 1993.

Leenheer, J.A., Comprehensive approach to preparative isolation and fractionation of dissolved organic carbon from natural waters and wastewaters, *Environ. Sci. Technol.*, 15, 578, 1981.

Li, Z.B. and Shuman, L.M., Heavy metal movement in metal-contaminated soil profiles, *Soil Sci.*, 161, 656, 1996.

Liang, B.C. et al., Characterization of water extracts of two manures and their adsorption on soils, *Soil Sci. Soc. Am. J.*, 60, 1758, 1996.

Maxin, C. and Kögel-Knabner, I., Partitioning of PAH to DOM: Implications on PAH mobility in soils, *Eur. J. Soil Sci.*, 46, 193, 1995.

McBride, M.B., *Environmental Chemistry of Soils*. Oxford University Press, New York, 1994.

McCarthy, J.F. and Zachara, J.M., Subsurface transport of contaminants, *Environ. Sci. Technol.*, 23, 496, 1989.

McCarthy, J.F. et al., Mobility of natural organic matter in a sandy aquifer, *Environ. Sci. Technol.*, 667, 1993.

McKnight, D.M. et al., Sorption of dissolved organic carbon by hydrous aluminum and iron oxides occurring at the confluence of Deer Creek with the Snake River, Summit County, Colorado, *Environ. Sci. Technol.*, 26, 1388, 1992.

Messori, L. et al., The pH dependent properties of metallotransferrins: A comparative study, *Biometals*, 10, 303, 1997.

Morley, G.F. and Gadd, G.M., Sorption of toxic metals by fungi and clay minerals, *Mycol. Res.*, 99, 1429, 1995.

Morrison, R.T. and Boyd, R.N., *Organic Chemistry*, Allyn and Bacon, Newton, MA, 1983, p. 680.

Myneni, S.C.B. et al., Imaging of humic substance macromolecular structures in water and soils, *Science*, 286, 1335, 1999.

Naidu, R. et al., Ionic strength and pH effects on surface charge and Cd sorption characteristics of soils, *J. Soil Sci.*, 45, 419, 1994.

Newman, M.E., Elzerman, A.W., and Looney, B.B., Facilitated transport of selected metals in aquifer material packed columns, *J. Contam. Hydrol.*, 14, 233, 1993.

Nodvin, S.C., Driscoll, C.T., and Likens, G.E., Simple partitioning of anions and dissolved organic carbon in a forest soil, *Soil Sci.*, 142, 27, 1986.

Oden, W.I., Amy, G.L., and Conklin, M., Subsurface interactions of humic substances with Cu (II) in saturated media, *Environ. Sci. Technol.*, 27, 1045, 1993.

Ohno, T. and Crannell, B.S., Green and animal manure-derived dissolved organic matter effects on phosphorus sorption, *J. Environ. Qual.*, 25, 1137, 1996.

Qin, C.N. and Mao, S.S., *Environmental Soil Science*, China Agricultural Press, Beijing, 1993, p. 42.

Qualls, R.G. and Haines, B.L., Geochemistry of dissolved organic nutrients in water percolating through a forest ecosystem, *Soil Sci. Soc. Am. J.*, 55, 1112, 1991.

Raber, B. and Kögel-Knabner, I., Influence of origin and properties of dissolved organic matter on the partition of polycyclic aromatic hydrocarbons (PAHs), *Eur. J. Soil Sci.*, 48, 443, 1997.

Riekerk, H. and Zasoski, R.J., Effects of dewatered sludge applications to a Douglas fir forest soil on the soil, leachate, and groundwater composition, in *Utilization of Municipal Sewage Effluent and Sludge on Forest and Disturbed Land*, Sopper, W.E. and Kerr, S.N., Eds., Pennsylvania State University Press, University Park, 1979, p. 35.

Stevenson, F.J. and Ardakani, M.S., Organic matter reactions involving micronutrients in soils, in *Micronutrients in Agriculture*, Mortvedt, J.J., Ed., Soil Science Society of America, Madison, WI, 1972.

Streck, T. and Richter, J., Heavy metal displacement in a sandy soil at the field scale: I. Measurements and parameterization of sorption, *J. Environ. Qual.*, 26, 49, 1997.

Swift, R.S., Organic matter characterization, in *Methods of Soil Analysis: Part 3, Chemical Methods*, Sparks, D.L., et al., Eds., Soil Science Society of America Society of Agronomy, Madison, WI, 1996, p. 1011.

Temminghoff, E.J.M., Van der Zee, S.E.A.T.M., and De Haan, F.A.M., Copper mobility in a copper-contaminated sandy soil as affected by pH and solid and dissolved organic matter, *Environ. Sci. Technol.*, 31, 1109, 1997.

Totsche, K.U., Danzer, J., and Kögel-Knabner, I., Dissolved organic matter-enhanced retention of polycyclic aromatic hydrocarbons in soil miscible displacement experiments, *J. Environ. Qual.*, 26, 1090, 1997.

Van der Watt, H.v.H., Sumner, M.E., and Cabrera, M.L., Bioavailability of copper, manganese, and zinc in poultry litter, *J. Environ. Qual.*, 23, 43, 1994.

Zachara, J.M. et al., Cadmium sorption on specimen and soil smectites in sodium and calcium electrolytes, *Soil Sci. Soc. Am. J.*, 57, 1491, 1993.

Zhou L.X. and Wong, J.W.C., Microbial decomposition of dissolved organic matter derived from organic wastes and its control during sorption experiment, *J. Environ. Qual.*, 29, 1852, 2000.

Zhou, L.X. and Wong, J.W.C., Effect of dissolved organic matters derived from sludge and composted sludge on soil Cu sorption, *J. Environ. Qual.*, 30(3) 2001.

Zhou L.X. et al., Fractionation and characterization of dissolved organic matter derived from sewage sludge and composted sludge, *Environ. Technol.*, 21, 765, 2000.

Zhou, L.X. et al., Fractionation and characterization of sludge bacterial extracellular polymers by FT-IR, ^{13}C-NMR, ^{1}H-NMR, *Water Sci. Tec.*, 44, 71, 2001.

Zsolnay, A. and Gorlitz, H., Water extractable organic matter in arable soils: Effects of drought and long-term fertilization, *Soil Biol. Biochem.*, 26, 1257, 1994.

11 Analytical Techniques for Characterizing Complex Mineral Assemblages: Mobile Soil and Groundwater Colloids

John C. Seaman, M. Guerin, B.P. Jackson, P.M. Bertsch, and J.F. Ranville

CONTENTS

11.1 Introduction 272
11.2 Light-Scattering Techniques for Colloid Characterization 274
11.2.1 Turbidimetric Methods 274
11.2.2 Dynamic Light Scattering 276
11.2.3 Laser Doppler Velocimetry and Particle Charge 278
11.3 Acoustic Spectroscopy 282
11.3.1 Acoustic Attenuation and Particle Sizing 282
11.3.2 Electroacoustics 284
11.4 Field Flow Fractionation 286
11.4.1 Sedimentation (Sd-FFF) and Flow-Field Flow Fractionation (Fl-FFF) 286
11.4.2 FFF Applications 289
11.5 Electron-Based Analysis Techniques 292
11.5.1 Scanning Electron Microscopy 292
11.5.2 Automated SEM Techniques: Removing Instrument and Operator Biases 296
11.5.3 Transmission Electron Microscopy 298
11.6 Other Analytical Methods 300
11.7 Conclusions 300
11.8 Acknowledgments 302
11.9 List of Symbols 303
11.10 List of Greek Symbols 303
References 304

1-56670-623-8/03/$0.00+$1.50

11.1 INTRODUCTION

Surface chemical reactions play a major role in controlling contaminant fate and transport in the environment. To better understand such processes, one often resorts to well-defined laboratory studies using mineral and organic standards or synthetic analogs as surrogates for the more complicated natural systems, either focusing on homogeneous systems or assuming the additivity of the major system components. In reality, such mixtures may display changes in particle size, surface area, and reactivity that differ from the individual surrogate components or the natural diagenetic environment that the investigator wishes to emulate.[1–5] For example, natural colloids observed in the electron microscope often appear irregularly eroded or coated with other mineral or organic phases and rarely resemble synthetic or pure mineral particles.[6,7] Complex mixtures and the presence of "surface coatings" or surface heterogeneities, often representing only a small fraction of the total suspension or matrix composition, can alter the reactivity of the more abundant components in ways that are difficult to quantify or predict based on the idealized systems.[8–11] Even common lab practices, such as homogenization and air-drying of soil materials can alter surface reactivity more than generally recognized.[12–14]

In recent years the study of mobile soil and groundwater colloids has received considerable attention because of concerns that such a vector may enhance the mobility of strongly sorbing contaminants, a process that is often referred to as "facilitated transport."[15,16] However, our ability to predict colloid movement and deposition is often confounded by the complexities of surface interactions in such dynamic, unstable systems. The lack of universally accepted analytical techniques and failure to realize instrumental limitations have made it difficult to compare and critically evaluate the results of different studies. Artifacts associated with groundwater sampling, filtration, and storage, and the dilute nature of most soil and groundwater suspensions further hamper characterization efforts.[17–21]

Not surprisingly, elevated concentrations of mobile groundwater colloids are generally associated with a disruption in the native hydrogeochemical environment, including those induced by artificial recharge, groundwater contamination, and even elevated flow rates associated with conventional sampling practices.[22,23] When precautions are taken to ensure that groundwater samples are representative of actual geohydrologic conditions within an aquifer, background or control wells outside the influence of the contamination source generally yield few mobile colloids.[24,25] However, artifactual colloids can be introduced during well construction or development (drilling fluids, bentonite, etc.),[26] result from changes in chemistry or redox due to inadequate sample preservation,[22,24,25] or become suspended from the immobile matrix by the shear forces associated with pumping.[17,22,23] Aggregation after sampling and membrane clogging can increase the efficiency of phase separation and reduce the average size and percentage of total suspended solids passing through the filter[18,19]; thus, the relative percentage of colloid-associated metals in filtered samples may not vary systematically with turbidity, that is, with colloid mass or concentration. In addition, larger size particles may contribute much of the colloidal mass, but reflect a smaller portion of the surface area available for contaminant sorption.[20]

Common colloidal materials found in subsurface environments include phyllosilicate clays, Al, Fe, and other metal oxyhydroxides, $CaCO_3$, microorganisms, and other biological debris. Field and laboratory studies have identified several mechanisms by which such materials can be mobilized in the environment: (1) clay dispersion due to changes in groundwater pH, ionic strength, and/or Na^+/Ca^{2+} ratios[27–32]; (2) manipulation of surface charge using a chemical dispersing agent[33,34]; (3) dissolution of carbonate or Fe-cementing agents resulting in the release and transport of silicate clays[35–38]; and (4) precipitation of colloidal particulates resulting from a change in groundwater chemistry.[24] In some instances, more than one of these mechanisms may be operative,[38] but essential in the development of such hypotheses is a thorough characterization of the composition and chemical nature of the colloidal suspension, including the inherent associations between various colloidal components, and their reactivity with respect to aggregation/filtration processes, as well as contaminant sorption properties.

Bulk quantification and characterization techniques, such as turbidity and chemical digestion/extraction methods, and certain commonly used instrumental characterization techniques [i.e., photon correlation spectroscopy (PCS), x-ray diffraction, electron microscopy, etc.] are extremely sensitive to the presence of artifactual colloids that are not inherently mobile within the soil or aquifer. Small sample sizes and the presence of organics and poorly ordered mineral phases can confound identification by x-ray diffraction, the primary method used by many in identifying clay minerals. Furthermore, discrete particle analysis techniques have confirmed that contaminants tend to be associated with specific colloidal types within a complex suspension and not generally distributed on all surfaces.[39] Chemical digestion may result in the overestimation of contaminant metals due to the dissolution of particulates that are not truly mobile, making it difficult to correlate elevated contaminant levels with a specific solution chemistry or sorptive colloidal fraction. Even nondestructive surface characterization methods, such as streaming potential, can yield macroscopic information about the surface charge of the immobile matrix that may not be indicative of surface chemical processes regulating colloidal deposition.[10,30]

This chapter will focus on a few key instrumental analysis methods that have wide application to the study of mobile colloids, including light-scattering methods (i.e., PCS), acoustic/electroacoustic methods, field flow fractionation (FFF), and electron microscopy (scanning electron microscope, SEM, and transmission electron microscope, TEM). Although the current chapter focuses on mobile colloids, the discussion is of general utility to any discipline in which the physicochemical characteristics of a suspension are of interest. The objective is to improve the quantitative nature of colloid characterization and description within an environmental context, and to ensure that the limitations of analytical techniques are fully recognized by environmental practitioners when the results are interpreted and reported in the literature. When appropriate, specific examples will be given illustrating the biases associated with certain widely applied analytical techniques. This text is not meant, however, to serve as a comprehensive discussion of the colloid transport literature, for which several excellent reviews have been published.[15,16,23]

11.2 LIGHT-SCATTERING TECHNIQUES FOR COLLOID CHARACTERIZATION

Various methods including sedimentation, centrifugation, zone-sensing and sequential filtration have been used to determine the concentration and particle size of submicron colloidal suspensions.[40–42] Such time-consuming methods may be sensitive to changes in particle size due to aggregation or the unforeseen alteration of solution chemistry during the sampling and analysis process. In contrast, light scattering provides a rapid noninvasive method of estimating particle size and concentration for dilute environmental suspensions. An extensive review of light scattering is beyond the scope of this chapter; thus, only a few qualitative aspects with respect to the characterization of environmental colloids are discussed. Those interested in an in-depth treatment of light-scattering methods and their application to the study of environmental colloids are directed to an excellent review by Schurtenberger and Newman.[43]

11.2.1 Turbidimetric Methods

When a light beam passes through a suspension, the dispersed particles scatter light away from the forward direction, thus reducing the intensity of the transmitted beam. Turbidity, the reduction in light intensity due to such scattering, is directly analogous to the Beer–Lambert relationship used in absorption spectrophotometry,[44,45]

$$I_l = I_o e^{-\tau l}$$

where τ is the turbidity or turbidimetric coefficient, analogous to the absorption coefficient, I_o is the incident beam intensity, and I_l is the remaining transmitted intensity after the beam passes through a sample of path length, l.[46]

Turbidimetric methods are often used to estimate the relative mass of suspended solids generated in laboratory column studies or present in surface- and groundwater samples.[7,17,29,38,47–50] In fact, turbidity is commonly used as an indicator when the chemistry within a monitoring well has stabilized during pumping so that a representative groundwater sample can be taken.

In many instances, researchers have simply used a UV/Vis spectrophotometer to estimate the colloid concentration, rather than a dedicated turbidimeter. For dilute suspensions, a linear relationship

$$-\log \frac{I_l}{I_o} = k_{turb} l c$$

for particle concentration, c, is usually observed where k_{turb} is a turbidimetric proportionality constant.

Nephelometric turbidimeters measure the radiant power, I_{sc}, of the scattered radiation at 90° from the incident light path, a scattering angle that is least sensitive to the presence of relatively few large particles. A calibration curve is obtained by

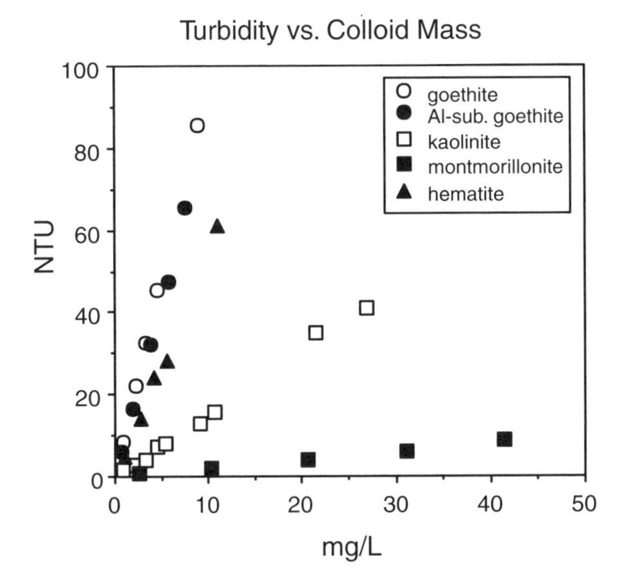

FIGURE 11.1 Turbidity of suspensions containing one of three synthetic Fe oxides displaying different particle sizes and morphologies: goethite (acicular, needle-like crystals 200+ nm in length); Al-substituted goethite (somatoidal crystals ~100 nm in length); and hematite (diamond-shaped crystals ~ 30 to 50 nm at the longest dimension), or the <2.0 μm fraction of kaolinite or montmorillonite.

simply relating the concentration of a given standard, usually a formazin suspension, to the sample scattering under carefully controlled conditions with results reported in nephelometric turbidity units:

$$I_{sc} = I_o I_{sc} c$$

Nephelometric turbidimeters are more accurate for measuring dilute suspensions and less sensitive to minor changes in instrumental design. Sensitivity increases with path length; however, linearity is sacrificed at higher suspension concentrations and self-quenching can result in anomalously low turbidity levels.[9] Obviously, dirty, scratched, or etched glassware, air bubbles, and vibration can all interfere with the accurate determination of turbidity.

Correlating turbidity with the actual mass of suspended particulates is often difficult because, in addition to the concentration of suspended solids, the size, shape, relative refractive index of the suspended particulates, and the wavelength of the incident radiation affect the light-scattering properties of the suspension.[42,46] To account for variations in scattering efficiency associated with different minerals, researchers often use reference minerals that are deemed to be representative of the mobile colloidal phase to calibrate instrumental response and estimate the mass of suspended colloids generated in column and groundwater studies.[17,29,37,51] In one case, Ryan and Gschwend[29] observed that the mass of suspended colloids generated in a laboratory column study was 5.1 to 11% less than calibration estimates using kaolinite, possibly resulting from the presence of more efficient scatterers. To illustrate the impact of such differences, the turbidity of three synthetic Fe oxides/oxyhydroxides and the <2μm fraction of kaolinite and montmorillonite suspended in deionized water was determined as a function of suspension concentration (Figure 11.1). Linear turbidity relationships with colloid mass were observed for each of the mineral suspensions. However, dramatic differences in the turbidity were observed for the two phyllosilicate clays compared to the Fe oxides, despite their large variation in size and morphology. Such differences likely reflect the higher refractive index for the Fe oxides (2.3 to 3.2) when compared to the phyllosilicates (~1.5).

11.2.2 Dynamic Light Scattering

The determination of particle size distributions for environmentally relevant suspensions is difficult due to their dilute nature, wide distribution of particle sizes (polydispersivity), and the large variation in particle morphologies.[22,37,43,52] Rayleigh light scattering occurs for particles much smaller than the wavelength of the light, the intensity of which is dependent on the wavelength ($1/\lambda^4$) and scattering angle (θ) with short wavelength radiation being scattered more than longer wavelengths.[42] For particle sizes comparable to the wavelength of light, multiple scattering events occur at different sites within a given particle, and the resulting scattering pattern becomes more complicated with an emphasis on forward scattering as particle size increases.[42,43] As Hunter[42] notes, even though a suspension of colloidal particulates is beyond the scope of Rayleigh theory, it demonstrates the strong dependence of scattering intensity (I) on particle mass (m^2), that is, particle size:

$$I_{sc(\theta)} \propto \frac{m^2 N_p}{\lambda^4}$$

where N_p is number of particles per unit volume, and λ is the wavelength of light.[42] For a spherical particle, the scattering intensity (I_{sc}) is dependent on the polarizability (α) of the particle:

$$I_{sc} \propto \alpha^2$$

where

$$\alpha = 4\pi\varepsilon_o r^3 \left(\frac{n^2 - 1}{n^2 + 2}\right)$$

The radius of the spherical particle is r, and the relative refractive index, n, is the ratio (n_p/n_o) of the refractive index for the particle (n_p) to that of the suspending media (n_o). As seen in Figure 11.1, particles with higher refractive indices scatter light more efficiently. The refractive index of polystyrene beads (1.6), generally used to standardize light-scattering instruments, is similar to the refractive index of phyllosilicate clays.

Light scattering from an intense, coherent, monochromatic beam, usually a He-Ne or Ar laser in most instruments, can be used to estimate colloidal particle size. Colloidal suspensions, due to the small particle size, are subject to Brownian motion resulting in local fluctuations in particle concentration and light scattering, the rate of which depends on the size (dispersion coefficient) of the particles. Thus, small particles that are more subject to Brownian motion will induce rapid transient concentration fluctuations while large, slowly moving particles will produce slow fluctuations in the scattering intensity.

PCS, also known as dynamic light scattering or quasi-elastic light scattering, quantifies the particle motion (Brownian motion) by measuring the signal intensity at a given instance and comparing that with signals obtained at successively longer time intervals, which is integrated over time [Figure 11.2(a)].[41–43] With little or no motion, the product of the signal will vary little with time; however, subsequent signals at different time intervals will vary dramatically if particles are moving rapidly. Scattering intensity measurements are taken at short time intervals relative to the diffusive motion of colloidal particles, resulting in high initial correlation that gradually disappears with longer time intervals.[21] Analysis of the decay function for signal correlation can yield the diffusion coefficient (D) for the suspended particulates:

$$D = \frac{kT}{3\pi\eta d}$$

where k is the Boltzmann constant, T is the absolute temperature, η is the viscosity, and d is the hydrodynamic diameter. The noninvasive nature of dynamic light scattering eliminates artifacts associated with particle isolation, such as centrifugation, filtration, or sample drying. However, PCS is very sensitive to contamination of larger particles, provides nondetailed size information, and determines the "effective" hydrodynamic diameter for nonspherical particles.[41,43] The detection limit (mg l^{-1}) for PCS is dependent on the particle size and scattering angle, as well as measurement duration, instrument sensitivity, and laser source.[43] Average size estimates may be heavily weighted in favor of larger particles; thus, fractionation, such as filtration, sedimentation, or centrifugation, designed to remove extremely large particles (>1μm) prior to PCS analysis may be necessary to resolve the size of smaller, more abundant colloids.[43,53]

Schurtenberger and Newman[43] emphasized that researchers must be more explicit in describing the data analysis techniques relating particle size to the measured autocorrelation function. To date, few studies have been successful at resolving multimodal size distributions for environmental suspensions.[9] Multi-angle PCS is critical for sizing environmental suspensions where no *a priori* knowledge of the particle size distribution is available to confirm multimodal size distributions, resolve artifacts associated with particle anisotropy and particle-particle interaction at high concentrations, and account for variations in scattering intensity as a function of particle size that are manifested as local scattering minima at specific angles.[43] Electron microscopy can be used to confirm PCS results; however, qualitative agreement between the two techniques is not surprising because of the limited resolution inherent to PCS, with an applicable size range that is consistent with the imaging and sizing capabilities of the SEM. This, combined with the qualitative nature of most SEM particle surveys, ensures that most investigators will observe results that are consistent with PCS. Care must be taken to ensure that particle associations observed in the electron microscope are typical of the particle state in suspension and do not reflect changes in aggregation induced by filtration or other sample preparation artifacts.

Changes in particle size with even limited storage suggest that timely analysis of environmental suspensions reduces artifacts associated with aggregation, changes in chemistry and biological activity.[20,21,43,53] The development of *in situ* PCS systems offers the ability to monitor particle size distribution and concentration of groundwater colloids without the necessity of altering the system during the sampling and handling processes before analysis.[25]

11.2.3 Laser Doppler Velocimetry and Particle Charge

Evaluating particle surface charge is critical to understanding the mechanisms of mobile colloid formation, stabilization, and physicochemical filtration in the environment, as well as other sorption phenomena. Without a significant electrostatic barrier to particle approach, smaller colloidal particles with higher diffusion coefficients collide more frequently and aggregate faster than larger particles, with discrete colloids or colloidal aggregates in the size range of 0.1 to 1.0 μm being most stable.[20] In addition, surface charge may be inconsistent with bulk suspension mineralogy due to the presence of organic or oxide surface modifiers.[8,9,34,54] For example, highly negative electrophoretic mobilities commonly observed for Fe/Al oxide-rich suspensions under pH conditions well below the reported point of zero charge (PZC; pH at which net charge is zero) have generally been attributed to organic coatings on the mobile oxide fractions.[9,37,49,54]

Colloidal particulates develop surface charge in one of two ways: either through isomorphic substitution within the mineral structure, which is insensitive to the external solution conditions (permanent charge), or from reactions of surface functional groups (e.g., surface hydroxyls associated with organics, edge sites on aluminosilicates, and metal oxyhydroxides) with adsorptive ions at the mineral/particulate-solution interface, which is subject to changes in the aqueous environment surrounding the particle (i.e., pH, ionic strength, etc.), and therefore considered "variable charge."[55,56]

The dilute nature of most environmental suspensions does not lend itself to conventional wet-chemical techniques for evaluating the surface charge of particulates, such as potentiometric titration or ion exchange methods. Potentiometric titrations, especially when applied to mixed, constant/variable-charge suspensions, are complicated by the presence of species other than H^+ and OH^- that act as potential determining ions (PDI) and various reactions that consume H^+ and OH^- without generating equivalent surface charge, such as exchangeable Al.[8,28,57,58] Ion exchange/extraction methods depend on the identity and concentration of the probe ion used to extract surface-associated species and yield little information related to overall colloidal stability. In contrast, electrophoretic methods can be used to evaluate the surface charge properties of dilute colloidal suspensions under the specific chemical conditions to which the colloids are subjected, thus reducing the errors and biases associated with altering the suspending solution or quantifying various poorly defined surface/solute reactions (ion exchange, mineral dissolution, Al hydrolysis, etc).

When an electrical field is applied to a suspension of charged particles, the particles migrate toward the electrode of opposite sign, reaching a terminal velocity in a matter of microseconds. The electrophoretic mobility (EM), u (μm cm s^{-1} V^{-1}), for a particle is defined as:

$$u = \frac{v_e}{E}$$

where v_e is the terminal velocity of the particle at a specified unit field strength, E (V cm^{-1}), with the sign being positive if the particles migrate from a region of high electrical potential to a region of low electrical potential.[55] A boundary is established between the strongly sorbed species and solvent that remains associated with the charged particle as it moves through the solution and the loosely sorbed diffuse species. The inner potential at the shear plane, known as the zeta potential, ζ, depends on the surface charge density of the particle at the shear plane and is indicative of the "effective charge" that particles and surfaces experience as they approach each other, that is, colloid stability.[42,55] Analysis of the solution chemistry (i.e., pH, ionic strength, solution composition, etc.) is critical to understanding the system, since the EM (i.e., zeta potential) is a function of the colloidal material and aqueous chemical environment. Such information can then be used to predict the effect of various solution–particle and particle–particle interactions on aggregation, flow, sedimentation, and filtration behavior.

Various equations have been derived for relating EM, u, to the zeta potential, ζ. Traditionally, the Smoluchowski equation,

$$\zeta = \frac{\varepsilon_o D u}{\eta}$$

has been used for soil clays where ε_o is the permitivity of a vacuum, D is the dielectric constant for water, and η is the viscosity of the solution. However, the validity of such an expression depends on a number of assumptions and the choice of molecular models used to represent the "plane of shear."[58,59] In many instances it may be more appropriate to simply report the measured mobility, u.

Electrophoretic instruments for analysis of colloidal suspensions can be divided into two basic classes: optical instruments for which the operator observes the migration of particles in a field using a microscope; and laser-based instruments that measure the Doppler shift in the frequency of the scattered light from particles moving in response to an electric field, that is, laser Doppler velocimetry (LDV). Analysis using optical instruments is slow and tedious, making it difficult to analyze unstable suspensions. Particle detection is limited by the resolving power of the light microscope and possibly biased by differences in particle size and the refractive index of various colloidal constituents of multicomponent suspensions. Therefore, electrophoretic results can be biased by the analysis of a relatively few discrete particles that may display the expected behavior.[60] Design improvements, such as laser illumination to improve particle resolution and rotating prism systems that measure the mobility of a field of particles, have addressed some of the inherent limitations of microscope-based systems.

In many respects, LDV instruments are superior to optical-based instruments, especially for polydisperse samples with a range of surface properties. Doppler broadening is generally evaluated at lower scattering angles to reduce the impact of inherent Brownian motion on the frequency shift, thus increasing instrument resolution.[60] Information about the particle size of the suspension can be obtained by measuring frequency broadening due to Brownian motion in the absence of the electric field. Verification of LDV results generally involves comparing frequency shifts at different scattering angles or different electric field strengths at a fixed angle; an alternating electrical field is used to avoid electrode polarization. However, Bertsch and Seaman[8] observed the disaggregation of colloidal particles that could impact charge characterization when repeatedly subjected to the alternating field without sufficient relaxation time between electrophoretic analyses. Operator bias associated with particle selection is eliminated and the mobility of a much larger population of particles can be rapidly determined, thus facilitating the analysis of colloidal samples that are inherently unstable. Typically, greater standard deviations in the measured mobilities are observed for LDV instruments, but this may reflect a more statistically relevant sample population that better accounts for actual mobility distributions.

Care must be taken to evaluate the influence of other electrokinetic phenomena occurring within the EM sample cell. When an electrical field is applied to the capillary containing a colloidal suspension, migration is observed for the suspending solution due to the osmotic flow of the counterions (electro-osmosis), as well as the particles (electrophoresis), resulting in a parabolic velocity distribution in colloid migration across the capillary that is the sum of the electrophoretic velocity and electro-osmotic flow (Figure 11.2(b)). Electro-osmosis is a consequence of the surface potential associated with the capillary walls, which induces a nonuniform distribution of solution ions within the tube. For example, cations associated with the capillary wall are attracted to the negative potential of the electrical field in a manner similar to a positively charged particle. However, there is a location within the tube, known as the stationary layer, where the net osmotic flow in either direction is zero.[42,61] Absolute measurements of EM should be taken in the stationary layer of the capillary tube. Unfortunately, the change in mobility as a function of minor changes in cell position is quite great in the region surrounding the stationary layer. Taking numerous measurements across the capillary (i.e., EM fingerprinting) may be an effective means of evaluating relative changes in surface charge under various solution conditions (i.e., pH, ionic strength, solution composition, etc.). Such an approach is also recommended to determine if significant particle settling has

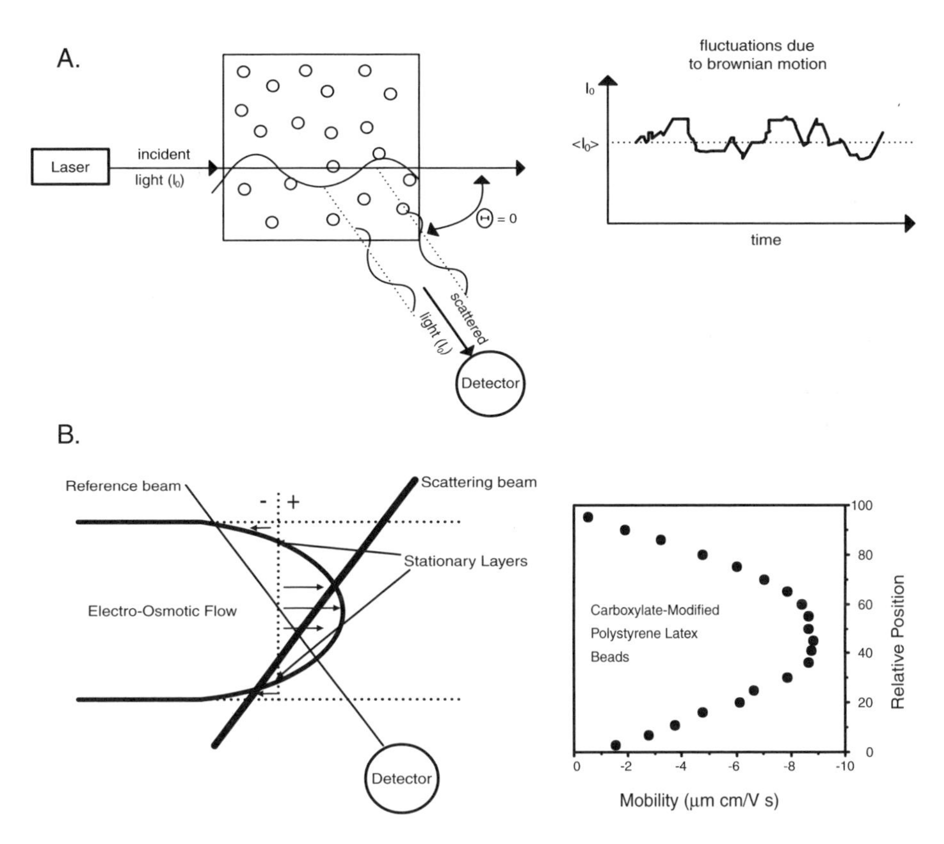

FIGURE 11.2 (a) Schematic representation of dynamic light scattering, and (b) laser Doppler velocimetry. (Part (a) modified from Buffle, J. and Leppard, G.G., *Environ. Sci. Technol.*, 29, 2176, 1995. With permission.)

occurred during EM analysis, which would tend to alter the symmetric nature of the parabolic velocity function as larger particles and aggregates settle away from the stationary layer located at the top of the capillary and accumulate on the bottom of the analysis cell. Particle segregation due to aggregation and settling may significantly skew results in favor of the smaller, more-stable fraction. Unfortunately, one is often interested in the EM of a suspension under conditions that are not conducive to colloidal stability, such as the pH region near the PZC for a given suspension. In such a case, electroacoustic methods discussed in a subsequent section may be appropriate. Recent advances in LDV instrumentation have focused on reducing/eliminating the need for optical alignment, and automating certain time-consuming aspects of analysis, such as incorporating autotitration systems to evaluate changes in charge as a function of pH.

Despite the poorly defined nature of the shear plane, EM has been commonly used as a noninvasive technique for evaluating various surface–surface and surface–solute reactions, such as evaluating the PZC for amphoteric minerals; specific sorption reactions for inorganics and organic compounds on mineral surfaces;

charge reversal for surfactant modified clays; changes in microbial cell membrane properties in response to contaminated environments; and cation demixing on 2:1 phyllosilicate clay minerals.[62–66]

Electrophoretic behavior, however, is not necessarily indicative of a colloidal sorption affinity for particular ionic species. For example, zeolite suspended in HDTMA, a commonly used cationic surfactant, will undergo charge reversal at high surface coverages resulting from the head-to-tail arrangement of the surfactant once the external cation exchange sites are filled, yielding a relatively stable suspension that displays a positive electrophoretic mobility (Figure 11.3).[111] However, the zeolite retains a high affinity for cations, especially Cs, due to the presence of internal exchange sites that are inaccessible to HDTMA and contribute little to the electrophoretic behavior. Internal exchange sites located within the open zeolite structure are analogous to surface heterogeneity and similar high affinity sorption sites associated with many soil minerals.

11.3 ACOUSTIC SPECTROSCOPY

11.3.1 Acoustic Attenuation and Particle Sizing

Recent advances have been made in the theory and application of acoustic and electroacoustic spectroscopies for measuring the particle size distribution (PSD) and ζ-potential of colloidal suspensions, respectively.[67–69] To date, the use of acoustics has been confined mainly to industrial applications, despite the clear potential for the technique to characterize colloids with environmental or agricultural significance.

Acoustic spectroscopy measures the speed and attenuation of sound waves interacting with a colloidal suspension. When a sound wave in the range of 1 to 100 MHz interacts with a colloidal suspension, the measured acoustic attenuation and

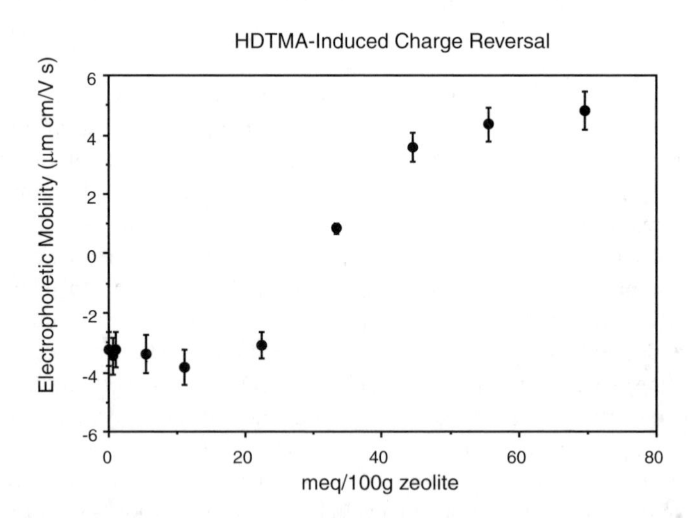

FIGURE 11.3 Electrophoretic mobility of zeolite as a function of HDTMA loading indicating external charge reversal. (Adapted from Sullivan, E.J., Hunter, D.B., and Bowman, R.S., *Environ. Sci. Technol.*, 32, 1948, 1998. With permission.)

sound speed can be theoretically related to PSD by accounting for viscous, thermal, scattering, intrinsic, electrokinetic, and structural losses.[70,71] The first two losses are the most significant as particles mainly interact with sound waves hydrodynamically through viscous losses and thermodynamically through temperature losses. Several different theoretical approaches and acoustically based instruments are available, such as the DT-1200, ESA-8000, Malvern Ultrasizer, and AcoustoSizer.[67,69,72]

When either of the main attenuation mechanisms (viscous or thermal) predominates, the other may be neglected. A quantity called the "viscous depth," δ_V (L), characterizes the decay distance of the shear wave from a particle's surface, while the "thermal depth," δ_t (L), is the penetration depth of the temperature wave into the liquid.[70] Each depth has a "critical frequency," comparable to the particle radius, corresponding to maximum attenuation. Thermal losses dominate in emulsions and low-density dispersions, so viscous losses may be neglected. Viscous losses are more sensitive to suspension concentration, while thermal losses may not become important until 30%-volume fractions. When viscous losses predominate, the so-called "long wavelength requirement" sets a lower limit of 10 nm on detectable particle size. The formulas for viscous depth and thermal depth, respectively, are

$$\delta_V = \sqrt{\frac{2\upsilon}{\omega}} \quad , \quad \delta_t = \sqrt{\frac{2\kappa}{\omega \rho_0 C_p^0}}$$

where υ (l t^{-1}) is the kinematic velocity; ω is the sound frequency (radian t^{-1}); κ(t l^2 mol^{-1}) is the thermal conductivity; ρ_0 (m l^{-1}) is the liquid density; and C^0_p (J T^{-1} mol^{-1} m^{-1}) is the specific heat of the liquid at constant pressure.[70]

Two models are available for interpreting attenuation spectra as a PSD in suspensions with chemically distinct, dispersed phases using the extended coupled phase theory.[68] Both models assume that the attenuation spectrum of a mixture is composed of a superposition of component spectra. In the "multiphase model," the PSD is represented as the sum of two log-normal distributions with the same standard deviation, that is, a bimodal distribution. The appearance of multiple solutions is avoided by setting a common standard deviation to the mean size of each distribution. This may be a poor assumption for the PSD (see section 11.3.2). The "effective medium" model assumes that only one "target phase" of a multidisperse system needs to be determined, while all other phases contribute to a homogeneous system, the so-called "effective medium." Although not complicated by the possibility of multiple solutions, this model requires additional measurements to determine the density, viscosity, and acoustic attenuation of the effective medium. The attenuation spectrum of the effective medium is modeled via a polynomial fit, while the target phase is assumed to have a log-normal PSD.[68] This model allows the PSD for mixtures of more than two phases to be determined.

Acoustic spectroscopy has several characteristics that make it useful. One clear advantage over light-scattering techniques is the ability to stir, pump, or otherwise physically agitate the sample during analysis, making the technique well suited to potentiometric titration and analysis of unstable suspensions. When the acoustic signal is measured as a function of the transmitter–receiver gap, it requires no calibration. A PSD can be calculated even when there is little density contrast between sample and fluid. Attenuation spectra are independent of electrical properties of the particle surface, and supply independent information about PSD, even in concentrated systems with several dispersed phases.[72] Thus, attenuation spectra can characterize PSD in uncharged dispersed systems, in highly conducting systems, as well as in systems with conducting particles. The theoretical success of characterizing viscous losses in concentrated dispersions with large density contrast gives acoustics an advantage over light-scattering methods in measuring PSD. Using well-characterized samples and commercially available instruments, acoustic spectroscopy can measure the mean of the PSD with a precision and accuracy of up to 1%, and the width of the PSD with an accuracy of up to 5%.[71]

There are several shortfalls in acoustic spectroscopy. Information about particle shape is lacking in the spectrum, and a substantial amount of physical and thermodynamic information may be needed to interpret acoustic spectra, including particle density, liquid density and viscosity, and the weight or volume fraction of the suspension.[73] Such information may not always be available for complex environmental suspensions. Also, relatively large sample requirements may restrict the use of acoustics to idealized laboratory systems.

Acoustic spectroscopy shows promise for distinguishing particle–particle interactions in concentrated suspensions (up to 30% by volume), as well as in polydisperse suspensions with chemically distinct phases. Although acoustic spectroscopy does not provide information regarding particle shape, it has an advantage over SEM for determining PSD as *in situ* measurements are made, so that the colloids are not subject to changes in particle–particle interactions during filtration and drying.

11.3.2 Electroacoustics

Electroacoustic spectroscopy measures either colloid vibration potential/current (CVP/CVI) or electrokinetic sonic amplitude (ESA), each of which is quantitatively related to mean ζ-potential. In response to an acoustic wave, the density contrast between the particle and the medium causes a displacement, or polarization, of the electrical double layer, creating a dipole moment (Figure 11.4) whose magnitude varies with the sound wave amplitude.[67] In superposition, the individual dipole moments give rise to the macroscopic alternating electric field measured as the CVP.[70] Conversely, the application of an alternating electric field produces an oscillating electrophoretic motion for particles with a nonzero ζ-potential, which generates a sound wave, the ESA effect, and the resulting acoustic field is measured.[67,69] The CVI is analogous to a sedimentation current, the current arising when the potential generated as charged particles settle under gravity is short-circuited between vertically placed electrodes, while in CVP, the alternating acoustic field supplies the acceleration instead of gravity.[74]

In electrokinetic phenomena such as electroacoustics, theoretical models need to consider the induced movement of charge within the electrical double layer (EDL), the "surface current", I_s, as well as the interaction of the outer portion of the double layer with the applied signal (acoustic or electric field) and with the liquid medium. Hydrodynamic flows generate surface current as liquid moving relative to the particle

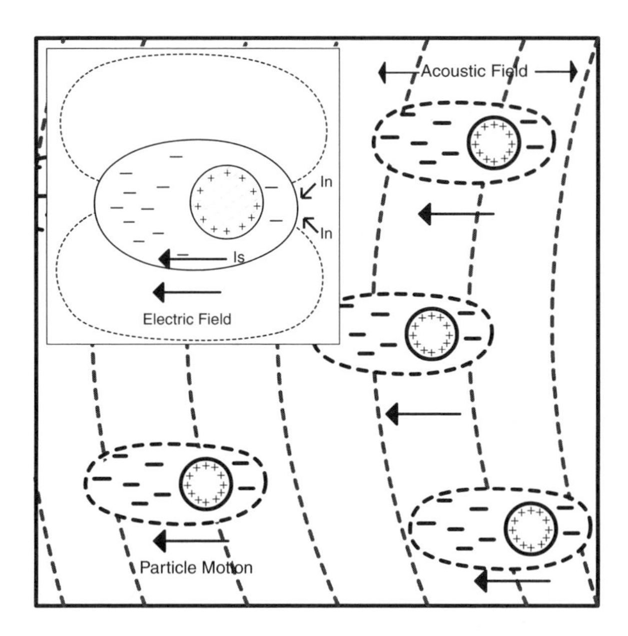

FIGURE 11.4 Dipole moments generated on charged particles in response to an applied field. Inset: location of CVI current, I_n, external to the electric double layer.

surface causes movement of charge in the outer portion of the electric double layer (EDL). A tangential electric field, E, is generated by the surface current, I_s, and the compensating current, I_n, is measured in CVI electroacoustics.[70] When particle size and/or electrolyte concentrations are small, surface conduction may be significant.[67]

Electroacoustic measurement of ζ-potential can be made even in concentrated suspensions (up to 40% by weight) where optical techniques fail. The interpretation of electroacoustic measurements is questionable when the EDL is thick and overlaps for neighboring particles, when the ζ-potential is large, or when anomalous conduction in the inner portion of the EDL is present. However, similar problems are encountered in theoretical analyses for all available techniques for measuring ζ-potential. Excluding electroacoustics, ζ-potential has typically been measured in extremely dilute colloidal suspensions by micro-electrophoresis, although suspension dilution may affect both particle size distribution and ζ-potential. Sedimentation potential measurements are also restricted to dilute suspensions in order to ensure free settling and uniform particle flow. Conversely, high solid concentrations of very large particles (e.g., sand) are required for electro-osmosis and streaming potential measurements due to the requirement of a tightly packed, immobile porous plug. In contrast, electroacoustic spectroscopy is applicable to a range of suspension concentrations.

Acoustic methods offer several advantages when compared to other comparable techniques: (1) applicable to concentrated suspensions; (2) less sensitive to particulate contamination; (3) better suited to polydisperse suspensions; (4) applicable to a wide size range; (5) well suited to automated potentiometric titrations and analysis of suspensions near their PZC and critical coagulation concentration (CCC) because of the ability to stir or pump the sample during analysis.

From an experimental standpoint, the ability to use concentrated suspensions offers obvious advantages when compared to other potentiometric and micro-electrophoretic techniques. Such conditions are more analogous to the solid/solution ratio encountered in a typical soil/sediment environment, but with less kinetic restrictions and system heterogeneity than encountered in the field. In addition, small aliquot suspensions can be removed during analysis for characterization using another technique. The large sample volumes and high colloidal loads, however, may restrict the use of acoustic techniques to idealized laboratory systems where sufficient sample is available.

11.4 FIELD FLOW FRACTIONATION

FFF is a separation technique that encompasses a range of procedures based on theory and subsequent instrumentation originally developed and advanced by Giddings.[75,76] FFF is a high-resolution chromatography-like technique applicable to the separation of macromolecules, colloids and particles encompassing a size range of 1 nm to 100 μm. In practice, a small volume of colloidal suspension (10–20 μL) is injected into the FFF channel, the concentration of which depends on the sensitivity of the online detector. The FFF channel is a thin (~0.02–0.05 cm), open rectangular channel through which a carrier solution is pumped by means of a peristaltic or HPLC pump. The breadth and length of the channel are generally on the order of 2 cm by 20 to 50 cm, respectively. The thinness of the channel ensures a laminar flow profile with the fastest flow vectors in the center and slowest flow at the channel walls. A force is applied across the thin dimension of the channel, and perpendicular to the direction of flow, driving molecules and colloidal particles against one of the channel walls (accumulation wall). Particles also diffuse back into the channel through Brownian motion; the extent of this diffusion being related to the molecular weight (M_w) or hydrodynamic diameter (d) of the particle. After sample injection the main channel flow is stopped for a period of time (relaxation time) to allow the particles to reach their equilibrium distribution from the channel wall without migrating down the channel. The equilibrium positions above the wall are based on the balance achieved between migration caused by the applied force and the back-diffusion of the particle (Figure 11.5(a)). After the relaxation time, the channel flow is resumed and particles whose equilibrium distance is higher in the channel (smaller particles) are carried in faster moving flow vectors, and, hence, are eluted first from the channel.

11.4.1 SEDIMENTATION (SD-FFF) AND FLOW-FIELD FLOW FRACTIONATION (FL-FFF)

The various FFF techniques arise through the different fields that are employed, including sedimentation, flow, electrical, and thermal fields. Of these techniques, sedimentation (Sd-FFF) (Figure 11.5(b)) and flow (Fl-FFF) (Figure 11.5(c)) have found the widest application in environmental studies. In Sd-FFF, the channel is

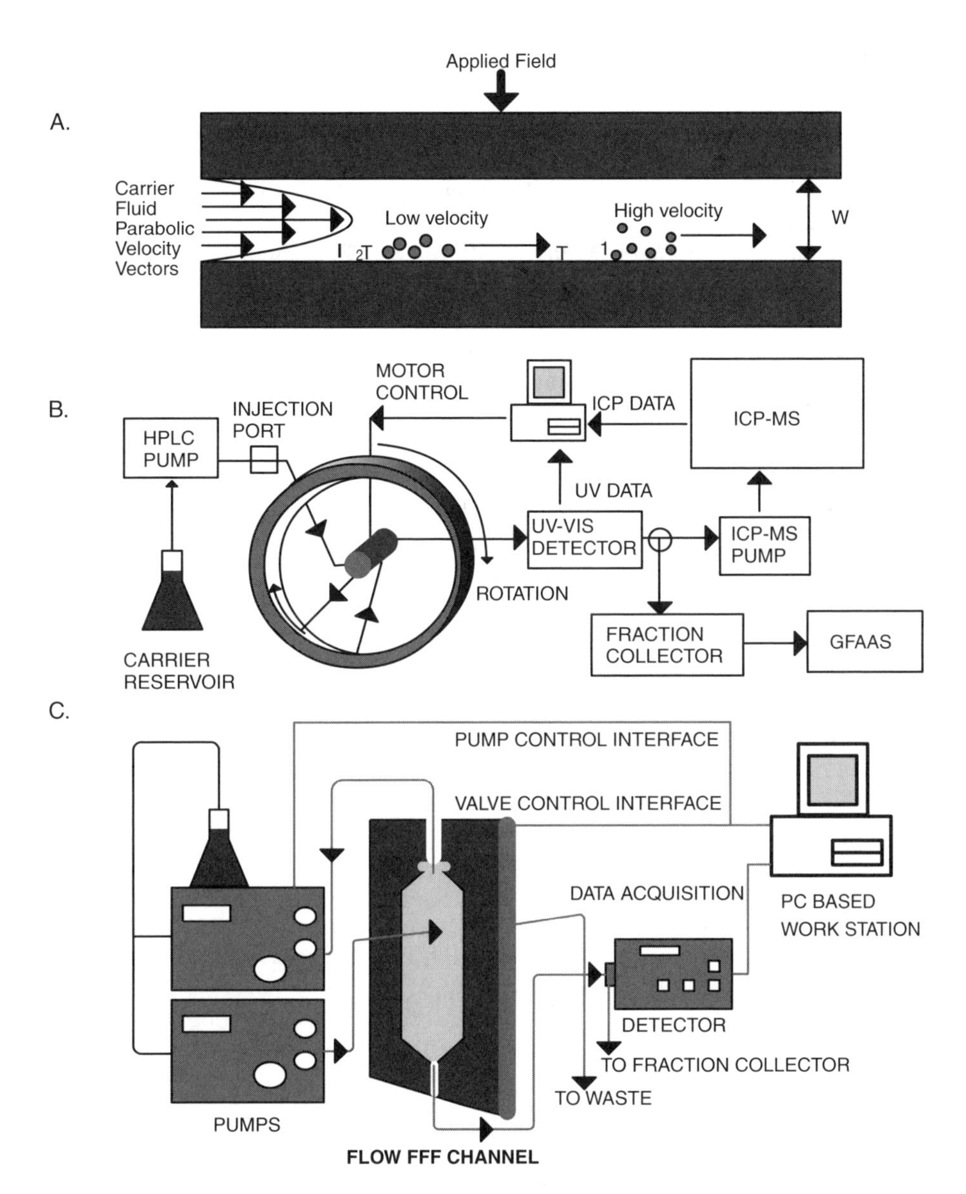

FIGURE 11.5 Schematic representation of the FFF. Schematic diagram of (a) Sd-FFF and (b) Fl-FFF systems coupled to ICP-MS and UV/Vis detector, respectively.

created by clamping a mylar spacer, with the channel cut out, between two concentric rings, which are then positioned within a centrifuge basket. The centrifugal forces used make Sd-FFF applicable to the separation of particles from about 30 nm to 100 μm, the retention volume of a particle being a function of the spherical diameter (d) and density (ρ_p). When operated in normal mode (see later description), that is, particles in the size range 30 nm to 1 μm, the theory of Sd-FFF is well defined mathematically and, when the gravitational field applied to effect the separation is constant, the retention ratio R (V_0/V_r) can be calculated from the following equation:

$$R = \frac{kT}{\pi w \Delta\rho_p \varpi^2 r d^3}$$

The relationship between R and d^3 affords Sd-FFF a high resolving power; however, it also means that the retention time of large particles can become unacceptably long when a constant field is employed. To overcome this, power programming is commonly used, where the gravitational field is decayed exponentially over time after an initial hold time at constant field strength.[77] Computer programs exist for optimizing run parameters and converting detector response and field strength as a function of analysis time into particle size distributions.

In Fl-FFF, the channel is created by placing a mylar spacer with the channel cut out between two porous frits. A membrane filter of a specific molecular weight cutoff is placed on one of the frits and acts as the accumulation wall to permit flow, without loss of particles. The applied force is then a perpendicular flow of the carrier solution across the porous frits. Fl-FFF is a versatile technique capable of separating macromolecules as small as roughly 1000 Da, in which case it is comparable to gel permeation (size exclusion) chromatography. However, Fl-FFF can also be applied to the separation of colloidal particles. In this case the hydrodynamic diameter of the colloidal particle is related to the retention volume, V_r, by the equation

$$V_r = \frac{\pi\eta w^2 v_c d_s}{2kT}$$

where w is the channel thickness, v_c is the cross-flow velocity, and d_s is the hydrodynamic diameter. For macromolecules, the usual practice is to create a calibration curve relating peak retention time to molecular weight, M_w. The diffusion coefficient, D, and can then be calculated from the equations below.

$$d_s = A' M_w^b$$

$$D = \frac{kT}{3\pi\eta d_s}$$

$$\log D = \log A'' - b \log M_w$$

A recent advance in Fl-FFF has been the introduction of asymmetric-flow FFF instrumentation. In asymmetric Fl-FFF, the upper channel wall is impermeable and the cross-flow rate is achieved by flow control of the cross-flow and channel flow. Upon sample injection channel flow is directed through both the channel inlet and outlet that allows for focusing of the sample and for preconcentration. For elution, channel flow is just introduced at the channel inlet.

The study of environmental colloids by FFF is complicated by a change in the main counter force opposing the applied force as particle size increases above about 0.8–1 μm. In normal mode FFF, the method previously described, the main force opposing the applied field is Brownian motion. Smaller particles have higher diffusion coefficients and therefore attain higher equilibrium distances from the accumulation wall and thus emerge first from the channel. However, as particle diameter increases, diffusion becomes less important compared to hydrodynamic lift forces. These forces are not well understood, but the net effect is to lift larger particles away from the accumulation wall. Also, for larger particles the actual particle diameter becomes comparable to the equilibrium cloud thickness for smaller particles. This is termed the steric effect and results in the opposite elution pattern to normal mode with larger particles eluting from the column first. The point at which the elution mode changes from normal to hyperlayer/steric mode is termed the inversion point and occurs around 1 μm. Steric-mode FFF has found numerous applications in the separation of particles >1 μm, such as in the separation of cells.[78] However, the steric effect complicates analysis of environmental suspensions and necessitates preseparation of particles >1 μm before FFF analysis in either normal or steric mode.

11.4.2 FFF Applications

Environmental applications of FFF fall into two broad groups. First, studies utilizing FFF with a nonspecific detector (i.e., UV spectrophotometer) to determine the particle size or molecular weight distribution of the sample. This approach has been applied extensively for the study of humic and fulvic acids in natural waters.[79–82] Fl-FFF is used exclusively for this application, in which it is essentially equivalent to gel permeation chromatography (GPC). As in GPC, the main analytical consideration is minimizing interactions between organic acids and the accumulation wall by the judicious choice of carrier solutions and membrane materials. Thang et al.[80] reported optimal recovery of humic substances using a 0.005-M Tris-buffer (trishydroxymethylaminomethane) as the carrier solution at pH 9.1 and utilizing a regenerated cellulose membrane. For molecular weight calculations, the system is first calibrated by running polystyrenesulfonate molecular weight standards to create a calibration curve. Caution is advised when calibrating because differences in structure between calibration standards and samples may introduce errors in molecular weight calculations of unknowns. UV detectors have been commonly used for larger colloids based on light scattering by the particles. Multi-angle light scattering has also been used for accurate molecular weight determinations in FFF studies of humic substances.[83,84]

The second major environmental application of FFF has been the use of an element-specific detector, usually in series with a UV detector, to provide elemental composition data along with the PSD. Graphite-furnace atomic absorption spectrometry has been used off-line on fractions collected from the FFF run. However, the multi-element detection, low detection limits and capability to function as an online detector have made inductively coupled plasma mass spectrometry (ICP-MS) the ideal detector for FFF.[85,86] The sample introduction system of the ICP-MS is able to efficiently transport micron-sized particles into the high-temperature plasma, where the particles are completely decomposed, atomized, and ionized. The advantage of this technique is that it allows for the major element mineralogy of the colloidal particles to be determined simultaneously with the UV-based PSD to provide element-based size distributions. Although the light-scattering response of a UV detector can be a complex function of particle size, agreement between the UV response and the element-specific response for environmental colloids has been generally observed. In most cases the sensitivity of the ICP-MS allows multiple elemental signals to be continuously monitored to determine minor and trace elements and, therefore, assess potential correlations between element distributions.

Sd-FFF-ICP-MS has been used to determine major mineral constituents and distributions of soil colloids to study trace metal adsorption to colloids and the effect of colloidal surface coatings on phosphate sorption.[87–90] Examples of the UV-based and ICP-MS–based fractograms for soil colloids are shown in Figure 11.6(a). The original surface soil was prefractionated by centrifugation in order to obtain a 0.2-0.8-μm size fraction. The example shows the element-based fractogram for a minor and trace element, manganese (Mn) and uranium (U), respectively. The fractogram was converted into an element-based particle size distribution (Figure 11.6(b)) and an element ratio distribution (Figure 11.6(c)) using a major element (Si). The element ratio shows changes in mineralogy that occur across the size range. In this case, it appears that U is uniformly distributed across the particle size range whereas a larger Mn-rich colloid, perhaps Mn oxide, appears to be present. Interpretations of element-based size distributions may prove useful for understanding trace metal behavior in soils.

Fl-FFF has also been coupled to ICP-MS for the determination of element size distributions of 28 elements, including C, in natural waters.[91] Hassellov et al.[91] further developed methodology for on-channel preconcentration that enables up to 50 ml of sample to be introduced onto the channel. This significantly enhances the effective detection limits of the technique, which can otherwise be problematic due to the low concentration of trace elements in natural waters, the dilution inherent in FFF analysis, and the small injection volume, typically 10 to 50 μl.

Clearly, FFF techniques are potentially useful in environmental analysis, as they can provide PSD analysis and, when coupled to a suitable detector, element sized distribution analysis. However, FFF theory for particle size determination does not account for particle–particle or particle–wall interactions, both of which will cause errors in the calculated particle size. Particle–wall interaction forces can be incorporated into FFF theory to allow for accurate particle sizing in low-ionic-strength carrier solutions if standards of the same material and of known size are available.[92] Particle interactions are minimized when the ionic strength of the carrier solution is near 10^{-3} M. Common carrier solutions employed in previous studies include dilute surfactant solutions (0.1 mM) sodium pyrophosphate, and 0.01 M sodium bicarbonate usually at high pH (>7). These conditions might be expected to change the primary PSD of the natural sample and confound interpretation of trace element–colloidal interactions and sorption behavior. Hence, interpretation of FFF results requires a consideration of the potential changes to surface charge that may result from the carrier solution.

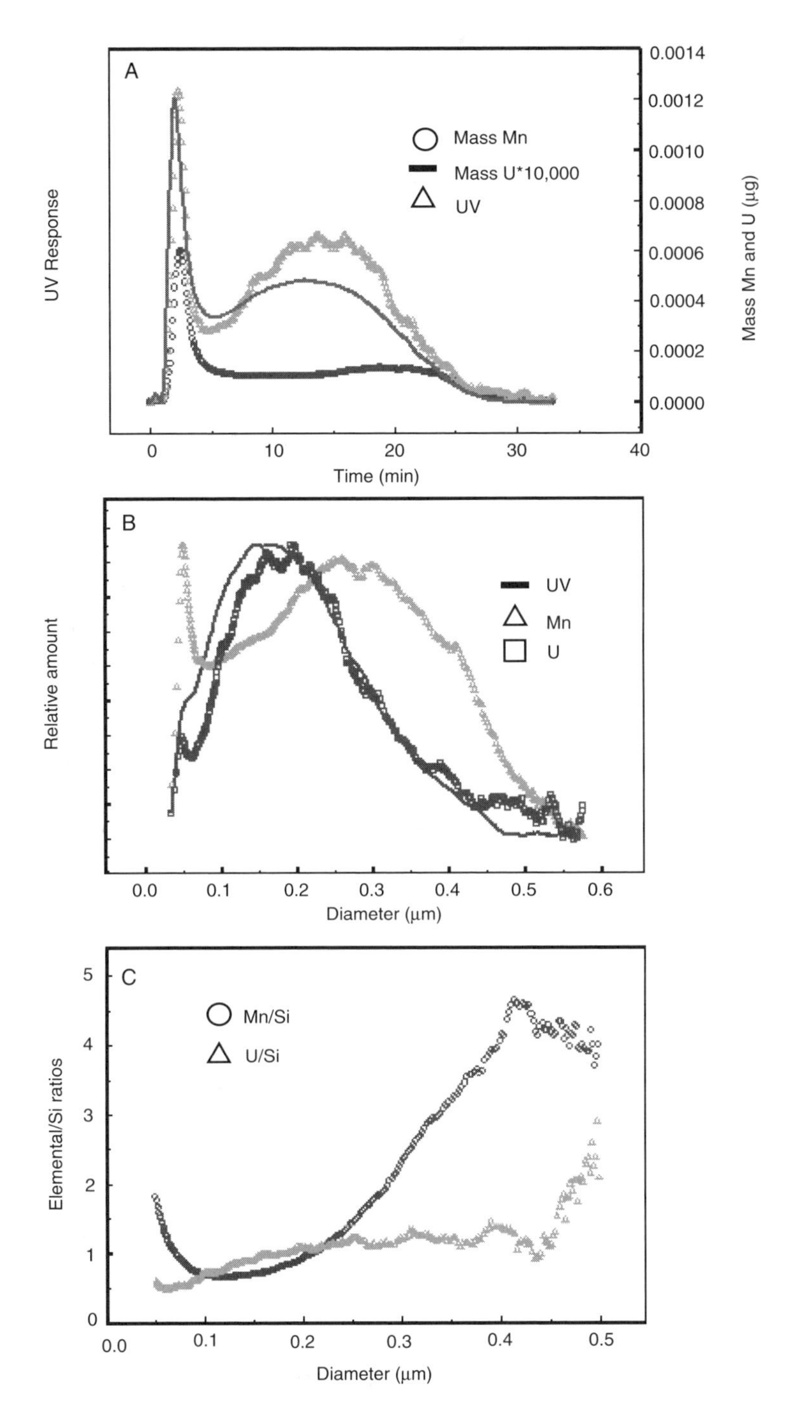

FIGURE 11.6 UV and ICP-MS-based Sd-FFF fractogram of a colloidal (0.2 to 0.8 μm) fraction: detector responses versus time (a); particle size distribution (UV response) and element-based size distribution for Mn and U (b); and element ratios for Mn/Si and U/Si (c).

11.5 ELECTRON-BASED ANALYSIS TECHNIQUES

11.5.1 SCANNING ELECTRON MICROSCOPY

A major advantage of the scanning electron microscope (SEM) when compared to other microscopic analysis methods is that it produces topographical images that are easily interpreted without an understanding of the theory behind the instrumentation, that is, colloids look as we think they should. SEM combined with energy dispersive spectrometry (EDS), that is, x-ray microanalysis, also commonly known as energy dispersive x-ray analysis, has been used as the primary discrete-particle analysis technique to characterize mobile groundwater colloids generated in field and laboratory studies, and colloidal particulates suspended in surface water as well.[7,24,29,30,35,37,39,51,93] A curious feature in many published micrographs from colloid transport studies is the common presence of particles that are much larger (>1 μm) than one would expect to be mobile in the subsurface environment. In many cases, the large particles may be artifacts associated with sampling, but their presence may also reflect resolution biases inherent in SEM analysis, and perhaps the tendency to fixate on interesting visual anomalies rather than regions of the sample that reflect the more mundane common particle morphology (i.e., aesthetic bias).

Typically, colloidal suspensions are deposited on polycarbonate filters immediately following sampling to reduce post-collection precipitation or aggregation artifacts and mimic the operational procedures used to define dissolved and particulate sample components. Obviously, a well-dispersed dilute sample should be used to avoid artifactual particle overlap. Filter sections are then mounted on SEM stubs and carbon or metal coated (Au/Pd) for compositional analysis and micrographic imaging, respectively. The tedious nature of operator controlled analysis of particulate samples in the SEM and the TEM has often precluded the acquisition of statistically valid information regarding suspension composition.

SEM images are generated by rastering a focused electron probe across the surface of the specimen while measuring various signals as a function of beam position. Elastic and inelastic electron scattering combine to limit the penetration of the beam within the sample, resulting in a region known as the "interaction volume," the dimensions of which (~1 μm^3) are greater than that of the focused probe.[94] Various signals responsible for image generation, that is, backscattered (BSE) and secondary electrons (SE), and "characteristic" x-rays indicative of sample composition are generated within this interaction volume (Figure 11.7(a)). An understanding of the size and shape of the interaction volume for a given specimen as influenced by the beam parameters (i.e., instrumentation) is critical in properly interpreting SEM images and the resolution capabilities of x-ray microanalysis.

For a given initial probe diameter, the energy of the electron beam (keV) has a strong influence on the relative size of the interaction volume, with greater electron penetration occurring as the beam energy increases. Electrons may travel significant lateral distances from the impact point before generating SE signals or escaping as BSEs, thus reducing image resolution by contributing to noise. Electron backscattering increases with atomic number (Z) to a degree, and thus, provides additional information about sample composition, as well as local topography.[94]

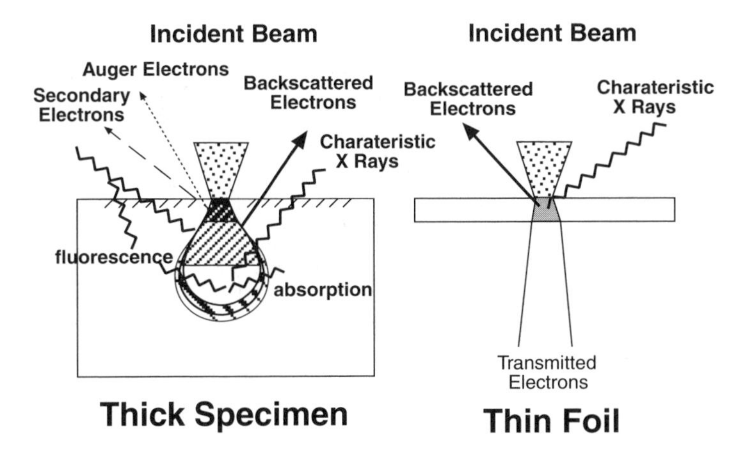

FIGURE 11.7 Electron/sample interactions for (a) conventional and (b) thin-foil mounting techniques in the SEM. (Adapted from Seaman, J.C., *Environ. Sci. Technol.*, 34, 187, 2000. With permission.)

Imaging at low electron voltages reduces beam penetration and ensures that the BSE and SE signals are more indicative of local surface features. However, a reduction in probe size necessary to achieve high resolution reduces the beam current within that probe, limiting the intensity of the BSE and SE signal (contrast) and the effective spot size that can be achieved for a given electron source. The brightness of an electron source, current density per solid angle, is an indicator of its ability to maintain current within a finely focused probe (Table 11.1); therefore, resolving power differs for various electron sources, a factor that most studies of environmental colloids fail to recognize when reporting instrumental operating conditions.

To illustrate the importance of emission source, synthetic goethite particles deposited on 0.1-μm pore-size polycarbonate filters (150,000 × mag.) were imaged using both a JEOL 6310 and a JEOL 6320F microscope equipped with thermoionic LaB_6 and field-emission (FE) electron sources, respectively (Figure 11.8). The

TABLE 11.1
Brightness and Stability of Electron Sources at 20 keV

Source Type	Brightness	Energy Spread (eV)	Current Stability (%)
Thermoionic			
Tungsten	10^5	1–3	1
LaB_6	10^6	1–2	1
Field Emission	10^8	0.3–1	2–5

Source: From Goldstein, J.I. et al., *Scanning Electron Microscopy and X-ray Microanalysis,* 2nd ed., New York, 1992. Reprinted with permission of Kluwer Academic/Plenum Publishers.

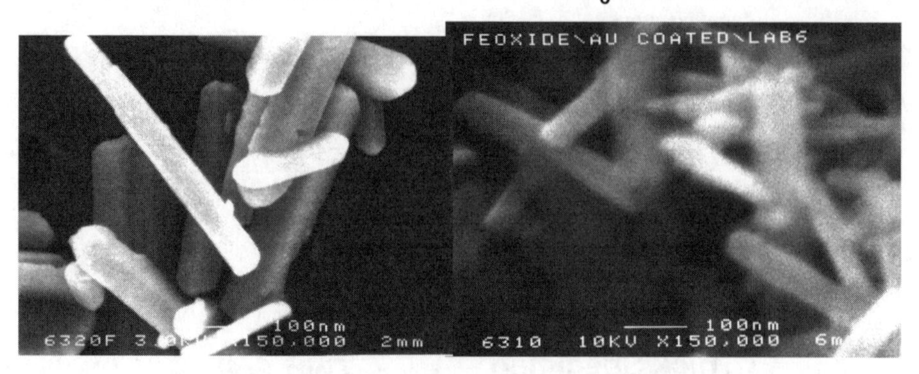

FIGURE 11.8 Electron micrographs of synthetic goethite imaged using two different electron sources, (a) field emission, and (b) LaB_6, at the same magnification level.

enhanced resolving power of the FE instrument is derived from both the finer spot size and the ability to image the sample at lower excitation voltages, 3 compared to 10 keV for the FE and LaB_6 instruments, respectively, with image signals originating from surficial regions of the sample closer to the initial probe location.

The influence of sample coating on SEM imaging and EDS analysis has frequently been neglected. A fraction of the beam electrons lose their initial high energy and are captured within the specimen. If the sample is a conductor, the charge will flow to ground; however, if the sample is a poor conductor, such as silicate clay minerals and polycarbonate filters, or if the path to ground is broken or inadequate due to poor contact between irregular-shaped particulates and the underlying substrate, excess charge accumulates. Sample charging can result in artifactual contrast features, beam aberrations that contribute to noise, and particle loss due to electrostatic repulsion with the incident beam.[94,95] Operation at low-beam energies reduces the potential for charging, but soil clays generally require a thin coating of a conductive substance to eliminate charging artifacts. Metal coating (e.g., Au/Pd) can increase both the SE and BSE signals for high-resolution imaging, but may interfere with performing microanalysis by increasing electron backscattering at the expense of specimen x-rays while producing x-rays characteristic of the metal coating. Therefore, a carbon coating that does not significantly interfere with x-ray generation is typically applied to nonconducting samples for EDS analysis. To illustrate this effect, mobile colloids generated in a series of column studies were deposited on a 0.1-μm pore size polycarbonate filter, a portion of which was mounted to two separate SEM stubs.[7,93] One stub was coated with evaporated carbon and the other was sputter-coated with Au/Pd prior to imaging with a Hitachi S 800 SEM equipped with a field emission electron source. The poor resolution observed for the carbon-coated sample illustrates an important limitation to the use of SEM for the analysis of colloids, with submicron particles becoming less distinct and appearing somewhat amorphous (Figure 11.9(a) and (b)).

Although the impact of sample coating and beam intensity on image resolution is quite obvious, there are more subtle implications to the x-ray microanalysis of

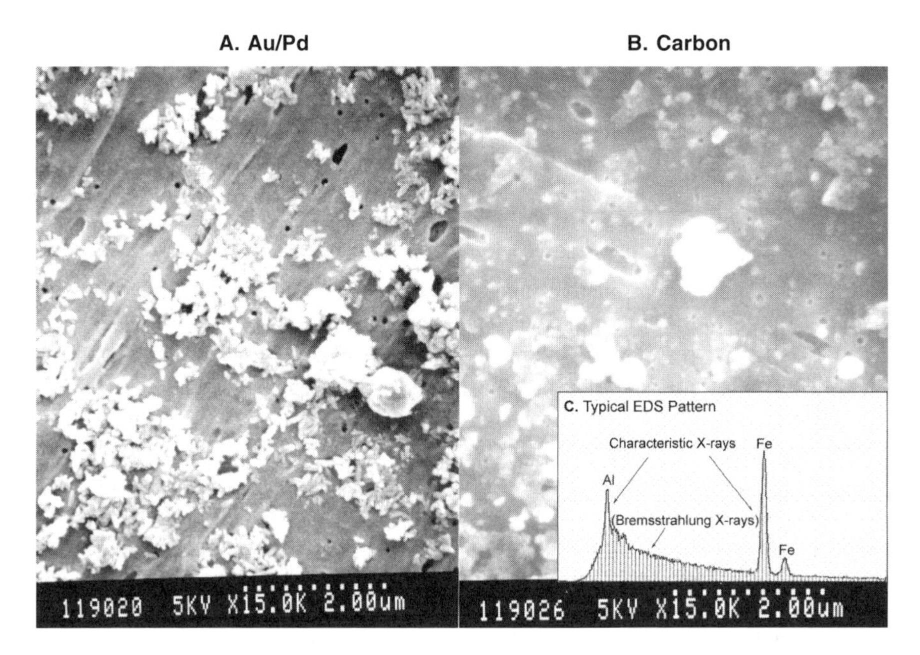

FIGURE 11.9 Electron micrograph images of mobile colloids collected in a series of column studies, deposited on polycarbonate filters, and then (a) metal (AuPd) or (b) carbon coated prior to imaging and EDS analysis with (c) a field-emission SEM.

submicron-sized particulates.[93,96] X-rays are produced in the SEM as a consequence of inelastic scattering events, with the resulting spectra consisting of two distinct components: "characteristic" x-rays indicative of the atomic composition of the beam–specimen interaction volume with sharply defined energy values, and a nonspecific background continuum from zero to the beam energy, known as Bremsstrahlung x-rays (Figure 11.9(c)). To generate an x-ray signal, the electron beam voltage must be greater than the x-ray energy for that sample element of interest (i.e., overvoltage). For example, x-ray lines indicative of the elements beryllium to uranium are found in the energy range of 0.1 to 14 keV.[94] Typically, a beam energy of 20 keV is used for EDS, which results in greater beam penetration and the generation of x-rays from other regions of the bulk sample or instrument chamber, yielding less x-ray intensity for smaller particles than would be predicted for a bulk material with the same composition and an increase in the Bremsstrahlung background with decreasing particle size.

Using a smaller electron probe does not improve resolution during microanalysis because electron scattering and penetration prior to x-ray generation is more a function of beam current.[94,97] Carbon sample coating necessary to perform x-ray analysis because of the higher potential for charging at 20 keV further limits visual and photomicrographic resolution, making it difficult to resolve small particles against the substrate or distinguish primary particles and aggregates. In essence, the resolution displayed in published micrographs of metal-coated samples is usually far superior to the resolution achieved during microanalysis. EDS patterns collected on C-coated samples cannot be directly related to published micrograph images. Therefore, x-ray microanalysis can be biased in favor of larger particles or aggregates, and elements representing multiple discrete solid phases may be incorrectly attributed to one mineral phase.

EDS analysis of environmental colloids has typically been restricted to qualitative identification of major elements with rough estimates of composition based on relative peak intensities.[7,9,24,37,39,98] With few exceptions,[7,24,39,93,99] x-ray spectra representing assigned particle categories are rarely included in the final reviewed publication, making it impossible to critically evaluate the degree of noise present within a spectra, suggest other mineral categories that account for observed element ratios, or assess potential resolution problems. Heavy filter loadings commonly observed in published photomicrographs are more likely to yield x-ray data indicative of multiple particles or aggregates.

Analysis schemes developed for identifying clay minerals in the TEM based on EDS spectra (e.g., Murdoch et al.[100]) are inappropriate for colloidal samples dispersed on polycarbonate filters due to complications associated with the various sample–beam–substrate interactions that differ dramatically from that of ideal samples or standards with smooth polished surfaces.[94,96,101,102] Correction procedures that account for the influence of particle size and morphology on x-ray spectra have been widely available for some time,[101,102] but these techniques have not been applied to the analysis of environmental particulates. To overcome the limitation of quantitative elemental analysis, some research groups have compared the x-ray spectra for sample colloids to the spectra for various minerals of similar size and composition under the same instrumental and sample preparation conditions to "calibrate" instrumental response.[7,24,93] Noting the resolution problems associated with SEM analysis of submicron colloids, several research groups have chosen TEM as the primary discrete particle analysis technique,[21,52,103,104] or have combined TEM analysis techniques, such as electron diffraction and x-ray microanalysis, to confirm conclusions drawn from SEM surveys.[7,93,105]

11.5.2 Automated SEM Techniques: Removing Instrument and Operator Biases

One method to increase spatial resolution during EDS analysis is to reduce the generation of x-rays by electrons that have passed through the specimen by using a modified analysis architecture similar to the TEM, known as thin-foil analysis (Figure 11.7(b)). A carbon foil supported on a metal grid, like a common TEM mount, is used as a sample mount in the SEM. Electrons passing through the sample are captured in a deep-hole graphite trap mounted below the sample, and thus, fail to contribute to the Bremsstrahlung background in the spectra.[93,94,96]

As an example of thin-foil analysis in the SEM, samples of the mineral standard kaolinite (<2 μm fraction, KGa-2, Source Clay Minerals Repository) were dispersed on TEM grids and analyzed using a JEOL JSM 6400 SEM under the control of an automated Noran Voyager EDS system equipped with a thin-window EDS detector.[93] For comparison, EDS spectra were also collected using the same instrument for the kaolinite size fractions deposited on polycarbonate filters, fixed to SEM sample stubs using carbon tape, and then carbon coated, a procedure that represents conventional colloid analysis in the SEM. One advantage of the thin foil method is that the BSE can be used to distinguish particles based on the atomic-weight contrast from the low-molecular-weight carbon substrate; thus, particles appear as bright featureless regions against the dark substrate background. The resulting high contrast image is ideal for automated particle recognition to estimate particle size and morphology using various image-analysis techniques. A reduction in the spectral background that is less dependent on particle size can be seen in Figure 11.10(a) for kaolinite. In contrast, a significant Bremsstrahlung background that increases with decreasing particle size can be seen for kaolinite dispersed on the polycarbonate grid (Figure 11.10(b)).

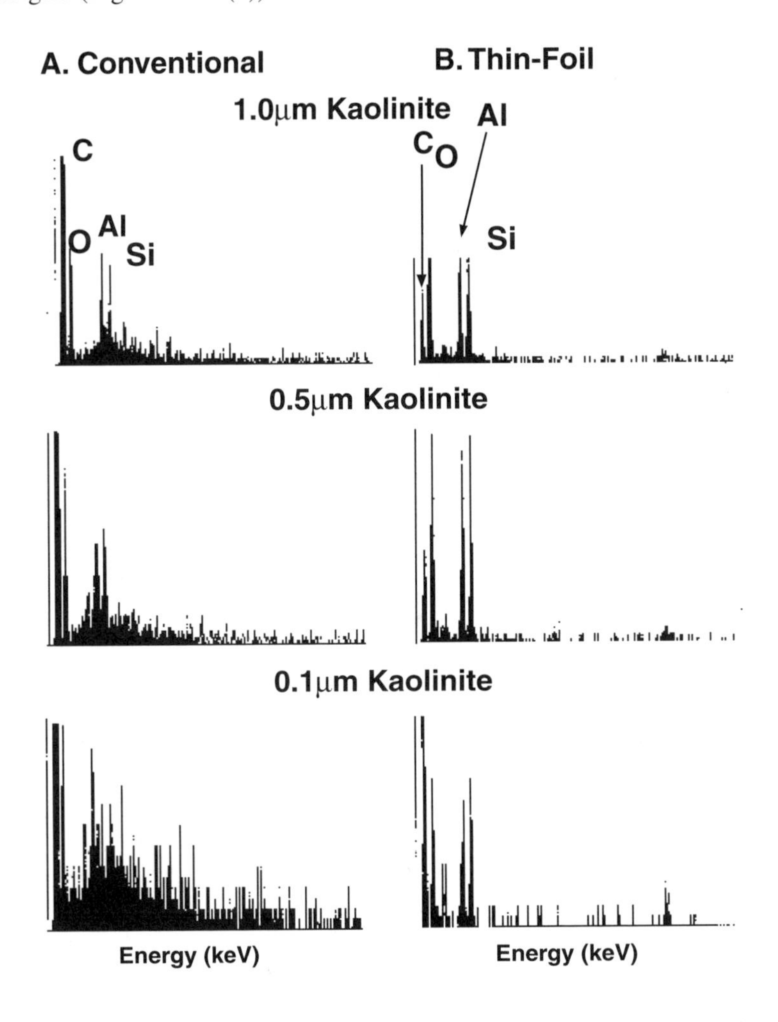

FIGURE 11.10 Conventional and thin foil EDS patterns for kaolinite analyzed using a LaB_6 SEM. (Adapted from Seaman, J.C., *Environ. Sci. Technol.*, 34, 187, 2000. With permission.)

Instead of a deep hole graphite trap, Laskin and Cowin[96] employed a copper foil with a small hole placed directly below the sample such that transmitted electrons pass through to the chamber floor, where they spread over such a wide angle that the foil serves to shield any residual signal from the electron and x-ray detectors located above the sample. The primary advantage of such a system is that it reduces the background carbon signal when analyzing the carbon content of particles or other low-molecular-weight compounds. Laskin and Cowin[96] also note that utilization of the SE signal may be necessary when performing automated analysis of low-atomic-number materials, as they may not yield sufficient BSE for particle recognition against the background.

Clearly, thin-foil analysis offers several advantages over conventional techniques when analyzing particulate samples in the SEM, such as (1) superior BSE contrast that improves particle recognition and particle sizing during automated analyses; (2) superior resolution during the microanalysis of submicron particles due to reduced spurious electron scattering; (3) greater signal to noise ratio during x-ray microanalysis of submicron particles (i.e., less Bremsstrahlung x-rays), improving the quantitative nature of EDS; (4) less beam damage for sensitive materials (i.e., carbon-, sulfur-, and phosphorus-containing compounds) due to reduced electron deposition within the sample; (5) reduces the need to correct x-ray data for the size and shape of small particles (i.e., improved elemental quantification); (6) requires little or no coating to reduce beam charging as less electron energy is deposited in the sample; and (7) particle recognition and analysis are similar for most SEM instruments, regardless of emission source.[93,94,96] A major drawback to the thin-foil method is the inability to collect high-quality images, a major reason why most researchers choose SEM over TEM techniques. Even so, characterization in the TEM is likely to be the best method for defining assigned particle categories by EDS and selective-area electron diffraction and identifying common aggregate types that may not be readily obvious in the SEM.[7]

11.5.3 Transmission Electron Microscopy

The transmission electron microscope can be used to determine the morphology of particles down to 10 nm, resolve crystal spacings and single-crystal electron diffraction patterns, and perform chemical analysis by measuring the x-ray spectra emitted by the specimen in the electron beam, that is, EDS. Thus, the TEM can provide valuable information concerning the size, shape, structure, and composition of submicron-size particulate specimens. Excitation voltages from 100 to 300 keV have typically been used for the study of colloidal particulates and clay minerals. Gilkes[106] provides an extensive review of the application of TEM to the study of soil minerals. Interaction of the electron beam with the specimen in the TEM produces the same x-ray signals observed in the SEM: characteristic x-rays indicative of sample composition overlying a continuum x-ray background (i.e., Bremsstrahlung radiation). The principal advantage of EDS in the TEM is the greater spatial resolution resulting from the smaller interaction volume generated for thin samples at high excitation voltages and lower background signal emitted by the relatively electron-transparent substrate (carbon film). Unlike the SEM, lateral interaction volumes for thin TEM samples approach the actual dimensions of the focused probe.

Bright-field, dark-field, and high-resolution images are commonly produced in the TEM; however, analysis and imaging require extremely thin specimens and yield images that are more difficult to interpret than SEM micrographs. Fortunately, the physicochemical filtration processes that limit the mobility of soil and groundwater colloids yield submicron particulates with dimensions along at least one axis that are generally compatible with TEM analysis without prior fixation and sectioning. In bright-field imaging, electrons scattered at higher angles by the sample, particularly those scattered by diffraction, are restricted by placing a small aperture in the focal plane of the objective lens, yielding an image of the transmitted electrons that appears as dark shadows on a bright background (Figure 11.11(a)). Areas of high electron density scatter a higher fraction of the electron beam and thus appear dark, while areas of low electron density appear light. Therefore, contrast in the specimen image reflects differences in composition, thickness, and diffraction conditions that control the degree of incident beam scattering. In contrast, the dark-field image is generated by positioning the objective aperture to exclude the transmitted beam and include one or more diffracted beams (Figure 11.11(b)). The resulting image reflects only portions of the crystal contributing to the diffracted beams and appears as bright regions of the specimen on a dark background. For high-resolution (phase contrast)

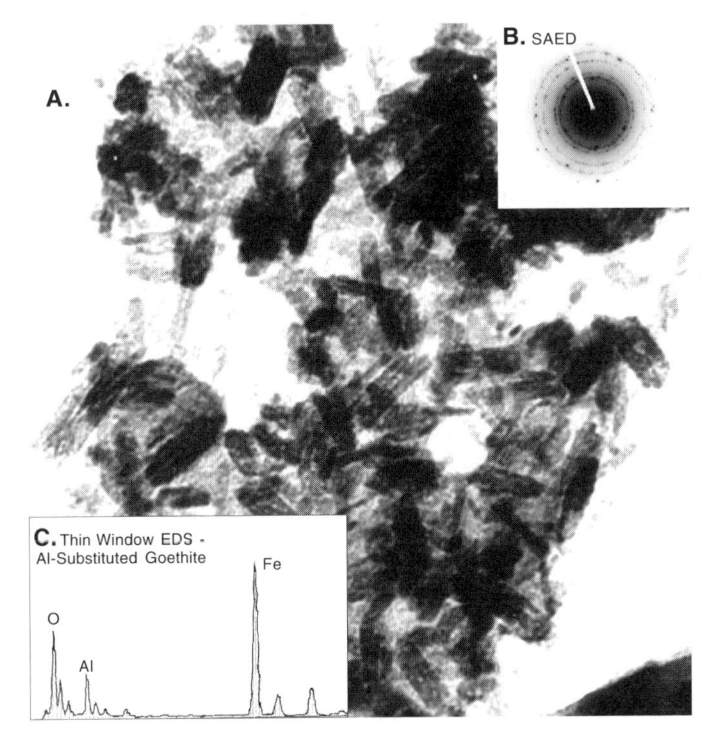

FIGURE 11.11 (a) TEM image, (b) selective-area electron diffraction pattern, and (c) EDS spectra for synthetic Al-substituted goethite.

images, several of the diffracted beams are recombined with the transmitted electrons to produce a periodic fringe image resolving the individual lattice plane. Electrons are more strongly diffracted than x-rays; thus, a very small crystalline sample, such as single colloidal particulates, yields diagnostic electron-diffraction patterns. However, the much shorter wavelength of electrons, 0.0037 nm for a 100-keV electron source compared to 0.15 nm for CuKα x-rays, results in diffraction angles that are much smaller than those observed for x-ray diffraction, typically $<1°$. Buffle and Leppard[21] outlined an extensive sample preparation sequence specifically developed for TEM analysis of environmental colloids. In addition to the general characterization of colloidal materials, TEM has been used to confirm PCS results for sized clay separates,[47] and the effectiveness of various size fractionation techniques.[105,107]

11.6 OTHER ANALYTICAL METHODS

A concise summary of the advantages and limitations for each of the analytical techniques discussed in this chapter is provided in Table 11.2. Jackson[108] provides a comprehensive practical discussion of numerous standard soil analysis procedures, including an extensive discussion of centrifugation procedures for fractionating soil clays. Also discussed in Jackson,[108] thermal gravimetric analysis can provide quantitative mineral analysis for samples containing appreciable amounts of goethite, gibbsite, and/or kaolinite, because of their well-defined thermal weight loss properties and the modest sample requirement, that is, a few milligrams.[7,9] For additional information regarding other analytical methods, see Amonette[109] for a very concise review of analytical methods that are commonly used to characterize clay minerals, and McCarthy and Degueldre[110] for a comprehensive list of analytical techniques that have been applied to the study of mobile soil and groundwater colloids, including a brief description of the advantages and limitations inherent to each technique.

11.7 CONCLUSIONS

Despite potential sampling artifacts, the study of mobile soil and groundwater colloids provides the unique opportunity to characterize complex mineral/organic assemblages found in nature with minimal disturbance when compared to other separation and fractionation schemes, such as the dispersion methods commonly used for particle size analysis. However, the inherent diversity and instability of environmental suspensions, with smaller particles often present in much greater numbers, but comprising less total mass than the larger particles, severely hampers characterization efforts. Many commonly used analytical techniques, including some of those discussed in this chapter, are inherently biased in favor of the larger materials that represent less of the reactive surface area for contaminant sorption, and in many instances may be artifacts of the actual sampling and characterization process. Coagulation, unforeseen changes in redox potential, and even microbial activity can induce chemical and physical changes to the system during sample collection and storage. Issues associated with sampling and storage, factors not addressed to any

TABLE 11.2
Summary of Advantages and Limitations inherent to Analytical Techniques Discussed in This Chapter

Instrumental Technique	Analytical Principle	Information Provided	Advantages/Limitations
Turbidity	Beam attenuation due to light scattering	Suspension concentration	Dilute suspensions Subject to quenching at high colloid loads
Dynamic light scattering (photon correlation spectroscopy, PCS)	Dynamic light-scattering diffusion coefficient	Particle size for dilute stable suspensions	Little/no sample prep Poor resolution for polydisperse suspensions
Laser doppler velocimetry	Doppler shift in scattered laser photons induced by charged particles in an electric field	Electrophoretic mobility/zeta potential	Dilute stable suspensions
Acoustic particle sizing	Acoustic attenuation	Particle size	Large sample volume/mass Requires significant information about the sample
Electroacoustics	Polarization of electrical double layer (EDL) in response to applied acoustic or electric field	Zeta potential	Ability to stir/agitate sample during analysis Applicable to concentrated suspensions
Scanning electron microscopy (SEM)	Imaging: backscattered and secondary electrons Energy-dispersive spectrometry (EDS): composition based on characteristic x-ray emission induced by electron beam	Colloid images (particle size and aggregate associations) and composition	Micrographic resolution dependent on electron beam source, excitation voltage, and sample coating Poor visual resolution during EDS analysis
Automated SEM thin-foil technique	Particle recognition by BSE EDS: composition based on characteristic x-ray emission induced by electron beam	Particle size, aggregate associations and composition	Greater spatial resolution Better suited than conventional SEM for automation Better signal-to-noise ratio for submicron particulates Subject to filtration artifacts

TABLE 11.2 (Continued)
Summary of Advantages and Limitations inherent to Analytical Techniques Discussed in This Chapter

Instrumental Technique	Analytical Principle	Information Provided	Advantages/Limitations
Transmission electron microscopy (TEM)	High keV electrons transmitted through sample	Imaging, particle size, composition by EDS Particle identification by electron diffraction	Tedious Subject to filtration artifacts Superior resolution (imaging and EDS) when compared to SEM Electron diffraction for individual particulates
Flow field flow fractionation (Fl-FFF)	Cross-flow–based particle separation	Size-based separation of suspensions Coupled to detector for compositional analysis of separates	Sample dilution during FFF
Sedimentation field flow fractionation (Sd-FFF)	Centrifuge-based particle separation	Coupled to detector for compositional analysis of separates	Sample dilution during FFF

great degree in this chapter, are just as important as the instrumental/analytical techniques that one employs.

In addition to sampling artifacts, analysis of colloidal suspensions is subject to numerous limitations, some of which are logistical in nature, such as the lack of general access to newer instrumentation and in some cases the expertise and experience to utilize it properly. Those with greater experience in a given instrumental technique often lack specific knowledge regarding the site in question or experimental objectives for which they are analyzing samples. Therefore, it is essential that environmental researchers develop at least a basic understanding of the theory behind a given analytical technique, even when they will not be directly involved in conducting the analysis.

11.8 ACKNOWLEDGMENTS

This research was supported by Financial Assistance Award Number DE-FC09–96SR18546 from the U.S. Department of Energy to The University of Georgia Research Foundation. We are grateful to Angel Kelsey-Wall, Jane Logan, and Jennifer Derrick for their assistance in manuscript preparation.

11.9 LIST OF SYMBOLS

c	Concentration
d	Hydrodynamic diameter
E	Electric field strength (V cm^{-1})
I_o	Incident beam intensity
I_l	Transmitted beam intensity
I_{sc}	Scattered beam intensity
k	Boltzmann constant
k_{turb}	Turbidimetric proportionality factor
k_{sc}	Scattering proportionality factor
l	Length
m	Mass
M_w	Molecular weight
mol	Mole
n	Relative refractive index ratio (n_p/n_o)
n_p	Refractive index for the particle
n_o	Refractive index of suspending media
N_p	Number of particles per unit volume
r	Spherical particle radius
R	Sd-FFF retention ratio
t	Time
T	Absolute temperature
u	Electrophoretic mobility (μm cm s^{-1} V^{-1})
v_e	Terminal velocity of particle in an electric field
v_r	Fl-FFF cross flow velocity
V	Volts
V_r	Retention volume
w	Channel thickness

11.10 LIST OF GREEK SYMBOLS

ρ_p	Particle density
ρ_o	Liquid density
λ	Wavelength of light
τ	Turbidimetric coefficient
α	Polarizability
θ	Scattering angle
κ	Thermal conductivity
δ_V	Viscous depth
δ_t	Thermal depth
η	Viscosity
ω	Sound frequency

REFERENCES

1. Zachara, J.M. et al., Cadmium sorption to soil separates containing layer silicates and iron and aluminum oxides, *Soil Sci. Soc. Am. J.*, 56, 1074, 1992.
2. Golden, D.C. and Dixon, J.B., Silicate and phosphate influence on kaolin-iron oxide interactions, *Soil Sci. Soc. Am. J.*, 49, 1568, 1985.
3. Anderson, P.R. and Benjamin, M.M., Surface and bulk characteristics of binary oxide suspensions, *Environ. Sci. Technol.*, 24, 692, 1990.
4. Anderson, P.R. and Benjamin, M.M., Effects of silicon on the crystallization and adsorption properties of ferric oxides, *Environ. Sci. Technol.*, 19, 1048, 1985.
5. Davis, J.A. and Kent, D.B., Surface complexation modeling in aqueous geochemistry, *Rev. Mineralogy*, 23, 177, 1990.
6. Leppard, G.G., Evaluation of electron microscopic techniques for the description of aquatic colloids, in *Environmental Particles*, Buffle, J. and van Leeuwen, H.P., Eds., Lewis Publishers, Ann Arbor, MI, 1992.
7. Seaman, J.C., Bertsch, P.M., and Strom, R.N., Characterization of colloids mobilized from southeastern coastal plain sediments, *Environ. Sci. Technol.*, 31, 2782, 1997.
8. Bertsch, P.M. and Seaman, J.C., Characterization of complex mineral assemblages: Implications for contaminant transport and environmental remediation, *Proc. Natl. Acad. Sci.*, 96, 3350, 1999.
9. Kaplan, D.I. et al., Soil-borne colloids as influenced by water flow and organic carbon, *Environ. Sci. Technol.*, 27, 1193, 1993.
10. Elimelech, M. et al., Relative insignificance of mineral grain zeta potential to colloid transport in geochemically heterogeneous porous media, *Environ. Sci. Technol.*, 34, 2143, 2000.
11. Song, L., Johnson, P.R., and Elimelech, M., Kinetics of colloid deposition onto heterogeneously charged surfaces in porous media, *Environ. Sci. Technol.*, 28, 1164, 1994.
12. Seaman, J.C., Bertsch, P.M., and Miller, W.P., Ionic tracer movement through highly weathered sediments, *J. Contam. Hydrol.*, 20, 127, 1995.
13. Boggs, M.J. and Adams, E.E., Field study in a heterogeneous aquifer 4. Investigation of adsorption and sampling bias, *Water Resour. Res.*, 28, 3325, 1992.
14. McMahon, M.A. and Thomas, G.W., Chloride and tritiated water flow in disturbed and undisturbed soil cores, *Soil Sci. Soc. Am. Proc.*, 38, 727, 1974.
15. Kretzschmar, R. et al., Mobile subsurface colloids and their role in contaminant transport, *Adv. Agron.*, 66, 121, 1999.
16. Ryan, J.N. and Elimelech, M., Colloid mobilization and transport in groundwater, *Colloids Surfaces A Physicochem. Eng. Aspects*, 107, 1, 1996.
17. Backhus, D.A. et al., Sampling colloids and colloid-associated contaminants in ground water, *Ground Water*, 31, 466, 1993.
18. Buffle, J., Perret, D., and Newman, M., The use of filtration and ultrafiltration for size fractionation of aquatic particles, colloids, and macromolecules, in *Environmental Particles*, Buffle, J. and van Leeuwen, H.P., Eds., Lewis Publishers, Ann Arbor, MI, 1992, chap. 5.
19. Honeyman, B.D. and Santschi, P.H., The role of particles and colloids in the transport of trace radionuclides and trace metals in oceans, in *Environmental Particles*, Buffle, J. and van Leeuwen, H.P., Eds., Lewis Publishers, Ann Arbor, MI, 1992, p. 379.
20. Buffle, J. and Leppard, G.G., Characterization of aquatic colloids and macromolecules. 1. Structure and behavior of colloidal materials, *Environ. Sci. Technol.*, 29, 2169, 1995.

21. Buffle, J. and Leppard, G.G., Characterization of aquatic colloids and macromolecules. 2. Key role of physical structures on analytical results, *Environ. Sci. Technol.*, 29, 2176, 1995.
22. Ryan, J.N., Groundwater Colloids in Two Atlantic Coastal Plain Aquifers: Colloid Formation and Stability, M.S. thesis, Massachusetts Institute of Technology, 1988.
23. Cusack, F. et al., Enhanced oil recovery—Three-dimensional sandpack simulation of ultramicrobacteria resuscitation in reservoir formation, *J. Gen. Microbiol.*, 138, 647, 1992.
24. Gschwend, P.M. and Reynolds, M.D., Monodisperse ferrous phosphate colloids in an anoxic groundwater plume, *J. Contam. Hydrol.*, 1, 309, 1987.
25. Ledin, A. et al., Measurement *in situ* of concentration and size distribution of colloidal matter in deep groundwater by photon correlation spectroscopy, *Water Res.*, 28, 1539, 1994.
26. McCarthy, J.F. and Wobber, F.J., *Manipulation of Groundwater Colloids for Environmental Restoration*, Lewis Publishers, Ann Arbor, MI, 1993.
27. Nightingale, H.I. and Bianchi, W.C., Ground-water turbidity resulting from artificial recharge, *Ground Water*, 15, 146, 1977.
28. Seaman, J.C., Bertsch, P.M., and Miller, W.P., Chemical controls on mobile colloid generation and transport in a sandy aquifer, *Environ. Sci. Technol.*, 29, 1808, 1995.
29. Ryan, J.N. and Gschwend, P.M., Effect of solution chemistry on clay colloid release from an iron oxide coated aquifer sand, *Environ. Sci. Technol.*, 28, 1717, 1994.
30. Bunn, R.A. et al., Mobilization of natural colloids from an iron oxide-coated sand aquifer: Effect of pH and ionic strength, *Environ. Sci. Technol.*, 36, 314, 2002.
31. McCarthy, J.F., McKay, L.D., and Bruner, D.D., Influence of ionic strength and cation charge on transport of colloidal particles in fractured shale saprolite, *Environ. Sci. Technol.*, 36, 3735, 2002.
32. Khaleel, R., Yeh, T.C.J., and Lu, Z., Upscaled flow and transport properties for heterogeneous unsaturated media, *Water Resour. Res.*, 38, 11/1, 2002.
33. Johnson, C.R. et al., Colloid mobilization in the field using citrate to remediate chromium, *Ground Water*, 39, 895, 2001.
34. Seaman, J.C. and Bertsch, P.M., Selective colloid mobilization through surface-charge manipulation, *Environ. Sci. Technol.*, 34, 3749, 2000.
35. Gschwend, P.M. et al., Mobilization of colloids in groundwater due to infiltration of water at coal ash disposal site, *J. Contam. Hydrol.*, 6, 307, 1990.
36. Ronen, D. et al., Characterization of suspended particles collected in groundwater under natural gradient flow conditions, *Water Resour. Res.*, 28, 1279, 1992.
37. Ryan, J.N. and Gschwend, P.M., Colloid mobilization in two Atlantic coastal plain aquifers: Field studies, *Water Resour. Res.*, 26, 307, 1990.
38. Swartz, C.H. and Gschwend, P.M., Mechanisms controlling release of colloids to groundwater in a southeastern coastal plain aquifer sand, *Environ. Sci. Technol.*, 32, 1779 , 1998.
39. Kaplan, D.I. et al., Application of synchrotron x-ray fluorescence spectroscopy and energy dispersive x-ray analysis to identify contaminant metals on groundwater colloids, *Environ. Sci. Technol.*, 28, 1186, 1994.
40. Rees, T.F., Comparison of photon correlation spectroscopy with photosedimentation analysis for the determination of aqueous colloid size distributions, *Water Resour. Res.*, 26, 2777, 1990.
41. Rees, T.F., A review of light-scattering techniques for the study of colloids in natural waters, *J. Contam. Hydrology*, 1, 425, 1987.

42. Hunter, R.J., *Introduction to Modern Colloid Science*, Oxford Science Publications, Oxford, 1993.
43. Schurtenberger, P. and Newman, M.E., Characterization of biological and environmental particles using static and dynamic light scattering, in *Environmental Particles*, Buffle, J. and van Leeuwen, H.P., Eds., Lewis Publishers, Ann Arbor, MI, 1993, p. 426.
44. Hunter, K., Application of active film multipliers in ICP-MASS spectrometry, 2nd Australian Symposium on Applied ICP-Mass Spectrometry, Brisbane, Australia, May 17-19, 1993.
45. Atkins, P.W., *Physical Chemistry*, 4th ed., W.H. Freeman, New York, 1990.
46. Ingle, J.D. and Crouch, S.R., *Spectrochemical Analysis*, Prentice Hall, Englewood Cliffs, NJ, 1988.
47. Noack, A.G., Grant, C.D., and Chittleborough, D.J., Colloid movement through stable soils of low cation-exchange capacity, *Environ. Sci. Technol.*, 34, 2490, 2000.
48. Swartz, C.H. and Gschwend, P.M., Field studies of *in situ* colloid mobilization in a southeastern coastal plain aquifer, *Water Resour. Res.*, 35, 2213, 1999.
49. Kaplan, D.I., Subsurface Mobile Colloids: Their Surface Characterization, Mineralogy, and Role in Contaminant Transport in a Coastal Plain Aquifer, Ph.D. thesis, University of Georgia, 1993.
50. McCarthy, J. and Shevenell, L., Obtaining representative ground water samples in a fractured and kartistic formation, *Ground Water*, 36, 251, 1998.
51. Ryan, J.N. and Gschwend, P.M., Effects of ionic strength and flow rate on colloid release: Relating kinetics to intersurface potential energy, *Colloid Interface Sci.*, 164, 21, 1994.
52. Ivanovich, M. et al., Natural analogue study of the distribution of uranium series radionuclides between the colloid and solute phases in the hydrogeological system of the Koongarra uranium deposit, Australia, United Kingdom Atomic Energy Authority Doc. No. AERE-R 12975, Oxfordshire, U.K., Jan. 1988.
53. Gallegos, C.L. and Menzel, R.G., Submicron size distributions of inorganic suspended solids in turbid waters by photon correlation spectroscopy, *Water Resour. Res.*, 23, 596, 1987.
54. Kretzschmar, R., Hesterberg, D., and Sticher, H., Effects of adsorbed humic acid on surface charge and flocculation of kaolinite, *Soil Sci. Soc. Am. J.*, 61, 101, 1997.
55. Sposito, G., *The Surface Chemistry of Soils*, Oxford University Press, New York, 1984.
56. Johnston, C.T. and Tombacz, E., Surface chemistry of soil minerals, in *Soil Mineralogy with Environmental Applications*, Dixon, J.B. and Schulze, D.G., Eds., Soil Science Society of America, Madison, WI, 2002, p. 37.
57. Parker, J.C. et al., A critical evaluation of the extentsion of zero point of charge (ZPC) theory to soil systems, *Soil Sci. Soc. Am. J.*, 43, 668, 1979.
58. Sposito, G., Characterization of particle surface charge, in *Environmental Particles*, Buffle, J. and van Leeuwen, H.P., Eds., Lewis Publishers, Ann Arbor, MI, 1992, chap. 7.
59. Harsh, J.B. and Xu, S., Microelectrophoresis applied to the surface chemistry of clay minerals, *Adv. Soil Sci.*, 14, 131, 1990.
60. Thompson, R.G., Practical zeta potential determination using electrophoretic light scattering, *Am. Lab.*, 24, 48, 1992.
61. Giese, R.F. and Oss, C.J.V., *Colloid and Surface Properties of Clays and Related Minerals*, Marcel Dekker, New York, 2001.
62. Bleam, W.F. and McBride, M.B., Cluster formation versus isolated-site adsorption: a study of Mn(II) and Mg(II) adsorption on boehmite and goethite, *J. Colloid Interface Sci.*, 103, 124, 1985.

63. Celi, L. et al., Effects of pH and electrolytes on inositol hexaphosphate interaction with goethite, *Soil Sci. Soc. Am. J.*, 65, 753, 2001.
64. Ognalaga, M., Frossard, E., and Thomas, F., Glucose-1-phosphate and myo-hexaphosphate adsorption mechanisms on goethite, *Soil Sci. Soc. Am. J.*, 58, 332, 1994.
65. Sullivan, E.J., Hunter, D.B., and Bowman, R.S., Topological and thermal properties of surfactant-modified clinoptilolite studied by tapping mode™ atomic force microscopy and high-resolution thermogravimetric analysis, *Clays Clay Min.*, 45, 42, 1997.
66. Xu, S. and Boyd, S.A., Cation exchange chemistry of hexadecyltrimethylammonium in a subsoil containing vermiculite, *Soil Sci. Soc. Am. J.*, 58, 1382, 1994.
67. Hunter, R.J., Recent development in the electroacoustic characterization of colloidal suspensions and emulsions, *Colloids Surfaces*, 141, 37, 1998.
68. Dukhin, A.S. and Goetz, P.J., Acoustic and electroacoustic spectroscopy for characterizing concentrated dispersions and emulsions, *Adv. Colloid Interface Sci.*, *92*, 73–132, 2001.
69. O'Brien, R.W., Cannon, D.W., and Rowlands, W.N., Electroacoustic determination of particle size and zeta potential, *J. Colloid Interface Sci.*, 173, 406, 1995.
70. Dukhin, A.S. and Goetz, P.J., Acoustic and electroacoustic spectroscopy, *Langmuir*, 12, 4336, 1996.
71. Dukhin, A.S. and Goetz, P.J., Characterization of aggregation phenomena by means of acoustic and electroacoustic spectroscopy, *Colloids Surfaces*, 144, 49, 1998.
72. Dukhin, A.S. and Goetz, P.J., Characterization of concentrated dispersions with several dispersed phases by means of acoustic spectroscopy, *Langmuir*, 16, 7597, 2000.
73. Dukhin, S.A., Installation Handbook and User Manual: Model DT-1200 Electroacoustic Spectrometer, Dispersion Technology Inc., Mount Kisco, NY, 2001.
74. Duhkin, A.S. et al., Electroacoustics for concentrated dispersions, *Langmuir*, 15, 3445, 1999.
75. Giddings, J.C., Field-flow fractionation: Analysis of macromolecular, colloidal, and particulate materials, *Science*, 260, 1456, 1993.
76. Giddings, J.C., Field flow fractionation, *Sep. Sci. Technol.*, 19, 831, 1984.
77. Williams, P.S. and Giddings, J.C., Power programmed field-flow fractionation: A new program form for improved uniformity of fractionating power, *Anal. Chem.*, 59, 2038, 1987.
78. Chianea, T., Assidjo, N.E., and Cardot, P.J.P., Sedimentation field-flow-fractionation: Emergence of a new cell separation methodology, *Talanta*, 51, 835, 2000.
79. Beckett, R., Jue, Z., and Giddings, J.C., Determination of molecular weight distributions of fulvic and humic acids using flow field-flow fractionation, *Environ. Sci. Technol.*, 21, 289, 1987.
80. Thang, N.M. et al., Application of the flow field flow fractionation (FFFF) to the characterization of aquatic humic colloids: evaluation and optimization of the method, *Colloids Surfaces*, 181, 289, 2001.
81. Zanardi-Lamardo, E., Clark, C.D., and Zika, R.G., Frit inlet/frit outlet flow field-flow fractionation: Methodology for colored dissolved organic material in natural waters, *Anal. Chim. Acta*, 443, 171, 2001.
82. Benedetti, M. et al., Field-flow fractionation characterization and binding properties of particulate and colloidal organic matter from the Rio Amazon and Rio Negro, *Org. Geochem.*, 33, 269, 2002.
83. Petteys, M.P. and Schimpf, M.E., Characterization of hematite and its interaction with humic material using flow field-flow fractionation, *J. Chromatogr. A*, 816, 145, 1998.

84. Kammer, F.v. and Forstner, U., Natural colloid characterization using flow-field-flow-fractionation followed by multi-detector analysis, *Water Sci. and Technol.*, 37, 173, 1998.
85. Beckett, R., Field-flow fractionation-ICP-MS: A powerful new analytical tool for characterizing macromolecules and particles, *Atomic Spectrosc.*, 12, 228, 1991.
86. Taylor, H.E. et al., Inductively coupled plasma-mass spectrometry as an element-specific detector for field-flow fractionation particle separation, *Anal. Chem.*, 64, 2036, 1992.
87. VanBerkel, J. and Beckett, R., Estimating the effect of particle surface coatings on the adsorption of orthophosphate using sedimentation field-flow fractionation, *J. Liq. Chrom. Rel. Technol.*, 20, 2647, 1997.
88. Chittleborough, D.J., Hotchin, D.M., and Beckett, R., Sedimentation field flow fractionation: A new technique for the fractionation of soil colloids, *Soil Sci.*, 153, 341, 1992.
89. Ranville, J.F. et al., Development of sedimentation field-flow fractionation-inductively coupled plasma mass-spectrometry for the characterization of environmental colloids, *Anal. Chim. Acta*, 381, 315, 1999.
90. Hassellov, M., Lyven, B., and Beckett, R., Sedimentation field-flow fractionation coupled online to inductively coupled plasma mass spectrometry: New possibilities for studies of trace metal adsorption onto natural colloids, *Environ. Sci. Technol.*, 33, 4528–, 1999.
91. Hassellov, M. et al., Determination of continuous size and trace element distribution of colloidal material in natural water by on-line coupling of flow field-flow fractionation with ICPMS, *Anal. Chem.*, 71, 3497, 1999.
92. Du, Q. and Schimpf, M.E., Correction for particle-wall interactions in the separation of colloids by flow field-flow fractionation, *Anal. Chem.*, 74, 2478, 2002.
93. Seaman, J.C., Thin-foil SEM analysis of soil and groundwater colloids: Reducing instrument and operator bias, *Environ. Sci. Technol.*, 34, 187, 2000.
94. Goldstein, J.I. et al., *Scanning Electron Microscopy and X-ray Microanalysis*, 2nd ed., Plenum Press, New York, 1992.
95. Goldstein, J.I., Williams, D.B., and Cliff, G., Quantitative x-ray analysis, in *Principles of Analytical Electron Spectroscopy*, Joy, D.C., Romig, A.D. Jr., and Goldstein, J.I., Eds., Plenum Press, New York, 1986.
96. Laskin, A. and Cowin, J.P., Automated single-particle SEM/EDX analysis of submicrometer particles down to 0.1μm, *Anal. Chem.*, 73, 1023, 2001.
97. Sawhney, B.L., Electron microprobe analysis, in *Methods of Soil Analysis: Part 1—Physical and Mineralogical Methods*, 2nd ed., Klute, A., Ed., American Society of Agronomy, Madison, WI, 1986.
98. Seaman, J.C., Physicochemical and Mineralogical Controls on Colloid Generation and Transport within the Highly Weathered Alluvial Sediments of the Upper Coastal Plain, Ph.D. thesis, University of Georgia, 1994.
99. VanPut, A. et al., Geochemical characterization of suspended matter and sediment samples from the Elbe River by EPXMA, *Water Res.*, 28, 643, 1994.
100. Murdoch, A., Zeman, A.J., and Sandilands, R., Identification of mineral particles in fine grained lacustrine sediments with transmission electron microscope and x-ray energy dispersive spectroscopy, *J. Sediment. Petrology*, 47, 244, 1977.
101. Armstrong, J.T., Methods of quantitative analysis of individual microparticles with electron beam instruments, *Scanning Electron Microscopy/1978*, 1, 455, 1978.
102. Armstrong, J.T. and Buseck, P.R., Quantitative chemical analysis of individual microparticles using the electron microprobe: theoretical, *Anal. Chem.*, 47, 2178, 1975.

103. Perret, D. et al., Electron microscopy of aquatic colloids: Non-perturbing preparation of specimens in the field, *Water Res.*, 25, 1333, 1991.
104. Davey, B.G. et al., Report on electron microscope studies of groundwater colloids from the Koongarra Uranium Ore Body, Northen Territory, Australia, in *Natural Analogue Study of the Distribution of Uranium Series Radionuclides Between the Colloid and Solute Phases in the Hydrogeological System of the Koongarra Uranium Deposit, Australia*, United Kingdom Atomic Energy Authority, Oxfordshire, U.K., 1987.
105. Perret, D. et al., Submicron particles in the Rhine River-1: Physico-chemical characterization, *Water Res.*, 28, 91, 1994.
106. Gilkes, R.J., Transmission electron microscope analysis of soil minerals, in *Quantitative Methods in Soil Mineralogy*, Amonette, J.E. and Zelazny, L.W., Eds., Soil Science Society of America, Madison, WI, 1994, p. 177.
107. Newman, M.E. et al., Submicron particles in the Rhine River. II. Comparison of field observations and model predictions, *Water Res.*, 28, 107, 1994.
108. Jackson, M.L., *Soil Chemical Analysis: Advanced Course*, 2nd ed., M.L. Jackson, Madison, WI, 1979.
109. Amonette, J.E., Methods for determination of mineralogy and environmental availability, in *Soil Mineralogy with Environmental Applications*, Dixon, J.B. and Schulze, D.G., Eds., Soil Science Society of America, Madison, WI, 2002, p. 153.
110. McCarthy, J.F. and Degueldre, C., Sampling and characterization of colloids and particles in groundwater for studying their role in contaminant transport, in *Environmental Particles*, Buffle, J. and van Leeuwen, H.P., Eds., Lewis Publishers, Ann Arbor, MI, 1993, chap. 6.
111. Sullivan, E.J., Hunter, D.B., and Bowman, R.S., Fourier transform raman spectroscopy of sorbed HDTMA and the mechanism of chromate sorption to surfactant-modified clinoptilolite, *Environ. Sci. Technol.*, 32, 1948, 1998.

12 Kinetic Modeling of Sulfate Transport in a Forest Soil

H. Magdi Selim, George R. Gobran, Ximing Guan, and Nicholas Clarke

CONTENTS

12.1 Introduction 311
12.2 Multireaction Kinetic Model 313
  12.2.1 Model Formulation 313
12.3 Sulfate Transport: Case Study 315
  12.3.1 Field Site Description and Soil Sampling 315
  12.3.2 Sequential Leaching 315
12.4 Modeling 317
  12.4.1 Equilibrium Sorption: Linear and Nonlinear 319
  12.4.2 Kinetic Sorption 323
  12.4.3 Model Predictions 323
  12.4.4 BC Layer 325
12.5 Summary 326
12.6 Acknowledgments 328
References 328

12.1 INTRODUCTION

Over the last three decades, two general approaches have been proposed in the literature for describing the interactions of sulfate in soils. The first approach is that of a chemical nature where thermodynamic interrelationships with speciation of cations and anions present in soil solution and the interaction with the soil surface are the major mechanisms. These models may be referred to as chemical models. Examples of such models include that of Cosby et al. (1986), Reuss and Johnson (1986), De Vries et al. (1994), among others. A common feature of these models is that both ion exchange and aluminum hydrolysis reactions are similar. Their capability of quantifying these processes varies according to whether the interactions are

1-56670-623-8/03/$0.00+$1.50

under batch conditions vs. when transport is included. In addition, transport mechanisms can be the simple mixing cell type or the convective-dispersive type where finite difference or element methods for solving simultaneous sets of equations are included (Cosby et al., 1986). A common theme of all these chemical models is the failure to accurately describe the retention and reactions of sulfate in soils. In addition, the sorption of the sulfate in all these chemical models is based on either Freundlich or Langmuir approaches. Both types are empirical where they are included simultaneously with the chemical models. In addition, both approaches assume that equilibrium is valid, that is, that the local equilibrium assumption is dominant. The literature is replete with evidence that sulfate reaction is time dependent (for a review, see Sparks, 1989).

Another class of models disregards the chemical thermodynamic relationships and focuses primarily on empirical approaches in which lumped parameters quantify the fate of sulfate in soils. Such lumped-parameters approaches can be looked at it terms of equilibrium, fully reversible kinetic approaches, and irreversible kinetic reactions. Hodges and Johnson (1987) tested the validity of several models for describing sulfate adsorption and desorption in a Cecil soil. Their test included first-order kinetics, and Elovich and other diffusion equation types. First-order reactions provided only adequate descriptions of results. They also found that sulfate desorption results were very different in terms of relations with soil solution in comparison to adsorption data. Such behavior is commonly referred to as "hysteresis," and may be a result of irreversible reactions as suggested by Hodges and Johnson (1987). Lack of irreversible retention of sulfate has been observed by Gobran et al. (1998b). This observation has also been noted in soils by Harrison et al. (1989) and on oxide mineral surfaces by Turner and Kramer (1992). It should also be noted that deviations between sorption and desorption could also be related to kinetic retention behavior (Gobran et al., 1998a, 1998b). Irreversible sorption/desorption and the extent of kinetic reactions has also been quantified using the pressure-jump relaxation method on goethite by Zhang and Sparks (1990), who found that sulfate adsorption occurs rapidly at initial reaction stages whereas desorption is a slow one and may be considered as a limiting step. Such findings have been observed for other inorganic and heavy metal species present in the soil solution (Selim, 1992).

Other modeling efforts include soil acidification models of the macroscopic type that account for the process of SO_4 sorption in different ways. These approaches, which assume equilibrium conditions to prevail, include the adsorption isotherm, solubility product, and anion exchange. Prenzel (1994) discussed the various limitations of the above approaches in their capability to account for changes in pH. Recently, Fumoto and Sverdrup (2000) used a constant capacitance approach to describe the pH dependency of SO_4 sorption isotherms in an andisol. Other modeling efforts of SO_4 isotherms were reported by Gustafsson (1995) in a spodosol. Such isotherm models are of the equilibrium type and include linear and Temkin types of models.

In this chapter we present a general-purpose transport model of the multireaction type. The model was successfully used to predict the adsorption as well as transport of several heavy metals in soils (Selim, 1992; Hinz and Selim, 1994; Selim and Amacher, 2001). Multireaction models are empirical and include linear and nonlinear equilibrium and reversible and irreversible retention reactions. A major feature of multireaction approaches is that they account for linear as well as nonlinear kinetic reactions of the consecutive as well as the concurrent types. Limitations of the multireaction approach are also presented. In order to illustrate the capability of the multireaction model, we present model predictions of SO_4 breakthrough results (BTCs) from a forest soil system. Specifically, the multireaction model was utilized to describe SO_4 BTCs from a sequential leaching experiment on forest (spodic) soil layers. Input pulse solutions of different $CaSO_4$ concentrations were used in the sequential leaching experiment. Various versions of the multireaction model (equilibrium and kinetic) were needed in order to describe effluent results from the various soil layers.

12.2 MULTIREACTION KINETIC MODEL

12.2.1 Model Formulation

The multireaction approach, often referred to as the multisite model, acknowledges that the soil solid phase is made up of different constituents (clay minerals, organic matter, iron, and aluminum oxides). Moreover, a heavy metal species is likely to react with various constituents (sites) via different mechanisms (Amacher et al., 1988). As reported by Hinz et al. (1994), heavy metals are assumed to react at different rates with different sites on matrix surfaces. Therefore, a multireaction kinetic approach is used to describe heavy metal retention kinetics in soils. The multireaction model used here considers several interactions of one reactive solute species with soil matrix surfaces. Specifically, the model assumes that a fraction of the total sites reacts rapidly or instantaneously with solute in the soil solution, whereas the remaining fraction reacts more slowly with the solute. As shown in Figure 12.1, the model includes reversible as well as irreversible retention reactions that occur concurrently and consecutively. We assumed that a heavy metal species is present in the soil solution phase, C (mg/L), and in several phases representing metal species retained by the soil matrix designated as S_e, S_1, S_2, S_s, and S_{irr} (mg/kg of soil). We further considered that the sorbed phases S_e, S_1, and S_2 are in direct contact with the solution phase (C) and are governed by concurrent reactions. Specifically, C is assumed to react rapidly and reversibly with the equilibrium phase (S_e) such that

$$S_e = K_e C^n \tag{12.1}$$

where K_e is a distribution coefficient (cm³/g), and n is the reaction order (dimensionless). Moreover, n represents a nonlinearity parameter which is commonly less than unity (Buchter et al. 1989). This parameter represents the heterogeneity of sorption sites having different affinities for heavy metal retention on matrix surfaces (Kinniburgh, 1986). The relations between C and the sorbed phases S_1 and S_2 were assumed to be governed by nonlinear kinetic reactions expressed as

$$\frac{\partial S_1}{\partial t} = k_1 \frac{\Theta}{\rho} C^m - k_2 S_1 \tag{12.2}$$

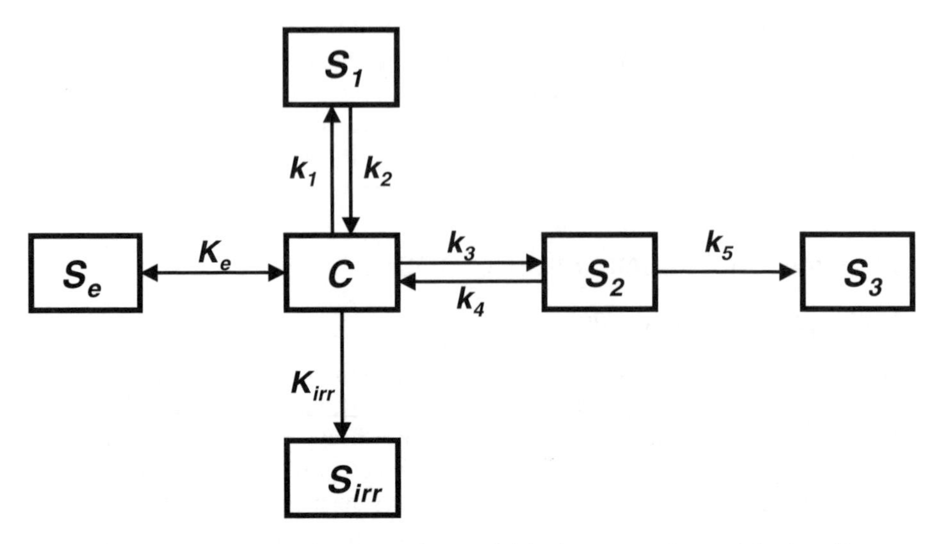

FIGURE 12.1 A schematic of multireaction model for heavy metals reactivity in soils.

$$\frac{\partial S_2}{\partial t} = \left[k_3 \frac{\Theta}{\rho} C^m - k_4 S_2\right] - k_5 S_2 \tag{12.3}$$

where ρ is soil bulk density (g/cm^3), Θ is water content (cm^3/cm^3), t is time (h), k_1 and k_2 are the rate coefficients (h^{-1}) associated with S_1, and m is the reaction order. Similarly, for the reversible reaction between C and S_2, k_3 and k_4 are the respective rate coefficients (h^{-1}). In the above equations, we assume n = m, since there is no known method for estimating n and/or m independently.

The multireaction model accounts for irreversible retention in two ways. First, as a sink term, Q, which represents a direct reaction between the solution phase C and S_{irr} (e.g., precipitation/dissolution or immobilization) as a first-order kinetic process, where k_{irr} is the associated rate coefficient (h^{-1}).

$$Q = \rho \frac{\partial S_{irr}}{\partial t} = k_{irr} \Theta C \tag{12.4}$$

Irreversible retention was also considered to be the result of a subsequent reaction of the S_2 phase into a less accessible or strongly retained phase S_3 such that,

$$\frac{\partial S_3}{\partial t} = k_5 S_2 \tag{12.5}$$

One may regard the slowly reversible phase S_3 as a consequence of rearrangement of solute retained by the soil matrix. Mechanisms associated with irreversible reactions include various types of surface precipitation that account for the formation or sorption of metal polymers on the surface, a solid solution or coprecipitate that involves co-ions dissolved from the sorbent, and a homogeneous precipitate formed on the surface composed of ions from the bulk solution or their hydrolysis products (Farley, Dzombak, and Morel, 1985). The continuum between surface precipitation and chemisorption is controlled by several factors including (1) the ratio of the number of sites to the number of metal ions in solution; (2) the strength of the metal-oxide bond; and (3) the degree to which the bulk solution is undersaturated with respect to the metal hydroxide precipitate. Such mechanisms are consistent with one or more irreversible reactions associated with our model presented in Figure 12.1.

12.3 SULFATE TRANSPORT: CASE STUDY

12.3.1 Field Site Description and Soil Sampling

Soil samples were collected from the Gårdsjön catchment, located at 58° 04' N and 12° 03' E on the west coast of Sweden. The area is covered by coniferous forest with Norway spruce (*Picea abies* (L.) Karst), which dominate the landscape. The area is glacial till, silt loam soil that is fairly thin. The soil is classified as a spodosol haplorthod, common in Sweden (Nilsson, 1988). A detailed description of the geology, soils, and vegetation of the Gårdsjön catchment is given by Olsson et al. (1985).

Soil samples were taken by horizon, that is, forest floor (FF), eluvial (E), and spodic (Bs and BC) horizons, within the control (F1) catchment. Although the uppermost mineral horizon is not well enough developed to be called either an E or A horizon, we classify it here as an E horizon. Samples were obtained by the irregular hole method (Blake and Hartge, 1986) so as to obtain the field bulk density. Moist composite field samples of humus (FF) were passed through a 5 mm-sieve, and mineral soils through a 2-mm sieve. The soils were then kept in black plastic bags at 4°C in field-moist condition during the course of the experiments.

12.3.2 Sequential Leaching

In order to simulate leaching in a natural forest soil, and due to the importance of the interaction between inorganic and organic soil constituents, a sequential leaching experiment was designed, which is illustrated in Figure 12.2. The sequential leaching technique is thought to resemble the field situation, in that the input solution passes through each horizon in turn on its way "down" through the soil. This means that the ions and other dissolved species leached from one horizon are included in the input to the underlying horizon. The leaching columns (4.7 × 5 cm) were made of PVC and were packed with soil from each mineral horizon (E, Bs, and BC). Disturbed soils were used in order to minimize problems due to soil heterogeneity. The bulk density of the soil column was equivalent to that in the field. Leaching solutions were prepared by mixing the FF layer with $CaSO_4$ solutions at two different input concentrations (C_o) of 0.005 M (SI) and 0.0005 M (SII). These concentrations

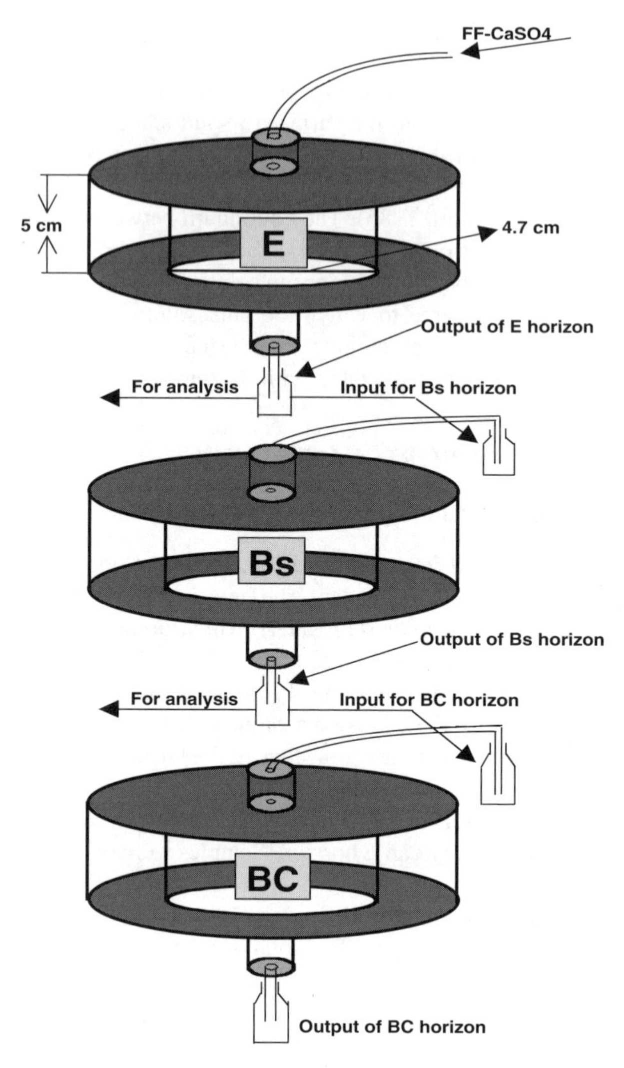

FIGURE 12.2 A schematic of the sequential leaching experiment through 5-cm soil columns from the E, Bs, and BC horizons (FF = forest floor).

were chosen in order to examine the impact of applying the neutral salt $CaSO_4$ to acid forest soils over a wide concentration range. The soil columns were leached with an amount of leaching solution equivalent to at least 1 year of throughfall in Gårdsjön. The concentration of SO_4-S in the SII treatment was similar to that of solution obtained by centrifuging the O horizon of Gårdsjön (Giesler et al., 1996). In the SI treatment, the total amount of S applied was about equivalent to the total amount of SO_4-S added in the form of $(NH_4)_2SO_4$ and S deposition in the $(NH_4)_2SO_4$ treatment in Skogaby, southwest Sweden during 4 years (Bergholm, 1994).

The proportions used in mixing the solutions (SI and SII) and the FF layer varied depending on the bulk density and the pore volume (PV) of the FF layer, which were 0.6 g/ml and 70 ml per column, respectively. As ten PVs were assumed to pass through the FF layer, we mixed the FF layer with the sulfate solutions in the proportion of 2 l of solution to 150 g of FF. After mixing, the leaching solutions were shaken at about 1-h intervals for a period of about 8 h. The solutions were then allowed to stand overnight, after which they were filtered. The chemical characteristics of the leaching solutions are given in Table 12.1.

The soil columns were leached with an amount of leaching solution equivalent to at least 1 year of throughfall at Gårdsjön. All experiments were carried out at a constant temperature of 15±1°C (arbitrarily chosen). Leaching was continuous, at conditions of near soil-water saturation with input solutions at a constant average velocity of 33 cm d^{-1}. This high leaching rate was used for two reasons: (1) the processes under investigation are known to be fast, and (2) to reduce any microbial activity that might occur when the time of contact between the soil and water is prolonged. The filtered leaching solutions were used to leach the soil column containing the E horizon. In order to obtain enough leachate sample at each collection from the first soil column (the E horizon) for analysis and for leaching the Bs horizon and then the BC horizon, we used nine, six, and two soil columns for the E, Bs, and BC horizons, respectively. At each collection, all leachate samples from soil columns of the same horizon were mixed to form a composite sample. Each composite sample from the E horizon was divided into two parts; one was used for analysis and the rest was used as the equivalent input to leach the lower (Bs) horizon and so on (see Figure 12.2), so that ten leachates were obtained from each horizon. The leachates were then filtered through a 0.45-μm membrane filter before analysis. The input leaching solutions and effluent solutions were analyzed for pH, electrical conductivity (EC), dissolved organic carbon (DOC), anions (Cl^-, SO_4^{2-} and NO_3^-) and cations (Ca, Mg, Na, K, total dissolved Al, Fe, Mn, and NH_4^+).

12.4 MODELING

In an effort to describe effluent results obtained from the different soil layers, we utilized various versions of the multireaction model described above. In principle, we based our efforts on the assumption of the miscible displacement approach that describes retention reactions of solutes during transport in porous media (Selim, 1992). Several simplifying assumptions were necessary in order to describe the SO_4 experimental data based on these models. Briefly, we tested the capability of the convection–dispersion (CD) equation to describe the mobility of applied sulfates in individual soil layers where steady-state conditions were assumed.

Experimental constraints were such that the assumptions of saturated and steady flow were not completely met, especially when the pulse was first introduced to each column. Our experimental conditions did not maintain water-saturated conditions initially and the applied pulse was introduced to moist but not fully water-saturated columns. As a result, effluent adjustments in the concentration of sulfate versus pore volume (V/V_o) for the various layers were made in order to reflect the late arrival of a wetting front (outflow) due to the unsaturated condition of the columns. Here V_o is the pore volume associated with individual soil layers. Our adjustments, which were based primarily on the initial moisture conditions for individual columns, were V/V_o

TABLE 12.1
Chemical Characteristics of the (Input) Leaching Solutions Used for the Soil Layers SI, SII, FFSI, and FFSII

Leaching Solution	pH	EC (mS/cm)	DOC (mg/l)	Ca	Cl	SO_4	Fe	Mn	Mg	Al	Na	K
				(mmolc/l)								
SI	6.36	0.86	0.00	9.57	0.00	9.61	—	—	—	—	—	—
SII	5.40	0.12	0.00	0.99	0.00	0.94	—	—	—	—	—	—
FFSI	3.22	1.03	5.80	7.30	0.03	9.51	0.001	0.018	0.82	0.06	0.28	0.23
FFSII	3.76	0.18	7.90	0.30	0.05	0.97	0.001	0.003	0.15	0.01	0.28	0.12

Note: EC = electrical conductivity; DOC = dissolved organic carbon.

of 0.5, 0.25, and 0.15 for the E, Bs, and BC soil layers, respectively. All other parameters used with the CD equations were based on our experimental condition (e.g., bulk density, flux, input concentrations, and so on).

The convective-dispersive transport equation for reactive solutes in porous media may be expressed as (Selim, 1992)

$$\Theta \frac{\partial C}{\partial t} + \rho \frac{\partial S}{\partial t} = \Theta D \frac{\partial^2 C}{\partial z^2} - v \frac{\partial C}{\partial z} \tag{12.6}$$

where D is the hydrodynamic dispersion coefficient (cm^2/h), z is soil depth (cm), and v is Darcy's water flux density (cm/h). In addition, S is the solute concentration associated with the solid phase (μg/g soil). Based on our multireaction model, the term S represents all phases of a solute except that which is present in the soil solution. Therefore S is the sum of all phases represented in Figure 12.1 (see also equations 12.1 through 12.5):

$$S = S_e + S_1 + S_2 + S_3 + S_{irr} \tag{12.7}$$

Therefore, we consider the ($\partial S/\partial t$) term in the CD equation to represent all reversible processes between the solution and the solid phases, as well as irreversible (sinks or sources) rates of reactions, that is, transformation reactions.

In the following sections, we restricted our analysis to a single species (SO_4), where the assumption that all other species were time and space invariant and SO_4 was the dominantly changing species in the system. This applicability and model validity will be carried for the two cases considered. In case I, the SO_4 input concentration (C_o) of 0.005 M was maintained, and in case II, the input (C_o) was 0.0005 M.

12.4.1 Equilibrium Sorption: Linear and Nonlinear

In order to describe effluent SO_4 from the top layer shown in Figure 12.3, we assumed a simple linear sorption to account for the retention in the transport equation. Specifically, we used the following linear (equilibrium) model,

$$S = S_e = K_d C \tag{12.8}$$

to describe the reactivity of SO_4 in the topsoil. Equation 12.8 is similar to equation 12.1 where n = 1. Therefore, we ignored all retention reactions illustrated in Figure 12.1 except for the equilibrium reaction of the linear type. Here K_d is a distribution coefficient and is a measure of the extent of sorption or affinity of SO_4 to the soil system (cm^3/g). The associated (dimensionless) retardation factor R can be expressed as

$$R = 1 + (\rho/\theta) K_d \tag{12.9}$$

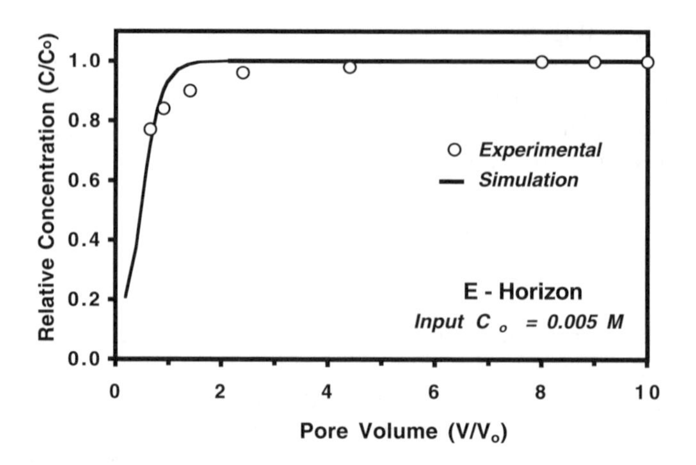

FIGURE 12.3 Experimental and simulated breakthrough results of SO_4 effluent concentrations vs. pore volume (BTCs) from the E horizon (column E-I, input SO_4 of 0.005 M). The simulation is based on best fit of the data when a linear-equilibrium sorption model was assumed.

For R = 1, the solute is considered nonreactive. Due to early arrival or breakthrough and the lack of a well-defined effluent front, our attempts to describe SO_4 results for the top layer (E-I) based on the linear equilibrium approach were not adequate. The simulation shown in Figure 12.3 is a result of the use of R less than unity, which implies negative sorption or ion exclusion (van Genuchten and Weirenga, 1976). Since a value for K_d and/or sorption isotherms for SO_4 were not independently measured, we cannot support such a finding. The estimated R value that provided the best fit of the BTC for the top layer E-I (where C_o = 0.005 M) was 0.59 with a standard error of 0.062 (r^2 = 0.815). We should also stress here that for the second soil column from the top layer (E-II) where low input concentration of SO_4 was applied (C_o = 0.0005 M), the results indicated a concentration in the effluent no different from that of the input solution (figure not shown). As a result, no attempts were made to describe effluent results for the E-II layer.

Our miscible-displacement modeling approach was modified in order to describe SO_4 effluent from the BS (as well as BC) layers. We adjusted the computer code to account for a variable concentration of the input pulse rather than a constant one as is commonly accepted in most column experiments and mathematical solutions. In all our simulations presented here, for each soil column, the SO_4 input concentrations from our experimental results were incorporated as inputs to the model. In addition, presentations of relative concentrations (C/C_o) were based on the respective C_o of the applied solution to the top layer (E).

Figure 12.4 shows the use of a linear equilibrium approach (equation 12.8) for several K_d values (ranging from 0.5 to 3 cm^3/g) in order to describe the BTC from the Bs-I column (C_o = 0.005 M). It is obvious that a linear approach failed to describe the shape of the BTC results. We therefore attempted to use the Freundlich approach of equation 12.1 in the transport equation where the dimensionless parameter n was allowed to vary from 0.5 to 4. As shown in Figure 12.5, the use of $n < 1$, which is

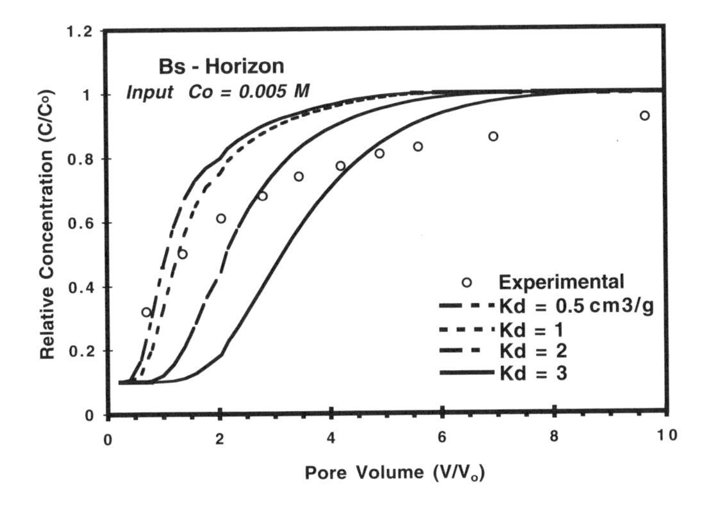

FIGURE 12.4 Experimental and simulated (solid and dashed curves) BTCs of SO_4 effluent concentrations from the Bs horizon (column Bs-I, input SO_4 (C_o) of 0.005 M). Simulations are for a range of K_d values where a linear-equilibrium sorption model was assumed.

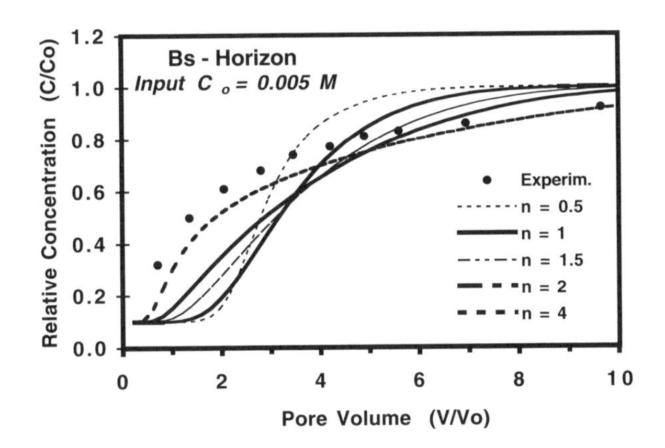

FIGURE 12.5 Experimental and simulated (solid and dashed curves) BTCs of SO_4 effluent concentrations from the Bs horizon (column Bs-I, input SO_4 (C_o) of 0.005 M). Simulations are for a range of n values where a nonlinear-equilibrium model was assumed.

commonly observed for most solutes, did not provide a good description of the shape of the SO_4 BTC (Selim, 1992). We were surprised, however, by the fact that when a larger n value ($n>>1$) was used, this resulted in some improvement in the description of the effluent data. Although the use of larger n value is unusual, recent work of Martinez (1998) showed isotherms of the convex type where $n>1$ is expected. However, others used a sigmoidal model to describe the initial convex shape of isotherms of selected heavy metals (Schmidt and Sticher, 1986). It was not surprising that a good fit of the data was obtained when a nonlinear least-squares optimization scheme was used with the CD equation. As shown in Figure 12.6, the solid curve

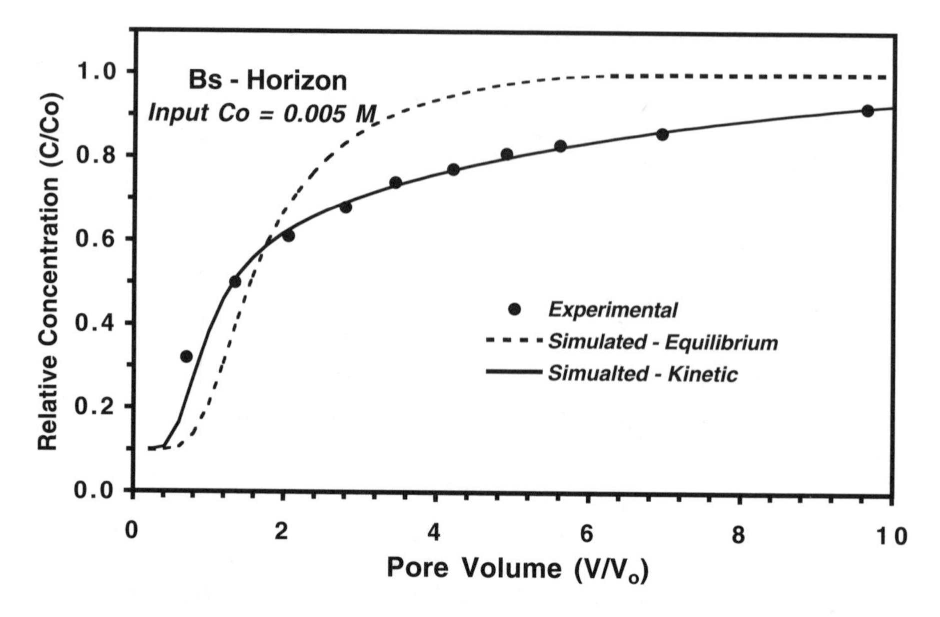

FIGURE 12.6 Experimental and simulated BTCs of SO_4 effluent concentrations from the Bs horizon (column Bs-I, input SO_4 (C_o) of 0.005 M). Simulations are based on best fit of the data when a linear equilibrium model (dashed curve) and a nonlinear equilibrium (solid curve) were used.

TABLE 12.2
Parameter Estimates, Standard Errors, Root Mean Square Errors, and Correlation Coefficients for Linear and Nonlinear Equilibrium Models for Various Columns

Column	Model r^2	RMSE	K_d (cm³/g)	SE (cm³/g)	K_e (cm³/g)	SE (cm³/g)	n	SE
Bs–I	0.948	0.155	0.232	0.1134	—	—	—	—
Bs–I	0.983	0.0361	—	—	0.802	0.117	5.896	0.846
Bs–II	0.962	0.0645	0.138	0.1065	—	—	—	—
BC–I	0.972	0.0902	0.655	0.0808	—	—	—	—
BC–I	0.996	0.0153	—	—	0.873	0.033	2.360	0.103
BC–II	0.958	0.0565	0.299	0.0643	—	—	—	—

Note: RMSE = root mean square error; SE = standard error.

is based on the following parameter estimates that provided the best fit of the BTC; $K_e = 0.802$ (cm³/g) and n = 5.89 ($r^2 = 0.983$). The use of a linear approach as shown by the dotted curve ($K_d = 0.232$ cm³/g, standard error = 0.11 cm³/g) gave a poor description of the effluent data (see Table 12.2).

12.4.2 Kinetic Sorption

Since the use of equilibrium (Freundlich) type with $n > 1$ is uncommon, we also attempted the kinetic reversible approach given by equation 12.2 to describe the effluent results from the Bs-I column. The use of equation 12.2 alone represents a fully reversible SO_4 sorption of the n-th order reaction where k_1 to k_2 are the associated rates coefficients (h^{-1}). Again, a linear form of the kinetic equation is derived if $m = 1$. As shown in Figure 12.7, we obtained a good fit of the Bs-I effluent data for the linear kinetic curve with $r^2 = 0.967$. The values of the reaction coefficients k_1 to k_2, which provided the best fit of the effluent data, were 3.42 and 1.43 h^{-1} with standard errors of 0.328 and 0.339 h^{-1}, respectively (see Table 12.3). Efforts to achieve improved predictions using nonlinear (m different from 1) kinetics was not successful (figures not shown). We also attempted to incorporate irreversible (or slowly reversible) reaction as a sink term (see equation 12.5) concurrently with first-order kinetics. A value of $k_{irr} = 0.0456$ h^{-1} was our best estimate, which did not yield improved predictions of the effluent results as shown in Figure 12.7.

12.4.3 Model Predictions

Thus far, the transport model, after incorporation of equilibrium or kinetic retention, was used in a calibration mode where, along with nonlinear least-squares approximation, a best fit of the model to the experimental BTC was attempted. This resulted in a set of model parameter estimates that provided the best fit of the BTC for a specific version of the model.

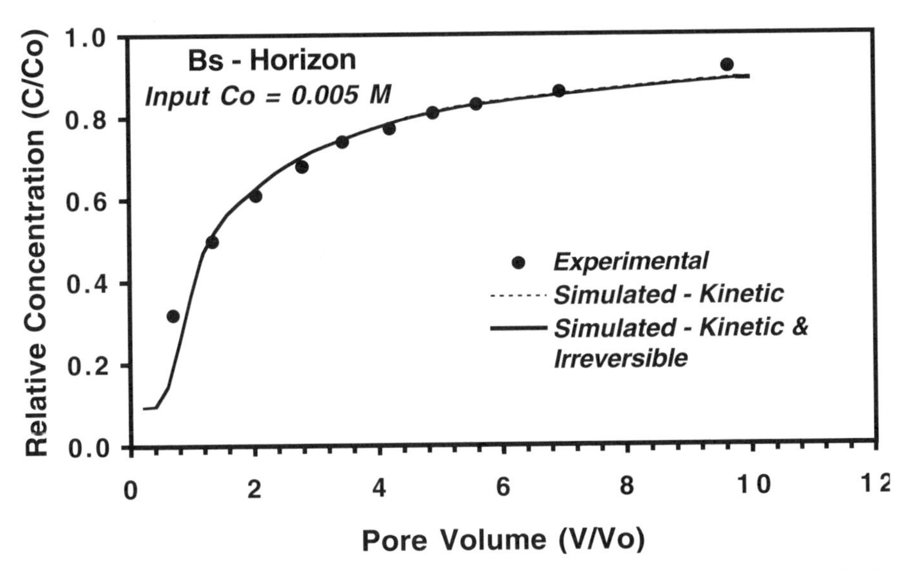

FIGURE 12.7 Experimental and simulated BTCs of SO_4 effluent concentrations from the Bs horizon (column Bs-I, input SO_4 (C_o) of 0.005 M). Simulations are based on best fit of the data when a kinetic model (dashed curve) and kinetic model with an irreversible reaction (solid curve) were used.

TABLE 12.3
Parameter Estimates, Standard Errors, Root Mean Square Errors, and Correlation Coefficients for First-Order Kinetic and Irreversible Models for Various Columns

Column	Model r^2	RMSE	k_1 (h^{-1})	SE (h^{-1})	k_2 (h^{-1})	SE (h^{-1})	k_{irr} (h^{-1})	SE (h^{-1})
Bs–I	0.967	0.0426	3.422	0.328	1.431	0.339	0.0456	0.224
Bs–II	0.979	0.0258	1.075	0.163	0.887	0.377	—	—
BC–I	0.964	0.0216	13.53	1.427	4.904	0.728	—	—
BC–II	0.831	0.0554	1.494	0.404	1.519	0.897	—	—

To test the capability of the model, we used the model in a predictive mode where no model fitting of measured BTC was carried out. Specifically, we utilized independently derived parameters from Bs-I to predict BTC effluent results from the Bs-II column. Specifically, we used the kinetic linear model (with $k_1 = 3.42\ h^{-1}$ and $k_2 = 1.43\ h^{-1}$), based on Bs-I where the input SO_4 (C_o) was 0.005 M to predict Bs-II where C_o of 0.0005 M was used. As illustrated by the solid curve in Figure 12.8, the predictions underestimated the SO_4 concentrations versus pore volume,

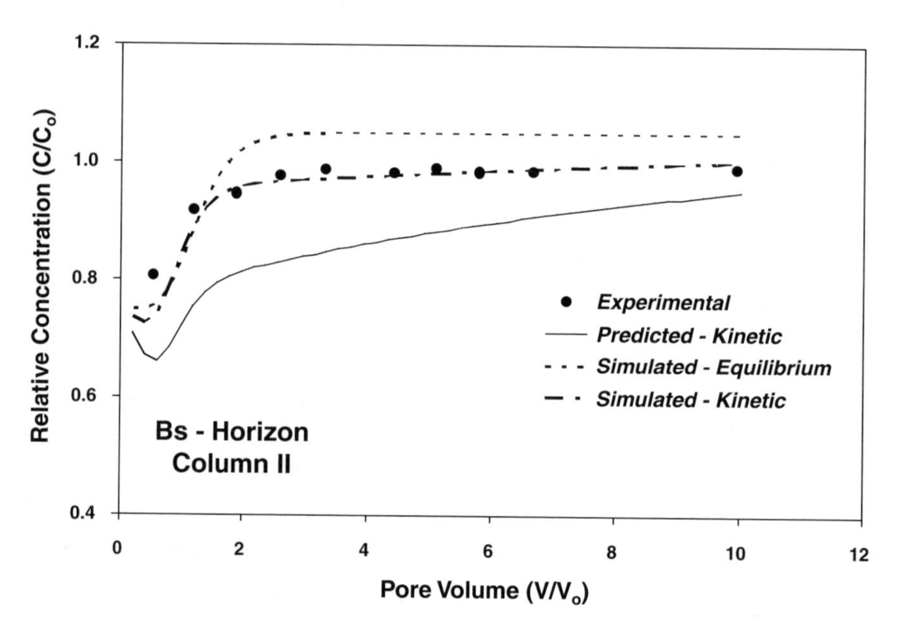

FIGURE 12.8 Experimental and simulated BTCs of SO_4 effluent concentrations from the Bs horizon (column Bs-II, input SO_4 (C_o) of 0.0005 M). Simulations are based on best fit of the data when a linear equilibrium model (dotted curve) and kinetic model (dashed curve) were used. The solid curve is a prediction based on the kinetic model.

which were indicative of higher reactivity or retention when a lower input concentration C_o was used. When the model was relaxed and used in a fitting mode in order to find best-fit predictions of Bs-II using the (linear) kinetic approach, once again adequate BTC predictions were obtained (see dashed curve in Figure 12.8). Best-fit parameters estimated were, as expected, different from those derived for the Bs-I column. Specifically, best-fit k_1 to k_2 values, which provided the best fit of Bs-II effluent data, were 1.07 and 0.89 h^{-1}, respectively (see Table 12.3). Such values for the rate coefficients were lower than those for the Bs-I column and are indicative of concentration-dependent reactions when the assumption of kinetic first-order sorption during transport is made. We also tested the linear equilibrium approach to model the Bs-II results and obtained less than adequate predictions as indicated by the dashed curve in Figure 12.8. The estimated value for K_d was 0.138 cm^3/g and r^2=0.963 (see Table 12.2). This K_d value was smaller than the 0.232 cm^3/g value obtained for the Bs-I column, which illustrates the dependency of sorption on the input concentration (C_o). This finding is consistent with assuming kinetic sorption rather than equilibrium sorption.

12.4.4 BC Layer

Attempts to describe the SO_4 effluent data from the BC layer (columns BC-I and BC-II) were carried out in a similar fashion to the above E and Bs layers. Our aim was to identify model versions that provide the best predictions of effluent results. Moreover, common features as well as trends of parameter estimates and retention characteristics among the different layers were sought. We described effluent data from the BC columns based on model fitting (or simulation) by use of the linear and nonlinear equilibrium approaches as well as first-order kinetics. The experimental results as well as model simulations are illustrated in Figures 12.9 to 12.11. For both BC columns, the worst simulations were obtained using the simple linear equilibrium model. Furthermore, the K_d value that provided the best fit of the data was higher for the BC-I column than for the BC-II column. Such a finding is consistent with that for the Bs layer and indicative of the dependency of sorption on the input concentration (C_o). The associated K_d values, their standard errors, and r^2 are given for the two columns in Table 12.2. In addition, the use of the nonlinear equilibrium model (equation 12.9) to describe BC-I provided an extremely good model fit of the effluent data with an r^2 of 0.996. Moreover, similar to the column from the second layer (Bs-I), the nonlinear parameter n was greater than 1. In fact, the best values were K_e = 0.873 cm^3/g and n = 2.36.

As clearly illustrated in Figures 12.9 and 12.11, good descriptions of the effluent results from BC-I and BC-II were achieved when the first-order kinetic model was implemented. Furthermore, increased sorption was realized for the higher input concentration (C_o = 0.005 M) of the BC-I column than for BC-II. This finding is based on the ratio of the parameters (k_1:k_2) that provided the best-fit model (see Table 12.3). For BC-I, the value of (k_1:k_2) was 2.76 compared to a value of 0.98 for BC-II. Such trends are consistent with the upper Bs layer and indicate sorption dependency on the dominant concentration within the soil column as influenced by the SO_4 input (C_o).

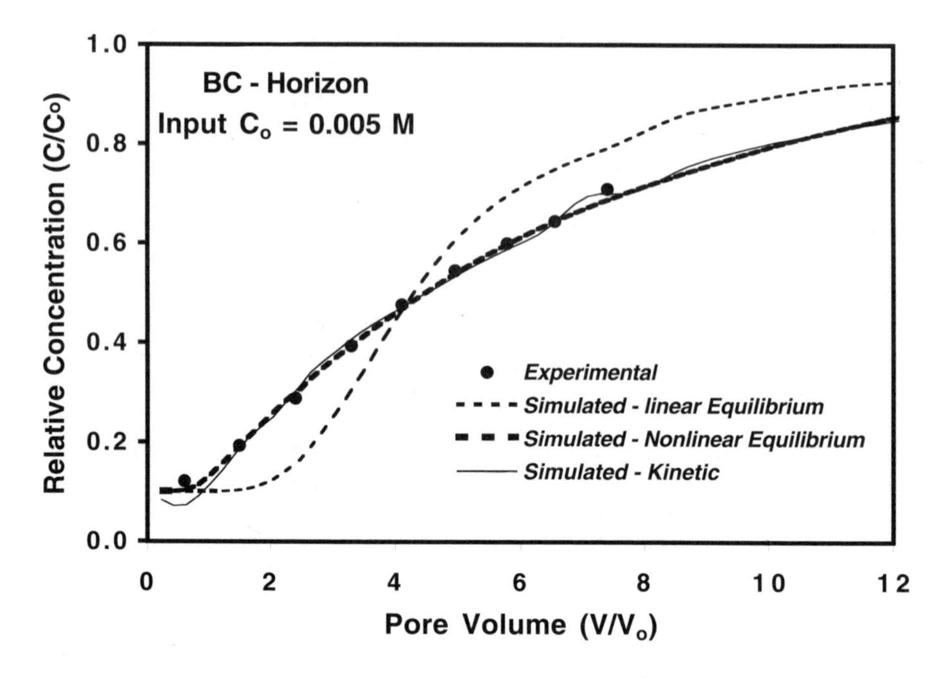

FIGURE 12.9 Experimental and simulated BTCs of SO_4 effluent concentrations from the BC horizon (column BC-I, input SO_4 (C_o) of 0.005 M). Simulations are based on best fit of the data when linear equilibrium (dotted curve), nonlinear equilibrium (dashed curve), and kinetic models (solid curve) were used.

In Figures 12.10 and 12.11, we illustrate a comparison of simulated (curve-fitted) versus predicted BTCs for the BC-II column. Clearly, regardless of whether a kinetic or equilibrium model was used, the predictions overestimated the extent of sorption and resulted in much-delayed BTCs. These predictions were obtained with independently derived parameters from the BC-I column for the equilibrium linear model as well as the kinetic model. It is obvious that the independently measured parameters for the high concentration were inadequate in describing BTC for the low concentration. Therefore, the reactivity or retention of SO_4 during transport is concentration dependent.

12.5 SUMMARY

Understanding sulfate transport and retention dynamics in forest soils is a prerequisite in predicting SO_4 concentration in the soil solution and in lake and stream waters. In this study, forest soil samples from the Gårdsjön catchments, Sweden, were used to study SO_4 transport in soil columns from the upper three soil horizons (E, Bs, and BC). The columns were leached using a sequential leaching technique. The input solutions were $CaSO_4$ equilibrated with forest floor material. Leaching behavior of SO_4 and concentration in the effluent were measured from columns from individual horizons. SO_4 was always retained in the Bs and BC horizons, while

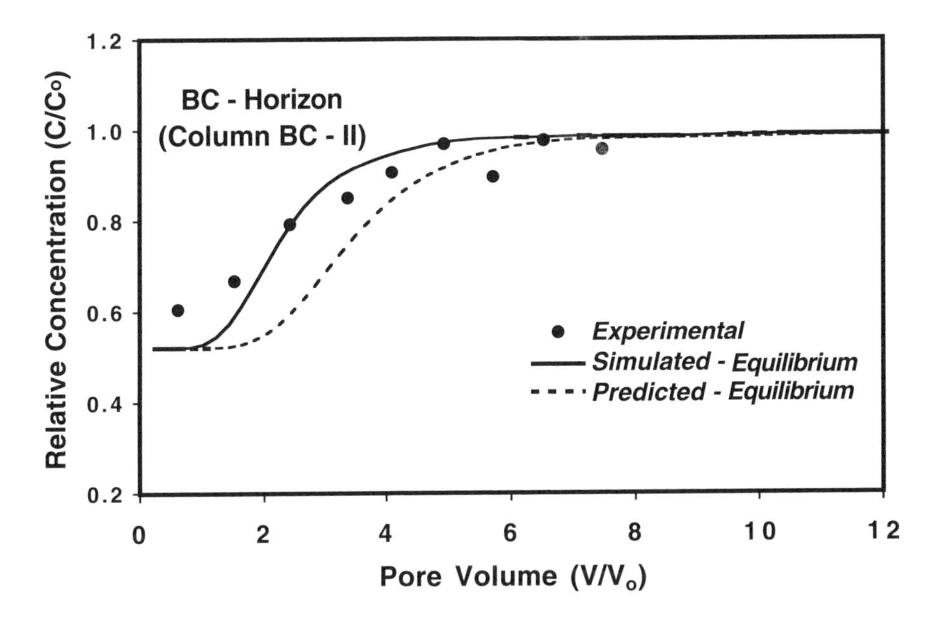

FIGURE 12.10 Experimental and simulated BTCs of SO_4 effluent concentrations from the BC horizon (column BC-II, input SO_4 (C_o) of 0.0005 M). Calculations are based on the linear equilibrium model in a curve-fitting mode (solid curve) and prediction mode (dashed curve).

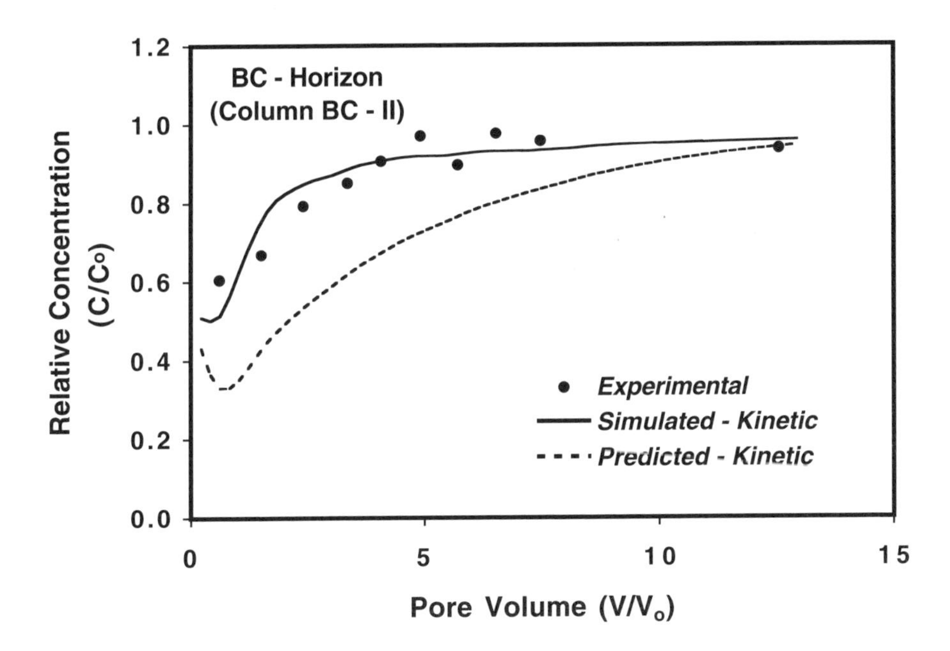

FIGURE 12.11 Experimental and simulated BTCs of SO_4 effluent concentrations from the BC horizon (column BC-II, input SO_4 (C_o) of 0.0005 M). Calculations are based on the kinetic model in a curve-fitting mode (solid curve) and prediction mode (dashed curve).

the pattern for the E horizon varied. Attempts were also made to model SO_4 breakthrough results based on miscible displacement approaches and the solute CD equation in porous media. Several retention mechanisms were incorporated into the CD equation to account for possible reversible and irreversible SO_4 reactions in individual soil layers. Our modeling efforts were less than adequate in describing the mobility of SO_4 in the top (E) horizon. Moreover, a linear equilibrium approach was generally inadequate for describing SO_4 sorption during transport in the Bs and BC horizons, whereas improved model descriptions were obtained when nonlinear equilibrium and first-order kinetic approaches were utilized. Moreover, based on model parameter estimates, the reactivity or retention of SO_4 during transport is concentration dependent. We conclude that sulfate retention during transport in this forest soil is most likely controlled by kinetic reactivity of SO_4 of the reversible and irreversible mechanisms.

12.6 ACKNOWLEDGMENTS

This work was financed in part by the Commission of the European Communities as part of the REPITISC (Role of Exchange Processes in Controlling Ion Transport in Forest Soils and Catchments) Project, and the National Swedish Environmental Protection Agency.

REFERENCES

Amacher, M.C., Selim, H.M., and Iskandar, I.K., Kinetics of chromium (VI) and cadmium retention in soils: A nonlinear multireaction model, *Soil Sci. Soc. Am. J.*, 52, 398, 1988.

Bergholm, J., Nutrient Flow in Soil and Soil Chemical Properties, Skogaby Results: 6, Department of Ecology and Environmental Research, Swedish University of Agricultural Sciences, Sweden, 1994.

Blake, G.R. and Hartge, K.H., Bulk density, in *Methods of Soil Analysis, Part 1, Physical and Mineralogical Methods*, 2nd ed., Klute, A., Ed., American Society of Agronomy, Madison, WI, 1986, p. 363.

Buchter, B. et al., Correlation of Freundlich K_d and n retention parameters with soils and elements, *Soil Sci.*, 148, 370, 1989.

Cosby, B.J et al., Modeling the effects of acid deposition: Control of long-term sulfate dynamics by soil sulfate adsorption, *Water Resour. Res.*, 22, 1283, 1986.

De Vries, W., Kros, J., and Van der Salm, C., The long term impact of various-emission deposition scenarious on Dutch forest soils, *Water Air Soil Pollut.*, 75, 1, 1994.

Farley K.J., Dzombak, D.A., and Morel, F.M.M., A surface precipitation model for the sorption of cations on metal oxides, *J. Colloid Interface Sci.*, 106, 226, 1985.

Fumoto, T. and Sverdrup, H., Modeling of sulfate adsorption on Andisols for implementation in the SAFE model, *J. Environ. Qual.*, 29, 1284, 2000.

Giesler, R. et al., Reversing acidification in a forested catchment in southwestern Sweden: Effects on soil solution chemistry, *J. Environ. Qual.*, 25, 110, 1996.

Gobran, G.R., Courchesne, F., and Dufresne, A., Relationships between sulfate retention and release, solution pH and DOC, in *Experimental Reversal of Acid Rain Effects: The Gårdsjön Soils*, Hultberg, H. and Skeffington, R.A., Eds., John Wiley & Sons, Ltd., England, 1998a, p. 207.

Gobran, G.R. et al., Description of sulfate adsorption-desorption and movement in a Swedish forest soil, *Water AirSoil Pollut.*, 108, 411, 1998b.

Gustafsson, J.P., Modeling pH-dependent sulfate adsorption in the Bs horizons of podsolized soils, *J. Environ. Qual.*, 24, 882, 1995.

Harrison, R.B., Johnson, D.W., and Todd, D.E., Sulphate adsorption and desorption reversibility in a variety of forest soils, *J. Environ. Qual.*, 18, 419, 1989.

Hinz, C. and Selim, H.M., Transport of Zn and Cd in soils: Experimental evidence and modelling approaches, *Soil Sci. Soc. Am. J.*, 58, 1316, 1994.

Hinz, C., Selim, H.M., and Gaston, L.A., Effect of sorption isotherm type on predictions of solute mobility in soil, *Water Resour. Res.*, 30, 3013, 1994.

Hodges, S.C. and Johnson, G.C., Kinetics of sulfate adsorption and desorption by Cecil soil using miscible displacement, *Soil Sci. Soc. Am. J.*, 51, 323, 1987.

Kinniburgh, D.G., General purpose adsorption isotherms, *Environ. Sci. Technol.*, 20, 895, 1986.

Martinez, C.E., Kleinschmeidt, A.W., and Tabatabai, M.A., Sulfate adsorption by variable chanrge soils: Effect of low-molecular–weight organic acids, *Biol. Fertil. Soils*, 26, 157, 1998.

Nilsson, S.I., Acidity properties in Swedish soils-regional paterns and implications for forest liming, *Scand. J. For. Res.*, 3, 417, 1988.

Olsson, B. et al., The Lake Gårdsjön area—physiographical and biological features, *Ecol. Bull.*, 37, 10, 1985.

Prenzel, J., Sulfate sorption in soils under acid deposition: Comparison of two modeling approaches, *J. Environ. Qual.*, 23, 188, 1994.

Reuss, J.O. and Johnson, D.W., *Acid Deposition and the Acidification of Soils and Waters*, Ecolological Studies 59, Springer-Verlag, Berlin, 1986.

Selim, H.M., Modeling the transport and retention of organics in soils, *Adv. Agron.*, 47, 331, 1992.

Selim, H.M. and Amacher, M.C., Sorption and release of heavy metals in soils: Nonlinear kinetics, in *Heavy Metal Release in Soils*, Selim, H.M. and Sparks, D.L., Eds., Lewis Publishers, Boca Raton, FL, 2001, p. 275.

Sparks, D.L., *Kinetics of Soil Chemical Processes*, Academic Press, San Diego, CA, 1989.

Schmidt, H.W. and Sticher, H., Long-term trend analysis of heavy metal content and translocation in soils, *Geoderma*, 38, 195, 1986.

Turner, L.J. and Kramer, J.R., Irreversibility of sulfate sorption on goethite and hematite, *Water Air Soil Pollut.*, 63, 23, 1992.

van Genuchten, M.Th. and Wierenga, P.J., Mass transfer studies in sorbing porous media I: Analytical solutions, *Soil Sci. Soc. Am. J.*, 40, 473, 1976.

Zhang, P. and Sparks, D.L., Kinetics and mechanisms of sulfate adsorption/desorption on goethite using pressure-jump relaxation, *Soil Sci. Soc. Am. J.*, 54, 1266, 1990.

13 Solubility of Fluoride in Semi-Arid Environments

K.J. Reddy, Michelle M. Patterson, J. Daniel Rodgers, Richard E. Jackson, and Barry L. Perryman

CONTENTS

13.1 Introduction 331
13.2 Case Study of F-Solubility in Semi-Arid Soil 334
13.3 Conclusions 343
13.4 Acknowledgments 347
References 347

13.1 INTRODUCTION

Fluoride (F^-) is an important element because it has a wide variety of applications and uses. It is utilized in the production of phosphate fertilizers, insecticides, and high-temperature plastics. Fluoride occurs naturally in soils. Additionally, fluorine is regarded as the most common phytotoxic pollutant. The most phytotoxic fluorine gases include hydrogen fluoride (HF) and silicon tetrafluoride (SiF_4) (National Academy of Science, 1974). The total F^- content of natural soils ranges between 200 and 300 mg/kg (Worl et al., 1973). The most important naturally occurring F^- minerals include fluorite (CaF_2), fluorapatite [$Ca_5(PO_4)_3F$], cryolite (Na_3AlF_6), and fluorophlogopite ($KMg_3AlSi_3O_{10}F_2$) (Elrashidi and Lindsay, 1986). In addition to these natural sources, F^- is deposited in soil through a variety of anthropogenic activities; such as mining, aluminum smelting, fertilizer production, burning fossil fuels, and long-term application of phosphate fertilizers (Drury et al., 1980; Polomski et al., 1982; Adriano, 1986; Rutherford et al., 1994; Arnesen and Krogstad, 1998; Vedina and Kreidman, 1999). Once F^- is released into soil water, it undergoes several geochemical processes including metal complexation, adsorption, and precipitation processes (Figure 13.1). These chemical processes in turn control the solubility, plant uptake, and mobility of F^- in the soil subsurface ecosystem. Fluoride is beneficial in small doses to animals and humans; however, in excess it becomes toxic. Relative to the risk of F^- contamination, our knowledge of F^- mineral dissolution and precipitation processes in natural soils, particularly in semi-arid soils, is very limited.

1-56670-623-8/03/$0.00+$1.50

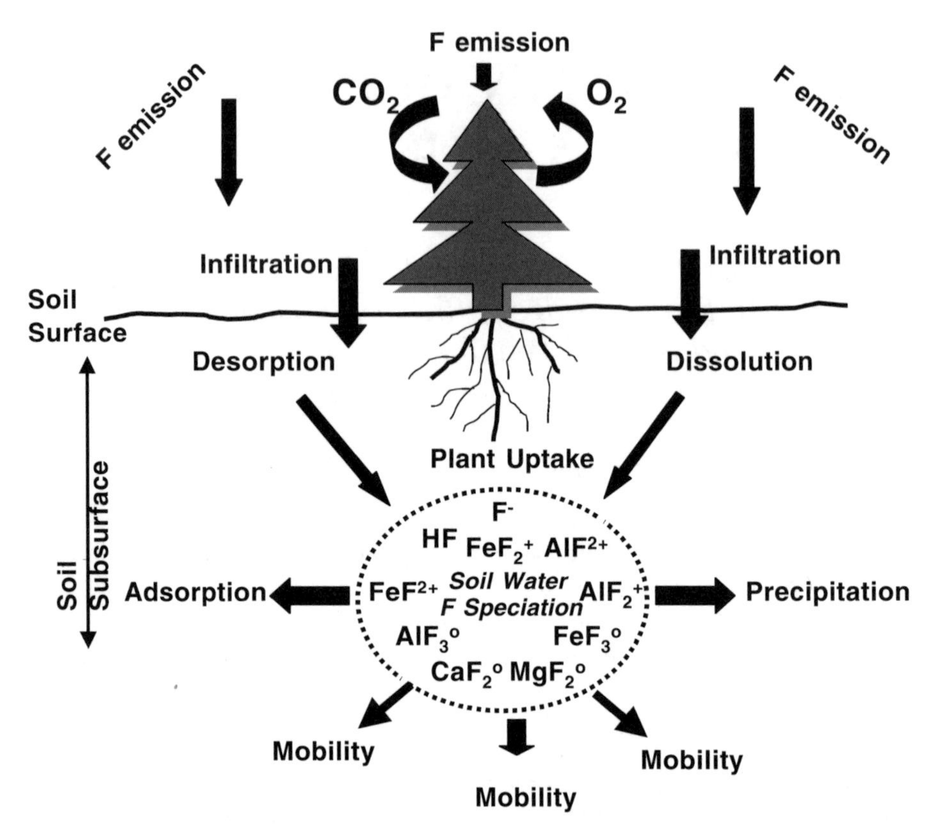

FIGURE 13.1 Hypothetical F^- geochemical processes in a soil subsurface system (not to scale).

In soils, F^- can be found in four major fractions: (1) dissolved in soil solution; (2) sorbed to Al, Fe, and Mn oxides and hydroxides and carbonates; (3) solid phases, such as fluorite and fluorophlogopite; and (4) associated with organic compounds. The solubility of F^- in soil solution is variable and is affected by pH, speciation, adsorption and desorption reactions, and dissolution and precipitation reactions (Luther et al., 1996). Acidic conditions and low calcium carbonate content are favorable to F^- solubility and can therefore enhance both root uptake (Weinstein and Alscher-Herman, 1982) and migration to surface and ground water (Smith, 1983). These conditions can lead to human, plant, and animal health issues. Soils that do contain appreciable amounts of calcium carbonate and are neutral to slightly alkaline conditions can fix F^- as insoluble calcium fluoride (CaF_2), and reduce its bioavailability and mobility (Kubota et al., 1982; Tracy et al., 1984; Reddy et al., 1993; Poulsen and Dudas, 1998).

As rainwater percolates through the soil profile, fluoride-bearing minerals dissolve and release F^- into soil water (Figure 13.1). Fluoride is also released into soil water from mineral surfaces through desorption processes. Studies have shown that F^- in soil water complexes with Al and Fe, which can influence plant uptake of F^-. Other studies have suggested that at higher F^- concentrations and strong acidic and alkaline conditions in soils cause precipitation of AlF_3 and CaF_2, and $[Ca_5(PO_4)_3F]$ minerals, respectively. However, in slightly acidic and neutral soils with low Ca concentrations, F^- solubility is controlled by surface adsorption to Al and Fe oxides and hydroxides.

Using a thermodynamic approach, Lindsay (1979) predicted that aluminum could form strong complexes with F^- in soil solution. Elrashidi and Lindsay (1985) further extended chemical equilibrium concepts of F^- in soils and presented information on solution complexes of F^- in soils. Results of this study suggested that aluminum and iron complexes (e.g., AlF^{2+}, AlF_2^+, AlF_3^0, and AlF_4^- and FeF^{2+}, FeF_2^+, and FeF_3^0) are the major solution species in strongly acid soils. In soils above pH 6, F^- is the predominant solution species. Peek and Volk (1986) examined composition and speciation of sodium-fluoride extracted soil solutions to determine the effects of different chemical components on F^- in ten soils. Results from this study suggested that aluminum complexed 72%, calcium 0.7%, and magnesium 2.4% of the total F^- in solution. The abovementioned findings were confirmed by Wenzel and Blum (1992), who investigated F^- speciation and mobility in slightly acidic soils contaminated by an aluminum smelter, and by Munns et al. (1992) in acid soils with pH ranging from 3.3 to 5.8.

Studies have shown that plants differ in their mechanisms of F^- tolerance (Horner and Bell, 1995; Vike and Habjorg, 1995). MacLean et al. (1992) and Takmaz-Nisancioglu and Davison (1988) found that the rate of F^- uptake at equivalent activities follows the order $HF > AlF^{2+} = AlF_2^+ > AlF_3 = AlF_4^- = F^-$. It was also suggested that F^- uptake by plants increased in the presence of Al due to positively charged AlF_x complexes, which are taken up by the plant roots more easily than negatively charged F^- ions. Recent studies by Stevens et al. (1997, 1998a, 1998b) support these conclusions.

Adsorption and desorption processes of F^- in soils and soil components has been studied extensively. Bar-Yosef et al. (1988) studied F^- sorption by montmorillonite and kaolinite and reported that F^- affinity was greatest in K-kaolinite and least with Ca-montmorillonite. These authors also concluded that the Langmuir model satisfactorily fit F^- sorption by these adsorbents. Fluoride sorption in soils has been described by both Langmuir and Freundlich isotherms (Omuetti and Jones, 1977; Chhabra et al., 1980; Morshina, 1980; Gupta et al., 1982). Peek and Volk (1985) reported that the Langmuir isotherm is applicable only for limited concentrations of F^- in soils.

Meeussen et al. (1996) used a surface complexation model and successfully predicted F^- sorption by pure goethite between pH 4 and 6. This study also extended the equilibrium-surface complexation model and predicted transport of F^- in a goethite-coated sand column at variable pH. Arnesen and Krogstad (1998) examined F^- sorption and desorption processes in soil polluted from an aluminum smelter in western Norway. Results in this study suggested that maximum sorption of F^- occurred between the pH range of 4.8 and 5.5 and the B-horizons sorbed more F^- than A_h-horizons. These findings were attributed to the higher content of Al-oxides and hydroxides in B-horizons compared to A_h-horizons.

Very few studies have examined precipitation and dissolution processes of F^- minerals in natural and contaminated soils. Lindsay (1979) developed solubility isotherms for different F^- minerals and predicted that CaF_2 is more stable than $KMg_3AlSi_3O_{10}F_2$ in soils below pH 6.6. Above pH 6.6 both minerals are more unstable than $Ca_5(PO_4)_3F$ in the presence of adequate phosphate. Elrashidi and Lindsay (1985) developed predictive models for F^- minerals and compared them to the F^- solubility measurements made on 109 soils. Results from these studies indicated that AlF_3(c) may control the solubility of F^- in strongly acid soils, CaF_2 in slightly acid soils, and $KMg_3AlSi_3O_{10}F_2$ in alkaline soils.

13.2 CASE STUDY OF F-SOLUBILITY IN SEMI-ARID SOIL

The major objective of this study was to examine soil F^- solubility and mobility to help explain F^- uptake by native rangeland species of semi-arid environments near a phosphate fertilizer plant in Wyoming. Phosphate fertilizer plant emissions can contain HF and SiF_4. Fluoride emissions are estimated at 35 tons yr^{-1}. The majority of the F^- (ca. 96%) is emitted from a gypsum tailings disposal pond covering nearly 104 ha. The F^- is in the form of HF, formed when the pond's liner heats in the sunlight. The remaining 4% is emitted as HF and SiF_4 entrained in a water vapor plume from the exhaust stacks of the production facility (Patterson, 2002). Fluoride from these sources can accumulate on the plant surface and tissues and eventually enter into the soil subsurface during leaf fall. Previous studies in this area suggested that native plant species such as Wyoming big sagebrush (*Artemisia tridentata* Nutt. ssp. *wyomingensis*) accumulated as much as 3000 $\mu g\ F^-\ g^{-1}$ through stomatal entry (Patterson et al., 2001).

The study site is located 7 km southeast of Rock Springs, Wyoming, 41°33' N, 109°07' W on property belonging to SF Phosphates, Ltd., and adjacent public property. The mean elevation is 2040 m, with short hills and some deep erosion gullies. The climate is arid; precipitation averages 23 to 25 cm yr^{-1}. Soils are dominated by aridisols, that is, ustic haplargids, fine to course loamy, mixed, frigid containing calcareous minerals from sandstone parent materials. Dominant vegetation consists of a mix of Wyoming big sagebrush (*Artemisia tridentata* Nutt. ssp. *wyomingensis*), greasewood (*Sarcobatus vermiculatus* (Hook.) Torr.), bluebunch wheatgrass (*Agrypyron spicatim* (Push) Scribn. and Smith), western wheatgrass (*Pascopyrum smithii* (Rydb.) A. Love), shadscale (*Atriplex confertifolia* (Torr. and Frem.) S. Wats.), and winterfat (*Ceratoides lanata* (Pursh) J.T. Howell) (Skinner et al., 1999; Stubbendieck et al., 1981).

We collected soil samples from the fluorine source and upwind (control) areas. Four 700-m transects were established—two downwind of the tailings pond (TP 1 and 2), one downwind of the production facility (PL), and one upwind of the tailings pond as a control. Soil samples were taken every 100 m from near an *A. tridentata* plant. This was done to determine if additional F^- input from decaying vegetation would deplete the carbonate buffering capacity of the surrounding soil. Soil samples were collected by depth, at 0–15 cm and 15–30 cm. Dried and sieved samples were extracted with distilled deionized water for 24 h and the supernatant collected after centrifugation was filtered through a 0.45-μm membrane. Each clear filtrate was divided into two subsamples; one was acidified to pH 3 with nitric acid and the other was left unaltered.

The unacidified samples were analyzed for pH, total alkalinity, F^-, chloride (Cl^-), nitrate (NO_3^-), phosphate (PO_4^{3-}), and sulfate (SO_4^{2-}). The acidified samples were analyzed for calcium (Ca^{2+}), magnesium (Mg^{2+}), sodium (Na^+), and potassium (K^+). The pH was measured with a pH combination electrode using a Hanna Instruments 8314. The alkalinity was measured as calcium carbonate ($CaCO_3$) by acid titration method. Fluoride was analyzed using an Orion 96–09 ion selective electrode (ISE). The $Cl,^-$ $NO_{3,}^-$ PO_4^{3-}, and SO_4^{2-} were measured by ion chromatography (IC) using a Dionex DX 500 chromatography system. An eluent of 2.25 mM sodium carbonate and 2.8 mM sodium bicarbonate was pumped at 2.0 ml min^{-1} through an AS-4 column system using a self-regenerating suppressor and an electrical conductivity detector. Calcium, Mg, and K were analyzed with ELAN 6100 ICP-MS (inductively coupled, plasma mass spectrometry). Sodium was analyzed with atomic absorption. All analyses followed standard analytical procedures of the American Public Health Association (1992). Speciation and potential solid phase control of total dissolved F^- in soil water extracts was estimated from the data using the MINTEQA2 program (Allison and Brown, 1992).

Soils in the study area are low in organic matter (average 1.5%) and total alkalinity averaged 11.7%. Selected soil water extract data for PL, TP 1 and 2, and control transects are shown in Tables 13.1 to 13.4. Dissolved concentrations of aluminum, iron, and phosphates were below the detection limit of 0.01 mg/l. The charge balance difference for most of the samples was less than 10%, suggesting a very good accounting of the positive and negative ions. The pH of soil water extracts was alkaline and ranged between 7.28 and 8.60 with a mean value of 8.01. The water extractable alkalinity varied from 50 to 460 mg/l as $CaCO_3$, with a mean value of 146.7 mg/l. This range of pH is expected for soils in this area, which are dominated by calcium carbonate.

Distribution of water extractable F^- in PL, TP 1 and 2, and control transects as a function of distance is shown in Figures 13.2 to 13.4. For the PL transect, dissolved F^- concentrations ranged between 0.69 and 4.0 mg/l for the 0- to 15-cm layer and 0.26 and 1.75 mg/l for the 15- to 30-cm layer. For the TP 1 and 2 transects, dissolved F^- concentrations were between 1.98 and 57 mg/l for the 0- to 15-cm layer and 1.08 and 55 mg/l for the 15- to 30-cm layer. There was one notable exception. At the 500-m point on the TP 2 transect, F^- was present at 57 mg/l in the 0- to 15-cm layer, and 55 mg/l in the 15- to 30-cm layer. Fluoride concentrations generally decreased with increasing distance as well as increasing depth from the source in the PL and TP transects. Soil F^- concentration were somewhat uniform in the control transect, quickly dropping to 0.5 mg/l, which we assumed to be background levels for these soils. As expected, both PL and TP transects had much higher dissolved F^- concentrations compared to the control transect.

Comparisons of F^-, Ca^{2+}, and alkalinity (as $CaCO_3$) concentrations by depth (0–15 cm vs. 15–20 cm) showed higher concentrations of all three components in the surface layer (Table 13.5 and 13.6). Paired t-tests were run for all three parameters to determine if the difference is significant. For F^-, the difference was significant for the two TP and the PL transects, but not the control transect. The control transect

TABLE 13.1
Chemical Characterization of Water Extracts for 0–15 cm (A) and 15–30 cm (B) Depths from Processing Plant Transect Soil Samples

Sample Distance (m)	pH		Ca (mg/l)		Mg (mg/l)		K (mg/l)		Alkalinity ($CaCO_3$)		F (mg/l)	
	A	B	A	B	A	B	A	B	A	B	A	B
0	7.36	8.00	104.9	84.17	15.35	22.16	31.64	59.56	165	270	3.25	1.75
100	7.81	8.03	61.03	53.05	8.53	5.99	8.46	6.84	150	140	1.93	1.0
200	8.1	8.11	44.48	36.88	7.29	7.41	5.37	2.39	100	95	0.69	0.26
300	8.17	8.15	79.00	69.2	27.59	12.89	37.71	13.81	125	145	1.25	1.5
400	8.13	8.02	95.64	57.82	19.76	18.37	11.23	8.19	250	155	1.68	1.08
500	8.15	7.99	52.98	43.5	4.29	3.74	4.34	3.72	115	100	1.85	1.5
600	7.94	8.04	79.1	59.41	24.72	19.26	50.37	50.89	195	130	4.0	0.8
700	8.18	7.97	64.83	101.7	7.87	8.04	6.87	5.57	175	130	1.93	1.09

TABLE 13.2
Chemical Characterization of Water Extracts for 0–15 cm (A) and 15–30 cm (B) Depths from Tailings Pond 1 Transect Soil Samples

Sample Distance (m)	pH		Ca (mg/l)		Mg (mg/l)		K (mg/l)		Alkalinity (CaCO3)		F (mg/l)	
	A	B	A	B	A	B	A	B	A	B	A	B
0	7.86	7.52	150.3	151.6	24.28	19.82	80.27	38.17	215	200	6.45	4.84
100	8.17	7.98	73.71	77.87	10.47	14.86	6.56	4.18	175	110	4.34	2.1
200	7.97	7.99	73.01	93.7	6.73	21.96	4.9	18.4	175	80	3.9	1.5
300	7.63	7.66	74.85	43.93	12.36	8.28	15.18	7.01	270	100	4.2	1.65
400	7.97	8.08	67.17	42.6	12.08	6.78	5.85	5.41	110	180	4.15	2.2
500	8.07	8.22	105.3	59.3	12.58	23.55	10.15	9.85	150	235	3.13	2.52
600	8.03	8.17	61.88	38.11	13.06	16.6	8.87	2.63	115	125	3.86	1.81
700	8.66	3.25	61.78	69.46	41.43	30.74	62.28	30.86	460	290	1.98	1.08

TABLE 13.3
Chemical Characterization of Water Extracts for 0–15 cm (A) and 15–30 cm (B) Depths from Tailings Pond 2 Transect Soil Samples

Sample Distance (m)	pH		Ca (mg/l)		Mg (mg/l)		K (mg/l)		Alkalinity ($CaCO_3$)		F (mg/l)	
	A	B	A	B	A	B	A	B	A	B	A	B
0	7.83	8.01	202.4	120.6	17.82	13.68	22.76	17.62	140	100	8.73	6.91
100	7.93	8.08	103.8	59.62	14.39	13.25	5.57	3.75	90	65	8.00	6.20
200	7.90	7.94	102.9	57.09	7.38	7.79	5.88	2.02	135	90	4.60	3.2
300	7.40	8.01	53.15	36.99	8.75	12.31	4.39	2.12	85	90	5.00	3.08
400	8.28	7.81	103.3	65.00	16.62	45.29	4.34	2.2	105	70	11.91	3.56
500	7.28	7.82	48.24	49.85	27.94	25.01	22.48	19.43	50	90	57.00	55.0
600	8.00	8.09	103.2	61.89	10.87	7.43	8.35	4.3	150	95	3.80	2.65
700	7.82	8.03	49.22	36.55	9.92	9.9	4.52	2.79	115	105	2.70	1.93

TABLE 13.4
Chemical Characterization of Water Extracts for 0–15 cm (A) and 15–30 cm (B) Depths from Upwind (Control) Transect Soil Samples

Sample Distance (m)	pH		Ca (mg/l)		Mg (mg/l)		K (mg/l)		Alkalinity ($CaCO_3$)		F (mg/l)	
	A	B	A	B	A	B	A	B	A	B	A	B
0	8.38	7.54	76.88	56.89	7.38	7.13	11.39	8.38	195	155	0.77	0.55
100	8.26	8.4	35.7	45.76	6.78	4.58	9.78	4.78	205	110	0.5	0.5
200	8.60	8.02	30.6	40.97	8.7	12.62	1.6	15.19	150	140	0.35	0.55
300	7.77	8.15	31.84	42.0	14.75	7.0	16.42	11.0	235	210	0.66	0.38
400	7.93	8.17	60.76	22.2	11.2	8.79	17.76	21.97	225	140	0.54	0.26
500	8.17	8.4	37.64	27.81	7.13	64.6	10.18	3.34	175	125	0.6	0.6
600	8.09	7.67	44.79	38.3	8.12	10.02	4.77	5.58	140	140	0.65	0.55
700	8.4	8.29	36.53	41.09	6.28	7.45	3.51	4.34	115	125	0.56	0.7

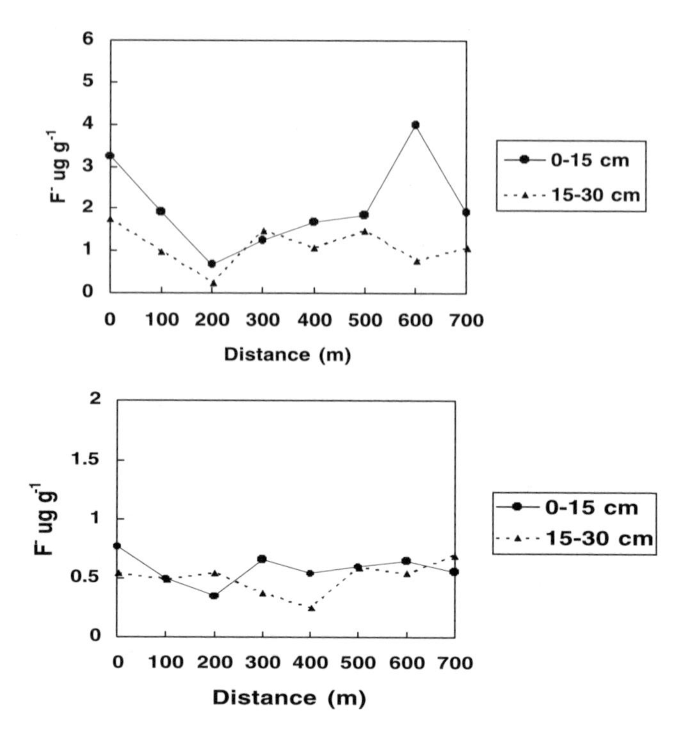

FIGURE 13.2 Distribution of water-extractable F^- in processing plant and control transects as a function of distance. Top figure, processing plant transect; bottom figure, control transect.

is subject to little F^- input, so this was anticipated. Calcium differed only for TP 2 transect, while alkalinity differed only in control transect. This variation may be caused by the relatively few samples analyzed in a heterogeneous environment.

Predicted chemical speciation of F^- in soil water extracts is presented in Tables 13.7 to 13.10. These results indicate that total dissolved F^- concentrations were dominated by free ion F^- (91–98%) followed by MgF_2^0 (<1–7%) and CaF_2^0 (<1–1.9%) ion pairs. These results are in agreement with previous studies by Elrashidi and Lindsay (1985) and Peek and Volk (1986). The relation between measured Ca^{2+} and F^- activities and CaF_2 solubility for PL and TP transects is shown in Figure 13.5. In this graph the solid line is the theoretical solubility isotherm for CaF_2. The experimental points are above the CaF_2 solubility isotherm, indicating that the system is oversaturated with respect to Ca^{2+} and F^- activities for the 0- to 15-cm layer. For the lower depth, Ca^{2+} and F^- activities approached saturation with respect to CaF_2 solubility.

The reported solubility product of CaF_2 is $10^{-10.98}$ (Allison and Brown, 1992). The model predicted values for TP and PL 0- to 15-cm soil water extracts between $10^{-8.29}$ and $10^{-11.52}$ with a mean of $10^{-10.46}$. For the 15- to 30-cm depth, predicted ion-activity product values ranged from $10^{-8.31}$ and $10^{-13.00}$ with a mean value of $10^{-11.02}$. These results indicate that near the source of fluorine, as soil depth increases, the F^- precipitates as CaF_2 mineral. The soils in the study area are alkaline and also contain sufficient amounts of dissolved Ca^{2+} (see Tables 13.1 to 13.3). Both alkaline

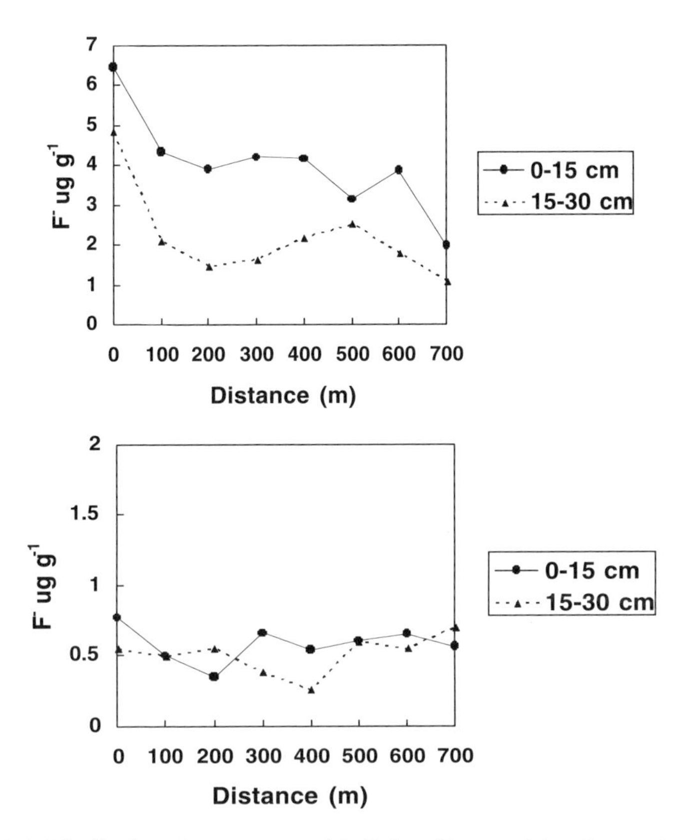

FIGURE 13.3 Distribution of water extractable F^- in tailing pond 1 and control transects as a function of distance. Top figure, tailing pond 1 transect; bottom figure, control transect.

pH and the availability of dissolved Ca^{2+} probably created favorable conditions for the precipitation of CaF_2. Comparison between measured Ca^{2+} and F^- activities for CaF_2 solubility for control transects (upwind) are shown in Figure 13.6. These results suggest that for the control soils, CaF_2 is not regulated by the solubility of F^- because Ca^{2+} and F^- activities for both depths are well below the CaF_2 solubility isotherm. Predicted CaF_2 ion activity products for control soil water extracts averaged $10^{-12.27}$ for 0–15 cm, and $10^{-12.42}$ for 15–30 cm. Thus, CaF_2 is probably not controlling the solubility of F^- in the upwind (control) soils.

Patterson et al. (2001) examined the uptake of F^- by Wyoming big sagebrush and western wheatgrass in the area where the PL and TP transects were made, and reported very little or no translocation of F^- from roots to plant leaves. The solubility estimates, combined with the observations of plant uptake, indicate that precipitation of CaF_2 in these soils may limit root uptake of F^-. Therefore, the soil geochemistry supports the Patterson (2002) findings of F^- plant uptake via stomatal entry.

We also examined the possibility of precipitation of other F^- insoluble minerals, such as fluorapatite in these soils. However, total dissolved phosphate concentrations in soil water extracts were extremely low and in most cases phosphate was well

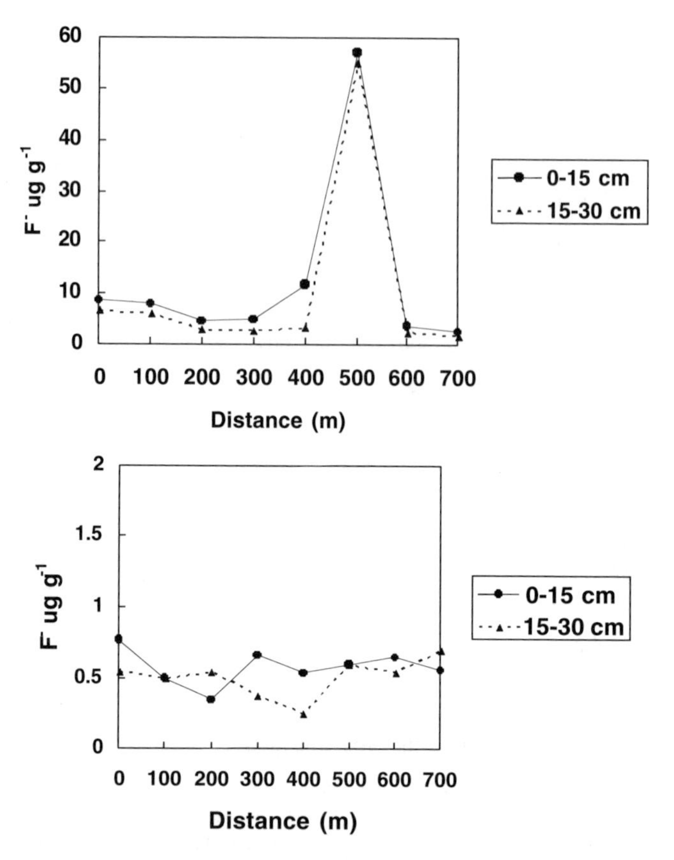

FIGURE 13.4 Distribution of water extractable F^- in tailing pond 2 and control transects as a function of distance. Top figure, tailing pond 2 transect; bottom figure, control transect.

TABLE 13.5
Comparisons of Mean Water Extractable Soil F^-, Ca^{2+}, and Alkalinity by Depth for Processing Plant and Upwind (Control) Transects

Depth (cm)	PL			Control		
	F^- (mg/l)	Ca^{2+} (mg/l)	Alkalinity ($CaCO_3$)	F^- (mg/l)	Ca^{2+} (mg/l)	Alkalinity ($CaCO_3$)
0–15	2.07	73.0	159	0.58	44.3	180
15–30	1.12	63.0	146	0.51	39.4	143
p-value	0.0360	0.2400	0.5400	0.3350	0.4500	0.0300

below the detection limit of 0.01 mg/l. In addition, we conducted an experiment in the laboratory to examine the precipitation kinetics of CaF_2 in the presence of Ca^{2+} and F^-. In this experiment, 150 ml of each NaF and $CaCl_2$ solutions were reacted in volumetric flask for 30 min, 1.0 h, 1.5 h, 2.0 h, and 3.5 h. After each reaction

TABLE 13.6
Comparisons of Mean Water Extractable Soil F^-, Ca^{2+}, and Alkalinity by Depth for Tailings Pond Transects

Depth (cm)	TP1			TP2		
	F^- (mg/l)	Ca^{2+} (mg/l)	Alkalinity ($CaCO_3$)	F^- (mg/l)	Ca^{2+} (mg/l)	Alkalinity ($CaCO_3$)
0–15	4.0	83.5	209	8.36	95.8	109
15–30	2.21	60.9	165	6.26	60.9	88
p-value	0.0002	0.2030	0.2500	0.0003	0.0060	0.1039

TABLE 13.7
Model Calculation of Predominant Solution Speciation of Fluoride in Soil Water Extracts for 0–15 cm (A) and 15–30 cm (B) Depths from Processing Plant Transect

Sample Distance (m)	F^- (%)		MgF^0_2 (%)		CaF^0_2 (%)	
	A	B	A	B	A	B
0	96.6	96.0	2.2	3.0	1.2	<1.0
100	97.8	98.2	1.4	1.0	<1.0	<1.0
200	98.1	98.1	1.3	1.4	<1.0	<1.0
300	95.6	97.1	3.6	2.0	<1.0	<1.0
400	95.9	96.2	3.0	3.0	1.1	<1.0
500	98.4	98.6	<1.0	<1.0	<1.0	<1.0
600	95.6	96.8	3.5	2.6	<1.0	<1.0
700	97.7	97.3	1.4	1.4	<1.0	1.4

time, solutions were filtered through a 0.45-micron filter and clear filtrates were analyzed for pH and F^-. The precipitate, separated from the solutions, was analyzed with x-ray diffraction (XRD) to identify the mineral phase. The analytical data from this study are shown in Table 13.11. These results suggest that as the reaction time increased, the F^- concentration decreased from 199 mg/l to 5.5 mg/l. The XRD analysis identified the precipitate as CaF_2 mineral. Both the solution data and the XRD analysis confirmed that CaF_2 can rapidly precipitate under normal conditions.

13.3 CONCLUSIONS

The results of this study are in agreement with previous studies. Lindsay (1979) suggested that precipitation of CaF_2 mineral could limit the solubility of F^- in calcareous soils. Reddy and Gloss (1993) studied geochemical speciation as related to the mobility of F^- in soil water extracts of semi-arid environments. In their study, surface and subsurface soil samples were extracted with distilled deionized water

TABLE 13.8
Model Calculation of Predominant Solution Speciation of Fluoride in Soil Water Extracts for 0–15 cm (A) and 15–30 cm (B) Depths from Tailings Pond 1 Transect

Sample Distance (m)	F^- (%)		MgF^0_2 (%)		CaF^0_2 (%)	
	A	B	A	B	A	B
0	95.7	95.0	2.9	3.1	1.4	1.9
100	97.5	97.0	1.6	2.1	<1.0	<1.0
200	97.8	96.2	1.2	2.8	1.0	<1.0
300	96.9	97.8	2.1	1.5	1.0	<1.0
400	97.1	98.1	2.0	1.2	<1.0	<1.0
500	96.7	95.3	2.0	3.9	1.3	<1.0
600	97.0	96.4	2.2	3.0	<1.0	<1.0
700	93.4	94.5	5.9	4.6	<1.0	<1.0

TABLE 13.9
Model Calculation of Predominant Solution Speciation of Fluoride in Soil Water Extracts for 0–15 cm (A) and 15–30 cm (B) Depths from Tailings Pond 2 Transect

Sample Distance (m)	F^- (%)		MgF^0_2 (%)		CaF^0_2 (%)	
	A	B	A	B	A	B
0	96.7	96.6	1.7	2.0	1.5	1.4
100	96.6	97.1	2.2	2.1	1.3	<1.0
200	97.5	97.9	1.2	1.3	1.3	<1.0
300	97.6	97.1	1.6	2.3	<1.0	<1.0
400	96.3	91.7	2.5	7.4	1.2	<1.0
500	95.3	95.7	4.1	3.7	<1.0	<1.0
600	97.0	97.7	1.7	1.4	1.3	<1.0
700	97.4	97.6	1.8	1.8	<1.0	<1.0

and water extracts were analyzed for pH, alkalinity, and dissolved cations and anions. Speciation and potential solid phases controlling the dissolved F^- were predicted with the GEOCHEM model. The study reported that in alkaline soil waters dissolved F^- concentrations were dominated by F^- species. Dissolved concentration of F^- was near saturation with respect to CaF_2.

In another study, Street and Elwali (1983) reacted three soils with calcium carbonate ($CaCO_3$) and CaF_2 for 6 weeks and measured the solubility of dissolved F^-. After 6 weeks of incubation, F^- solubility in these soils approached saturation with respect to CaF_2. In a lysimeter experiment, Tracy et al. (1984) studied alfalfa grown on a calcareous soil and irrigated with water high in F^-. Dissolved F^- concentrations in soil water extracts and leachates were near the solubility of CaF_2. Murray and Lewis

TABLE 13.10
Model Calculation of Predominant Solution Speciation of Fluoride in Soil Water Extracts for 0–15 cm (A) and 15–30 cm (B) Depths from Upwind (Control) Transect

Sample Distance (m)	F^- (%)		MgF^0_2 (%)		CaF^0_2 (%)	
	A	B	A	B	A	B
0	97.6	97.8	1.3	1.3	1.1	<1.0
100	98.2	98.4	1.3	<1.0	<1.0	<1.0
200	97.9	97.1	1.6	2.3	<1.0	<1.0
300	96.9	98.1	2.6	1.3	<1.0	<1.0
400	97.2	97.9	2.0	1.7	<1.0	<1.0
500	98.1	98.3	1.3	1.3	<1.0	<1.0
600	97.8	97.5	1.5	1.9	<1.0	<1.0
700	98.2	97.9	1.2	1.4	<1.0	<1.0

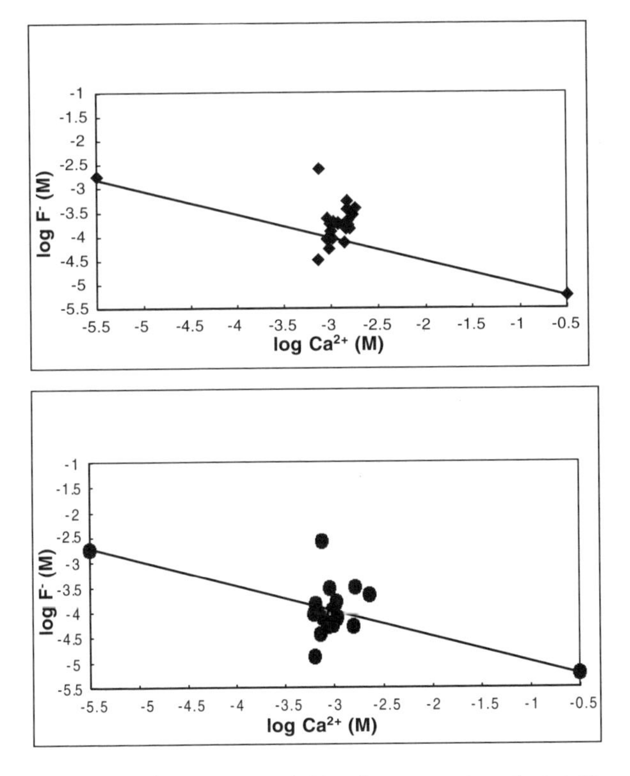

FIGURE 13.5 Measured Ca^{2+} versus F^- activities for processing plant, tailing pond 1 and 2 boil water extracts. Solid line is the theoretical solubility line for CaF_2. Top, 0–15 cm depth; bottom, 15–30 cm depth.

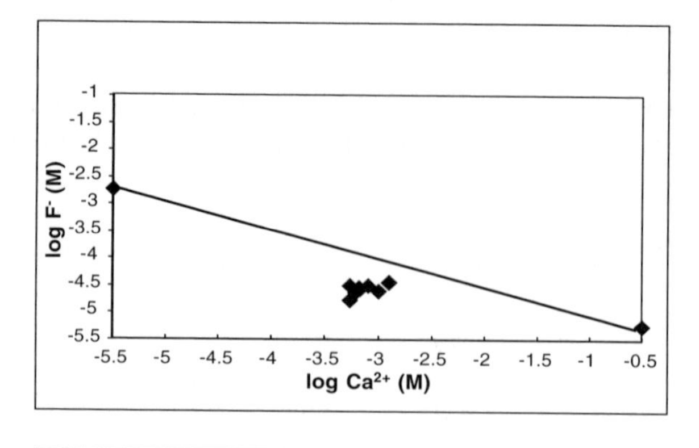

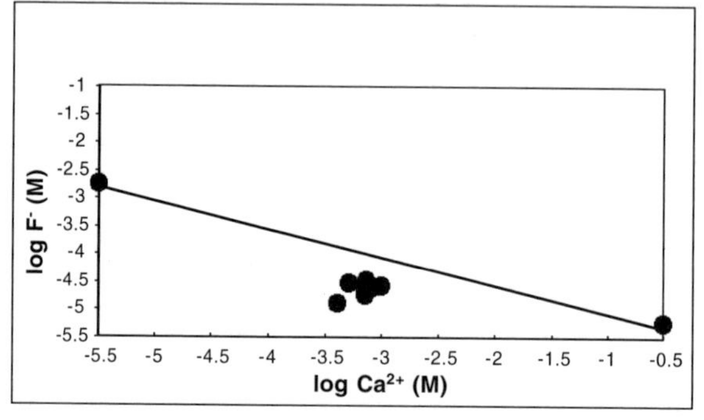

FIGURE 13.6 Measured Ca^{2+} versus F^- activities for water extracts from upwind (control) soil. Solid line is the theoretical solubility line for CaF_2. Top, 0–15 cm depth; bottom, 15–30 cm depth.

TABLE 13.11
Analytical Data from NaCl and $CaCl_2$ Experiment

Reaction Time	pH	F^- mg/l
Initial	6.85	190
30 minutes	6.31	11.5
1.0 hour	6.33	8.5
2.0 hours	6.38	6.5
3.5 hours	6.41	5.5

(1985) reported that $CaCO_3$ is the most effective mineral in removing F^- from phosphogypsum-process water. Luther et al. (1996) reacted phosphogypsum leachates containing F^- with calcareous and noncalcareous soils and reported precipitation of CaF_2 in calcareous soils. Poulsen and Dudas (1998) examined attenuation of F^- in phosphogypsum-process waters by calcareous soil. Results from these studies suggest that calcareous soil effectively immobilizes F^- through precipitation processes.

Overall results of this study suggest that dissolved F^- concentrations in semi-arid alkaline soils, near a phosphate fertilizer facility with adequate amounts of $CaCO_3$, are near saturation with respect to CaF_2. However, in control soils (upwind) dissolved F^- concentrations were highly undersaturated with respect to CaF_2. Perhaps, as suggested by Elrashidi and Lindsay (1985), in neutral and alkaline soils, $KMg_3AlSi_3O_{10}F_2$ is more stable than CaF_2; thus $KMg_3AlSi_3O_{10}F_2$ may be controlling the solubility of F^- in control (upwind) soils. However, we could not explore this possibility because of very low levels (<0.01 mg/l) of dissolved Al^{3+} concentrations in soil water extracts.

13.4 ACKNOWLEDGMENTS

Funding for this project was provided by the SF Phosphates, Ltd., Rock Springs, Wyoming. We thank Darin Howe and Bill Lew of SF Phosphates for their assistance and help in the project. We also thank Amber Jackson, Marji Patz, Rachel Shorma, Junran Li, Mary Knight, Kari Fink, and Mary Fortier for their help in plant and soil sampling and laboratory analyses.

REFERENCES

Adriano, D.C., *Trace Elements in the Environment*, Springer-Verlag, New York, 1986.

Allison, J.D. and Brown, D.S., MINTEQA2: A Geochemical Assessment Model for Environmental Systems, Environmental Research Laboratory, Office of Research and Development, U.S. Environmental Protection Agency, Athens, GA, 1992.

American Public Health Association, Standard Methods for the Examination of Water and Wastewater, American Public Health Association, Washington, D.C., 1992.

Arnesen, A.K.M. and Krogstad, T., Sorption and desorption of fluoride in soil polluted from the aluminum smelter at Ardal in western Norway, *Water, Air, Soil Pollut.*, 103, 357, 1998.

Bar-Yosef, B. Afik, I., and Rosenberg, R., Fluoride sorption by montmorillonite and kaolinite, *Soil Sci.*, 145, 194, 1988.

Chhabra, R.A., Singh, A., and Arbol, I.P., Fluoride in sodic soils, *Soil Sci. Soc. Am. J.*, 44, 33, 1980.

Drury, J.S., Ensminger, J.T., and Hammons, A.S., Review of Environmental Effects of Pollutants: 9, Fluoride. publication no. ORNL/EISS-85, National Technical Information Service, Springfield, VA, 1980.

Elrashidi, M.A. and Lindsay, W.L., Solubility relationships of fluorine minerals in soils, *Soil Sci. Soc. Am. J.*, 49, 1133, 1985.

Elrashidi, M.A. and Lindsay, W.L., Chemical equilibria of fluoride in soils: A theoretical development, *Soil Sci.*, 141, 274, 1986.

Gupta, R.K., Chhabra, R., and Arbol, I.P., Fluorine adsorption behavior in alkali soils: Relative roles of pH and sodicity, *Soil Sci.*, 133, 364, 1982.

Horner, J.M. and Bell, J.N.B., Effects of fluoride and acidity on early plant growth, *Agric. Econ. Environ.*, 52, 205, 1995.

Kubota, J., Naphan, E.A., and Oberly, G.H., Fluoride in thermal spring water and in plants of Nevada and its relationship to fluorosis in animals, *J. Range Manage.*, 35, 188, 1982.

Lindsay, W.L., *Chemical Equilibria in Soils*, John Wiley & Sons, New York, 1979.

Luther, S.M. et al., Fluoride sorption and mineral stability in an Alberta soil interacting with phosphogypsum leachate, *Can. J. Soil Sci.*, 76, 83, 1996.

MacLean, D.C., Hansen, K.S., and Schneider, R.E., Amelioration of aluminum toxicity I wheat by fluoride, *New Phytology*, 121, 81, 1992.

Meeussen, J.C.L. et al., Predicting multicomponent adsorption and transport of fluoride at variable pH in a goethite-silica sand system, *Environ. Sci. Technol.*, 30, 481, 1996.

Morshina, T.N., Fluoride cycles in an estuarine ecosystem, *Total Environ.*, 17, 223, 1980.

Munns, D.N., Helyar, K.R., and Conyers, M., Determination of aluminum activity from measurements of fluoride in acid soil solutions, *J. Soil Sci.*, 43, 441, 1992.

Murray, L.C. and Lewis, B.G., Phosphogypsum waste anion removal by soil minerals, *J. Environ. Eng.*, 111, 681, 1985.

National Academy of Science, Effects of Fluoride in Animals, National Academy of Science, Washington, D.C., 1974.

Omuetti, J.A.I. and Jones, R.L., Fluoride adsorption by Illinois soils, *J. Soil Sci.*, 28, 564, 1977.

Patterson, M.M. et al., Assessment of Ambient, Plant Tissue, and Soil Fluoride Concentrations Associated with Phosphate Fertilizer Plant, Proceedings of Biogeochemistry of Trace Elements Conference, University of Guelph, Guelph, Ontario, Canada, 2001.

Patterson, M.M., Distribution of fluoride in vegetation and soils near a fertilizer production facility in southwest Wyoming, M.S. thesis, University of Wyoming, Laramie, WY, 2002, p. 74.

Peek, D.C. and Volk, V.V., Fluoride sorption and desorption in soils, *Soil Sci. Soc. Am. J.*, 49, 583, 1985.

Peek, D.C. and Volk, V.V., Composition and speciation of sodium fluoride extracted soil solution, *Commun. Soil Sci. Plant Anal.*, 17, 741, 1986.

Polomski, J., Fluhler, H., and Blaser, P., Accumulation of airborne fluoride in soils, *J. Environ. Qual.*, 11, 457, 1982.

Poulsen, L. and Dudas, M.J., Attenuation of cadmium, fluoride, and uranium in phosphogypsum process water by calcareous soil, *Can, J. Soil Sci.*, 78, 351, 1998.

Reddy, K.J. and Gloss, S.P., Geochemical speciation as related to the mobility of F, Mo, and Se in soil leachates, *Appl. Geochem.*, S2, 159, 1993.

Rutherford, P.M., Dudas, M.J., and Samek, R.A., Environmental impacts of phosphogypsum, *Sci. Total Environ.*, 149, 1, 1994.

Skinner, Q.D. et al., Grasses of Wyoming, Agricultural Experiment Station Research Journal, 202, University of Wyoming, Laramie, 1999.

Smith, F.A., Fluorides in everyday life, in *Fluorides: Effects on Vegetation, Animals and Humans*, Shupe, J.L., Peterson, H.B., and Leone, N.C., Eds., Paragon Press, Inc., Salt Lake City, UT, 1983, p. 7.

Stubbendieck, J., Hatch, S.L., and Butterfield, C.H., *North American Range Plants*, 5th ed., University of Nebraska Press, Lincoln, 1981.

Stevens, D.P., McLaughlin, M.J., and Alston, A.M., Phytotoxicity of the aluminum-fluoride complexes and their uptake from solution culture by *Avena sativa and Lycopersicon esculentum, Plant Soil*, 192, 81, 1997.

Stevens, D.P., McLaughlin, M.J., and Alston, A.M., Phytotoxicity of fluoride ion and its uptake from solution culture by *Avena sativa and Lycopersicon esculentum, Plant Soil*, 200, 119, 1998a.

Stevens, D.P., McLaughlin, M.J., and Alston, A.M., Phytotoxicity of hydrogen fluoride and fluoroborate and their uptake from solution culture by *Avena sativa and Lycopersicon esculentum, Plant Soil*, 200, 175, 1998b.

Street, J.J. and Elwali, A.M.O., Fluorite solubility in limed acid sandy soils, *Soil Sci. Soc. Am. J.*, 47, 483, 1983.

Takmaz-Niasncioglu, S. and Davison, A.W., Effects of aluminum on fluoride uptake by plants, *New Phytology*, 109, 149, 1988.

Tracy, P.W., Robbins, C.W., and Lewis, G.C., Fluorite precipitation in a calcareous soil irrigated with high fluoride water, *Soil Sci. Soc. Am. J.*, 48, 1013, 1984.

Vedina, O. and Kreidman, J., Fluoride distribution in burozems of Moldova, *Fluoride*, 32, 71, 1999.

Vike, E. and Habjorg, A., Variation in fluoride content and leaf injury on plants associated with three aluminum smelters in Norway, *Sci. Total Environ.*, 163, 25, 1995.

Weinstein, L.H. and Alscher-Herman, R., Physiological responses of plants to fluorine, in*Effects of Gaseous Air Pollution in Agriculture and Horticulture*, Unsworth, M.H. and Ormrod, D.P., Eds., Butterworths, London, 1982, p. 139.

Wenzel, W. and Blum, W.E.H., Fluoride speciation and mobility in F-contaminated soils, *Soil Sci.*, 153, 357, 1992.

Worl, R.G., Van Alstine, R.E., and Shawe, D.R., Fluorine, in U.S. Mineral Resources, Brobst, D.A. and Pratt, W.P., Eds., Geological Survey Professional Paper 820, U.S. Department of the Interior, Washington, D.C., 1973.

Index

A

Ab initio methods/modeling, 116, 127, 128, 137-138, 153-154, 163
Acoustic attenuation/particle sizing, 282-284
Acoustic spectroscopy, 282-286
ADI (Alternating-direction-implicit) method, 57-61
Adsorption, definition, 52
Adsorption entropy, 149-150
Advanced Photon Source (APS), 215
AES and speciation, 192
Agrypyron spicatim (bluebunch wheatgrass), 334
Al^{3+}-complexed humic acid
  molecular modeling, 134-137, 138
  pyridine interaction, 136-137
Al atoms/oxides and Zn, 207, 209, 213-214, 218, 221-222
Alternating-direction-implicit (ADI) method, 57-61
Anionic models, 118-119
Aperture measures, ORNL study, 11
Argonne National Lab, Chicago, 215
Artemisia tridentata wyomingensis (Wyoming big sagebrush), 334, 341
Atriplex confertifolia (shadscale), 334
Auger electron spectroscopy, 192

B

Bacillus subtilis
  Mn oxidation, 170, 172
  Mn oxide/metal scavenging, 175
Backscattered electron images, *see* BSEs
Batch method
  ion adsorption, 94
  proton adsorption quantification, 93-94
Beasley soil, 28, 35
Becke 3-parameter exchange functional, 119
Bedrock core samples (ORNL)
  mineralogical characteristics, 15
  photo, 5
Beer-Lambert relationship, 274
Benzoic acids, 121-123, 124
Bioceramic prosthetic implants, 146, 152
Biogenic Mn oxidation
  abiotic Mn oxidation vs., 170
  cell concentration factor, 171, 173
  Cu factor, 171, 172-174
  rate of, 171-175
Biogenic Mn oxides
  Fe mixture/metal adsorption, 179-180
  Fe oxides comparison, 176-178, 181
  natural biofilms/trace metal adsorption, 180-183
  overview, 169-170
  oxidation kinetics, 170-175
  Pb adsorption, 176-179, 181-183
  significance, 169, 183
  trace metal adsorption, 175-179
Biosolids
  colloid metal transport, 39-45, 46
  increased use, 26
Biotite, 14-15
Birnessite, 209, 210, 219
Bluebunch wheatgrass (*Agrypyron spicatim*), 334
Blue Mountain, Pennsylvania, 195-196
Boltzmann distribution of velocities, 120
Bond-valence principle, 86-90
Boundary conditions, MRTM model, 56-57
Br, gas tracer, 6-10, 11-13, 14, 16, 17, 19-20
Bremsstrahlung x-rays, 295, 297
Brownian motion, 277, 280, 286
BSEs
  electron microscopy, 292, 297
  Zn contamination/speciation study, 198, 199, 200, 205
Buckingham potential, 120

C

C++ programming language, 61, 67
Carbonate content, biosolid colloids, 40-45, 46
Cation exchange capacity (CEC) measurements, 94
Cayuga Lake, New York, 181
Cell membranes
  molecular composition, 147, 163
  oxides effect on, 146-147
Ceratoides lanata (winterfat), 334
Charge of solid particles classification, 94
Chelated radionuclide transport (ORNL)
  dissociation, 13-16
  modeling, 18-21
  oxidation, 16-18

retardation coefficients, 11
Chelated reactive tracers, 4
Chemical compound, general characteristics, 249
Chemical transport models, 242
Classical mechanical simulations, 116-117, 119-121
Clay structure, *see* Phyllosilicate clay
Clay-water interface
representations, 100, 101, 102
significance, 79-80
Colloid mass vs. turbidity, 275
Colloid-mediated metal transport, *see also* Mobile colloid characterization
colloid breakthrough, 36-37
colloid surface charges, 32
humic materials, 26
metal saturation sequence, 29
metal specificity, 32-33
mineral controls
biosolids, 39-45
metal contamination, 27-34
Pb contamination, 34-39
modeling, 26
nuclear test cavities, 26
overview, 26-27, 46-47
remediation purposes, 27, 46-47
Colloid vibration potential/current (CVP/CVI), 284, 285
COMPASS method
description, 121
use, 127, 128, 129, 131, 132, 133
Contaminant transport, *see also* Radioactive waste transport
storm events, 17
three dimensional simulation
conclusions, 76-77
MRTM model/ADI method, 55-61
overview, 54-55
results, 69-76
system implementation, 61-69
Convection-dispersion (CD) equation, 317, 319
Corundum, 151
Coulombic forces, 119
C programming language, 61, 67
CRAFLUSH, 10-13, 18-21
Crank-Nicholson algorithm, 57-61
Criteria Maximum Concentration (CMC), Hg, 71, 72, 74
Criterion Continuous Concentration (CCC), 71, 72, 74
Critical frequency, acoustic attenuation, 283
Crystallochemical model, 86
Cu
biosolid colloids, 39-40, 43-45, 46
colloid-mediated transport, 30, 32-34
contamination, 245

D

Dekalb stony loam soils, 196
Diatoms and silica, 152-153, 161, 162
Dirichlet-type boundary condition, 57
Dissolved organic matter, *see* DOM
DLVO model, 148
DMol program, 117
DOM
biodegradability, 253-254
biodegradation kinetics, 254
Cd sorption, 255, 256, 257, 258-259
composting, 248, 249
Cu plant uptake, 263-264
Cu release, 261-262
Cu sorption, 255, 256-259, 260-261
definition, 246
effects on ocean, 134, 135
fractionation/characterization, 247-251
green manure biodegradability, 253-254
green manure/Cu release, 261, 262
green manure fractionation/characterization, 247-248, 249, 250, 251
hydrophobic/hydrophilic fractions, 247-249
metal bio-availability, 262-264
metal dissolution, 261-262
metal sorption, 254-261
metal sorption/pH, 257, 258-259
metal sorption/sludge concentration, 260-261
metal sorption/soil types, 254-259
metal transport facilitation, 246, 247
molecular-size fractions, 249-251
overview, 246-247, 265-266
peat biodegradability, 253-254
peat fractionation/characterization, 247-248, 249, 250, 251
peat sorption, 251-252
pig manure biodegradability, 253
pig manure fractionation/characterization, 247-248, 249, 250, 251
pig manure sorption, 251-252
plant metal uptake, 262-264
research needs, 266
rice residue fractionation/characterization, 247-248, 249, 250, 251
rice residue sorption, 251-252
sewage sludge biodegradability, 253
sewage sludge/Cu release, 261, 262
sewage sludge fractionation/characterization, 247-248, 249, 250, 251
sewage sludge sorption, 251-252
sorption in soil, 251-253

sorption parameters/distribution coefficients, 252
use of, 245-246, 261, 264
Zn sorption, 255, 256-257, 258-259
Donnan gel, 102
Donnan-like model
clay-water interface representation, 102-103
illite soils, 105
DTPA, metal speciation determination, 191
Dynamic light scattering, 276-278
Dynamic/quasi-elastic light scattering, 277-278

E

EDS
SEM colloid characterization, 292, 294, 295-296, 297, 298
speciation, 192-193
TEM, 298-299
EDTA
metal speciation determination, 191
tracers transport/dissociation, 13-16
tracers transport/oxidation, 16-18
Electrical double layer (EDL), 99, 284-285
Electroacoustics, 284-286
Electrokinetic sonic amplitude (ESA), 284
Electron correlation, molecular modeling, 119
Electroneutrality condition/equation, 94
Electron microprobe analysis, 192, 193, 198, 216
Electro-osmosis, 280
Electrophoresis, 280
Electrophoretic mobility (EM)
definition, 279
mobile colloids, 279, 280-282
EM fingerprinting, 280-281
EMPA and speciation, 192, 193, 198, 216
Energy dispersive spectrometry, *see* EDS
Enterprise Java Beans, 64, 68
EPSs (extracellular polymeric substances), 152
Equilibrium (Freundlich) sorption, 54
Equilibrium model, retention, 52
Erythrocytes (red blood cells), 146
Eukaryotic blood cells, 146
Eukaryotic cell membranes, 147
EXAFS
data analysis, 202-205
sequential extraction, 209-214
shell fitting parameters, 208
spectroscopy, 202-209
Zn reference parameters, 206
Zn speciation, 193, 195
Exchange-free energy, 148-149
Extended portion (EXAFS), 193

F

Fe
ORNL study, 14-16, 21
Zn contamination/speciation study, 202, 209, 216, 218, 219, 221
Field flow fractionation, 286-291
Fluoride
anthropogenic sources, 331
geochemical processes overview, 332
health issues, 331, 332
natural sources, 331
plant uptake/effects, 331, 332-333
Fluoride solubility
aluminum interactions, 333, 334
controlling factors, 332-333
Fluoride solubility study
Ca, 340-341, 342, 343, 345, 346-347
fluoride chemical speciation, 340, 343, 344, 345
fluoride concentrations, 335, 336-339, 340, 341, 342, 343, 346-347
semi-arid soil, 334-347
site description, 334, 335
soil description, 335
water extract chemical characterization, 336-339
FORTRAN programming language, 61, 67
Fourier-transform-ion-cyclotron resonance-mass spectrometry, 115
Fractured zones, bedrock, 2-3
Fracture spacing, ORNL study, 11
Franklinite, 205, 206, 207, 210, 211-212, 218, 219, 222
Freundlich approach, 312, 320, 323
Freundlich metal distribution coefficients, 28
Freundlich solute retention, 55
Froth floatation, 146
Fulvic acid molecular modeling
benzene/interactions comparison, 129, 131-133
charging/solvation effects, 127-129, 130-131, 138
prediction limitations, 132-134
pyridine interactions, 129, 131-133
Fulvic acid studies and FFF, 289

G

Galerkin finite element technique, 54
GAMESS program, 117, 154
Gas tracers (ORNL) breakthrough profiles, 6-9
Gaussian 94 program, 154
Gaussian 98 program, 117, 154
Gaussian functions, 118

GEOCHEM model, 344
GIAO method, 154
Gibbsite, 209, 210, 213, 214, 218, 219
Gibbsite surface group, 83, 90
Gouy-Chapmann and Stern model, 99, 108-109
Gouy-Chapmann equation, 109
Greasewood (*Sarcobatus vermiculatus*), 334
Groundwater
  chelated metals, 15
  monitoring, 2-4
  ONRL description, 2

H

H+ adsorption charge density, 94
He, gas tracer, 6-9, 12
Heavy metal contamination
  anthropogenic sources, 188
  consequences, 188
  humic materials, 26-27
  metal form significance, 188, 189
  natural sources, 188
  toxicity factors/risk assessment, 188, 189, 220
  U.S., 26
Herbicides, colloid-mediated transport, 26
Hg
  EPA criteria, 71
  three dimensional model study, 71-76
High-performance computing model, 64, 65, 66
High-performance parallel computing, 63
Histidine-Si complexes, 159-160
Hopscotch algorithm, 54
Humic acid, *see also* Al^{3+}-complexed humic acid
  FFF studies, 289
  models, 115
Hydrogen bonding, 119-120
HyperChem 5.0 use, 128
HyperChem program, 117
"Hysteresis," 312

I

Illite platelet representation, 104, 105
Illite soil
  proton adsorption, 97
  proton adsorption modeling, 104-105
  structure, 81, 82
Illitic colloids
  colloid elution, 29, 30, 31, 32
  Pb contamination study, 36-37, 38-39
Inversion point, FFF, 289
Ion adsorption
  measuring, 94
  Solution and Electrostatic (SE) model, 147-148
Ion adsorption charge density, 94
Ion adsorption/desorption, *see also* Proton adsorption
  mineral-water interface, 79-80
  significance, 79
Isolated layer model
  clay-water interface representation, 101
  description/equations, 107-109
Isomorphic substitutions, phyllosilicate clay, 81, 105

J

J2EE technologies
  business tier, 68
  client tier, 66-67
  overview, 64-66
  resource tier, 68-69
  web tier, 67-68
Java 2 Enterprise Edition, *see* J2EE technologies
Java applets, 66-67
Java Native Interface technology, 61-69
Java Plug-in, 67
Java servlets, 64-66, 67
Java technologies, 61-69
Java Virtual Machine (JVM), 67-68
JNI technology, 61-69

K

Kaolinitic colloids, 29, 30, 31, 32
Kaolinitic soil
  charge balance test, 95
  ion/proton adsorption, 98
  proton adsorption modeling, 105-106
  structure, 81
Kinetic models
  multireaction model, 313-315, 317
  retention, 52-53

L

Laidig stony loam soils, 196
LaJara soils
  cation exchange coefficients, 236
  chemical/physical properties, 232
  reduction/cation exchange model, 233-234, 236-237, 239-241
Langmuir isotherm predictions, 179-180
Langmuir method, 312, 333
Laser doppler velocimetry (LDV), 279-282
Layers, phyllosilicate clay, 80-82

Lee-Yang-Parr correlation functional, 119
Lennard-Jones (6-12) potential, 120
Leptothrix discophora
  Mn oxidation, 170-174
  Mn oxides/metal scavenging, 175-176
Lewis acids, 86
Lewis bases, 86
Light-scattering techniques, 274-282
Line source contamination, 71, 72, 74
Liposomes, 146
Lithiophorite, 214, 216-217
Long wavelength requirement, 283
Loradale soil, 28, 34, 35
Lysosomes, 146, 148

M

MacMolplt program, 154
Macrophages, 146, 148
Matrix zones (bedrock), 2-3
Maury soils, 28, 34, 39, 40
Medical geology, 146
Membranolytic oxides
  adsorption entropy, 149-150
  description, 146
  exchange-free energy, 148-149
  thermodynamic basis, 148-152
Metal contamination
  analytical models, 53
  contaminant/soil properties, 52, 53
  models overview, 52-55
  numerical models, 53-54
  soil migration, 52
  sources of, 52, 245-246
Metallurgic activity impact determination, 191
Metal speciation
  components, 188-189
  microenvironment factor, 194-195
Metal speciation approaches
  analytical techniques, 192-195
  data analyses, 194
  microspectroscopic approaches, 194-195
  sequential extraction methods, 191-192
  single extraction methods, 191
  synchrotron-based methods, 193-194
  technique combining, 195
  total metal concentration, 191
Michaelis-Menten kinetics, 171-172
Microfocused XAFS and speciation, 194, 195
Microspectroscopic methods, 194-195
Mineral-water interface, 79-80
Mining impact determination, 191
Mn, *see also* Biogenic Mn oxides
  cycling, 169
  Mn oxides/trace metal associations, 230
  Zn contamination/speciation study, 202, 209, 216, 219, 221
Mobile colloid characterization, *see also* Colloid-mediated metal transport
  acoustic attenuation/particle sizing, 282-284
  acoustic methods advantages, 285-286
  acoustic spectroscopy, 282-286
  acoustic spectroscopy advantages, 283-284
  acoustic spectroscopy shortfalls, 284
  contaminant fate, 272-273
  dynamic light scattering, 276-278
  electroacoustics, 284-286
  electron-based techniques, 292-300
  environmental applications, 289-291
  FFF, 286-291
  FFF applications, 289-291
  fractogram, 291
  laser doppler velocimetry, 279-282
  light-scattering techniques, 274-282
  overview, 272-273, 300-302
  particle charge, 278-282
  PSD, 283-284
  sedimentation/flow-field flow fractionation, 286-289
  SEM, 292-296
  SEM automated techniques, 296-298
  technique advantages, 300, 301-302
  technique comparison, 301-302
  technique limitations, 300, 301, 302
  TEM, 298-300
  turbidimetric methods, 274-276
Mogote soils
  cation exchange coefficients, 236
  chemical/physical properties, 232
  reduction/cation exchange model, 233-234, 236-241
Molecular dynamic (MD) simulations
  clay structural charges, 92, 100
  proton adsorption, 99
Molecular modeling, *see also* SOM chemistry
  overview, 114-116
  problems, 117
  QM/MM models, 116-117
  time involved, 117
  types, 116-117
Møller-Plesset method, 119
Monoprotic protonating-deprotonating surface groups, 99
Monte Carlo (MC) simulations
  clay structural charges, 92, 100
  proton adsorption, 99
Montmorillonite soil
  proton adsorption, 95-96, 97, 98
  proton adsorption modeling, 103-104
  structure, 81, 82

Montmorillonitic colloids
colloid elution, 29, 30, 31, 32
Pb contamination study, 36-37, 38, 39
Morse potential, 120
MRTM model (three dimension)
ADI method, 54-61
description/formulas, 55-61
one/three dimensional comparison, 69-76
online computing resources, 61-69
visualization aspect, 76-77
Multiple nonreactive tracer transport
fracture regime, 6-7, 10-11
matrix diffusion, 6-9, 10-13
modeling, 10-13
ORNL study, 6-13
preferential flow, 6-9
Multireaction models
retention mechanisms, 52-53
sulfate transport, 312-315, 317
Museum of Natural History, Washington, D.C., 204
MUSIC model, 86-90, 100

N

National Soil Survey Center, 28, 31
National Synchrotron Light Source (NSLS), 202
Natural organic matter, *see* NOM
Ne, gas tracer, 6-9, 12
Nephelometric turbidimeters, 274-276
Neumann-type no-flow boundary condition, 57
New Jersey Zinc Company, 196
NMR spectroscopy, 115, 153-159
NOM
Al^{3+} complexation, 134-136
complexity, 138
effects on metals, 134
molecular modeling, 127-128, 129

O

Oak Ridge National Laboratory, *see* ORNL
Octahedral sheets, clay, 80-82, 90
One-dimensional Richards equation, 54
Operator bias
laser Doppler velocimetry, 280
SEM, 296-298
Oracle servers, 68
ORNL
contaminant transport study, 2-22
radioactive waste disposal, 2
Waste Area Grouping 5, 2-3
Oxides
bacterial adhesion, 147
effects on cell membrane, 146-147
membranolysis, 148-152

P

Palmerton smelting plant, 195-196; *see also* Zn contamination/speciation study
Parallel Quantum Solutions program, 117
Pascopyrum smithii (Western wheatgrass), 334, 341
Pauling bond strength, 148
Pauling's principle of electroneutrality, 86
Pb
biosolid colloids, 26, 39-40, 43-45, 46
colloid desorption, 34-35, 37-39
colloid-mediated transport, 30, 32-34
Ph/desorption, 39
surface charge/desorption, 38-39
Zn contamination/speciation study, 198, 221
PC-based Linux clusters, 117
PCS (dynamic/quasi-elastic light scattering), 277-278
pH
biosolid colloids, 40, 42, 43
DOM/metal sorption, 257, 258-259
membranolytic oxides, 148-149
Mn oxidation, 170, 171, 173, 183
Pb adsorption/Mn oxide, 179
proton adsorption (clay), 95-99, 106
seawater and Si, 161, 162
Si-organic complexes, 159
soil reduction/metal solubility, 230-231, 242
Zn contamination/speciation study, 221-222
Phospholipid-oxide surface interactions, 146
Phospholipids
adhesion to oxides, 147-148
membrane structure, 149
membranolytic oxides, 148-151, 163
molecular structure, 147
Phthalic acids, 121-123, 124, 126
Phyllosilicate clay, *see also* Proton adsorption (clay)
flocs, 82, 83, 84, 90
gel particles, 82, 102
layer surface, 82-83
mineral classification, 80-81
platelets, 82, 90
structural charges, 91-92, 105, 107
structure, 80-82
surface groups, 83, 84
Platelets (clay), aspect ratio, 105
Point source (two) contamination Hg model, 71, 74-76
Point source contamination Hg model, 71, 72, 73
Pople, John, 117

Porosity, ORNL study, 11
Portable Batch System (PBS), 69
Potentiometric titration, 93
Pre-edge/near-edge (XANES), 193
Pressure-jump relaxation method, 312
Principal component analyses, 202, 203
Prokaryotic cell membranes/walls
composition, 147
oxide effects on, 151-152
Proton adsorption, 80
Proton adsorption (clay)
basal/edge surfaces, 83, 90-93, 99, 100-102, 105
bond-valence principle, 86-90
calculation parameters, 104
crystallochemical model, 86
electric potential mapping, 90
electrostatic component, 90-93
mathematics, 83, 85
measuring, 93-99
metal oxides vs., 106
modeling, 99
model selection, 99-103
MUSIC model, 86-90, 100
overview, 83, 107
Pauling's principle of electroneutrality, 86
pH, 95-99, 106
structural charges, 91-92, 105, 107
surface groups, 88
surfaces, 84
surfaces/H bonds, 88
Pseudomonas putida, 170, 172

Q

Q-CHEM program, 117
QM/MM models, 116-117
Quantum mechanical calculations methods, 115-119
Quartz effects, 147, 151, 152
Quasi-elastic/dynamic light scattering, 277-278

R

Radial structure functions (RSFs), 202, 207, 213, 217
Radioactive waste transport
fractured bedrock, 1-22
overview, 1-2, 21-22
remediation strategies, 2, 21-22
Radiolaria and silica, 152-153
Rayleigh light scattering, 276
Remediation
colloid-mediated metal transport, 27, 46-47
radioactive waste transport, 2, 21-22
web-based simulation, 77
Respiratory disorders and oxides, 146, 148
Retention mechanism conceptualization, 52-55
RSFs (radial structure functions), 202, 207, 213, 217
Rutile, 151

S

Salicyclic acids, 121-126
"Salting-out effect," 134
Sarcobatus vermiculatus (greasewood), 334
Scanning electron microscopy, *see* SEM
Schrodinger equation, 118
SCRF method, 154
Seawater pH and silica, 161, 162
Secondary mass spectroscopy, 192
Sedimentation/flow-field flow fractionation, 286-289
SEM
automated techniques, 296-298
electron source, 293-294
mobile colloid characterization, 292-296
speciation, 192, 193
thin-foil method, 293, 296, 297, 298
Zn sorption, 190
Semi-empirical (PM3) methods, 128, 131-132
SE model
ion adsorption, 147-148
membranolysis, 148, 151, 152
Serine-Si complexes, 159-161
Shadscale (*Atriplex confertifolia*), 334
Sheets, phyllosilicate clay, 80-82
Shrouts soil, 28, 35
Si-amino acid complexes, 159-162
Silica and cell membranes, 151, 152
Silica bioceramics, 152
Silica deposition vesicles, diatoms, 161, 162
Silicatein enzyme, 159
Silicic acid and water, 162, 163
Silicosis, 146
Siloxane group, 83, 90
SIMS and speciation, 192
Si-organic complexes
approach to studying, 153-154
monocyclic Si-polyalcohol complexes, 158
NMR anisotropic shifts, 155-156
NMR isotropic shifts, 155-156
NMR shifts, 153-159
overview, 162-163
Si-amino acid complexes, 159-162
silicon cycle, 152-153
Si-polyalcohol complexes, 154-159
spirocyclic Si-polyalcohol complexes, 157

SIP (Strongly implicit procedure), 54
Si-polyalcohol complexes, 154-159
Si-serine complexes, 160-161
Smectite (montmorillonite) group, 81, 82
Smelting facilities, *see* Zn contamination/speciation study
Soil acidification models, 312
Soil organic matter, *see* SOM
Soil reduction/metal solubility
  Ca cation exchange, 235
  Ca prediction, 236, 237, 238
  cation exchange coefficients, 235-236
  cation exchange model, 235
  cation exchange model limitations, 242
  Cd, 230-231
  chemical transport models, 242
  electrolyte concentration, 231
  Fe oxides, 230
  metal concentrations, 234
  Mg cation exchange, 235
  Mg prediction, 236, 238
  Mn cation exchange, 235
  Mn factors controlling, 231-232
  Mn oxides, 230
  Mn prediction, 236-237, 238-239
  model/experimental predictions, 236-239
  Ni, 231, 234
  Ni cation exchange, 235
  Ni prediction, 239-240, 241
  overview, 230, 242-243
  Pb, 230-231
  pH, 230-231, 242
  redox potential, 230, 233
  reduction model, 233-235
  Sr cation exchange, 235
  Sr prediction, 236, 237, 238-239
  Sr release, 234
  Zn cation exchange, 235
  Zn prediction, 239-240, 241
  Zn release, 234
Soils complexity, 113-115
Solution and Electrostatic model, *see* SE model
SOM adsorption significance, 115
SOM chemistry
  Al^{3+} addition, 123, 125-127
  Al^{3+}-complexed humic acid, 134-137, 138
  classical mechanical simulations, 116-117
  classical mechanical simulations description, 119-121
  fulvic acid, 127-134, 138
  overview, 113-114
  QM/MM models, 116-117
  quantum mechanical calculations approach, 115-116
  quantum mechanical calculations description, 117-119
  quantum mechanical calculations overview, 116-117
  simple organic acids, 121-127, 137-138
  UV resonance Raman frequencies, 123, 124, 125-126
Spartan program, 117
Sphalerite, 206, 207, 210, 211-212, 219, 221
Spirocyclic Si-polyalcohol complexes, 157
Spodosol haplorthod soil study, 315
Sponges and silica, 152-153, 159, 160
Stacks, phyllosilicate clay, 81, 82
Static dielectric constant, 151
Stern-Gouy-Chapman model, 99, 108-109
Stern layer, 108-109
Stevenson structure, 135, 137
Stishovite, 151, 163
Storm events and contaminant transport, 17
Strongly implicit procedure (SIP), 54
Structural charge density, 94
Sulfate interactions/soil
  approaches to studying, 311-313
  BC layer, 325-326, 327
  BTC, 320-322, 324, 327
  case study, 315-317
  chemical models, 311-312
  empirical approaches, 312-313
  equilibrium sorption, 319-322, 323
  Freundlich approach, 312, 320, 323
  "hysteresis," 312
  irreversible retention, 312, 314-315, 328
  kinetic sorption, 323, 325
  Langmuir approach, 312
  leaching experiment, 315-317, 318
  linear equilibrium approach, 320-322, 325, 327
  miscible displacement approach, 317, 320
  modeling, 317-326, 327
  model predictions, 323-325
  multireaction kinetic model, 313-315, 317
  multireaction models, 312-313
  nonlinear equilibrium approach, 322
  overview, 326, 328
  pressure-jump relaxation method, 312
  reversible retention, 313, 315, 328
  soil acidification models, 312
Sun SuperMSPARC processors, 68
Suwannee fulvic acid, 128-134
Synchrotron-based methods, 193-194

T

Teichoic acids, 151
TEM

EDS, 298-299
mobile colloid characterization, 298-300
Temperature and Mn oxidation, 170, 171, 173
TEOS (tetreethoxyorthosiliate), 159
Tetrahedral sheets, clay, 80-82, 90
Tetreethoxyorthosiliate (TEOS), 159
Thermal depth, acoustic attenuation, 283
Thermal gravimetric analysis, 300
Three-dimensional architecture, 62-63, 76
Three dimensional flow simulation, *see also* Web-based simulation
concentration curves, 70-71, 76
conclusions, 76-77
Hg contamination, 71-76
MRTM model/ADI method, 54-61
overview, 54-55
results, 69-76
system implementation, 61-69, 76
visualization aspect, 76-77
Three-dimensional method of characteristics (MOC3D), 54
TMA^+
adsorption entropy, 149-150
membranolytic oxide effects on, 148, 151
molecular composition, 147
Total proton adsorption, 96
Tracer strategy, ORNL study, 3-5
Traditional computing model, 63, 65
Transmission electron microscopy, *see* TEM
Turbidimetric methods, 274-276
Turbidity vs. colloid mass, 275

U

Unix-based workstations, 118
U.S. Department of Agriculture National Soil Survey Center, 28, 31
U.S. Department of Energy (DOE), *see also* ORNL
radioactive waste disposal, 1-2
U.S. EPA
heavy metal contamination, 26, 188
Hg standards, 71
Superfund National Priority List, 188, 196
UV resonance Raman frequencies, 123, 124, 125-126

V

Van der Waal's forces, 119-120, 136, 137
Vermiculite
proton adsorption, 98
structure, 81, 82
Viscous depth, acoustic attenuation, 283

W

Waste contamination overview, 26
Wastewater treatment plants biofouling, 146
Waynesboro soil, 28
Web-based simulation, *see also* Three dimensional flow simulation
high-performance computing model, 64, 65, 66
J2EE technologies, 64-69
one-dimensional architecture, 61-62
overview, 61-63
programming language overview, 62
three-dimensional architecture, 62-63, 76
traditional computing model, 63, 65
visualization aspect, 76-77
Western wheatgrass (*Pascopyrum smithii*), 334, 341
Winterfat (*Ceratoides lanata*), 334
Wyoming big sagebrush (*Artemisia tridentata wyomingensis*), 334, 341

X

XAFS
microfocused XAFS, 194
speciation, 193-194
Zn contamination study, 202-205, 207-209
Zn sorption, 190
XANES, speciation, 193, 195
XPS, 192
X-ray absorption fine structure spectroscopy, *see* XAFS
X-ray fluorescence spectroscopy, 192
X-ray photoelectron spectroscopy, 192
XRD
speciation, 192
Zn contamination/speciation study, 197-198
Zn sorption, 190
XRF spectroscopy and speciation, 192

Z

Zeta potential, 279, 282, 284, 285
Zeta potential measurements, 94, 104
Zn
biosolid colloids, 39-40, 43-45, 46
colloid-mediated transport, 30, 32-34
contamination and pH, 189
contamination sources, 189
diffusion limiting, 220
leaching, 219-220
toxicity, 189

Zn contamination/speciation study
  Al atoms/oxides, 207, 209, 213-214, 218, 221-222
  backscattered electron images (BSEs), 198, 199, 200, 205
  birnessite, 209, 210, 219
  Cd, 221
  desorption, 219-220
  EMPA use, 198, 216
  μ-EXAFS, 217-219
  EXAFs data analysis, 202-205
  EXAFs/sequential extraction, 209-214
  EXAFS shell fitting parameters, 208
  EXAFS spectroscopy, 202-209
  EXAFS Zn reference parameters, 206
  Fe, 202, 209, 216, 218, 219, 221
  franklinite, 205, 206, 207, 210, 211-212, 218, 219, 222
  gibbsite, 209, 210, 213, 214, 218, 219
  LCF, 202, 203, 209, 210, 223
  lithiophorite, 214, 216-217
  Mn, 202, 209, 216, 219, 221
  octahedral coordination, 207, 211-212
  overview, 190, 221-223
  Pb, 198, 221
  percent metals, 203
  pH, 221-222
  radial structure functions (RSFs), 202, 207, 213, 217
  sampling method, 196-197
  sequential extractions, 198, 201-202
  significance, 221-222
  site description, 195-199
  sphalerite, 206, 207, 210, 211-212, 219, 221
  subsurface soil, 207, 209, 218-219
  surface soil, 205, 207, 217-218
  synchotron-μ-XRF/SXRF, 215-217, 223
  tetrahedral coordination, 205, 207, 211-212, 219
  XAFS, 202-205, 207-209
  XRD analysis, 197-198
  Zn leaching, 219-220
Zn-EXAFS (bulk), 205
Zn-EXAFS (normalized), 204
Zn K-edge EXAFS spectra of soil, 202